OXFORD MATHEMATICAL MONOGRAPHS

OXFORD MATHEMATICAL MONOGRAPHS

A. Belleni-Moranti: *Applied semigroups and evolution equations*
M. Rosenblum and J. Rovnyak: *Hardy classes and operator theory*
J. W. P. Hirschfeld: *Finite projective spaces of three dimensions*
A. Pressley and G. Segal: *Loop groups*
J. C. Lennox and S. E. Stonehewer: *Subnormal subgroups of groups*
D. E. Edmonds and W. D. Evans: *Spectral theory and differential operators*
Wang Jianhua: *The theory of games*
S. Omatu and J. H. Seinfeld: *Distributed parameter systems: theory and applications*
J. Hilgert, K. H. Hofmann, and J. D. Lawson: *Lie groups, convex cones, and semigroups*
S. Dineen: *The Schwarz lemma*
B. Dwork: *Generalized hypergeometric functions*
S. K. Donaldson and P. B. Kronheimer: *The geometry of four-manifolds*
T. Petrie and J. Randall: *Connections, definite forms, and four-manifolds*
R. Henstock: *The general theory of integration*
D. W. Robinson: *Elliptic operators and Lie groups*
A. G. Werschulz: *The computational complexity of differential and integral equations*
P. N. Hoffman and J. F. Humphreys: *Projective representations of the symmetric groups*
I. Györi and G. Ladas: *The oscillation theory of delay differential equations*
J. Heinonen, T. Kilpelainen, and O. Martio: *Non-linear potential theory*
B. Amberg, S. Franciosi, and F. de Giovanni: *Products of groups*
M. E. Gurtin: *Thermomechanics of evolving phase boundaries in the plane*
I. Ionescu and M. Sofonea: *Functional and numerical methods in viscoplasticity*
N. Woodhouse: *Geometric quantization* 2nd edition
U. Grenander: *General pattern theory*
J. Faraut and A. Koranyi: *Analysis on symmetric cones*
I. G. Macdonald: *Symmetric functions and Hall polynomials* 2nd edition
B. L. R. Shawyer and B. B. Watson: *Borel's methods of summability*
D. McDuff and D. Salamon: *Introduction to symplectic topology*
M. Holschneider: *Wavelets: an analysis tool*
Jacques Thévenaz: *G-algebras and modular representation theory*
Hans-Joachim Baues: *Homotopy type and homology*
P. D. D'Eath: *Black holes: gravitational interactions*
R. Lowen: *Approach spaces: the missing link in the topology–uniformity–metric triad*
Nguyen Dinh Cong: *Topological dynamics of random dynamical systems*

The Geometry of Four-Manifolds

S.K. DONALDSON
The Mathematical Institute, Oxford

AND

P.B. KRONHEIMER
Merton College, Oxford

CLARENDON PRESS · OXFORD

Oxford University Press, Great Clarendon Street, Oxford OX2 6DP
Oxford New York
Athens Auckland Bangkok Bogota Bombay
Buenos Aires Calcutta Cape Town Dar es Salaam
Delhi Florence Hong Kong Istanbul Karachi
Kuala Lumpur Madras Madrid Melbourne
Mexico City Nairobi Paris Singapore
Taipei Tokyo Toronto Warsaw
and associated companies in
Berlin Ibadan

Oxford is a trade mark of Oxford University Press

Published in the United States
by Oxford University Press Inc., New York

First published 1990
First published in paperback 1997

British Library Cataloguing in Publication Data
The geometry of four-manifolds
I. Topological spaces: Manifolds
I. Donaldson, S. K. II. Kronheimer, P. B.
514.223

Library of Congress Cataloging in Publication Data
Donaldson, S. K.
The geometry of four-manifolds/S. K. Donaldson and P. B. Kronheimer.
p. cm.—(Oxford mathematical monographs)
Includes bibliographical references.
1. Four-manifolds (Topology) I. Kronheimer, P. B. II. Title.
III. Series
QA613.2.D66 1990 514'.3—dc20 89-77530
ISBN 0 19 850269 9

Printed in Great Britain by
Biddles Ltd, Guildford and King's Lynn

CONTENTS

PREFACE

This book grew out of two lecture courses given by the first author in Oxford in 1985 and 1986. These dealt with the applications of Yang–Mills theory to 4-manifold topology, which, beginning in 1982, have grown to occupy an important place in current research. The content of the lectures was governed by two main aims, and although the treatment of the material has been expanded considerably in the intervening years, some of the resulting structure is preserved in the present work. The primary aim is to give a self-contained and comprehensive treatment of these new techniques as they have been applied to the study of 4-manifolds. The second aim is to bring together some of the developments in Yang–Mills theory itself, placed in the framework of contemporary differential and algebraic geometry. Leaving aside the topological applications, ideas from Yang–Mills theory—developed by many mathematicians since the late 1970's—have played a large part in fixing the direction of modern research in geometry. We have tried to present some of these ideas at a level which bridges the gap between general text books and research papers.

These two aims are reflected in the organization of the book. The first provides the main thread of the material and begins in Chapter 1 with the mysteries of 4-manifold topology—problems which have been well-known in that field for a quarter of a century. It finishes in the last chapters, when some of these problems are, in part, resolved. On the way to this goal we make a number of detours, each with the purpose of expounding a particular area of interest. Some are only tangentially related, but none are irrelevant to our principal topic. It may help the reader to signpost here the main digressions.

The first is in Chapter 3, which deals for the most part with the description of instanton solutions on the 4-sphere; some of the facts which emerge are an ingredient in later arguments (in Chapters 7 and 8, for example) and serve as a model for more general results, but their derivation is essentially independent from the rest of the book. Chapter 6 is concerned with the proof of a key theorem which provides a route from differential to algebraic geometry. This result underpins calculations in Chapters 9 and 10, but it could be taken on trust by some readers. In Chapter 7, only the last section is central to the subject matter of the book, and the main topological results can be obtained without the rather lengthy analysis which it contains. The reader who wants only to discover how Yang–Mills theory has been applied to 4-manifold topology might want to read only Chapter 1, the first part of Chapter 2, and Chapters 4, 5, 8, and 9.

The ten chapters are each reasonably self-contained and could, to a large extent, be read as individual articles on different topics. In general we have tried to avoid duplicating material which is readily available elsewhere.

Almost all of the results have appeared in research papers but we have spent some time looking for different, or simplified, proofs and for a streamlined exposition. Where other books already cover a topic in detail, we have tried to keep our treatment brief. While we hope that readers with a wide range of backgrounds will be able to get something useful from the book, we have assumed a familiarity with a definite body of background material, well represented in standard texts, roughly equivalent to first-year graduate courses in topology, differential geometry, algebraic geometry, and global analysis. The pre-requisites in analysis are summarized in the appendix; for the other subjects we hope that the references given will enable the reader to track down what is needed.

There are notes at the end of each chapter which contain a commentary on the material covered. Nearly all the references have been consigned to these notes. We feel that this streamlines the main text, although perhaps at the cost of giving precise references at all points. We have tried to acknowledge the original sources for the ideas and results discussed, and hasten to offer our apologies for any oversights in this regard.

Turning away from the content of the book, we should now say what is missing. First, although the subject of Yang–Mills theory, as an area of mathematical research, is rooted in modern physics, we have not discussed this side of the story except in passing. This is not to deny the importance of concepts from physical theories in the topics we treat. Indeed, throughout the last decade this area in geometry has been continually enriched by new ideas from that direction, and it seems very likely that this will continue. We are not, however, the right authors to provide an account of these aspects. Secondly, we have not given an exhaustive treatment of all the results on 4-manifolds which have been obtained using these techniques, nor have we tried to bring the account up-to-date with all the most recent developments; this area is still very active, and any such attempt would inevitably be overtaken by events. We hope that, by concentrating on some of the central methods and applications, we have written a book which will retain its value. Finally, while we have tried to give a thorough treatment of the theory from its foundations, we feel that there is still considerable scope for improvement in this respect. This holds both for a number of technical points and also, at a more basic level, in the general ethos of the interaction between Yang–Mills theory and 4-manifold topology. The exploratory drive of the early work in this field has not yet been replaced by any more systematic or fundamental understanding. Although the techniques described here have had notable successes, it is at present not at all clear what their full scope is, nor how essential they are to the structure of 4-manifolds. Looking to the future, one might hope that quite new ideas will emerge which will both shed light on these points and also go further in revealing the nature of differential topology in four dimensions. In any case, we hope that this book will help the reader to appreciate the fascination of these fundamental problems in geometry and topology.

It is a pleasure to record our thanks to a number of people and institutions for their help in the writing of this book. We are both indebted to our common doctoral supervisor, Sir Michael Atiyah, who originally suggested the project and has been a great source of encouragement throughout. Together with Nigel Hitchin, he also introduced us to many of the mathematical ideas discussed in the book. We have learnt a tremendous amount from Cliff Taubes and Karen Uhlenbeck, whose work underpins the analytical side of the theory, and from Werner Nahm. We should also like to take this opportunity to record the significant contribution that discussions in 1981 with Mike Hopkins and Brian Steer made in the early development of this subject.

The first author wishes to thank Nora Donaldson and Adriana Ortiz for their encouragement and help with the typing of the manuscript, and The Institute for Advanced Study, All Souls College and The Mathematical Institute, Oxford, for support. The second author is grateful for hospitality and support provided by Balliol College, The Institute for Advanced Study, the Mathematical Sciences Research Institute and the United States' National Science Foundation. Finally, we should like to thank the staff at Oxford University Press for their patience in awaiting the manuscript, and for their efficiency in the production of this book.

Oxford S.K.D.
March 1990 P.B.K.

1

FOUR-MANIFOLDS

This chapter falls into three parts. In the first we review some standard facts about the geometry and topology of four-manifolds. In the second we discuss a number of results which date back to the 1960's and before; in particular we give an account of a theorem of Wall which accurately portrays the limited success, in four dimensions, of the techniques which were being used to such good effect at that time in the study of high-dimensional manifolds. This discussion sets the scene for the new developments which we will describe in the rest of this book. In the third section we summarize some of the main results on the differential topology of four-manifolds which have sprung from these developments. The proofs of these results are given in Chapters 8, 9, and 10. The intervening chapters work, with many digressions, through the background material required for these proofs.

This first chapter has an introductory nature; the material is presented informally, with many details omitted. For thorough treatments we refer to the sources listed in the notes at the end of the chapter.

1.1 Classical invariants

1.1.1 Homology

In this book our attention will be focused on compact, simply connected, differentiable four-manifolds. The restriction to the simply connected case certainly rules out many interesting examples: indeed it is well known that any finitely presented group can occur as the fundamental group of a four-manifold. Furthermore, the techniques we will develop in the body of the book are, in reality, rather insensitive to the fundamental group, and much of our discussion can easily be generalized. The main issues, however, can be reached more quickly in the simply connected case. We shall see that for many purposes four-manifolds with trivial fundamental group are of beguiling simplicity, but nevertheless the most basic questions about the differential topology of these manifolds lead us into new, uncharted waters where the results described in this book serve, at present, as isolated markers.

After the fundamental group we have the homology and cohomology groups, $H_i(X; \mathbb{Z})$ and $H^i(X; \mathbb{Z})$, of a four-manifold X. For a closed, oriented four-manifold, Poincaré duality gives an isomorphism between homology and cohomology in complementary dimensions, i and $4 - i$. So, when X is simply connected, the first and third homology groups vanish and all the homological information is contained in H_2. The universal coefficient theorem for cohomology implies that, when H_1 is zero, $H^2(X; \mathbb{Z}) = \mathrm{Hom}(H_2(X; \mathbb{Z}), \mathbb{Z})$ is a free abelian group. In turn, by Poincaré duality, the homology group $H_2 = H^2$ is free.

There are three concrete ways in which we can realize two-dimensional homology, or cohomology, classes on a four-manifold, and it is useful to be able to translate easily between them (this is standard practice in algebraic geometry). The first is by *complex line bundles*, complex vector bundles of rank 1. On any space X a line bundle L is determined, up to bundle isomorphism, by its Chern class $c_1(L)$ in $H^2(X; \mathbb{Z})$ and this sets up a bijection between the isomorphism classes of line bundles and H^2. The second realization is by smoothly embedded two-dimensional oriented surfaces Σ in X. Such a surface carries a fundamental homology class $[\Sigma]$ in $H_2(X)$. Given a line bundle L we can choose a general smooth section of the bundle whose zero set is a surface representing the homology class dual to $c_1(L)$. Third, we have the de Rham representation of real cohomology classes by differential forms.

Let X be a compact, oriented, simply connected four-manifold. (The choice of orientation will become extremely important in this book.) The Poincaré duality isomorphism between homology and cohomology is equivalent to a bilinear form:

$$Q\colon H_2(X; \mathbb{Z}) \times H_2(X; \mathbb{Z}) \longrightarrow \mathbb{Z}.$$

This is the *intersection form* of the manifold. It is a unimodular, symmetric form (the first condition is just the assertion that it induces an isomorphism between the groups H_2 and $H^2 = \mathrm{Hom}(H_2, \mathbb{Z})$). We will sometimes write $\alpha . \beta$ for $Q(\alpha, \beta)$, where $\alpha \in H_2$, and also $Q(\alpha)$ or α^2 for $Q(\alpha, \alpha)$. Geometrically, two oriented surfaces Σ_1, Σ_2 in X, placed in general position, will meet in a finite set of points. To each point we associate a sign ± 1 according to the matching of the orientations in the isomorphism,

$$TX = T\Sigma_1 \oplus T\Sigma_2,$$

of the tangent bundles at that point. The intersection number $\Sigma_1 . \Sigma_2$ is given by the total number of points, counted with signs. The pairing passes to homology to yield the form Q. Going over to cohomology, the form translates into the cup product:

$$H^2(X) \times H^2(X) \longrightarrow H^4(X) = \mathbb{Z}.$$

Thus the form is an invariant of the oriented homotopy type of X (and depends on the orientation only up to sign). In terms of de Rham cohomology, if ω_1, ω_2 are closed 2-forms representing classes dual to Σ_1, Σ_2, the intersection number $Q(\Sigma_1, \Sigma_2)$ is given by the integral:

$$\int_X \omega_1 \wedge \omega_2.$$

To see this correspondence between the integration and intersection definitions one chooses forms ω_i supported in small tubular neighbourhoods of the surfaces. Locally, near an intersection point, we can choose coordinates (x, y, z, w) on X so that Σ_1 is given by the equations $x = y = 0$, and Σ_2 by $z = w = 0$. For the dual forms we can take:

$$\omega_1 = \psi(x, y)\mathrm{d}x\mathrm{d}y, \quad \omega_2 = \psi(z, w)\mathrm{d}z\mathrm{d}w,$$

where ψ is a bump function on $\mathbb{R}^2$, supported near $(0, 0)$ and with integral 1. The 4-form $\omega_1 \wedge \omega_2$ is now supported near the intersection points, and for each intersection point we can evaluate the contribution to the total integral in the coordinates above:

$$\int \psi(x, y)\psi(z, w)\,\mathrm{d}x\mathrm{d}y\mathrm{d}z\mathrm{d}w = \pm 1$$

depending on orientations.

If we choose a basis for the free abelian group H_2, the intersection form is represented by a matrix with integer entries. The matrix is symmetric, and has determinant equal to ± 1 (this is the unimodular condition—a matrix with integer entries has an inverse of the same kind if and only if its determinant is ± 1). As we will explain below, the form on the integral homology contains more information than that on the corresponding real vector space $H_2(X; \mathbb{R})$. The latter is of course classified up to equivalence by its rank—the second Betti number b_2 of the manifold—and *signature*. Following standard notation we write

$$b_2 = b^+ + b^-, \tag{1.1.1}$$

where b^+, b^- are the dimensions of maximal positive and negative subspaces for the form on H_2. (In the familiar way we can identify the bilinear form with the associated quadratic form $Q(x)$.) The *signature* τ of the oriented four-manifold is then defined to be the signature of the form:

$$\tau = b^+ - b^-.$$

1.1.2 Some elementary examples

(i) The four-sphere S^4 has zero second homology group and so all intersection numbers vanish.

(ii) The complex projective plane $\mathbb{CP}^2$ is a simply connected four-manifold whose second homology is $\mathbb{Z}$. The standard generator is furnished by the fundamental class of a projective line $\mathbb{CP}^1 \subset \mathbb{CP}^2$. (The projective line is, of course, diffeomorphic to a two-sphere—the 'Riemann sphere'.) Two lines meet in a point and the conventional orientation is fixed so that this self-intersection number is 1. Thus the intersection form is represented by the 1×1 matrix (1). We write $\overline{\mathbb{CP}}^2$ for the same manifold equipped with the opposite orientation; so this manifold has intersection form (-1). (Note that there is no orientation reversing diffeomorphism of $\mathbb{CP}^2$.)

(iii) In the product manifold $S^2 \times S^2$ standard generators for the homology are represented by the embedded spheres $S^2 \times \{pt\}$ and $\{pt\} \times S^2$. These spheres intersect transversely in one point in the four-manifold and each has zero self-intersection. The intersection matrix is $\begin{pmatrix} 0 & 1 \\ 1 & 0 \end{pmatrix}$.

(iv) We can think of $S^2 \times S^2$ as being obtained from the trivial line bundle $S^2 \times \mathbb{C}$ by compactifying each fibre separately with a 'point at infinity'. More generally we can do the same thing starting with any complex line bundle over S^2. The line bundles are classified by the integers, via their first Chern class, so we get a sequence of four-manifolds M_d, $d \in \mathbb{Z}$. In each case H_2 is two dimensional; we can take generators to be the class of a two-sphere fibre and the zero section of our original bundle. Then the intersection matrix is

$$Q_d = \begin{pmatrix} 0 & 1 \\ 1 & d \end{pmatrix}.$$

Now it is easy to see that there are only two diffeomorphism classes realized by these manifolds; M_d is a diffeomorphic to $M_0 = S^2 \times S^2$ if d is even and to M_1 if d is odd. This is because the integer d detects the homotopy class of the transition function for the original line bundle in $\pi_1(S^1) = \mathbb{Z}$, while the manifold M_d, as the total space of a two-sphere bundle, depends only on the image of this in $\pi_1(SO(3)) = \mathbb{Z}/2$. It follows of course that the quadratic forms above depend, up to isomorphism, only on the parity of d, which one can readily verify by a suitable change of basis. All the forms have $b^+ = b^- = 1$; however the forms for d odd and d even are *not* equivalent over the integers, so M_1 is not diffeomorphic to $S^2 \times S^2$. The two non-equivalent standard models are:

$$\begin{pmatrix} 1 & 0 \\ 0 & -1 \end{pmatrix} \quad \begin{pmatrix} 0 & 1 \\ 1 & 0 \end{pmatrix}. \tag{1.1.2}$$

We say an integer quadratic form Q is of *even type* if $Q(x)$ is even for all x in the lattice, and that the form is of *odd type* if it is not of even type. Then we see that the form Q_d is even if and only if d is even.

(v) For any two four-manifolds X_1, X_2 we can make the *connected sum* $X_1 \# X_2$. If X_1, X_2 are simply connected, so is the connected sum; $H_2(X_1 \# X_2)$ is the direct sum of the $H_2(X_i)$ and the intersection form is the obvious direct sum. Starting with the basic building blocks above, we can make many more four-manifolds: for example by taking sums of copies of $\mathbb{CP}^2$ with appropriate orientations we get manifolds $l\mathbb{CP}^2 \# m\overline{\mathbb{CP}}^2$ with forms:

$$\mathrm{diag}(\underbrace{1, \dots, 1}_{l}, \underbrace{-1, \dots, -1}_{m}) = l(1) \oplus m(-1).$$

In fact the manifold $\mathbb{CP}^2 \# \overline{\mathbb{CP}}^2$ is diffeomorphic to M_1 of (iv). One can see this by thinking of the Hopf fibration $S^1 \to S^3 \to S^2$. The complement of a small ball in $\mathbb{CP}^2$ can be identified with the disc bundle over S^2 (a line in $\mathbb{CP}^2$) associated with this circle bundle. When we make the connected sum we glue two of these disc bundles, with opposite orientations, along their boundary spheres to get the S^2 bundle considered in (iv).

1.1.3 Unimodular forms

How far do these examples go to cover the possible unimodular forms? It turns out that the algebraic classification of unimodular *indefinite* forms is rather simple. Any odd indefinite form is equivalent over the integers to one of the $l(1) \oplus m(-1)$ and any even indefinite form to one of the family $l\begin{pmatrix} 0 & 1 \\ 1 & 0 \end{pmatrix} \oplus m E_8$, where E_8 is a certain positive definite, even form of rank 8 given by the matrix:

$$E_8 = \begin{pmatrix} 2 & 0 & -1 & & & & & \\ 0 & 2 & 0 & -1 & & & & \\ -1 & 0 & 2 & -1 & & & & \\ & -1 & -1 & 2 & -1 & & & \\ & & & -1 & 2 & -1 & & \\ & & & & -1 & 2 & -1 & \\ & & & & & -1 & 2 & -1 \\ & & & & & & -1 & 2 \end{pmatrix}. \tag{1.1.3}$$

In other words, indefinite, unimodular forms are classified by their rank, signature and type. (This is the *Hasse–Minkowski* classification of indefinite forms.) Thus we have found, so far, four-manifolds corresponding to all the odd indefinite forms but only the forms $l\begin{pmatrix} 0 & 1 \\ 1 & 0 \end{pmatrix}$ in the even family.

The situation for definite forms is quite different. For each fixed rank there are a finite number of isomorphism classes, but this number grows quite rapidly with the rank—there are many exotic forms, E_8 being the prototype, not equivalent to the standard diagonal form. In fact, up to isomorphism, there is just one even positive-definite form of rank 8, two of rank 16, namely $E_8 \oplus E_8$ and E_{16}, and five of rank 24, including $3E_8$, $E_8 \oplus E_{16}$ and the Leech lattice.

Notice that we only consider above those definite forms whose rank is a multiple of eight: this is due to the following algebraic fact. For any unimodular form Q, an element c of the lattice is called *characteristic* if

$$Q(c, x) = Q(x, x) \bmod 2$$

for all x in the lattice; then if c is characteristic we have

$$Q(c, c) = \text{signature}(Q) \mod 8. \tag{1.1.4}$$

If Q is even the element 0 is characteristic, and we find that the signature must be divisible by 8. (Note that characteristic elements can always be found, for any form.)

1.1.4 The tangent bundle: characteristic classes and spin structures

In general one obtains invariants of smooth manifolds, beyond the homology groups themselves, as characteristic classes of the tangent bundle. For an oriented four-manifold X the characteristic classes available comprise the Stiefel–Whitney classes $w_i(TX) \in H^i(X; \mathbb{Z}/2)$ and the Euler and Pontryagin classes $e(X)$, $p_1(TX) \in H^4(X; \mathbb{Z}) = \mathbb{Z}$. The second Stiefel–Whitney class w_2 can be obtained from the mod 2 reduction of the intersection form by the Wu formula:

$$Q(w_2(TX), \alpha) = Q(\alpha, \alpha) \mod 2, \tag{1.1.5}$$

for all $\alpha \in H^2(X; \mathbb{Z}/2)$. This is especially easy to see when X is simply connected. Then any mod 2 class is the reduction of an integral class and so can be represented by an oriented embedded surface Σ. We have:

$$\langle w_2(TX), [\Sigma] \rangle = \langle w_2(T\Sigma \oplus \nu_\Sigma), [\Sigma] \rangle = \langle w_2(T\Sigma) + w_2(\nu_\Sigma), [\Sigma] \rangle,$$

where ν_Σ is the normal bundle. The Wu formula follows for, on the oriented two-plane bundles $T\Sigma$ and ν_Σ, the class w_2 is the mod 2 reduction of the Euler class; $e(\nu_\Sigma)$ is the self-intersection number $\Sigma.\Sigma$ of Σ, and $e(T\Sigma)$ is the Euler characteristic $2 - 2\cdot\text{genus}(\Sigma)$, which is even. It is in fact the case that for any oriented four-manifold w_1 and w_3 are both zero. This is trivial for simply connected manifolds and we see that in this case the Stiefel–Whitney classes give no extra information beyond the integral intersection form.

The Euler and Pontryagin classes of a four-manifold can both be obtained from the rational cohomology ring. For the Euler class we have the elementary formula

$$e(TX) = \Sigma(-1)^i b_i,$$

the alternating sum of the Betti numbers b_i. The Pontryagin class is given by a deeper formula, the Hirzebruch Signature Theorem in dimension 4,

$$p_1(TX) = 3\tau(X) = 3(b^+ - b^-). \tag{1.1.6}$$

So in sum we see that all the characteristic class data for a simply connected four-manifold is determined by the intersection form on H_2.

In any dimension $n > 1$ the special orthogonal group $SO(n)$ has a connected double cover $\text{Spin}(n)$. If V is a smooth oriented n-manifold with a Riemannian metric, the tangent bundle TV has structure group $SO(n)$. The Stiefel–Whitney class w_2 represents the obstruction to lifting the structure group of TV to Spin(n). Such a lift is called a 'spin structure' on V. If $w_2 = 0$ a spin structure exists and, if also $H^1(X; \mathbb{Z}/2) = 0$, it is unique. In particular a simply connected four-manifold has a spin structure if and only if its intersection form is even, and this spin structure is unique.

A special feature, which permeates four-dimensional geometry, is the fact that Spin(4) splits into a product of two groups: $\mathrm{Spin}(4) = SU(2) \times SU(2)$. One way to understand this runs as follows. Distinguish two copies of $SU(2)$ by $SU(2)^+$, $SU(2)^-$ and let S^+, S^- be their fundamental two-dimensional complex representation spaces. Then $S^+ \otimes_{\mathbb{C}} S^-$ has a natural Hermitian metric and also a complex symmetric form (the tensor product of the skew forms on S^+, S^-). Together these define a real subspace $(S^+ \otimes S^-)_{\mathbb{R}}$, the space on which the symmetric form is equal to the metric. The symmetry group $SU(2)^+ \times SU(2)^-$ acts on $S^+ \otimes S^-$, preserving the real subspace, and this defines a map from $SU(2)^+ \times SU(2)^-$ to $SO(4)$ which one can verify to be a double cover. In the same way a spin structure on a four-manifold can be viewed as a pair of complex vector bundles S^+, S^-—the spin bundles—each with structure group $SU(2)$, and an isomorphism $S^+ \otimes S^- = TX \otimes \mathbb{C}$, compatible with the real structures. (We will come back to spin structures in Chapter 3.)

1.1.5 Self-duality and special isomorphisms

The splitting of Spin(4) is related to the decomposition of the 2-forms on a four-manifold which will occupy a central position throughout this book. On an oriented Riemannian manifold X the $*$ operator interchanges forms of complementary degrees. It is defined by comparing the natural metric on the forms with the wedge product:

$$\alpha \wedge *\beta = (\alpha, \beta)\, d\mu \tag{1.1.7}$$

where $d\mu$ is the Riemannian volume element. So, on a four-manifold, the $*$ operator takes 2-forms to 2-forms and we have $** = 1_{\Lambda^2}$. The *self-dual* and *anti-self-dual* forms, denoted Ω^+_X, Ω^-_X respectively, are defined to be the ± 1 eigenspaces of $*$, they are sections of rank-3 bundles Λ^+, Λ^-:

$$\Lambda^2 = \Lambda^+ \oplus \Lambda^-, \quad \alpha \wedge \alpha = \pm|\alpha|^2\, d\mu, \quad \text{for } \alpha \in \Lambda^{\pm}. \tag{1.1.8}$$

Reverting to the point of view of representations, the splitting of Λ^2 corresponds to a homomorphism $SO(4) \to SO(3)^+ \times SO(3)^-$. But $SU(2)$ can be identified with Spin(3) and the whole picture can be expressed by

$$\begin{array}{ccc} \mathrm{Spin}(4) = & SU(2)^+ \times SU(2)^- & = \mathrm{Spin}(3)^+ \times \mathrm{Spin}(3)^- \\ \downarrow & \downarrow & \\ SO(4) & \longrightarrow SO(3)^+ \times SO(3)^-. & \end{array} \tag{1.1.9}$$

Over a four-manifold the $*$-operator on two-forms, and hence the self-dual and anti-self-dual subspaces, depend only on the conformal class of the

Riemannian metric. It is possible to turn this around, and regard a conformal structure as being defined by these subspaces. This is a point of view we will adopt at a number of points in this book. Consider first the intrinsic structure on the six-dimensional space $\Lambda^2(U)$ associated with an oriented four-dimensional vector space U. The wedge product gives a natural indefinite quadratic form q on U, with values in the line Λ^4. A choice of volume element makes this into a real-valued form. Plainly this form has signature 0; a choice of conformal structure on U singles out maximal positive and negative subspaces Λ^+, Λ^- for q. Note in passing that the null cone of the form q on Λ^2 has a simple geometric meaning—the rays in the null cone are naturally identified with the oriented two-planes in U. On the other hand, given a metric, this set of rays can be identified with the set of pairs $(\omega^+, \omega^-) \in \Lambda^+ \times \Lambda^-$ such that $|\omega^+| = |\omega^-| = 1$. So we see that the Grassmannian of oriented two-planes in a Euclidean four-space can be identified with $S^2 \times S^2$.

Now, in the presence of the intrinsic form q one of the subspaces, say Λ^-, determines the other; it is the annihilator with respect to q. The algebraic fact we wish to point out is that for any three-dimensional negative subspace $\Lambda^- \subset \Lambda^2$ there is a unique conformal structure on U for which this is the anti-self-dual subspace. (Note that the discussion depends on the volume element in Λ^4 only through the orientation; switching orientation just switches Λ^+ and Λ^-.) This is a simple algebraic exercise: it is equivalent to the assertion that the representation on Λ^2 exhibits $SL(U) = SL(4, \mathbb{R})$ as a double cover of the identity component of $SO(\Lambda^2, q) = SO(3, 3)$. This is another of the special isomorphisms between matrix groups. (The double cover $SO(4) \to SO(3) \times SO(3)$ considered before can be derived from this by taking maximal compact subgroups.)

For purposes of calculation we can exploit this representation of conformal structures as follows. Fix a reference metric on U and let Λ_0^+, Λ_0^- be the corresponding subspaces. Any other negative subspace Λ^- can be represented as the graph of a unique linear map,

$$m : \Lambda_0^- \longrightarrow \Lambda_0^+, \tag{1.1.10}$$

such that $|m(\omega)| < |\omega|$ for all non-zero ω in Λ_0^- (see Fig. 1). Thus there is a bijection between conformal structures on U and maps m from Λ_0^- to Λ_0^+ of operator norm less than 1. We can identify the new subspace Λ^- with Λ_0^-, using 'vertical' projection, and similarly for Λ^+. Then if α is a form in Λ^2, with components (α^+, α^-) in the old decomposition, the self-dual part with respect to the new structure is represented by:

$$(1 + mm^*)^{-1} \, (\alpha^+ + m\alpha^-). \tag{1.1.11}$$

This discussion goes over immediately to an oriented four-manifold X; given a fixed reference metric, we can identify the conformal classes with bundle maps $m : \Lambda^- \to \Lambda^+$ with operator norm everywhere less than 1.

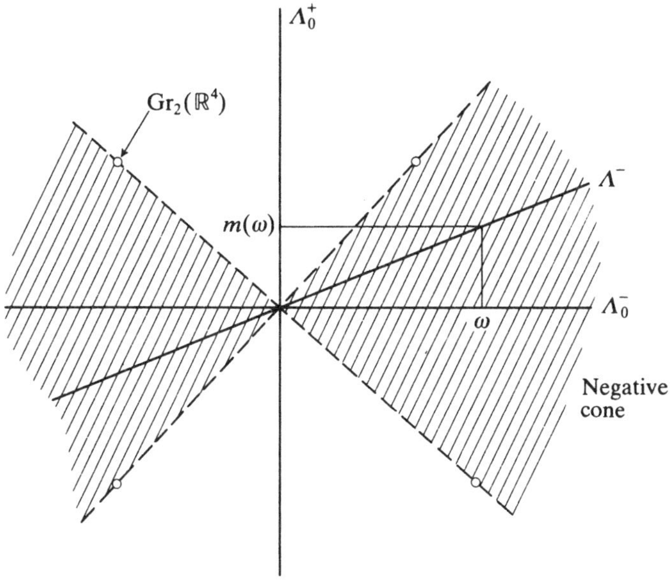

Fig. 1

1.1.6 Self-duality and Hodge theory

On any compact Riemannian manifold the Hodge Theory gives preferred representatives for cohomology classes by harmonic differential forms. Recall that one introduces the formal adjoint operator,

$$\mathrm{d}^*:\Omega^p \longrightarrow \Omega^{p-1}, \tag{1.1.12}$$

associated with the intrinsic exterior derivative by the metric, so that

$$\int (\mathrm{d}\alpha, \beta) = \int (\alpha, \mathrm{d}^*\beta); \tag{1.1.13}$$

in the oriented case $\mathrm{d}^* = \pm * \mathrm{d} *$. The Hodge theorem asserts that a real cohomology class has a unique representative α with:

$$\mathrm{d}\alpha = \mathrm{d}^*\alpha = 0. \tag{1.1.14}$$

For a compact, oriented four-manifold there is an interaction between the splitting of Λ^2 and the Hodge theory, which will be central to much of the material in this book. First, the harmonic two-forms are preserved by the $*$ operator (which interchanges $\ker \mathrm{d}$ and $\ker \mathrm{d}^*$), so given a metric we get a decomposition,

$$H^2(X; \mathbb{R}) = \mathscr{H}^+ \oplus \mathscr{H}^-, \tag{1.1.15}$$

into the self-dual and anti-self-dual (ASD) harmonic 2-forms. It follows immediately from the definition that these are maximal positive and negative subspaces for the intersection form Q, viewed now as the wedge product on de Rham classes. So:

$$\dim \mathscr{H}^+ = b^+, \quad \dim \mathscr{H}^- = b^-. \tag{1.1.16}$$

These spaces of $\pm$ self-dual harmonic forms can be obtained by a slightly different procedure. Over our four-manifold X, with a Riemannian metric, let

$$\mathrm{d}^+ : \Omega^1_X \longrightarrow \Omega^+_X \tag{1.1.17}$$

be the first-order differential operator formed from the composite of the exterior derivative $\mathrm{d}: \Omega^1 \to \Omega^2$ with the algebraic projection $\frac{1}{2}(1 + *)$ from Ω^2 to Ω^+. We then have a three-term complex:

$$\Omega^0_X \xrightarrow{\mathrm{d}} \Omega^1_X \xrightarrow{\mathrm{d}^+} \Omega^+_X. \tag{1.1.18}$$

This is, roughly speaking, half of the ordinary de Rham complex.

Proposition (1.1.19). *If X is compact the cohomology groups of this complex can be naturally identified with $H^0(X; \mathbb{R}), H^1(X; \mathbb{R})$ and $\mathscr{H}^+_X$, in dimensions* 0, 1, 2 *respectively.*

This is a simple exercise in Hodge theory. The assertion for H^0 is trivial. For the middle term, observe that for any α in Ω^1,

$$\int_X (|\mathrm{d}^+\alpha|^2 - |\mathrm{d}^-\alpha|^2)\,\mathrm{d}\mu = \int_X \mathrm{d}\alpha \wedge \mathrm{d}\alpha = \int_X \mathrm{d}(\alpha \wedge \mathrm{d}\alpha) = 0, \tag{1.1.20}$$

by Stokes' theorem. So if $\mathrm{d}^+\alpha = 0$ we must have $\mathrm{d}^-\alpha = 0$, hence $\ker \mathrm{d}^+/\mathrm{im}\,\mathrm{d} = \ker \mathrm{d}/\mathrm{im}\,\mathrm{d} = H^1(X; \mathbb{R})$, as required. For the last term, observe that $\mathscr{H}^+_X \subset \Omega^+_X$ is orthogonal, in the L^2 sense, to the image of d^+. So we have an induced injection from $\mathscr{H}^+_X$ to $\Omega^+_X/\mathrm{im}\,\mathrm{d}^+$. To show that this is surjective we write an element ω^+ of Ω^+_X in its Hodge decomposition:

$$\omega^+ = h + \mathrm{d}\alpha + \mathrm{d}^*\gamma = h + \mathrm{d}\alpha + *\mathrm{d}\beta,$$

where h is a harmonic form and $\beta = *\gamma$.
Then, since $*\omega^+ = \omega^+$, the uniqueness of the Hodge decomposition gives that $\alpha = \beta$, and $h = *h$, so

$$\omega^+ = h + \mathrm{d}^+(2\alpha)$$

and $h \in \mathscr{H}^+_X$ represents the same class in $\Omega^+_X/\mathrm{im}\,\mathrm{d}^+$.

1.1.7 Complex surfaces

A complex surface is a complex manifold of complex dimension two—that is, a smooth four-manifold S with an atlas of charts whose transition functions are holomorphic on domains in $\mathbb{C}^2 = \mathbb{R}^4$. The tangent bundle of a complex surface acquires the structure of a complex vector bundle (with fibre $\mathbb{C}^2$) and therefore has Chern classes $c_i(S) = c_i(TS) \in H^{2i}(S;\mathbb{Z})$. There is a standard convention fixing the orientation of a complex manifold and these Chern classes are related to the characteristic classes considered above by the formulae:

$$c_2(S) = e(S),$$

$$c_1(S) = w_2(S) \bmod 2,$$

$$c_1(S)^2 = p_1(S) + 2e(S) = 3\tau(S) + 2e(S). \qquad (1.1.21)$$

So the new cohomological data is the class c_1, a preferred integral lift of w_2, whose cup square is prescribed by the cohomology of S. (Notice that c_1 is a characteristic element for the intersection form, as defined in Section 1.1.3.) Now these characteristic classes are defined for any *almost complex* four-manifold, i.e. a four-manifold with a complex structure on its tangent bundle, not necessarily coming from a system of complex charts. The same formulae relating the Chern classes to the invariants of X hold, but for *almost* complex structures there is in addition a simple classification theory. This is because the existence of almost complex structures is purely a matter of bundle, and so homotopy, theory. We may always choose metrics compatible with a bundle structure, so an almost complex structure can be viewed as a reduction of the structure group of the tangent bundle from $SO(4)$ to $U(2) \subset SO(4)$. Such reductions correspond to sections of a bundle with fibre $SO(4)/U(2)$. This can be set up more explicitly in terms of our discussion of the 2-forms in Section 1.1.5. The double cover of $SO(4)$ induces a cover of $U(2)$ by $S^1 \times SU(2)$. In the decomposition of Spin(4) this $SU(2)$ corresponds to $SU(2)^-$ and the S^1 to a standard one-parameter subgroup in $SU(2)^+$. Passing down to the 2-forms, we see that an almost complex structure corresponds to a reduction of the structure group of Λ^+ from $SO(3)$ to S^1. Put in a different way, it corresponds to a decomposition

$$\Lambda^+ = \underline{\mathbb{R}} \oplus E \qquad (1.1.22)$$

where $\underline{\mathbb{R}}$ is the trivial real line bundle. (This is the piece spanned by the 2-form associated with a Hermitian metric; the decomposition (1.1.22) will have paramount importance in Chapter 6.) It follows then that almost complex structures correspond to homotopy classes of sections of the two-sphere bundle over X formed from the unit spheres in Λ^+.

From this starting point it is a comparatively routine matter to show that if we are given any characteristic element c in $H^2(X; Z)$ with $c^2 = 3\tau + 2e$ then there is an almost complex structure on x with $c_1 = c$. Thus the existence of almost complex structures comes down to a purely arithmetical question about the intersection form. Note that, in the simply connected case, a necessary condition for the existence of an almost complex structure on X is that $b^+(X)$ be *odd*, (since, by (1.1.4), $c_1^2 = \tau \bmod 8$.) For a (simply connected) complex surface S we have, indeed, the formula:

$$b^+ = 1 + 2p_g(S), \tag{1.1.23}$$

where $p_g(S)$ is the *geometric genus* (the dimension of the space of holomorphic 2-forms). This is a version of the 'Hodge index theorem'. (In fact the formula holds if the first Betti number of S is even. In the Kähler case the formula can be obtained by decomposing the space $\mathscr{H}^+$ of self-dual harmonic forms into the holomorphic forms, of real dimension $2p_g$, and a one-dimensional piece spanned by the Kähler form. Thus the formula is a global manifestation of the bundle decomposition (1.1.22).)

Complex surfaces arise most often as complex projective varieties—smooth submanifolds of $\mathbb{CP}^N$ cut out by homogeneous polynomial equations. (It is known from the theory of surfaces that any simply connected compact complex surface can be deformed into—hence is diffeomorphic to—such an algebraic surface.) These give us a large supply of examples of four-manifolds, with which we shall be specially concerned in this book. We will illustrate the usefulness of characteristic classes by computing the homotopy invariants in two simple families of complex surfaces.

(i) *Hypersurfaces in* $\mathbb{CP}^3$: We consider a smooth hypersurface S_d of degree d in $\mathbb{CP}^3$—zeros of a homogeneous polynomial of degree d in four variables. The diffeomorphism type does not depend on the polynomial chosen—we could take for example the hypersurface $z_0^d + z_1^d + z_2^d + z_3^d = 0$. The Lefschetz hyperplane theorem tells us, in general, that the homotopy groups of a hypersurface of complex dimension n in projective space agree with those of the ambient space up to dimension $n - 1$, so S_d is simply connected like $\mathbb{CP}^3$. The cohomology of $\mathbb{CP}^3$ is freely generated, as a group, by $1, h, h^2, h^3$ where $h \in H^2(\mathbb{CP}^3; \mathbb{Z})$ is the *hyperplane class*: the class dual to a hyperplane $\mathbb{CP}^2 \subset \mathbb{CP}^3$. To say that S_d has degree d means that its dual class in $\mathbb{CP}^3$ is $d.h$ and that its normal bundle is the restriction of H^d, where H is the Hopf line bundle on $\mathbb{CP}^3$, with first Chern class h. We can obtain the Chern classes of S_d from the Whitney product formula:

$$c(T\mathbb{CP}^3|_{S_d}) = c(TS_d \oplus H^d) = c(TS_d)c(H^d).$$

(Here c denotes the total Chern class, $c = 1 + c_1 + c_2 + \ldots.$) We need to know that $c(T\mathbb{CP}^3) = (1 + h)^4$; then we can invert the formula to calculate:

$$c_1(TS_d) = (4 - d)h|_{S_d}, \quad c_2(TS_d) = (6 - 4d + d^2)h^2|_{S_d}.$$

Then, using the fact that h^2 is d times the fundamental class on S_d and the various formulae above, one obtains:

$$b_2(S_d) = d(6 - 4d + d^2) - 2, \quad \tau(S_d) = \tfrac{1}{3}(4 - d^2)d \tag{1.1.24}$$

and

$$w_2(TS_d) = dh (\text{mod } 2).$$

Consider small values of d. When $d = 1$ the surface is a hyperplane $\mathbb{CP}^2$, one of our elementary manifolds discussed in Section 1.1.2. We have also met the four-manifolds S_2 and S_3 before. To see this we must appeal to some classical projective geometry. First, the quadric surface S_2 contains two rulings by lines in $\mathbb{CP}^3$ and these give a diffeomorphism $S_2 = S^2 \times S^2$ (thinking of S^2 as $\mathbb{CP}^1$), which can easily be written down explicitly. For S_3 our formulae give $b^+ = 1$, $b^- = 6$ and we have indeed that S_3 is diffeomorphic to $\mathbb{CP}^2 \# 6\overline{\mathbb{CP}}^2$. In geometry one knows that S_3 is a *rational surface*, admitting a birational equivalence $f: \mathbb{CP}^2 \to S_3$. The map f is undefined at six points in $\mathbb{CP}^2$, at which it has local singularities of the form $f(z, w) = (z, z/w)$. In general if x is a point in a complex surface S the 'blow-up' $\tilde{S}$ of S at x is another complex surface, obtained from S by replacing x with the projective line $P(TS_x)$. It is easy to show that, as a differentiable manifold, the blow-up $\tilde{S}$ can be identified with the connected sum $S \# \overline{\mathbb{CP}}^2$. Now blowing up the six points in $\mathbb{CP}^2$ has the effect of removing the indeterminacy in f, which induces a smooth map from $\mathbb{CP}^2 \# 6\overline{\mathbb{CP}}^2$ to S_3.

We mention here that complex geometry can be used to show that the manifolds $\mathbb{CP}^2 \# 2\overline{\mathbb{CP}}^2$ and $(S^2 \times S^2) \# \overline{\mathbb{CP}}^2$ are diffeomorphic (and so that we certainly do not have 'unique factorization' into connected sums). For this we consider $S^2 \times S^2$ as the quadric S_2 and $\mathbb{CP}^2$ as a plane in $\mathbb{CP}^3$. Fix a point p in S_2, then define the projection map π from $S_2 \setminus p$ to $\mathbb{CP}^2$ by letting $\pi(q)$ be the point of intersection of the line $\overline{pq}$ with the plane. This map does not extend over p but if we blow up p we get a smooth map from $S^2 \times S^2 \# \overline{\mathbb{CP}}^2$ to $\mathbb{CP}^2$. In the other direction, there are two lines in S_2 which are collapsed to a pair of points in $\mathbb{CP}^2$ by π, and if we blow up these points we can define an inverse map (a typical line in $\mathbb{CP}^3$ meets S_2 in two points); and in this way we find the required diffeomorphism.

The situation changes when we consider surfaces of degree $d \geq 4$. From the point of view of complex geometry we now have 'irrational' surfaces, and we cannot expect to use the same techniques to describe them as being built out of the elementary manifolds by connected sums. What our formulae do give are the intersection forms. We see that S_d is spin if and only if d is even, and the intersection form is indefinite for $d > 1$. So by the classification theorem the intersection form must be

$$\lambda_d(1) \oplus \mu_d(-1)$$

for d odd, with

$$\lambda_d = \tfrac{1}{3}(d^3 - 6d^2 + 11d - 3), \quad \mu_d = \tfrac{1}{3}(d-1)(2d^2 - 4d + 3), \tag{1.1.25}$$

and $\quad l_d \begin{pmatrix} 0 & 1 \\ 1 & 0 \end{pmatrix} \oplus m_d(-E_8)$

for d even, with

$$l_d = \tfrac{1}{3}(d^3 - 6d^2 + 11d - 3), \quad m_d = (1/24)d(d^2 - 4). \tag{1.1.26}$$

(Note the power of the classification theorem; it would be a formidable task to actually exhibit bases of two-cycles in S_d meeting in these intersection patterns.)

The surface S_4 has played an important role in four-manifold topology. It is an example of a '$K3$ surface'. (A $K3$ surface is a compact, simply connected complex surface with $c_1 = 0$. All $K3$ surfaces are diffeomorphic to S_4 but they cannot all be realized as complex surfaces in $\mathbb{CP}^3$.) We note that the intersection form is

$$3\begin{pmatrix} 0 & 1 \\ 1 & 0 \end{pmatrix} \oplus 2(-E_8), \tag{1.1.27}$$

and this is the smallest of the forms appearing in the family S_d which contain the 'exotic' summand E_8.

(ii) *Branched covers*: For our second family we begin with a smooth complex curve (Riemann surface) B of degree $2p$ in the plane $\mathbb{CP}^2$. Then we construct a surface R_p which is a double cover of $\mathbb{CP}^2$ branched along B. So we have an analogue, in two complex variables, of the familiar picture of a Riemann surface as a branched cover of the Riemann sphere. Precisely, we fix a section s of the line bundle $H^{\otimes 2p} = H^{2p}$ over $\mathbb{CP}^2$ cutting out B and define R_p to be the subspace of the total space of $H^p \to \mathbb{CP}^2$ defined by the equation $\xi^2 = s$. The projection map in H^p induces a map $\pi\colon R_p \to \mathbb{CP}^2$ which is two-to-one away from B.

A version of the Lefschetz theorem shows that R_p is simply connected. To compute its invariants we first use the formula (which can be derived by a simplex-counting argument) for the Euler characteristic of a branched cover:

$$e(R_p) = 2e(\mathbb{CP}^2) - e(B).$$

The branch curve B is a Riemann surface whose genus is

$$g(B) = (p-1)(2p-1),$$

hence

$$e(R_p) = 2(p-1)(2p-1) + 4.$$

Next we look at $c_1(R_p)$. This is minus the Chern class of the line bundle $\Lambda^2 T^* R_p$ of holomorphic two-forms. If ψ is a local non-vanishing holomorphic form on $\mathbb{CP}^2$ the lift $\pi^*(\psi)$ has a simple zero along $\pi^{-1}(B) = B$. This implies that

$$\begin{aligned} c_1(\Lambda^2 T^* R_p) &= \pi^*(c_1(\Lambda^2 T^* \mathbb{CP}^2)) + [\pi^{-1}(B)] \\ &= \pi^*(c_1(\Lambda^2 T^* \mathbb{CP}^2) + \tfrac{1}{2}B). \end{aligned}$$

Now $\Lambda^2 T^* \mathbb{CP}^2$ is isomorphic to H^{-3} and we deduce that

$$c_1(R_p) = (p-3)\pi^*(h); \tag{1.1.28}$$

so

$$c_1(TR_p)^2 = 2(c_1(H^{p-3}))^2 = 2(p-3)^2.$$

Putting our calculations together we get

$$b^+(R_p) = p^2 - 3p + 3, \quad b^-(R_p) = 3p^2 - 3p + 1. \tag{1.1.29}$$

Moreover we see that R_p is spin (i.e. $\pi^*((p-3)h)$ is 0 mod 2) precisely when p is odd, so we can obtain the intersection forms as before.

The family of surfaces R_p displays the same general behaviour as the family S_d. When p is 1 we get $S^2 \times S^2$ again, and when p is 2 we get another rational surface diffeomorphic to $\mathbb{CP}^2 \# 7\overline{\mathbb{CP}}^2$. When p is 3 we get a $K3$ surface, and for $p \geq 4$ we get irrational surfaces of 'general type'. We shall study the surface R_4 in more detail in Chapter 10.

1.2 Classification results obtained by conventional topological methods

It should be clear now that a central question in four-manifold theory is this: to what extent is a simply connected four-manifold determined by its intersection form? We have seen that the form contains all the homological information and the characteristic class data, and that questions of spin and almost complex structures can be settled knowing the intersection form alone. But what of the differential topology of the four-manifold?

Of course there is also a complementary question: which forms are realized by compact four-manifolds? In this section we will set down some results in this direction which are obtained by standard topological methods.

1.2.1 Homotopy type

The following theorem was deduced by Milnor (1958) from a general result of Whitehead (1949):

Theorem (1.2.1). *The oriented homotopy type of a simply connected, compact, oriented four-manifold X is determined by its intersection form.*

To understand this fact, start by removing a small ball B^4 from X. The punctured manifold has homology groups $H_2(X \setminus B^4) = H_2(X)$, $H_i(X \setminus B^4) = 0$ for $i = 1, 3, 4$. By the Hurewicz theorem, the generators of H_2 can be represented by maps $f_i : S^2 \to X \setminus B^4$. We thus obtain a map

$$f = \bigvee f_i : \bigvee S^2 \longrightarrow X \setminus B^4$$

from a 'wedge' (or one-point union) of two-spheres which induces isomorphisms of all homology groups and is therefore a homotopy equivalence. So, up to homotopy, X is obtained by attaching a four-cell to a wedge of two-spheres,

$$X = \left(\bigvee S^2\right) \cup_h e^4, \tag{1.2.2}$$

by an attaching map $h : S^3 \to \bigvee S^2$. The homotopy type of X is determined by the homotopy class of h, so the theorem comes down to the calculation of

$\pi_3(\bigvee S^2)$, which can be tackled in a number of ways. Notice that for any h we can construct a four-dimensional space X_h, which is not necessarily a manifold but which has a 'fundamental homology class' in H_4. Thus we can associate with any h a quadratic form via the cup product in X_h. The result we need then is that the homotopy classes of maps from S^3 to the wedge of two-spheres are in one-to-one correspondence with the symmetric matrices (expressing the quadratic form relative to the preferred basis for H^2).

One can see the symmetric matrix associated with h more directly as follows: we can suppose that h is smooth on the open subset in S^3, which maps to the complement of the vertex in the wedge. Then by Sard's theorem generic points x_i in the two-spheres are regular values of h and the preimages $h^{-1}(x_i)$ are smooth, compact one-dimensional submanifolds K_i of S^3. The off-diagonal entries in the matrix are given by the total *linking numbers* of the K_i in S^3 (note that K_i need not be connected). We also have a trivialization of the normal bundle of K_i in S^3, induced by the derivative of h. This trivialization has a winding number (use the trivialization to construct a 'parallel' copy K_i' of K_i, and take the linking number of K_i with K_i'.) These winding numbers form the diagonal entries of the matrix. With this interpretation it is not hard to show by a direct geometric argument that the matrix determines the homotopy class of h. (This is a generalisation of the Pontryagin–Thom construction.)

1.2.2 Manifolds with boundary

At this point we mention a related construction which gives information about the converse question of the realization of forms. Starting with any symmetric matrix M we can certainly find a link, made up of components $K_i = S^1 \subset S^3$, with linking numbers the off-diagonal entries of M. We use the diagonal entries to choose trivializations of the normal bundles of the K_i, as above, so we get a 'framed link'. Now think of S^3 as the boundary of B^4 and make a new four-manifold-with-boundary Y as follows. For each K_i we take a 'handle'

$$H_i = D^2 \times D^2$$

with a chosen boundary component $D^2 \times S^1$. Then we cut out a tubular neighbourhood N_i of K_i in S^3 and glue the H_i to $B^4 \backslash (\bigcup N_i)$ along the $D^2 \times S^1$ using the given trivialization of the normal bundle. This gives a four-manifold-with-boundary Y. Strictly we have to 'straighten corners' to give the boundary of Y a smooth structure. The whole construction becomes much clearer if one thinks of the two-dimensional analogue in which the integral framings are replaced by elements of $\mathbb{Z}/2$—see Fig. 2.

It is easy to see that Y is simply connected and that the two-dimensional homology of Y has a basis of elements associated with the K_i. The matrix M now reappears as the intersection matrix of Y, or equivalently as the

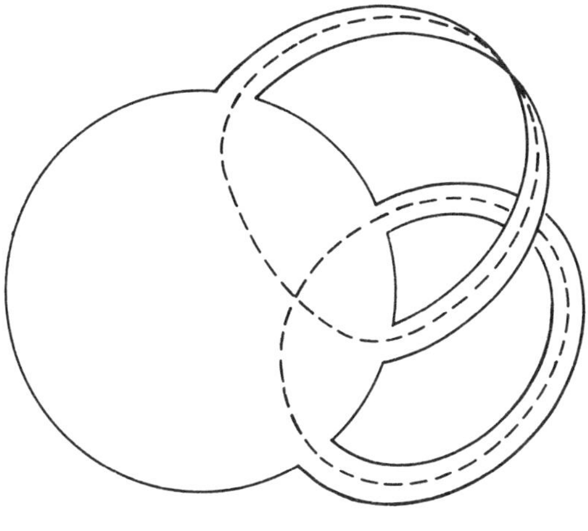

2-manifold with intersection matrix

$$\begin{pmatrix} 1 & 1 \\ 1 & 0 \end{pmatrix} \quad (\text{mod } 2)$$

Fig. 2

composite of the Lefschetz duality $H_2(Y) = H^2(Y, \partial Y)$ with the pairing $H^2(Y, \partial Y) \otimes H_2(Y) \to \mathbb{Z}$. If M is unimodular, one deduces from the exact homology sequence of $(Y, \partial Y)$ that the closed three-manifold ∂Y is a *homology three-sphere*, i.e. that $H_1(Y, \mathbb{Z}) = 0$. So we have:

Theorem (1.2.3). *For any unimodular form Q there is a simply connected four-manifold-with-boundary, having intersection form Q, and boundary a homology three-sphere.*

Alternatively we can add a cone on the boundary to get a closed homology four-manifold, with one singular point and cup-product form Q. This is essentially a specific version of the space X_h constructed, at the level of homotopy type, in Section 1.2.1, using the map h associated with Q.

One gets some insight into the problems of existence and uniqueness from the discussion above. The diffeomorphism type of Y depends, *a priori*, on the link $\bigcup K_i$, which could be very tangled and complicated, while the intersection form only detects the simplest invariant, the linking matrix.

1.2.3 Stable classification and cobordism

We now turn to questions on the classification of smooth manifolds up to diffeomorphism, and the following result of Wall (1964*b*):

Theorem (1.2.4). *If X, Y are simply connected, smooth, oriented four-manifolds with isomorphic intersection forms, then for some $k \geq 0$ there is a diffeomorphism $X \# k(S^2 \times S^2) = Y \# k(S^2 \times S^2)$.*

Here $k(S^2 \times S^2)$ denotes the connected sum of k copies of $S^2 \times S^2$. We shall outline first the proof of a weaker statement than this: that for some k, l, k', l' there is a diffeomorphism

$$X \# k(S^2 \times S^2) \# l(\mathbb{CP}^2 \# \overline{\mathbb{CP}}^2) = Y \# k'(S^2 \times S^2) \# l'(\mathbb{CP}^2 \# \overline{\mathbb{CP}}^2).$$

One can view this theorem in a broader context. In high dimensions Smale proved the generalized Poincaré conjecture, together with other classification theorems, using the 'h-cobordism' theorem. As we shall explain in Section 1.2.4, the proof of the h-cobordism theorem breaks down in four dimensions. Wall's argument follows the same pattern of proof, and the 'stable' classification he obtains (relative to the operation of connected sum with $S^2 \times S^2$) can be viewed as that part of the high-dimensional manifold theory which remains valid for smooth four-manifolds. It is possible to isolate precisely the point at which the proof of the h-cobordism theorem fails in four dimensions and this is the aspect we want to explain.

The procedure Wall follows is this: we know that X and Y are cobordant; that is, there is an oriented five-manifold W with (oriented) boundary the disjoint union $X \cup \bar{Y}$. (This follows from Thom's cobordism theory: the only oriented cobordism invariant of a four-manifold is the signature.) We try to modify W to a product cobordism $X \times [0, 1]$, and so deduce that the ends X, Y are diffeomorphic. The basic notion which enters at a number of points in the story is that of 'surgery'. We have already met a version of this in the construction of the manifold-with-boundary above. In general we exploit the fact that the manifolds $S^i \times B^{j+1}$ and $B^{i+1} \times S^j$ have the same boundary $S^i \times S^j$. So if U is an i-sphere, smoothly embedded in an ambient manifold V of dimension $i + j + 1$, and with trivial normal bundle, we can cut out a neighbourhood of U in V and glue back a copy of $B^{i+1} \times S^j$ to obtain a new manifold, the result of 'surgery along U'. (To make this unambiguous we have to specify a trivialization of the normal bundle of U.)

Surgery is intimately connected with Morse theory, and more generally with the variation of the structure of manifolds defined by generic one-parameter families of equations (cf. the material in Chapter 4, Section 4.3). In our problem we choose a Morse function on W, a smooth map $f: W \to [0, 1]$ which is 0 on X and 1 on Y, with only isolated, nondegenerate critical points. We fix a Riemannian metric on W, so we get a gradient vector field grad f on W. If f has no critical points then grad f has no zeros and the paths of the associated gradient flow all travel from Y to X; in that case these paths will define a diffeomorphism $W = X \times [0, 1]$ as desired. In general, the level set $Z_t = f^{-1}(t)$ is a smooth four-manifold whenever t is not one of the finite set of critical values of f, i.e., when Z_t does not contain a critical point. The diffeomorphism type of Z_t changes only when t crosses a critical value, and the change is precisely by a surgery. Around a critical point we can choose coordinates so that f is given by the quadratic function

$$f(x_1, \ldots, x_5) = c - (x_1^2 + \ldots + x_\lambda^2) + (x_{\lambda+1}^2 + \ldots + x_5^2)$$

for a constant c. Here λ is the *index* of the critical point. (The existence of such coordinates is the content of the Morse lemma.) Then one can see explicitly that Z_t changes by surgery along a sphere of dimension $\lambda - 1$ as t increases through c. (Here we assume that there is only one critical point in $f^{-1}(c)$. When $\lambda = 0$ the effect is to create a new S^4 component in Z_t.) We can illustrate this by considering the analogue for a function of three variables, see Fig. 3.

There now begins a process of modifications to W and to the Morse function f, rearranging and cancelling critical points. First, we may obviously suppose that W is, like X and Y, connected. Then it is not hard to see that we can choose f to have no critical points of index 0, i.e. local minima. We start with any Morse function and then remove the minima by cancelling with critical points of index 1. Again, the picture in lower dimensions (Fig. 4), of a cobordism between one-manifolds, illustrates the idea. Next we argue that we can choose W to be simply connected, like X and Y. This involves another use of surgery. Starting with any W, we represent a system of generators for $\pi_1(W)$ by disjoint, embedded loops. The normal bundles of these loops are trivial since W is oriented, so we can perform surgery on all of these loops to obtain a new manifold which is easily seen to be simply connected. Now a more complicated argument shows that if W, X, Y are all simply connected

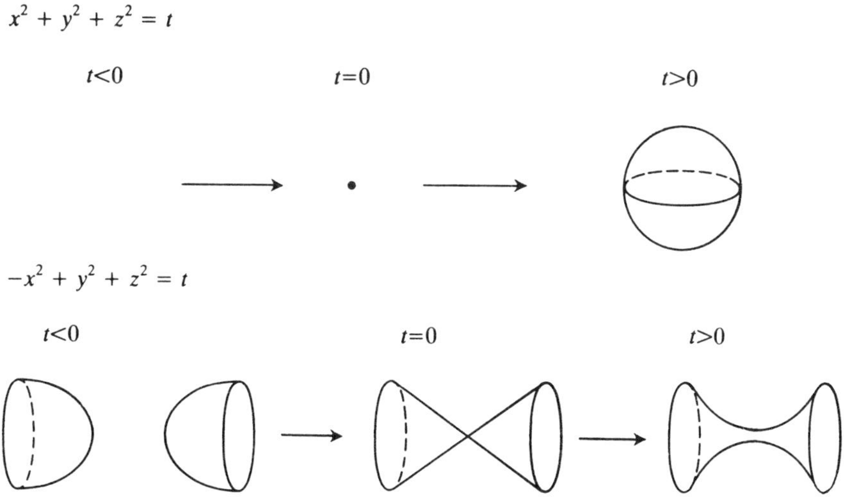

Fig. 3

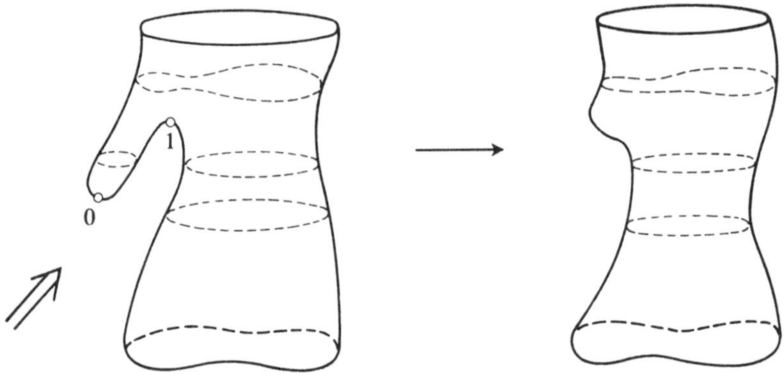

Fig. 4

we can choose f to have no critical points of index 1 or 0. Symmetrically (replace f by $1 - f$) we can remove critical points of index 4 or 5, and we need not introduce new critical points of index 0 or 1 in the process. So we get down to the situation where f has only critical points of index 2 and 3. Finally one shows that the points of index 2 can be arranged to lie 'below' all those of index 3, separated by the level set $Z_{1/2}$, say.

Given the existence of such a cobordism and Morse fuction, the result we are after follows immediately from a consideration of the effect of passing the critical levels, the surgeries, on the global structure of the level sets. Consider again the situation in a lower dimensional example, where W has dimension 3; the level sets are surfaces and we pass a critical point of index 1, performing surgery on a 0-sphere (i.e. a pair of points). Globally, one of three things can happen. The first possibility is that the two points lie in different components of $Z_{c-\varepsilon}$; then the result is that these components come together in $Z_{c+\varepsilon}$ making a connected sum. If on the other hand the two points can be joined by an arc in $Z_{c-\varepsilon}$ then $Z_{c+\varepsilon} = Z_{c-\varepsilon} \# (S^1 \times S^1)$ or $Z_{c-\varepsilon} \# K$, where K is the Klein bottle, according to how a neighbourhood of the arc is twisted—see Fig. 5. Returning to four-manifolds, the analogue of the first possibility does not occur, since all level sets are simply connected. The four-dimensional versions of the other possibilities can occur, the roles of $S^1 \times S^1$ and K being played by the two S^2 bundles over S^2, namely $S^2 \times S^2$ and $\mathbb{CP}^2 \# \overline{\mathbb{CP}^2}$. It follows then that there is a diffeomorphism

$$Z_{1/2} = X \# k(S^2 \times S^2) \# l(\mathbb{CP}^2 \# \overline{\mathbb{CP}^2})$$

where $k + l$ is the number of critical points of index 2. Replacing f by $1 - f$ we see that $Z_{1/2}$ is similarly related to Y and this establishes the weaker form of Wall's result stated.

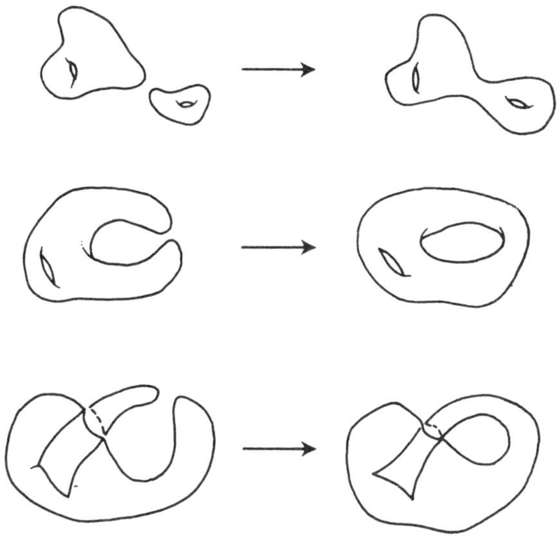

Fig. 5

1.2.4 h-cobordisms; embedded surfaces and the Whitney Lemma

To prove the stronger result (1.2.1), involving only $S^2 \times S^2$s, one must work a little harder. We 'killed' the fundamental group of W by surgeries; Wall goes further and shows that the relative homology group $H_2(W, X; \mathbb{Z})$ can also be killed. When this is done the inclusions of X and Y in W are both homotopy equivalences. A cobordism of this kind is called an 'h-cobordism' so what Wall establishes is the following:

Proposition (1.2.5). *Two simply connected four-manifolds with isomorphic intersection forms are h-cobordant.*

Then one shows that the twisted bundles $\mathbb{CP}^2 \# \overline{\mathbb{CP}}^2$ can be avoided by considering w_2 and spin structures (analogous to w_1 and orientations in the two-dimensional case).

What prevents us from showing that X and Y are actually diffeomorphic? For simply connected manifolds *of dimension* 5 *or more* we have Smale's h-cobordism theorem: if X, Y are h-cobordant then they are diffeomorphic. More precisely, any h-cobordism W between them is a product, i.e. there is a Morse function on W with no critical points. Were this theorem to be true in dimension four one could deduce from the proposition above that a simply connected four-manifold is determined up to diffeomorphism by its intersection form; but the proof of the h-cobordism theorem breaks down in four dimensions. We will now explain the reason for this. (The failure of this proof does not, of course, imply by itself that the h-cobordism theorem does not hold in four dimensions.)

Suppose then that, in line with the discussion above, we have an h-cobordism W and function f with, for simplicity, just one critical point, p, of index 2 and one, q, of index 3, separated by $Z_{1/2}$. We would like to cancel these to obtain a function without critical points. Now the gradient vector field grad f defines a flow on W, and every point flows as $t \to +\infty$ to X or to a critical point, and as $t \to -\infty$ to Y or to a critical point. In the proof of the h-cobordism theorem one shows that p and q can be cancelled if there is exactly one flow line running from q (at $t = -\infty$) to p (at $t = +\infty$)—compare Fig. 6. The points in $Z_{1/2}$ which flow down to p form an embedded two-sphere S_- and symmetrically the points which flow up to q as $t \to -\infty$ form a two-sphere S_+. So we can cancel p and q if S_+, S_- meet in exactly one point in $Z_{1/2}$ (and we are assuming that the intersection is transverse). On the other hand the fact that W is an h-cobordism implies, by straightforward homology theory, that in any case the *algebraic* intersection number of S_+, S_- (adding up intersection points with signs) is 1. The crucial point then is this: if there is an isotopy (a one parameter family of self-diffeomorphisms) of $Z_{1/2}$, moving S_+ to a sphere S'_+ whose geometric intersection with S_- *agrees* with its algebraic intersection, then we can modify f correspondingly to satisfy the gradient flow criterion and hence cancel p with q.

The Whitney lemma bears on precisely this issue: the comparison of geometric and algebraic intersection numbers. Suppose in general that P, Q are submanifolds of complementary dimensions in an ambient simply connected manifold M. Suppose P and Q intersect transversely but geometric

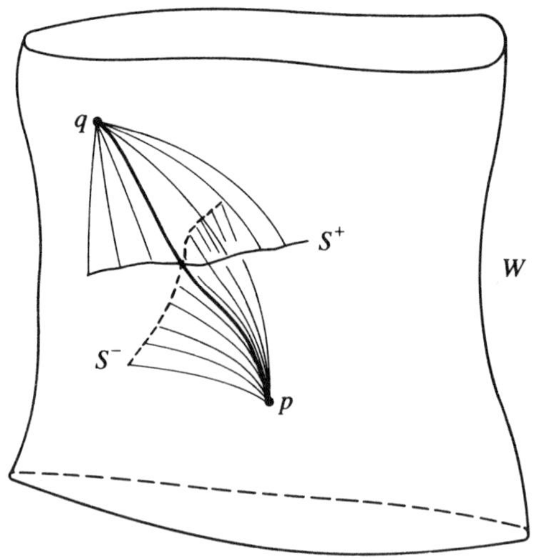

Fig. 6

and algebraic intersections are different, so there are intersection points x, y of opposite sign. We assume P, Q are connected and choose arcs α, β in P, Q respectively joining the intersection points. Since M is simply connected the composite loop $\alpha \cup \beta$ is inessential and we can try to find an embedded disc D, with boundary $\alpha \cup \beta$ but otherwise disjoint from P and Q. If such a 'Whitney disc' can be found it can be used to guide an isotopy of M, moving P say, to cancel the intersection points. Figure 7, in dimension three, should give an idea of the general construction. (More precisely, we also need a condition on the normal bundle of D in M.)

Now if P and Q both have codimension three or more, a Whitney disc can be found by straightforward general position arguments—a generic two-disc is embedded and does not meet a codimension three submanifold. By a more involved argument one can get at the case when one of P, Q has codimension at least three, which will automatically be the case if M has dimension five or more. So in high dimensions we can rather generally cancel intersection points and this fact lies at the heart of the h-cobordism theorem and thus of high-dimensional manifold theory. However if M is a four-manifold and P, Q are surfaces there are problems—our discs may have unwanted self-intersections or meet P, Q in interior points (and there is another problem with the normal bundle condition). Trying to remove these extra intersections puts us back at essentially the same problem we started with. Thus the point where the proof of the h-cobordism theorem fails is that the spheres S_+, S_- fall outside the range of dimensions covered by the Whitney lemma. Wall's theorem shows what can be salvaged from this failure: the 'stable' classification in which all but the obstinate index 2 and 3 critical points are removed.

We now move to the complementary question of the existence of smooth four-manifolds with a given intersection form. It has long been known that

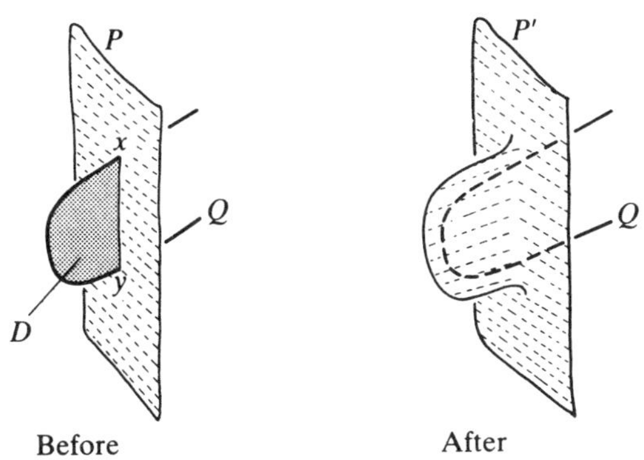

Fig. 7

not all forms can be realized in this way; a constraint is provided by a deep theorem of Rohlin:

Theorem (1.2.6). *The signature of a smooth, compact, spin four-manifold is divisible by* 16.

Here the spin condition is just that w_2 be zero, which as we have seen is equivalent in the simply connected case to the form being even. This factor 16 should be contrasted with the arithmetical factor 8 given by (1.1.5). We see in particular that E_8 cannot be the intersection form of a smooth four-manifold.

The same issue, the cancellation of intersection points of surfaces in four-manifolds, that we encountered in the discussion of the *h*-cobordism theorem enters in this complementary question. For example suppose we want to construct a four-manifold X with form $2\begin{pmatrix} 0 & 1 \\ 1 & 0 \end{pmatrix} \oplus 2(-E_8)$ (not ruled out by Rohlin's theorem). The obvious approach, in the light of our discussion of the $K3$ surface $K = S_4$, whose form has three $\begin{pmatrix} 0 & 1 \\ 1 & 0 \end{pmatrix}$ summands, is to try to find a connected sum decomposition $K = X \# (S^2 \times S^2)$. We can do this if we can find a pair of embedded two-spheres in K meeting transversely in exactly one point. (Make X by gluing a four-ball to the complement of tubular neighbourhoods of the spheres in K.) Once again, from our knowledge of the intersection form, there are immersed spheres with the correct algebraic intersection numbers, but we are lacking a procedure for removing unwanted intersection points. (We mention here a general result of Freedman and Taylor (1977), similar in spirit to (1.2.2): any direct sum decomposition of the intersection form of a simply connected four-manifold X can be realized by a 'generalized connected sum' $X = Y_1 \cup_Z Y_2$ where Y_1, Y_2 are four-manifolds with common boundary a homology three-sphere Z.)

1.3 Summary of results proved in this book

We have now sketched the background in four-manifold theory against which we can set off the results proved in this book. The results bear on the twin questions of the existence and uniqueness of smooth four-manifolds with given intersection forms. They can be summarized by saying that the classification of smooth, simply connected, oriented four-manifolds up to diffeomorphism is revealed to be very *different* from the classification of unimodular forms. Large classes of forms cannot be realized as intersection forms in this way and, on the other hand, there are many examples of distinct manifolds sharing the same forms. Thus the *h*-cobordism theorem does indeed fail in four dimensions. Our results show that the unwanted intersection points of surfaces in four-space cannot be avoided. While the intersection form gives a complete picture of the classical algebraic topology of a simply-connected four-manifold, there are additional subtleties in four-manifold theory arising from the fact that the intersection form does not capture the essence of the differential topology of these surfaces within the ambient, four-dimensional, space.

1.3.1 Realisation of forms

On the question of existence we shall prove, in Chapter 8:

Theorem (1.3.1). *The only negative definite forms realized as the intersection forms of smooth, simply connected, compact four-manifolds are the standard diagonalizable forms $n(-1)$.*

Thus none of the 'exotic' forms—multiples of E_8, E_{16}, the Leech lattice etc.—will actually arise from smooth, closed four-manifolds. There are various different ways of thinking of this result, each suggesting natural generalizations. In one direction, we know that all forms arise as intersection forms of four-manifolds with homology three-sphere boundaries. Our theorem says that for the exotic forms this boundary is never a three-sphere. It is natural to ask if one can say more about the three-manifolds which bound exotic forms. This is not however a question which we shall pursue in this book. Instead we will consider extensions to indefinite forms; for example Theorem (1.3.1) asserts that the form $2(-E_8)$ is not realized by a closed, smooth (simply connected) four-manifold, but what about the forms $2(-E_8)\oplus l\begin{pmatrix} 0 & 1 \\ 1 & 0 \end{pmatrix}$? Our remarks on instability under connected sum are relevant here, since we know that when $l \geq 3$ the form is realized by $K \# (l-3)(S^2 \times S^2)$. On the other hand, if we have proved that the form is not realized for one value of l this certainly implies the same assertion for smaller values. So we can regard the search for an extension of (1.3.1) to indefinite forms as a search for a version of the proof which is partially stable with respect to connected sum with $S^2 \times S^2$. This is the point of view we shall take in the second half of Chapter 8, where we will prove:

Theorem (1.3.2). *If the form $n(-E_8)\oplus m\begin{pmatrix} 0 & 1 \\ 1 & 0 \end{pmatrix}$ is realized by a smooth, compact, simply connected four-manifold then if $n > 0$ we must have $m \geq 3$.*

This result is satisfactory in that it gives the expected critical number 3 of $\begin{pmatrix} 0 & 1 \\ 1 & 0 \end{pmatrix}$ summands when $n = 2$; but it leaves us with the question of which forms are realized for higher values of n. For all known examples we have $m \geq 3n/2$ (note that n must be even by Rohlin's theorem). That is, connected sums of copies of K give the 'best' way known to represent forms. But our results fall far short of proving that the inequality $m \geq (3/2)n$ holds in general.

1.3.2 New invariants of smooth four-manifolds

We now turn to the question of uniqueness. In Chapter 9 we will define new invariants of smooth four-manifolds. These will be defined for any simply connected, oriented, four-manifold X with b^+ odd and not less than 3. The invariants are a sequence of distinguished *polynomials* in the second cohomology group of X,

$$q_k \in s^d(H^2(X; \mathbb{Z})), \tag{1.3.3}$$

of degree $d = 4k - 3(b^+ + 1)$, for sufficiently large integers k. These polynomials are invariant, up to sign, under diffeomorphisms of X. In general terms they are reminiscent of the Pontryagin classes $p_i \in H^{4i}(V; \mathbb{Z})$ of a smooth manifold V but, as we shall see, our invariants are something quite new, detecting phenomena beyond the reach of the standard topological methods.

Wall's theorem shows that interesting new invariants must be 'unstable'. As we shall see, our invariants do indeed have this property; we have:

Theorem (1.3.4). *If the four-manifold X is a smooth, oriented, connected sum $X = X_1 \# X_2$ and if $b^+(X_1)$, $b^+(X_2)$ are both strictly positive then $q_k(X) = 0$ for all k.*

In particular taking the connected sum with even a single $S^2 \times S^2$ kills the new invariants. The proof of Theorem (1.3.4), and its relation to other results of Wall, is described in Chapter 9.

In the opposite direction we shall show that the invariants do not always vanish. The condition that b^+ be odd is satisfied by any simply connected complex surface, and in this case the invariants are non-trivial.

Theorem (1.3.5). *If S is a compact, simply connected complex surface with $b^+ \geq 3$ then $q_k(S) \neq 0$ for sufficiently large k.*

This fact is discussed in Chapter 10 where we consider in special detail the 'double plane' R_4 and give a partial calculation of two of the invariants. This calculation makes use of ideas developed throughout all the earlier chapters.

The two theorems above show that, by and large, complex surfaces cannot be completely decomposed into connected sums. Consider for example the surface R_4: it has the same intersection form as the manifold $7\mathbb{CP}^2 \# 37\overline{\mathbb{CP}}^2$ but, by our theorems, the invariants vanish in one case but not the other. So we have:

Corollary (1.3.6). *R_4 is not diffeomorphic to $7\mathbb{CP}^2 \# 37\overline{\mathbb{CP}}^2$.*

We state this particular result separately because our explicit calculations allow us to avoid the use of some of the theory involved in the proof of (1.3.5), which we will not cover in full detail in this book. However, granted this theory, we see that special results like (1.3.6) understate the case; we have indeed:

Proposition (1.3.7). *For any simply connected complex surface S with $b^+(S) > 3$ there is a smooth four-manifold $X(S)$, homotopy equivalent but not diffeomorphic to S, nor to any complex surface.*

Here the manifold $X(S)$ is constructed as a connected sum to have the same intersection form as S (see Chapter 10).

From any of these examples we deduce, using Wall's result (1.2.5), the failure of the h-cobordism theorem in dimension four, as mentioned above.

Corollary (1.3.8). *There are h-cobordisms between simply connected four-manifolds which are not diffeomorphic to products.*

In fact we shall give a comparatively simple explicit proof of this assertion in Chapter 9, using a calculation for $K3$ surfaces.

Note that Proposition (1.3.7) can be viewed as saying that there are new obstructions for some manifolds (the connected sums $X(S)$) to admit complex structures. So, in contrast to the almost complex structures discussed in Section 1.1.7, the existence of a *complex* structure on a simply connected four-manifold is a delicate issue, beyond considerations of homotopy type and bundle theory alone.

1.3.3 Geometry: topological manifolds and homeomorphisms

It might be misleading if we did not point out explicitly here that the bulk of the material in this book lies within the realm of geometry, specifically geometrical aspects of Yang–Mills theory. These geometrical techniques will then be applied to obtain the differential-topological results mentioned above. It is precisely this departure from standard techniques which has led to the new results, and at present there is no way known to produce results such as these which does not rely on Yang–Mills theory. The geometrical ideas involved span a range between differential and algebraic geometry, the latter accounting for the special position of complex algebraic surfaces in the whole theory.

By way of contrast, we finish this chapter by mentioning briefly some facts that might have fitted more naturally into Section 1.2, but which we have postponed in order to present the discussion in roughly historical order. Throughout this chapter we have been discussing smooth manifolds and their classification up to diffeomorphism. One can also look at *topological* manifolds and their classification up to *homeomorphism*. Large parts of the foundations of this theory are technically much harder than in the smooth case, since one cannot appeal directly to transversality arguments. However in high dimensions topologists were able to develop a classification theory, involving an extension of the h-cobordism theorem, whose results followed those of the smooth theory quite closely. Until the early 1980s the classification of topological four-manifolds rested, stuck on the same basic questions which we have described in the smooth case. The work of Freedman completely changed this picture. Freedman (1982) gave a complete classification theorem for compact, simply connected topological four-manifolds by showing that in the topological category the *h*-cobordism theorem *does hold* for four-manifolds. For example his classification asserts that there is just one topological four-manifold, up to homeomorphism, for each even unimodular form. In the topological case the classification of manifolds is essentially the same as the classification of forms. This is of course the exact opposite of the conclusion we have reached in our discussion of smooth manifolds, and we see that there is a radical divergence between topology and differential topology in dimension four. This contrast has led to a number of corollaries, notably the existence of 'exotic $\mathbb{R}^4$s'—smooth manifolds homeomorphic but not diffeomorphic to $\mathbb{R}^4$. In the body of this book we shall only be concerned with smooth manifolds; the results of the topological theory can serve however as some justification for our preoccupation with ideas from geometry, showing that conventional manifold-theory techniques are unlikely, by themselves, to be adequate for the understanding of smooth four-manifolds.

Notes

Our main aim in this Chapter has been to present the theory of four-manifolds as it appeared *circa* 1980. Useful contemporary references are the survey article by Mandelbaum (1980) and the problem list (Kirby, 1978*b*).

Section 1.1.1

For the construction of a four-manifold with prescribed fundamental group see Markov (1960). The relations between the different geometrical representations of homology classes are developed by, for example, Griffiths and Harris (1978).

Section 1.1.3

For the classification of indefinite forms see Serre (1973) and Husemoller and Milnor (1973).

Section 1.1.4

General references for characteristic classes are Milnor and Stasheff (1974) and Husemoller (1966). The fact that $w_3(X)$ is zero for any orientable four-manifold is proved by Hirzebruch and Hopf (1958); this is equivalent to the existence of an integral lift of $w_2(X)$ or, geometrically, to the existence of a spin^c structure. For information on the spin representation in four dimensions see, for example, Atiyah *et al.* (1978*b*) and Salamon (1982).

Section 1.1.5

The representation of conformal classes by ASD subspaces is developed by Donaldson and Sullivan (1990). It is a direct analogy of the classical representation of conformal structures in two dimensions by Beltrami differentials.

Section 1.1.6

Expositions of the Hodge theory are found in Hodge (1989), de Rham (1984), Warner (1983), Wells (1980) and Griffiths and Harris (1978).

Section 1.1.7

A comprehensive general reference for the theory of complex surfaces is Barth *et al.* (1984). For the existence of almost complex structures see Hirzebruch and Hopf (1958) and Matsushita (1988). The article by Mandelbaum (1980) contains a wealth of information on the topology of complex surfaces.

The fact that a simply connected surface can be deformed into an algebraic one follows from classification theorems of Kodaira; in fact we only need assume that the first Betti number is even (Kodaira 1963).

Section 1.2.1

The original proof of Theorem (1.2.1) is given by Milnor (1958); see also Whitehead (1949). The simple result (1.2.3) leads on to the 'Kirby calculus' for manipulating handle descriptions of four-manifolds. See Kirby, (1978*a*) and Mandelbaum (1980). Explicit handle descriptions of non-diffeomorphic, homeomorphic four-manifolds with boundary are given by Gompf (1990).

Sections 1.2.2 and 1.2.3

Theorems (1.2.4) and (1.2.5) are proved by Wall (1964*b*). For the proof of the *h*-cobordism theorem we refer to Smale (1964) and Milnor (1965). The original reference for the Whitney lemma is Whitney (1944). A full account in the piecewise-linear setting is given in the book by Rourke and Sanderson (1982). For other classification theorems in high dimensions see, for example, Wall (1962).

The original proof of Rohlin's theorems is in Rohlin (1952); see also Kervaire and Milnor (1958). A geometric proof of a more general result is given by Freedman and Kirby (1978). There were extensive searches for a connected sum decomposition of a $K3$ surface. The result on generalized connected sums is proved by Freedman and Taylor (1977).

Section 1.3

We have not attempted to give a comprehensive survey of results on four-manifolds proved using Yang–Mills theory; for other surveys see Donaldson (1987*c*), Friedman and Morgan (1988*b*). The original proof of (1.3.1) is given by Donaldson (1983*b*); see also Freed and Uhlenbeck (1984). The result holds without the restriction to simply connected manifolds; see Fintushel and Stern (1984), Furuta (1987), Donaldson (1987*b*), Fintushel and Stern (1988). There are also versions for orbifolds (Fintushel and Stern 1985; Furuta 1990).

The extension of the theory to certain non-compact four-manifolds, including manifolds with boundary, was begun by Taubes (1986). The key new ingredients are the unitary representations of the fundamental group of the boundary, leading naturally to the 'Floer homology groups' of a three-manifold defined by Floer (1989). These are in turn related to the Casson invariant of a three-manifold (Akbulut and McCarthy 1990; Taubes 1990). For the relations between the Floer homology and four-manifolds see Atiyah (1988) and Donaldson *et al.* (1990).

The result (1.3.2) on indefinite forms is proved by Donaldson (1986). The assumption on the fundamental group can be weakened but not removed entirely; see the discussion in the introduction to that paper.

Section 1.3.2

The polynomial invariants (1.3.3) were introduced by Donaldson (1990*a*), which contains also Theorems (1.3.4) and (1.3.5). For many more detailed results see Friedman *et al.* (1987) and Friedman and Morgan (1989).

The main development which we do not mention in the text involves manifolds with $b^+ = 1$. Invariants can be defined in this case, but they have a more complicated form, see Donaldson (1987*a*), Kotshick (1989), Mong (1989) and Okonek and Van de Ven (1989). Using such invariants it has been shown that there are infinitely many, non-diffeomorphic, simply connected four-manifolds (Dolgachev surfaces) with intersection form $(1) \oplus 9(-1)$; see Friedman and Morgan (1988*a*) and Okonek and Van de Ven (1986). They also detect distinct manifolds with form $(1) \oplus 8(-1)$ (Kotshick 1989); at the time of writing this is the example of such phenomena with 'smallest' homology which is known.

Concerning our remark on almost-complex versus complex structures—in the non-simply-connected case there are comparatively straightforward examples of this phenomenon; see Barth *et al.* (1984).

Section 1.3.3

For the theory of topological manifolds in high dimensions see Kirby and Siebenmann (1977). Freedman's results appeared in Freedman (1982); see also Freedman and Quinn (1990). It is now known that there are uncountable families of exotic $\mathbb{R}^4$s; see Gompf (1983, 1985) and Taubes (1986). The results for smooth manifolds obtained using Yang–Mills theory can be proved under weaker smoothness hypotheses. They remain true for four-manifolds with Lipschitz or even quasiconformal structures; see Donaldson and Sullivan (1990).

2

CONNECTIONS

In this chapter we begin our study of Yang–Mills theory—the theory of connections—which makes up the core of this book. Classical differential geometry considers, by and large, connections on the tangent bundle of a manifold, the most important being the Levi–Civita connection defined by a Riemannian metric. The key feature of Yang–Mills theory is that it deals with connections on auxiliary bundles, not directly tied to the geometry of the base manifold. After reviewing standard definitions and notation in Section 2.1, we go on to give proofs of two fundamental theorems: the integrability theorem for holomorphic structures and Uhlenbeck's theorem on the existence of Coulomb gauges. These are essentially local results. We regard them as extensions of the elementary fact that connections with curvature zero are locally trivial.

2.1 Connections and curvature

2.1.1 Bundles and connections

We recall briefly the different definitions of connections, referring to standard texts for more details.

Let G be a Lie group. A principal G-bundle P over a smooth manifold X is a manifold with a smooth (right) G action and orbit space $P/G = X$. We demand that the action admit local product structures, i.e. is locally equivalent to the obvious action on $U \times G$, where U is an open set in X. Then we have a fibration $\pi: P \to X$. We say that P has structure group G.

Three useful ways of defining a connection on such a bundle are:

1. As a field of 'horizontal subspaces' $H \subset TP$ transverse to the fibres of π. That is, for each p in P we have a decomposition:

$$TP_p = H_p \oplus T(\pi^{-1}(x)),$$

where $\pi(p) = x$. The field of subspaces is required to be preserved by the action of G on P.

2. As a 1-form A on P with values in the Lie algebra $\mathfrak{g}$ of G; i.e. a section of the bundle $T^*P \otimes \mathfrak{g}$ over P. Again we require this to be invariant under G, acting by a combination of the given action on P and the adjoint action on $\mathfrak{g}$. Also, A should restrict to the canonical right-invariant form on the fibres.

3. For any linear representation of G, on $\mathbb{C}^n$ or $\mathbb{R}^n$, we get a vector bundle E over X associated with P and the representation. The fibres of E are copies of

$\mathbb{C}^n$ or $\mathbb{R}^n$ respectively. Conversely, given a vector bundle E we can make a principal bundle. For example, if E is a complex n-plane bundle we get a principal bundle P with structure group $GL(n, \mathbb{C})$ by taking the set of all 'frames' in E. A point in the fibre of P over $x \in X$ is a set of basis vectors for E_x. Additional algebraic structure on E yields a principal bundle with smaller structure group. For example, if E is a complex vector bundle with a Hermitian metric we get a principal $U(n)$ bundle of orthonormal frames in E. For the classical groups, (automorphisms of a vector space preserving some linear algebraic structure), the concepts of principal and vector bundles are completely equivalent. Now given a vector bundle E (real or complex) a connection on the frame bundle can be defined by a *covariant derivative* on E, that is a linear map:

$$\nabla : \Omega_X^0(E) \longrightarrow \Omega_X^1(E). \tag{2.1.1}$$

Here we introduce the notation $\Omega_X^p(E)$ to denote sections of $\Lambda^p T^* X \otimes E$—the p-forms with values in E. The map ∇ is required to satisfy the Leibnitz rule: $\nabla(f.s) = f\nabla s + \mathrm{d}f.s$, for any section s of E and function f (real or complex as appropriate) on X. To understand this definition, consider a tangent vector v to X at a point x. If s is a section of E we can make the contraction $\langle \nabla s, v\rangle \in E_x$ which is to be thought of as the derivative of s in the direction v at x. If E has additional algebraic structure we can require the covariant derivative to be compatible with this; for example if E has a metric, a compatible connection is one for which $\nabla(s, t) = (\nabla s, t) + (s, \nabla t)$, for all sections s, t.

One sees the equivalence of these three viewpoints as follows: to go from (2) to (1) we define the horizontal space H_p to be the kernel of A_p, regarded as a linear map $A_p : TP_p \to \mathfrak{g}$. To go from (3) to (1) we first observe that ∇ is a local operator; that is, if two sections s_1, s_2 agree on an open set U in X, so also do $\nabla s_1, \nabla s_2$. This follows from the Leibnitz rule by considering ϕs_i where ϕ is a cut-off function. Then we say that a local section σ of the frame bundle (i.e. a collection of local sections $s_1, \ldots s_n$ of E) is horizontal at x in X if all the ∇s_i vanish at x. Finally we define H_p to be the tangent space to a horizontal section σ through p, regarded as a submanifold of P. It is a simple matter to verify that these constructions can be inverted, so that the three give equivalent definitions of a connection—at least for the classical groups.

Our standard practice in this book will be to work with vector bundles, usually complex vector bundles, which we normally denote by E. This will cause no real loss of generality and will simplify notation at a number of points. However, at a few points where the more abstract principal bundle formulation gives extra insight we shall feel free to shift to this setting and introduce a principal bundle P. In fact for all our detailed work we can restrict ourselves to the two groups $SU(2)$ and, more occasionally, $SO(3)$. Thus we will be considering bundles with metrics and with fixed trivializations of their top exterior powers. We will write $\mathfrak{g}_E$ (or $\mathfrak{g}_P$ where appropriate) for the bundle of Lie algebras associated to the adjoint representation, so $\mathfrak{g}_E$ is a real subbundle of End $E = E \otimes E^*$. If the structure group is $SU(2)$, for example, then $\mathfrak{g}_E$ consists of skew-adjoint, trace-free endomorphisms of the rank-two complex vector bundle E. We will take the approach (3) (see above) in terms of a covariant derivative on E as a working definition of a connection, but we will write this, slightly illogically, as ∇_A using A to denote the connection. The connection on E induces one on $\mathfrak{g}_E$ and we will also denote the covariant derivative on $\mathfrak{g}_E$ by ∇_A.

Let us note here four facts about connections. First on a trivial bundle $\underline{\mathbb{C}}^n = \mathbb{C}^n \times X$ there is a standard product connection whose covariant derivative is ordinary differentiation of vector-valued functions. Second, connections are covariant objects: if $f: X \to Y$ is a smooth map and A is a connection on a bundle E over Y then there is an induced connection $f^*(A)$ on the bundle $f^*(E)$ over X. Third, connections yield (and can indeed be defined by) a parallel transport of bundle elements along paths in the base space. If $\gamma: [0, 1] \to X$ is a smooth path with end points x, y and E is a bundle with connection A over X then the parallel transport is a linear map,

$$T_\gamma: E_x \longrightarrow E_y. \tag{2.1.2}$$

In terms of horizontal subspaces this is defined by choosing a frame p in E_x, viewed as a point in the principal frame bundle, then finding the unique horizontal lift $\tilde{\gamma}$ of γ beginning at p, i.e. a path in P such that $\pi\tilde{\gamma} = \gamma$ and whose derivative vectors always lie in the horizontal subspaces. Then we get a frame $\tilde{\gamma}(1)$ in E_y and T_γ is the map which takes p to this new frame. In terms of covariant derivatives, we pull back A to get a connection on the bundle $\gamma^*(E)$ over $[0, 1]$. Then any vector in $E_x = (\gamma^*(E))_0$ has a unique extension to a 'covariant constant' section of $\gamma^*(E)$ (that is, with zero covariant derivative) and we evaluate this section at 1 to define T_γ.

As the fourth property in our list we note the familiar fact that 'the difference of two connections is a tensor'. Suppose A is a connection on E and a is an element of $\Omega^1_X(\mathfrak{g}_E)$, a bundle-valued one-form. Then the operator $\nabla_A + a$ is again a covariant derivative and so defines a new connection $A + a$. Here a acts algebraically on $\Omega^0_X(E)$ via the contraction

$$\Omega^0_X(E) \times \Omega^1_X(\text{End}\, E) \longrightarrow \Omega^1_X(E). \tag{2.1.3}$$

(The restriction to $\mathfrak{g}_E \subset \text{End}\, E$ ensures that the new connection is compatible with the structure group of E.) Conversely the difference of two connections on E is defined as an element of $\Omega^1_X(\mathfrak{g}_E)$. Thus the space $\mathscr{A}$ of all connections on E is an infinite-dimensional affine space, modelled on $\Omega^1_X(\mathfrak{g}_E)$.

We will now get a firmer grip on these ideas by studying connections locally, in trivializations of the bundle. Suppose that U is an open set in X over which E is trivial and fix a trivialization $\tau: E|_U \to \underline{\mathbb{C}}^n$. Then over U we can compare a connection A on E with the product connection on $\underline{\mathbb{C}}^n$, via τ. In terms of covariant derivatives, we write:

$$\nabla_A = d + A^\tau. \tag{2.1.4}$$

(We use the symbols ∇ and d interchangeably for the derivative on functions.) Here the 'connection form' A^τ is a $\mathfrak{g}$-valued 1-form (a matrix of 1-forms) over U. The meaning is that we identify local sections of E with (column) vector valued functions, using τ, and the covariant derivative is given by the indicated combination of ordinary differentiation and multiplication by A^τ. Still more explicitly, if we have local coordinates x_i on U we write

$$\nabla_A = \sum \nabla_i \, dx_i, \tag{2.1.5}$$

where the 'covariant derivative in the x_i direction' ∇_i is

$$\nabla_i = \frac{\partial}{\partial x_i} + A_i^\tau \tag{2.1.6}$$

for matrix-valued functions A_i^τ. In terms of our principal bundle P of frames, we interpret τ as a local section of P; then A^τ is the pull back $\tau^*(A)$ of the 1-form defining the connection by approach 2.

While the connection matrices A^τ give a concrete description of a connection, it is important to emphasize that they depend on the choice of trivialization τ. To understand this dependence let us return to the invariant point of view and suppose that $u: E \to E$ is an automorphism of E, respecting the structure on the fibres and covering the identity map on X. The set of these automorphisms form a group $\mathcal{G}$ which we call the *gauge group* of E. We have a pointwise exponential map exp: $\Omega^0(\mathfrak{g}_E) \to \mathcal{G}$. The gauge group acts on the set of connections by the rule:

$$\nabla_{u(A)} s = u \nabla_A (u^{-1} s). \tag{2.1.7}$$

We can expand the left-hand side as: $u\nabla_A u^{-1} = \nabla_A - (\nabla_A u)u^{-1}$, where the covariant derivative of u is formed by regarding it as a section of the vector bundle End (E). So

$$u(A) = A - (\nabla_A u)u^{-1}. \tag{2.1.8}$$

Now let u be an automorphism of the trivial bundle $\underline{\mathbb{C}}^n$, i.e. a smooth map from the base space to $U(n)$. If τ is a trivialization of E, then $(u\tau)$ is a new trivialization, and for a connection A we have

$$\begin{aligned} A^{u\tau} &= A^\tau - \{(\mathrm{d} + [A^\tau, \,])u\}u^{-1} \\ &= A^\tau - \{\mathrm{d}u + A^\tau u - uA^\tau\}u^{-1} \\ &= uA^\tau u^{-1} - (\mathrm{d}u)u^{-1}. \end{aligned} \tag{2.1.9}$$

Thus A^τ and $A^{u\tau}$ are different connection matrices representing the same connection. The choice of bundle trivialization is sometimes called the choice of a gauge. When working locally we will sometimes be rather imprecise about distinguishing connections and connection matrices, and may suppress the superscript τ. The transformation formula $A \to uAu^{-1} - (\mathrm{d}u)u^{-1}$ can be viewed equivalently as describing the effect of a change of trivialization on the matrices representing a fixed connection, or as the action of a bundle automorphism on the connections on a bundle, viewed in a fixed trivialization. Of course this picture is very familiar; it is just the same as in Euclidean geometry, say, where one can either rotate the coordinate axes in a fixed space or rotate the space in fixed coordinates. The main difference is that our symmetry group of gauge transformations is infinite dimensional.

However one looks at it, we can say that a connection on a bundle over X is given by the following set of explicit data. First, the bundle can be defined by a cover X_α of X and transition functions

$$u_{\alpha\beta}: X_\alpha \cap X_\beta \longrightarrow U(n), \tag{2.1.10}$$

such that $u_{\alpha\beta} = u_{\beta\alpha}^{-1}$ and $u_{\alpha\gamma} = u_{\alpha\beta}u_{\beta\gamma}$ on $X_\alpha \cap X_\beta \cap X_\gamma$. The connection is specified by matrix valued one-forms A_α on the open sets X_α satisfying

$$A_\beta = u_{\alpha\beta} A_\alpha u_{\alpha\beta}^{-1} - (\mathrm{d}u_{\alpha\beta})u_{\alpha\beta}^{-1} \tag{2.1.11}$$

on $X_\alpha \cap X_\beta$.

2.1.2 Curvature and differential operators

In addition to the covariant derivative there are various other differential operators defined by a connection which we will often use in this book. These extend operators defined on forms, or other tensors, to bundle valued tensors, in just the same way as a covariant derivative extends ordinary differentiation.

First, extending the ordinary de Rham complex,

$$\Omega_X^0 \xrightarrow{\mathrm{d}} \Omega_X^1 \xrightarrow{\mathrm{d}} \dots \xrightarrow{\mathrm{d}} \Omega_X^p \xrightarrow{\mathrm{d}} \Omega_X^{p+1} \dots,$$

we have exterior derivatives

$$\mathrm{d}_A: \Omega_X^p(E) \longrightarrow \Omega_X^{p+1}(E). \tag{2.1.12}$$

These are uniquely determined by the properties:

(i) $\mathrm{d}_A = \nabla_A$ on $\Omega_X^0(E)$,

(ii) $\mathrm{d}_A(\omega \wedge \theta) = (\mathrm{d}_A\omega) \wedge \theta + (-1)^p \omega \wedge \mathrm{d}_A\theta$,

for $\omega \in \Omega_X^p$, $\theta \in \Omega_X^q(E)$.

(From now on we will use ∇_A, d_A interchangeably to denote the covariant derivative on sections of E.)

In contrast with the ordinary exterior derivative, it is *not* true in general that $\mathrm{d}_A\mathrm{d}_A$ is zero. Instead the Leibnitz rule tells us that this composite is an algebraic operator (commuting with multiplication by smooth functions) which can be used to define the curvature F_A of the connection. Thus:

$$d_A d_A s = F_A s, \tag{2.1.13}$$

where $F_A \in \Omega^2_X(\mathfrak{g}_E)$. (When it is more convenient we will write $F(A)$ for F_A.) The curvature varies with the connection; we have

$$(d_A + a)(d_A + a)s = d_A d_A s + (d_A a)s + (a \wedge a)s,$$

so

$$F_{A+a} = F_A + d_A a + a \wedge a. \tag{2.1.14}$$

Here $a \wedge a$ denotes the combination of wedge product with multiplication on $\mathfrak{g}_E \subset \operatorname{End} E$. (From the point of view of principal bundles it is better to write this as $\frac{1}{2}[a, a]$, where we combine the antisymmetric wedge product on 1-forms with the antisymmetric Lie bracket in $\mathfrak{g}_E$ to obtain a quadratic form, $a \to [a, a]$.) Similarly, in a local trivialization the curvature is given in terms of the connection matrix A^τ by a matrix of 2-forms:

$$F^\tau_A = dA^\tau + A^\tau \wedge A^\tau. \tag{2.1.15}$$

To understand the definition more clearly, consider the situation in a local trivialization and with local coordinates on X. The curvature matrix $F^\tau_A = \sum F_{ij} dx_i dx_j$ has components F_{ij} which are the commutators of the covariant derivatives in the different coordinate directions:

$$F_{ij} = [\nabla_i, \nabla_j] = \left[\frac{\partial}{\partial x_i} + A^\tau_i, \frac{\partial}{\partial x_j} + A^\tau_j\right] \tag{2.1.16}$$

$$= \frac{\partial A^\tau_j}{\partial x_i} - \frac{\partial A^\tau_i}{\partial x_j} + [A^\tau_i, A^\tau_j]. \tag{2.1.17}$$

Here we mention a more geometric definition of curvature. In the coordinate system let $S_{ij}(\delta)$ be the square in the x_i, x_j plane with corners at $(0, 0)$, $(\delta, 0)$, $(0, \delta)$, (δ, δ) and let $T_{ij}(\delta)$ be the result of parallel transport around $S_{ij}(\delta)$—an automorphism of the fibre of E over 0. Clearly $T_{ij}(\delta)$ tends to the identity as δ tends to 0. The curvature gives an approximation to this parallel transport, in that we have

$$T_{ij}(\delta) = 1 + F_{ij}.\, \delta^2 + O(\delta^4) \quad \text{as } \delta \to 0.$$

(Note that the usual proof of the symmetry of partial derivatives can be thought of as using parallel transport around such a square.)

An important point to note is that curvature transforms as a tensor under bundle automorphisms:

$$F_{u(A)} = u F_A u^{-1}. \tag{2.1.18}$$

In particular the set of connections with curvature zero, called *flat* connections, is preserved by $\mathcal{G}$. We will have more to say about these in the next section.

It is convenient to note here the infinitesimal versions of (2.1.8) and (2.1.14). Let A_t be a smooth one-parameter family of connections, with time derivative

$a \in \Omega^1_X(\mathfrak{g}_E)$ at $t = 0$. (For example, we might have $A_t = A_0 + ta$.) Then, by (2.1.14), the time derivative of the curvatures $F(A_t)$, at $t = 0$, is $d_{A_0}a$. In other words, the derivative of the curvature, viewed as a map from $\mathscr{A}$ to $\Omega^2_X(\mathfrak{g}_E)$, is

$$(DF)_{A_0} = d_{A_0}: \Omega^1_X(\mathfrak{g}_E) \to \Omega^2_X(\mathfrak{g}_E). \tag{2.1.19}$$

Similarly, let u_t be a one-parameter family of gauge transformations, with u_0 the identity. The derivative $\mathrm{d}u/\mathrm{d}t|_{t=0}$ can be viewed naturally as a section ξ of $\mathfrak{g}_E$. Let A_0 be a connection and define a one-parameter family by $A_t = u_t(A_0)$. Then, differentiating (2.1.8) we find that the derivative of A_t at 0 is $-\nabla_A \xi = -\mathrm{d}_A \xi$. In other words, the derivative of the action of the gauge group on $\mathscr{A}$, at A_0, is

$$-\mathrm{d}_{A_0}: \Omega^0_X(\mathfrak{g}_E) \longrightarrow \Omega^1_X(\mathfrak{g}_E). \tag{2.1.20}$$

To complete the picture we consider the curvature of the family $u_t(A_0)$. This is given by the composite of (2.1.19) and (2.1.20) as $-\mathrm{d}_{A_0}\mathrm{d}_{A_0}\xi$, which according to our definition of curvature is $-[F_{A_0}, \xi] = [\xi, F_{A_0}]$. This agrees with the differentiated form of (2.1.18), since the bracket is the derivative of conjugation.

Finally we note the *Bianchi identity*: for any connection A we have

$$\mathrm{d}_A F_A = 0. \tag{2.1.21}$$

The verification of this identity is an easy exercise. (A highbrow proof exploits the naturality of the constructions under the action of the diffeomorphism group of X.)

To get more differential operators we need to have extra structure on X. Suppose X has a Riemannian metric, so we have Euclidean inner products on all the bundles $\Lambda^p T^*X$ and also on the other tensors. In addition we have the Levi–Civita connection on the tangent bundle of X. If A is a connection on an auxiliary bundle E over X we obtain covariant derivatives:

$$\nabla_A: \Omega^p_X(E) \longrightarrow \Gamma(\Lambda^p T^*X \otimes T^*X \otimes E). \tag{2.1.22}$$

These are defined by combining the Levi–Civita connection on the form component with ∇_A on the E component. The operators d_A can be obtained from these by the wedge product $\Lambda^p \otimes \Lambda^1 \to \Lambda^{p+1}$. We also have formal adjoint operators:

$$\nabla^*_A: \Gamma(\Lambda^p T^*X \otimes T^*X \otimes E) \longrightarrow \Omega^p_X(E), \tag{2.1.23}$$

$$\mathrm{d}^*_A: \Omega^{p+1}_X(E) \longrightarrow \Omega^p_X(E). \tag{2.1.24}$$

The second of these, for example, is characterized by the equation:

$$\int_X (\mathrm{d}_A\phi, \psi)\, d\mu = \int_X (\phi, \mathrm{d}^*_A\psi)\, \mathrm{d}\mu,$$

for forms ϕ, ψ, at least one of which has compact support.

To summarize this chapter so far, we have three basic objects: gauge transformations or bundle automorphisms, connections and curvature tensors. Locally these can be represented by, respectively, matrix-valued functions, 1-forms and 2-forms. Let us interpose a word about our notation. Until recently differential geometers commonly used ω and Ω to denote connection and curvature matrices. The A, F notation is derived from mathematical physics, beginning with electromagnetism. Consider a complex vector bundle E of rank 1, with a Hermitian metric. The structure group is the circle $U(1)$ which is Abelian and whose Lie algebra can be identified with $\mathbb{R}$ (or, more conveniently, with $i\mathbb{R}$). So, in a local trivialization, a connection on E is represented by an ordinary 1-form A and the curvature is the ordinary 2-form $F = \mathrm{d}A$. The non-linear $[A, A]$ term is absent. Moreover we can write any gauge transformation locally as $u = \exp(i\chi)$, and this acts on the connections simply by $A \to A - i\mathrm{d}\chi$. In electromagnetism the base space X is space–time and we interpret the electromagnetic four-vector potential, conventionally written A, as a connection on such a bundle. Then the curvature $\mathrm{d}A = F$ is the electromagnetic field, a tensor whose six components can be identified with those of the electric and magnetic field vectors if we choose a 'time' vector in TX. For pure magnetostatics we can take X to be $\mathbb{R}^3$. Then A and F can be identified, using the metric, with the magnetic vector potential A and vector field $B = \nabla \times A$. The fact that A is only defined up to gauge transformations corresponds to the older notion that the magnetic potential does not have a direct physical meaning. However it is now generally accepted that the right mathematical formulation of electromagnetism does indeed involve a connection on a $U(1)$ bundle, and this is the starting point for the Yang–Mills theories in physics, which led in turn to the mathematical ideas we describe in this book.

2.1.3 Anti-self-dual connections over four-manifolds

Suppose now that X is an oriented Riemannian four-manifold. We have already met in Chapter 1 the decomposition of the two-forms on X into self-dual and anti-self-dual parts:

$$\Omega^2_X = \Omega^+_X \oplus \Omega^-_X .$$

This splitting extends immediately to bundle-valued 2-forms and in particular to the curvature tensor F_A of a connection on a bundle E over X. We write

$$F_A = F^+_A \oplus F^-_A \in \Omega^+_X(\mathfrak{g}_E) \oplus \Omega^-_X(\mathfrak{g}_E), \tag{2.1.25}$$

where $\Omega^{\pm}_X(\mathfrak{g}_E) = \Gamma(\Lambda^{\pm}_X \otimes \mathfrak{g}_E)$. We call a connection *anti-self-dual* if $F^+_A = 0$. Of course we also have *self-dual* connections with $F^-_A = 0$, and the notions are interchanged by reversing the orientation of X; but, as we shall see, the anti-self-dual assumption fits better with standard conventions. We will usually abbreviate anti-self-dual to ASD. Note that, like the decomposition into self-dual and anti-self-dual forms, this is a *conformally invariant* notion. It depends only on the conformal class of the Riemannian metric on the base space.

Explicitly, on Euclidean space $X = \mathbb{R}^4$ with connection matrices A_i, the ASD equation $F_A^+ = 0$ comprises the system of partial differential equations (PDEs):

$$\begin{aligned} F_{12} + F_{34} &= 0, \\ F_{14} + F_{23} &= 0, \\ F_{13} + F_{42} &= 0, \end{aligned} \tag{2.1.26}$$

where $F_{ij} = [\nabla_i, \nabla_j] = \dfrac{\partial A_j}{\partial x_i} - \dfrac{\partial A_i}{\partial x_j} + [A_i, A_j]$.

2.1.4 *Bundle theory and characteristic classes*

Two interpretations of these ASD equations will be important for us, one topological and one geometrical. For the first we have to recall the rudiments of the Chern–Weil representation of characteristic classes. We begin by digressing slightly and considering complex line bundles. Let L be a Hermitian line bundle over a manifold X. As in Section 2.1.2 the curvature of a connection A on L is a purely imaginary 2-form which we write as $-2\pi i\phi$. So ϕ is a real 2-form, which is closed by the Bianchi identity (2.1.21). It therefore defines a de Rham cohomology class $[\phi]$ in $H^2(X; \mathbb{R})$. Consider a second connection $A' = A + a$; we have $F' = F + \mathrm{d}a$, so $[\phi'] = [\phi]$. So we obtain a cohomology class which is independent of the choice of connection, and thus depends only on the bundle L. It is well known that this class is just the first Chern class $c_1(L)$ which classifies L (cf. Section 1.1.1), or rather the image of this class in the real cohomology. More generally, for any complex vector bundle E, with a connection A, the first Chern class $c_1(E)$ is represented by $(i/2\pi)\mathrm{Tr}(F_A)$.

Now, in the same way, consider the 4-form $\mathrm{Tr}(F_A^2)$ defined by a connection on a Hermitian bundle E. This is again a closed form whose de Rham cohomology class depends only on E, not the particular connection. (The general Chern–Weil theory considers de Rham cohomology classes represented by invariant polynomials in the curvature—here the relevant polynomial is the negative definite form $\xi \to \mathrm{Tr}(\xi^2)$ on $\mathfrak{g}$.) Direct proofs of these assertions make good exercises in the notation set up above. The second assertion, for example, follows from the identity:

$$\mathrm{Tr}(F_{A+a}^2) - \mathrm{Tr}(F_A^2) = \mathrm{d}\{\mathrm{Tr}(a \wedge \mathrm{d}_A a + \tfrac{2}{3} a \wedge a \wedge a)\}. \tag{2.1.27}$$

Again, these forms represent a standard topological characteristic class of E. For a complex vector bundle E we have

$$[(1/8\pi^2)\mathrm{Tr}(F_A^2)] = c_2(E) - \tfrac{1}{2}c_1(E)^2 \in H^4(X), \tag{2.1.28}$$

where c_1, c_2 are the Chern classes. We have chosen this normalization since the most important case for us will be when we have bundles with structure group $SU(r)$ (especially when $r = 2$): then the trace of the curvature and c_1 are zero and $(1/8\pi^2)\operatorname{Tr}(F_A^2)$ represents the basic four-dimensional class c_2. When X is a closed, oriented, four-manifold we identify $H^4(X)$ with the integers and then write, for $SU(r)$ connections,

$$c_2(E) = \frac{1}{8\pi^2}\int_X \operatorname{Tr}(F_A^2) \in \mathbb{Z}. \tag{2.1.29}$$

To see this topological invariant explicitly consider the case of $SU(2)$ bundles over the four-sphere. A bundle E can be trivialized over the upper and lower 'hemispheres' separately, and is determined up to isomorphism by the homotopy class of the resulting transition function, regarded as a map from the equatorial three-sphere to the structure group. But $SU(2)$ can itself be identified with a three-sphere (cf. Section 3.1.1), so the transition function gives a map $u\colon S^3 \to S^3$. With appropriate orientation conventions the integer invariant $c_2(E)$ is just the degree of this map.

We now introduce the anti-self-dual condition. Observe that on the Lie algebra $\mathfrak{u}(n)$ of skew adjoint matrices $\operatorname{Tr}(\xi^2) = -|\xi|^2$. Combining this with the definition of the splitting of the 2-forms (1.1.8) we get

$$\operatorname{Tr}(F_A^2) = -\{|F_A^+|^2 - |F_A^-|^2\}\,\mathrm{d}\mu \tag{2.1.30}$$

where $\mathrm{d}\mu$ is the Riemannian volume element. In particular, a connection is ASD if and only if

$$\operatorname{Tr}(F_A^2) = |F_A|^2\,\mathrm{d}\mu$$

at each point.

We now integrate (2.1.28) over the closed four-manifold X to get

$$8\pi^2 c_2(E) = \int_X \operatorname{Tr}(F_A^2)\,\mathrm{d}\mu = \int_X |F_A^-|^2\,\mathrm{d}\mu - \int_X |F_A^+|^2\,\mathrm{d}\mu. \tag{2.1.31}$$

The significance of this for Yang–Mills theory is that the absolute value of $8\pi^2 c_2$ gives a lower bound on the *Yang–Mills functional*: the square of the L^2 norm of the curvature

$$\|F_A\|^2 = \int_X |F_A|^2\,\mathrm{d}\mu = \int_X |F_A^-|^2\,\mathrm{d}\mu + \int_X |F_A^+|^2\,\mathrm{d}\mu.^\dagger \tag{2.1.32}$$

When c_2 is positive this bound is achieved precisely for the ASD connections:

$$\|F_A\|^2 = 8\pi^2 c_2(E) \Leftrightarrow A \text{ is ASD}. \tag{2.1.33}$$

† In general, throughout this book, a norm symbol without further qualification will denote an L^2 norm.

In particular ASD connections are solutions of the *Yang–Mills equations.* These are the Euler–Lagrange equations for the functional $\|F_A\|^2$ on the space of connections $\mathscr{A}$, which take the form:

$$d_A^* F_A = 0. \tag{2.1.34}$$

In this book we shall be almost exclusively concerned with the more special ASD equations, but the information about the Yang–Mills functional that we get from the formulae above will play a vital role. Let us note here the important fact that, like the ASD condition, the Yang–Mills functional in four dimensions is *conformally invariant*. In general, in dimension d, if we scale the Riemannian metric by a factor c, the pointwise norm on 2-forms scales by c^{-2}, while the volume form scales by c^d. So an integral $\int |F|^2\, d\mu$ transforms to $\int c^{d-4}|F|^2\, d\mu$, which is invariant precisely when d is 4. This conformal invariance in four dimensions is another facet of the relation above between the Yang–Mills density $|F|^2$ and the intrinsic 4-form $\mathrm{Tr}(F^2)$.

In the discussion above we have fixed attention on $SU(r)$ connections, but this is purely for simplicity of notation. Indeed any compact group G admits an invariant, definite, inner product on its Lie algebra. This can be defined by taking the trace-square in some faithful, orthogonal representation of the Lie algebra. Such an invariant form gives rise to a characteristic number for G-bundles over compact, oriented four-manifolds for which the whole discussion above goes through. The only question is how best to normalize these topological invariants. The case we will need, beyond the complex Hermitian bundles considered above, is that of real vector bundles, with structure group $SO(r)$, and in particular $SO(3)$. We will first recall briefly the relevant bundle theory.

The standard four-dimensional characteristic class for a real orthogonal bundle is the Pontryagin class:

$$p_1(V) = -c_2(V \otimes \mathbb{C}) \in H^4(X; \mathbb{Z}). \tag{2.1.35}$$

In addition, such a bundle has a Stiefel–Whitney class $w_2(V)$ in $H^2(X; \mathbb{Z}/2)$ (in Section 1.1.4 we developed this theory for the tangent bundle of the four-manifold). The Stiefel–Whitney class satisfies

$$w_2(V)^2 = p_1(V) \mod 4. \tag{2.1.36}$$

(The reader will recall here the fact that the cup square of mod 2 classes has a lift to $\mathbb{Z}/4$ coefficients, the Pontryagin square.) By a theorem of Dold and Whitney (1959) the isomorphism classes of $SO(3)$ bundles over a four-manifold are in one-to-one correspondence with pairs (p_1, w_2) satisfying (2.1.36). The groups $SO(3)$ and $SU(2)$ are locally isomorphic: there is a two-fold covering homomorphism from $SU(2)$ to $SO(3)$ (Section 1.1.4). Thus with any $SU(2)$ bundle E we can associate an $SO(3)$ bundle V. In fact the homomorphism from $SU(2)$ to $SO(3)$ is just the adjoint representation of $SU(2)$ on its Lie algebra and the bundle V is what we have denoted by $\mathfrak{g}_E$. The characteristic classes are related by:

bundle V is what we have denoted by $\mathfrak{g}_E$. The characteristic classes are related by:

$$p_1(\mathfrak{g}_E) = -4c_2(E). \tag{2.1.37}$$

The $SO(3)$ bundles which arise from $SU(2)$ bundles in this way, i.e. those which admit a lifting of the structure group to $SU(2)$, are precisely those for which w_2 is zero (necessarily then, p_1 is divisible by 4). For the purposes of local differential geometry, $SO(3)$ connections and $SU(2)$ connections are completely equivalent. Globally, for simply connected base spaces X, a connection on $\mathfrak{g}_E$ determines a unique connection on E. So there is really not much difference between working with the structure groups $SU(2)$ and $SO(3)$. The only difference is that with $SO(3)$ connections we have the additional flexibility to choose w_2, and this can be extremely useful, as we shall see.

Suppose now E is a complex vector bundle with structure group $U(2)$, so we do not impose a trivialization of $\Lambda^2 E$. The bundle of Lie algebras $\mathfrak{g}_E$ splits into

$$\mathfrak{g}_E = \mathfrak{g}_E^{(0)} \oplus \underline{\mathbb{R}}, \tag{2.1.38}$$

corresponding to the trace-free and central endomorphisms. So we again get an $SO(3)$ bundle $\mathfrak{g}_E^{(0)}$. The characteristic classes are related by

$$p_1(\mathfrak{g}_E^{(0)}) = c_1(E)^2 - 4c_2(E); \quad w_2(\mathfrak{g}_E^{(0)}) = c_1(E) \mod 2. \tag{2.1.39}$$

Conversely, given an $SO(3)$ bundle, if $w_2(V)$ can be lifted to an integral class c then V can be obtained from a $U(2)$ bundle, with first Chern class c. In particular, this lift to $U(2)$ can always be made for bundles over a simply connected four-manifold. From this standpoint the group $SO(3)$ more naturally appears in the guise of the projective unitary group $PU(2)$. (It is an easy exercise to deduce the theorem of Dold and Whitney in the case when the base space is simply connected from the classification mentioned in Section 1.1.1 of complex line bundles. One begins by choosing an integral lift of w_2 and constructing an $SO(3)$ bundle of the form $\underline{\mathbb{R}} \oplus L$.)

Returning to the Chern–Weil theory, we make the following conventions for vector bundles over a compact oriented four-manifold. We take as basic characteristic number:

$$\begin{aligned} \kappa(E) &= c_2(E) && \text{for } SU(r) \text{ bundles } E, \\ &= c_2(E) - \tfrac{1}{2}c_1(E)^2 && \text{for } U(r) \text{ bundles } E, \\ &= -\tfrac{1}{4}p_1(V) && \text{for } SO(r) \text{ bundles } V. \end{aligned} \tag{2.1.40}$$

We write the Chern–Weil formula:

$$\kappa(E) = \frac{1}{8\pi^2}\int_X \operatorname{Tr}(F_A^2), \tag{2.1.41}$$

with the understanding that, in the $SO(r)$ case, the trace of F_A^2 is defined by the spin representation of the Lie algebra; e.g. for $SO(3)$ we identify the Lie algebra of $SO(3)$ and $SU(2)$ and use the fundamental representation of the latter. As an immediate consequence of (2.1.31) and (2.1.39) we have:

Proposition (2.1.42). *If a bundle E over a compact, oriented Riemannian four-manifold admits an ASD connection then $\kappa(E) \geq 0$, and if $\kappa(E) = 0$ any ASD connection is flat.*

2.1.5 Holomorphic bundles

The second interpretation of the ASD condition has to do with complex structures. We leave the world of four dimensions for a moment and consider a general complex manifold Z. A holomorphic vector bundle $\mathscr{E}$ over Z is a complex manifold with a holomorphic projection map $\pi: \mathscr{E} \to Z$ and a complex vector space structure on each fibre $\mathscr{E}_z = \pi^{-1}(z)$, such that the data is locally equivalent to the standard product bundle. Alternatively we can say that a holomorphic bundle is a bundle defined by a system of holomorphic transition functions:

$$g_{\alpha\beta}: Z_\alpha \cap Z_\beta \longrightarrow GL(n, \mathbb{C}). \tag{2.1.43}$$

A holomorphic bundle has a preferred collection (more precisely, a sheaf) $\mathcal{O}(\mathscr{E})$ of local sections—the local holomorphic sections. We can multiply holomorphic sections by holomorphic functions, so $\mathcal{O}(\mathscr{E})$ is a sheaf of modules over the structure sheaf $\mathcal{O}_Z$ of local holomorphic functions on Z. It is an easily seen fact that this gives a complete correspondence between holomorphic vector bundles and locally free $\mathcal{O}_Z$ modules.

We will now cast these ideas in more differential-geometric form, introducing a differential operator $\bar{\partial}_{\mathscr{E}}$ on sections of $\mathscr{E}$. To define this operator we first recall that, on the complex manifold Z, the complexified de Rham complex (Ω_Z^*, d) splits into a double complex $(\Omega_Z^{p,q}, \partial, \bar{\partial})$ with $\mathrm{d} = \partial + \bar{\partial}$ and

$$\partial: \Omega^{p,q} \longrightarrow \Omega^{p+1,q}; \quad \bar{\partial}: \Omega^{p,q} \longrightarrow \Omega^{p,q+1}. \tag{2.1.44}$$

(In local holomorphic co-ordinates z_λ we write forms in terms of $\mathrm{d}z_\lambda$, $\mathrm{d}\bar{z}_\lambda$, and $\Omega^{p,q}$ consists of forms with 'p dzs and q d$\bar{z}$s'.) Then a complex valued function f on an open set in Z is holomorphic if and only if $\bar{\partial} f = 0$. Now for any complex vector bundle E over Z we write $\Omega_Z^{p,q}(E)$ for the E-valued (p, q)-forms. Given a holomorphic structure $\mathscr{E}$ on E, as defined above, there is a linear operator

$$\bar{\partial}_{\mathscr{E}}: \Omega_Z^{0,q}(E) \longrightarrow \Omega_Z^{0,q+1}(E) \tag{2.1.45}$$

uniquely determined by the properties:

(i) $\bar{\partial}_{\mathscr{E}}(f.s) = (\bar{\partial} f)s + f(\bar{\partial}_{\mathscr{E}} s)$.
(ii) $\bar{\partial}_{\mathscr{E}} s$ vanishes on an open subset $U \subset Z$ if and only if s is holomorphic over U.

We construct $\bar{\partial}_{\mathscr{E}}$ as follows. Property (i) implies that the operator is local, so it suffices to work in a local holomorphic trivialization. In such a trivialization sections of $\mathscr{E}$ are represented by vector valued functions and we define $\bar{\partial}_{\mathscr{E}}$ on these by the ordinary $\bar{\partial}$ operator, acting on the separate components. This satisfies (i) and (ii). To see explicitly that it is independent of the local holomorphic trivialization consider two different trivializations related by a holomorphic map g into $GL(n, \mathbb{C})$; then for a vector valued function s we have

$$\bar{\partial}(gs) = (\bar{\partial}g)s + g(\bar{\partial}s) = g(\bar{\partial}s), \tag{2.1.46}$$

since g is holomorphic. Thus the $\bar{\partial}$ operator defined in terms of components transforms tensorially under holomorphic changes of trivialization.

The defining properties of $\bar{\partial}_{\mathscr{E}}$ are clearly analogous to those of a covariant derivative. Indeed if we are given a connection (not necessarily unitary) and covariant derivative $d_A = \nabla_A$ on any smooth complex vector bundle E we can decompose $\Omega^1_Z(E)$ into $\Omega^{1,0}_Z(E) \oplus \Omega^{0,1}_Z(E)$ and get corresponding components:

$$d_A = \partial_A \oplus \bar{\partial}_A : \Omega^0_Z(E) \longrightarrow \Omega^{1,0}_Z(E) \oplus \Omega^{0,1}_Z(E). \tag{2.1.47}$$

Extending this analogy, let us consider 'partial connections' on a C^∞ bundle E, i.e., operators

$$\bar{\partial}_\alpha : \Omega^0_Z(E) \longrightarrow \Omega^{0,1}_Z(E), \tag{2.1.48}$$

which satisfy the Leibnitz rule (i) above. In a local C^∞ trivialization τ of E such an operator can be expressed as

$$\bar{\partial}_\alpha = \bar{\partial} + \alpha^\tau, \tag{2.1.49}$$

where our notation follows (2.1.4). Thus α^τ is a matrix of (0, 1) forms. As for the covariant exterior derivative, the operator extends to the bundle-valued $(0, q)$ forms, and $\bar{\partial}^2_\alpha$ is an algebraic operator, Φ_α say:

$$\Phi_\alpha \in \Omega^{0,2}_Z(\operatorname{End} E). \tag{2.1.50}$$

In a local trivialization we have:

$$\Phi^\tau_\alpha = \bar{\partial}\alpha^\tau + \alpha^\tau \wedge \alpha^\tau, \tag{2.1.51}$$

and in local complex coordinates on the base space

$$\Phi_{\lambda\mu} = \left[\frac{\partial}{\partial \bar{z}_\lambda} + \alpha_\lambda, \frac{\partial}{\partial \bar{z}_\mu} + \alpha_\mu \right].$$

The operators $\bar{\partial}_{\mathscr{E}}$ obtained from a holomorphic bundle clearly satisfy $\bar{\partial}^2_E = 0$ and from this point of view the cohomology groups of $\mathscr{E}$, which we denote by $H^p(\mathscr{E})$, are defined as the 'Dolbeault cohomology'

$$H^*(\mathscr{E}) = \ker \bar{\partial}_E / \operatorname{im} \bar{\partial}_E. \tag{2.1.52}$$

All of the above is merely notation. The significant fact which we want to introduce is the 'integrability theorem', to be proved in Section 2.2, which gives a criterion for a partial connection to arise from a holomorphic structure on E. Explicitly, this means that any point z of Z is contained in a neighbourhood K over which there is a trivialization τ of E such that $\alpha^\tau = 0$. For, in such a trivialization, the solutions of $\bar{\partial}_\alpha s = 0$ are just the holomorphic vector functions, so we see that the sheaf of local solutions to this equation is locally free over $\mathcal{O}_Z$, and we have a holomorphic bundle. (Another way to express this is that any two trivializations in which α^τ vanishes differ by a holomorphic map into $GL(n, \mathbb{C})$, and this gives us a system of holomorphic transition functions.) We call a partial connection $\bar{\partial}_\alpha$ *integrable* if these local trivializations exist, i.e. if it comes from a holomorphic structure on E.

Theorem (2.1.53). *A partial connection $\bar{\partial}_\alpha$ on a C^∞ complex vector bundle over a complex manifold Z is integrable if and only if $\bar{\partial}_\alpha^2 = \Phi_\alpha$ is zero.*

The point is that for a general partial connection $\bar{\partial}_\alpha$ there may be no solutions to the equation $\bar{\partial}_\alpha s = 0$ whatsoever: the integrability condition $\bar{\partial}_\alpha^2 = 0$ is the necessary and sufficient condition for the existence of the maximal number of independent solutions to this equation.

We now bring the discussion back to connections and the ASD condition. As we have noted above, a connection A on E defines a partial connection $\bar{\partial}_A$. Conversely, we can look at connections compatible with a given operator $\bar{\partial}_\alpha$. If we have a holomorphic structure $\mathscr{E}$, we say that a connection A is compatible with the structure $\mathscr{E}$ if $\bar{\partial}_A = \bar{\partial}_{\mathscr{E}}$. (The condition can be expressed more geometrically as follows: the principal $GL(n, \mathbb{C})$-bundle P of frames in $\mathscr{E}$ is a complex manifold and the connection is compatible with the holomorphic structure if the horizontal subspaces are complex subspaces of the tangent bundle of P.)

Now given any connection A over Z we can decompose the curvature F_A according to the type:

$$F_A = F_A^{2,0} + F_A^{1,1} + F_A^{0,2}.$$

It is clear from the definitions that the component $F_A^{0,2}$ gives $\bar{\partial}_A^2$ (i.e. $F^{0,2}$ is the tensor denoted Φ above). So the integrability theorem (2.1.53) implies that the connection is compatible with a holomorphic structure precisely when $F_A^{0,2} = 0$.

We now introduce Hermitian metrics, through the following fundamental lemma.

Lemma (2.1.54). *If E is a complex vector bundle over Z with a Hermitian metric on the fibres, then for each partial connection $\bar{\partial}_\alpha$ on E there is a unique unitary connection A such that $\bar{\partial}_A = \bar{\partial}_\alpha$.*

The proof is very easy: we can work in a local unitary trivialization in which the partial connection is represented by a matrix of $(0, 1)$ forms α^τ. The connection matrix A^τ of one-forms must satisfy

$$A^\tau = -(A^\tau)^*$$

(the unitary condition), and have (0, 1) component α^τ (the compatibility condition). These uniquely determine A^τ as

$$A^\tau = \alpha^\tau - (\alpha^\tau)^*. \tag{2.1.55}$$

(The conjugate transpose of a matrix of (0, 1) forms is a matrix of (1, 0) forms.) In particular, if $\mathscr{E}$ is a holomorphic bundle with a Hermitian metric, there is a unique connection on $\mathscr{E}$ compatible with both structures. The curvature of a unitary connection is skew adjoint, so $F^{0,2} = -(F^{2,0})^*$. Thus in sum we have:

Proposition (2.1.56). *A unitary connection on a Hermitian complex vector bundle over Z is compatible with a holomorphic structure if and only if it has curvature of type* (1, 1), *and in this case the connection is uniquely determined by the metric and holomorphic structure.*

For calculational purposes another approach to relation (2.1.54) between connections, metrics and $\bar\partial$ operators is often useful. Given a holomorphic structure we work in a local *holomorphic* trivialization of the bundle, by sections s_i. A Hermitian fibre metric is represented in this trivialization by a self-adjoint matrix h, with

$$h_{ij} = (s_i, s_j).$$

Then in this trivialization the compatible connection is given by the matrix of (1, 0) forms $h^{-1}(\partial h)$. The curvature is given by the matrix of (1, 1) forms $\bar\partial(h^{-1}(\partial h))$.

Fix attention now on a complex surface Z—a complex manifold of complex dimension 2—with a hermitian metric on its tangent bundle. Forgetting the complex structure we obtain an oriented Riemannian four-manifold (using the standard orientation convention that if e_1, e_2 is a complex basis for a tangent space then e_1, Ie_1, e_2, Ie_2 is an oriented real basis). We have then two decompositions of the complexified 2-forms on Z: first the decomposition into bi-type,

$$\Omega^2 = \Omega^{2,0} \oplus \Omega^{1,1} \oplus \Omega^{0,2},$$

and second the decomposition into self-dual and anti-self-dual parts,

$$\Omega^2 = \Omega^+ \oplus \Omega^-.$$

We have already mentioned in Section 1.1.7 the relation between these decompositions. The complex structure and metric together define a (1, 1) form ω, by the rule

$$\omega(\xi, \eta) = (\xi, I\eta).$$

So we can decompose $\Omega^{1,1}$ into parts:

$$\Omega^{1,1} = \Omega_0^{1,1} + \Omega^0 \cdot \omega,$$

where $\Omega_0^{1,1}$ consists of forms pointwise orthogonal to ω. The algebraic fact we need now is:

Lemma (2.1.57). *The complexified self-dual forms over Z are*

$$\Omega^+ = \Omega^{2,0} \oplus \Omega^0\omega \oplus \Omega^{0,2}$$

and the complexified anti-self-dual forms are

$$\Omega^- = \Omega_0^{1,1}.$$

Taking real parts we get the decomposition (1.1.22) of the real self-dual forms into a one-dimensional piece spanned by ω and a real two-dimensional bundle, (which can be identified with the complex line bundle $\Lambda^{0,2}T^*Z$).

The proof of (2.1.57) is straightforward checking: in the model space $\mathbb{C}^2$, with complex coordinates $z_1 = x_1 + ix_2$, $z_2 = x_3 + ix_4$, the (0, 2) forms are spanned by

$$\mathrm{d}\bar{z}_1\mathrm{d}\bar{z}_2 = (\mathrm{d}x_1\mathrm{d}x_3 - \mathrm{d}x_2\mathrm{d}x_4) - i(\mathrm{d}x_2\mathrm{d}x_3 + \mathrm{d}x_1\mathrm{d}x_4),$$

and the metric form is:

$$\omega = \mathrm{d}x_1\mathrm{d}x_2 + \mathrm{d}x_3\mathrm{d}x_4.$$

The real and imaginary parts of $\mathrm{d}\bar{z}_1\mathrm{d}\bar{z}_2$ and the metric form give the standard basis for the self-dual forms.

We now bring the discussion to its fulfillment. For any connection A over Z put

$$\hat{F}_A = (F_A, \omega), \tag{2.1.58}$$

the component of the (1, 1) part of the curvature along the metric form. Then, combining (2.1.57) with the integrability theorem, we have:

Proposition (2.1.59). *If A is an ASD connection on a complex vector bundle E over the Hermitian complex surface Z then the operator $\bar{\partial}_A$ defines a holomorphic structure on E. Conversely if $\mathscr{E}$ is a holomorphic structure on E, and A is a compatible unitary connection, then A is ASD if and only if $\hat{F}_A = 0$.*

To sum up, in the presence of a complex structure on the base space, the ASD condition splits naturally into two pieces, one of which has a simple geometric interpretation as an integrability condition. It is instructive to see this splitting concretely in local coordinates. For simplicity, suppose we are working with the flat Euclidean metric on $\mathbb{C}^2$. Then the three ASD equations (2.1.26) decompose into:

$$[\nabla_1 + i\nabla_2, \nabla_3 + i\nabla_4] = 0 \quad \text{(the integrability condition)} \tag{2.1.60}$$

$$[\nabla_1, \nabla_2] + [\nabla_3, \nabla_4] = 0 \quad \text{(the condition } \hat{F} = 0\text{)}. \tag{2.1.61}$$

Another suggestive way of writing the equations uses the operators $D_1 = \nabla_1 + i\nabla_2$, $D_2 = \nabla_3 + i\nabla_4$ (essentially the components of $\bar{\partial}_A$) and their formal adjoints, e.g. $D_1^* = -\nabla_1 + i\nabla_2$. The equations are:

$$[D_1, D_2] = 0, \tag{2.1.60)'$$

$$[D_1, D_1^*] + [D_2, D_2^*] = 0. \tag{2.1.61)'$$

2.2 Integrability Theorems

2.2.1 Flat connections

We begin with the fundamental integrability theorem for connections, defined initially over the hypercube $H = \{x \in \mathbb{R}^d \,|\, |x_i| < 1\}$.

Theorem (2.2.1). *If E is a bundle over H and A is a flat connection on E there is a bundle isomorphism taking E to the trivial bundle over H and A to the product connection.*

We can prove this, in a procedure that will be used again later, as follows. We can choose any initial trivialization and represent our connection by matrices A_i (the superscript denoting the trivialization will be omitted). The hypothesis that A is flat asserts that the covariant derivatives $\nabla_i = (\partial/\partial x_i) + A_i$ in the different coordinate directions commute.

We want to show that there is a gauge transformation $u: H \to U(r)$ such that $u\nabla_i u^{-1} = \partial/\partial x_i$ for all i. This is clearly analogous to the simultaneous diagonalization of commuting matrices. To find u we suppose, inductively, that the required condition holds for the first p indices ($p < d$), i.e. $A_i = 0$ for $i = 1, \ldots, p$. It suffices to show that we can then find a gauge transformation h such that $h\nabla_i h^{-1} = \partial/\partial x_i$ for $i = 1, \ldots, p+1$. For then repeating this d times we get the desired gauge transformation, i.e. new trivialization. Now the equations for h that we wish to satisfy are:

(a) $$\frac{\partial h}{\partial x_i} = 0 \qquad \text{for } i = 1, \ldots, p,$$

(b) $$\frac{\partial h}{\partial x_{p+1}} + hA_{p+1} = 0.$$

Equation (b) is a linear ordinary differential equation (ODE) for h in the x_{p+1} variable. By the standard theory of ODEs there is a unique solution, for fixed $x_i, i \neq p+1$, with the initial condition $h(x_1, \ldots, x_p, 0, x_{p+2}, \ldots) = 1$. Moreover the solution is smooth in the variables x_i, regarded as parameters in the ODE. Now our hypotheses assert that

$$\left[\frac{\partial}{\partial x_i}, \frac{\partial}{\partial x_{p+1}} + A_{p+1}\right] = \frac{\partial A_{p+1}}{\partial x_i} = 0 \quad \text{for } i \leq p.$$

So A_{p+1} is independent of the first p variables. By uniqueness, the solution h

with the given initial conditions is also independent of these variables, so (a) is satisfied. Moreover if the connection matrices A_i are in the Lie algebra of $U(r)$ (the skew adjoint matrices), we have $\partial(h^* h)/\partial x_{p+1} = 0$, so h is a unitary gauge transformation and the proof is complete.

Let us now make some remarks about this theorem. First the geometric meaning of the proof is clear. We construct trivializations by successive parallel transports. First we choose a framing for the fibre E_0. Then we parallel transport this along the x_1 axis to trivialize E there. Next we transport along the lines in the x_2 direction to extend this trivialization to a square in the $x_1 x_2$ plane, and then transport in the x_3 direction to extend to a three-dimensional cube, and so on. Our proof can be viewed as saying that if we construct a trivialization of the bundle by this explicit procedure—a procedure for 'fixing the choice of gauge'—using a flat connection, the resulting connection matrices all vanish. We could equally well have used other procedures based on parallel transport along other families of lines. For example the rays '$\sigma = \text{constant}$' in generalized polar coordinates (r, σ), $\sigma \in S^{d-1}$. In the latter case we have connection matrices, written in a handy abbreviated notation, A_r, A_σ. We again choose a frame for E_0 and extend to a trivialization by parallel transport along the rays. In this trivialization we have $A_r = 0$ by construction. The curvature condition $F_{r\sigma} = 0$ asserts then that A_σ is independent of r. On the other hand, by considering the coordinate singularity at $r = 0$ one sees that $A_\sigma \to 0$ as $r \to 0$. So we conclude that A_σ also vanishes.

Next observe that in the Abelian case of a rank-one bundle, the statement of the theorem is precisely the Poincaré lemma for closed 1-forms: $dA = F = 0 \Rightarrow A = d\chi$, where we take the gauge change $u = \exp(\chi)$. Our proofs reduce to the standard proofs of the Poincaré lemma, which depend of course on the contractability of the base space H. The use of polar or cartesian coordinates corresponds to different explicit contractions. In terms of principal bundles our proof asserts that the family of horizontal subspaces defining the connection are *integrable* (or *involutive*) if the curvature vanishes. This can be proved directly using the Frobenius theorem. The horizontal subspaces define a horizontal foliation in the total space of the principal bundle P and parallel transport is given by moving in the leaves of this foliation. One sees from this that for a flat connection the parallel transport $T_\rho : E_x \to E_y$ depends only upon the homotopy class of the path ρ between x and y. In particular, considering loops we get, for any flat connection on a bundle with structure group G, a holonomy representation

$$r_A : \pi_1(X, x_0) \to \mathrm{Aut}(E_{x_0}) = G. \tag{2.2.2}$$

Then one can easily prove:

Proposition (2.2.3). *The gauge equivalence classes of flat G-connections over a connected manifold X are in one-to-one correspondence with the conjugacy classes of representations* $\pi_1(X) \to G$.

In the case of complex line bundles this classification theorem can be extended to all connections. Let L be a Hermitian complex line bundle over a manifold X and $\mathscr{C}_L \subset \Omega_X^2$ be the set of closed 2-forms representing $c_1(L)$. We have seen in section 2.1.4 that the normalized curvature form gives a map, f say, from the space $\mathscr{A}$ of unitary connections on L to $\mathscr{C}_L$. This map is surjective, since any connection can be changed by an imaginary 1-form a, and this changes f by $(i/2\pi)\mathrm{d}a$. On the other hand f is constant on the gauge equivalence classes, the orbits of $\mathscr{G}$ in $\mathscr{A}$. Suppose connections A_1, A_2 have the same curvature form; then their difference $a = ib = A_1 - A_2$ is closed, and b defines a cohomology class in $H^1(X; \mathbb{R})$. Changing A_1 by a gauge transformation $u: X \to U(1)$, with A_2 fixed, changes b by $i\mathrm{d}u\, u^{-1}$. If u can be written as an exponential $\exp(i\xi)$ this is just $-\mathrm{d}\xi$. Now the maps which can be written as exponentials are just the null-homotopic maps. The homotopy classes of maps from X to $U(1) = S^1$ may be identified with $H^1(X; \mathbb{Z})$ and one sees that the class of b in the '*Jacobian torus*'

$$J_X = H^1(X; \mathbb{R})/H^1(X; \mathbb{Z}) \tag{2.2.4}$$

is unchanged by gauge transformations acting on A_1. In sum we obtain a description of the space $\mathscr{B} = \mathscr{A}/\mathscr{G}$ of gauge equivalence classes of connections on L in the form of a fibration:

$$J_X \longrightarrow \mathscr{B} \longrightarrow \mathscr{C}_L. \tag{2.2.5}$$

(If $c_1(L)$ is zero then $\mathscr{B}$ has a group structure, induced from tensor product of line bundles, and this is an exact sequence of groups.)

As a special case we have:

Proposition (2.2.6). *If $H^1(X; \mathbb{R}) = 0$ and L is a line bundle over X then for any 2-form ω representing $c_1(L)$ there is a unique gauge equivalence class of connections with curvature $-2\pi i\omega$.*

2.2.2 *Proof of the integrability theorem for holomorphic structures*

Let us now take up the main business of this section: the proof of Theorem (2.1.53). This is very similar to the elementary proof above.

We consider a complex of operators $\bar{\partial}_\alpha: \Omega^{0,p}(E) \to \Omega^{0,p+1}(E)$, satisfying the Leibnitz rule and with $\bar{\partial}_\alpha^2 = 0$. We want to show that these define a holomorphic structure on E. The problem is purely local and we can work on an arbitrarily small neighbourhood of a given point. So, in line with the discussion above we can suppose that Z is a polydisc $K(1) = \{|z_\lambda| < 1\} \subset \mathbb{C}^d$. Then we can choose a trivialization of E, as a smooth bundle, and represent the operator in the form $\bar{\partial} + \alpha$, for a matrix of (0, 1)-forms α over $K(1)$. Our hypothesis is that $\bar{\partial}\alpha + \alpha \wedge \alpha = 0$, and we want to show that there is a smaller polydisc $K(r) = \{|z_\lambda| < r\}$ and a 'complex gauge transformation'

$$g: K(r) \longrightarrow GL(n, \mathbb{C})$$

with $g\alpha g^{-1} - (\bar{\partial} g)g^{-1} = 0$ on $K(r)$. More explicitly still, the operator has components $(\partial/\partial \bar{z}_\lambda) + \alpha_\lambda$, the hypothesis is that these components commute, and we want to show that there is a g with

$$g\left(\frac{\partial}{\partial \bar{z}_\lambda} + \alpha_\lambda\right)g^{-1} = \frac{\partial}{\partial \bar{z}_\lambda}.$$

We begin with the special case when the base space is of one complex dimension. Then the integrability condition is vacuous. We have a single coordinate which we denote z, and we write $\alpha = \rho\, \mathrm{d}\bar{z}$, say. So ρ is a matrix function on the unit disc D in $\mathbb{C}$. We want to solve the equation

$$\frac{\partial g}{\partial \bar{z}} - g\rho = 0, \tag{2.2.7}$$

with g invertible, and we are content with a solution in an arbitrarily small neighbourhood of 0. We incorporate this latter freedom by a rescaling procedure. For $r < 1$ let $\delta_r : \mathbb{C} \to \mathbb{C}$ be the map $\delta_r(z) = rz$. Then we are free to replace the matrix of forms α in our problem by $\delta_r^*(\alpha)$, since this just corresponds to a change of local coordinate. In other words, we are free to replace the function ρ in (2.2.7) by

$$r\rho(rz).$$

In particular, we can suppose that:

$$N = \sup|\rho(z)| = \|\rho\|_\infty$$

is as small as we please. Similarly, since we need only solve the equation in a neighbourhood of 0, we can multiply ρ by a cut-off function ψ, equal to 1 near 0, without changing the problem. If $\psi = 1$ on the $\frac{1}{2}$-disc about 1 and we solve the problem for $\rho'(z) = \psi(z)r\rho(rz)$ we can transform back to get a solution of the original equation over the $\frac{1}{2}r$-ball. These observations mean that we can suppose that the matrix function ρ is defined over all of $\mathbb{C}$, is smooth and supported in say the unit ball, and that $N = \|\rho\|_\infty$ is as small as we want (and the same holds for any standard norm of ρ). When $N < 1$ we will solve (2.2.7) over all of $\mathbb{C}$ by the familiar contraction mapping principle.

To set up the problem we write $g = 1 + f$, so we need to solve $\partial f/\partial \bar{z} = (1 + f)\rho$. Recall that the Cauchy kernel $-(1/2\pi i z)$ is a fundamental solution of the Cauchy–Riemann equation on $\mathbb{C}$. If θ is a compactly supported function (or, here, a matrix-valued function) on $\mathbb{C}$ then

$$\frac{\partial}{\partial \bar{z}}(L\theta) = \theta$$

where:

$$(L\theta)(w) = -\frac{1}{2\pi i}\int_{\mathbb{C}} \frac{\theta(z)}{z - w}\, \mathrm{d}\mu_z. \tag{2.2.8}$$

So if f satisfies the integral equation $f = L(\alpha + f\alpha)$, it will indeed satisfy the differential equation. If also $\|f\|_\infty$ is small then $g = 1 + f$ will be an invertible matrix giving a solution to our problem. Finally elliptic regularity for the $\bar{\partial}$ operator (or, what is essentially the same, estimates for the integral operator L) implies that any bounded solution f of the integral equation is smooth, so we can work in the Banach space $L^\infty(\mathbb{C})$.

Now it is clear that for any disc D in $\mathbb{C}$ of radius 1

$$\int_D \left|\frac{1}{z}\right| \mathrm{d}\mu_z \leq \int_0^1 \frac{1}{r} \cdot 2\pi r \,\mathrm{d}r = 2\pi.$$

Thus, using the fact that ρ is supported in the disc $|z| < 1$, we have:

$$\|L(h.\rho)\|_\infty \leq N\|h\|_\infty \quad \text{for any } h.$$

So, if $N < 1$, the map T defined by $T(g) = L(\rho + g.\rho)$ is a contraction mapping from L^∞ to itself. Thus T has a unique fixed point f, which yields a solution to our equation. The norm of f is bounded by $\sum N^i \| L(\rho)\|_\infty$ and so we can make $1 + f$ invertible by choosing N small. This completes the proof in the one-dimensional case. Notice that we can regard the argument as an application of the implicit function theorem to the function of two variables h, ρ give by $F(h, \rho) = h - L(\rho + h\rho)$: cf. Appendix A3.

To go to the many-dimensional case we need one last observation about the solution constructed above. If ρ depends on some additional parameters $\rho = \rho(z; \xi, \eta)$, but is always small enough for the contraction method to apply, we get a family of particular solutions g depending on these parameters. If ρ is *holomorphic* in ξ and *smooth* in η then g will be also. This is the standard addendum to the implicit function theorem in complex Banach spaces, since F is holomorphic in the variables ρ, h.

Following the scheme introduced in the proof of (2.2.1) we now turn to the general problem, with $\alpha = \sum \alpha_\lambda \mathrm{d}\bar{z}_\lambda$, and suppose that α_λ vanishes for $\lambda = 1, \ldots, p$. This condition is preserved by automorphisms g which are holomorphic in $z_1, \ldots, z_p$. On the other hand the hypothesis gives that $\rho = \alpha_{p+1}$ is holomorphic in these variables. We can dilate the z_{p+1} coordinate and multiply by a fixed cut-off function $\psi(|z_{p+1}|)$ to get a new matrix function,

$$\rho' = r.\psi.\rho(z_1, \ldots, rz_{p+1}, \ldots, z_d),$$

for a small constant r; and ρ' is still holomorphic in the first p variables. Then solve the equation

$$\frac{\partial h}{\partial \bar{z}_{p+1}} = h\rho' \qquad (z_{p+1} \in \mathbb{C})$$

with the z_λ variables ($\lambda \neq p + 1$)—regarded as parameters in the equation—running over a compact polydisc. We can choose r so small that the discussion of the one-dimensional case above applies to give a solution h, and this is holomorphic in the first variables. Finally, reversing the scale change we made in the z_{p+1} direction and restricting to a sufficiently small polydisc, h yields a new trivialization in which the α_λ vanish for $\lambda = 1, \dots, p+1$. The proof is completed by induction as before.

It is probably not necessary to point out now the similarities between our theorems (2.1.53) and (2.2.1) and their proofs. In each case we treat the problem by reducing to the one-dimensional situation, with the other variables regarded as parameters. The main difference is that in the complex setting the one variable problem already involves a partial rather than ordinary differential equation, and this is why we restrict ourselves to smaller and smaller polydiscs. However, our method of solving the partial differential equation—conversion to an integral equation to which the contraction mapping principle applies—is of course the same as the standard method of proving existence of solutions to ordinary differential equations.

Notice also that in the case of bundles of rank one our integrability theorem (2.1.53) reduces to a weak form of the '$\bar{\partial}$–Poincaré lemma' for (0, 1)-forms; i.e. $\bar{\partial}\alpha = 0 \Rightarrow \alpha = \bar{\partial}\eta$, when we take $g = \exp(\eta)$. Again, our proof reduces in the rank 1 case exactly to the standard proof of this lemma. So theorem (2.1.53) can be regarded as an extension of the basic integrability theorem to complex variables, which fits in naturally with familiar constructions for differential forms. In the next section we will look at another kind of extension of the integrability theorem which, in the same spirit, fits in with the ideas of Hodge theory for differential forms.

2.3 Uhlenbeck's Theorem

2.3.1 Gauge fixing

We have seen that a flat connection can be represented, at least locally, by the zero connection matrix in a suitable choice of gauge or bundle trivialization. It is natural to ask then whether a connection with small curvature can be represented by correspondingly small connection matrices in an appropriate gauge. We would like to have a canonical choice of the optimal connection matrix, some way of 'fixing the choice of gauge'. Consider for example our ASD equation over an oriented Riemannian four-manifold X, expressed in terms of a connection matrix A^τ, which we just denote by A. To write the equation compactly we introduce the differential operator,

$$d^+ : \Omega^1_X \longrightarrow \Omega^+_X,$$

of Section 1.1.6. Then the ASD equation is:

$$d^+ A + (A \wedge A)^+ = 0,$$

where of course the + superscript denotes the self-dual part. Now for *linear, elliptic* differential equations we can call on a substantial body of results, the most important of which are summarized in the Appendix. If D is an elliptic operator of order k and s is a solution of the equation $\mathrm{D}s = 0$ over a domain Ω in $\mathbb{R}^d$ we have *a priori* inequalities: the norm of s in any one function space—for example the L^2 norm—controls the norms of all derivatives on an interior domain. This leads to *compactness* properties—a sequence of solutions s_i with $\|s_i\| < C$ has a subsequence converging in C^∞ on interior domains. Now our ASD equation is non-linear (except for the special case of rank-one bundles) but more to the point it is not elliptic, i.e. the highest order part d^+ is not an elliptic operator. This is clear on abstract grounds from the invariance of the equation under gauge transformations. We can make solutions $A = -(\mathrm{d}u)u^{-1}$, gauge equivalent to the trivial one $A = 0$, whose higher derivatives are not controlled by, say, the L^2 norm. Plainly what we want is some way of removing this gauge invariance and in this section we shall explain how this is done.

In the course of our discussion of the integrability theorem we have already seen one method of fixing a choice of gauge for a connection defined over, say, the unit ball in $\mathbb{R}^d$. We choose a frame for the fibre over 0 and spread it out to trivialize the bundle by parallel transport along radial lines. In terms of connection matrices there is a representative A for a connection with $A_r = 0$, unique up to $A \mapsto uAu^{-1}$ for a *constant* unitary gauge transformation u. In this gauge we can indeed estimate the connection matrix by the curvature. In polar coordinates we have

$$\frac{\partial A_\sigma}{\partial r} = F_{r\sigma},$$

and

$$A_\sigma \longrightarrow 0 \quad \text{as} \quad r \longrightarrow 0.$$

Hence $|A_\sigma(r, \sigma)| \leq r \sup |F_{r\sigma}|$ or, transferring back to Cartesian coordinates: $|A_i(\underline{x})| \leq |\underline{x}| \sup |F_A|$. This is satisfactory up to a point but what we really require are estimates starting from the L^2 norm of F_A, because, as we have seen above, this is the measure of curvature which fits in most naturally with the ASD, and general Yang–Mills, equations. Even more important, the radial gauge fixing condition $A_r = 0$ does not combine happily with PDEs and elliptic theory.

We can see the way ahead by looking again at the linear case of rank-one bundles. Consider, for simplicity, connections on the trivial $U(1)$ bundle over a compact, simply connected manifold X. Then any gauge transformation can be written as $u = \exp(i\chi)$ for a real-valued function χ on X and these act on the connection 1-forms by $A \mapsto A - i\mathrm{d}\chi$. A choice of optimal representative is supplied by the Hodge theory. Given any A we can choose χ such that $\tilde{A} = A - i\mathrm{d}\chi$ satisfies the equation:

$$\mathrm{d}^* \tilde{A} = 0.$$

Moreover $\tilde{A}$ is unique, and χ is uniquely determined up to a constant. These assertions follow immediately from the Fredholm alternative for the Laplacian $\Delta = d^*d$ on Ω^0_X. We can solve the equation $\Delta f = g$ if and only if g is orthogonal to the kernel of the formal adjoint operator. But Δ is self-adjoint and its kernel consists exactly of the constant functions (for if $\Delta s = 0$ then $\int_X (\mathrm{d}s, \mathrm{d}s) = \int_X (s, \Delta s) = 0$, so $\mathrm{d}s = 0$). So the condition is just that $\int_X g = 0$. Now $\int_X \mathrm{d}^* A = 0$ by Stokes' Theorem so the equation $\mathrm{d}^*\mathrm{d}\chi = -i\mathrm{d}^* A$ can indeed be solved. Moreover this choice of representative fits in well with elliptic theory and estimates in Sobolev spaces. The operator

$$\mathrm{d}^* + \mathrm{d} \colon \bigoplus_i \Omega_X^{2i+1} \to \bigoplus_i \Omega_X^{2i}$$

is elliptic, its kernel decomposes according to degree and so if, as we suppose, $H^1(X)$ is zero, all the 1-forms are orthogonal to the kernel. So elliptic theory gives inequalities

$$\| A \|_{L^2_k} \leq \text{const.} (\| \mathrm{d}^* A \|_{L^2_{k-1}} + \| \mathrm{d} A \|_{L^2_{k-1}}).$$

In particular if A satisfies $\mathrm{d}^* A = 0$ then $\| A \|_{L^2_k} \leq \text{const.} \| F_A \|_{L^2_{k-1}}$, a 'gain' of one derivative, measured in L^2. In the case when X is four dimensional we can consider also the operator

$$\mathrm{d}^* + \mathrm{d}^+ \colon \Omega^1 \longrightarrow \Omega^0 \oplus \Omega^+. \tag{2.3.1}$$

This is an elliptic operator, analogous to the Cauchy–Riemann operator in two dimensions. Sometimes, for brevity, we shall denote this operator by δ.

So the linear version, $\mathrm{d}^+ A = 0$, of the ASD equation, *combined* with the gauge fixing condition $\mathrm{d}^* A = 0$, yields an elliptic system of equations $\delta A = 0$.

While we have been discussing above the case of a closed base manifold, similar ideas can be applied on manifolds with boundary or on complete manifolds, given appropriate boundary or decay conditions. For domains in $\mathbb{R}^3$ this is all standard classical procedure in the theory of magnetism. The vector potential is normalized by the condition $\mathrm{d}^* A = 0$ (or $\mathrm{div}(\mathbf{A}) = 0$ in vector notation). This is called the choice of the 'Coulomb gauge'. (For general electromagnetic fields, defined over four-dimensional space–time, it is more appropriate to use the 'Lorentz gauge' condition $\mathrm{d}^*_{(4)} A = 0$, where $\mathrm{d}^*_{(4)}$ is the formal adjoint of d defined by the pseudo-Riemannian metric on space–time. To satisfy this condition we have to solve a wave equation for χ.)

Now the motivation for the gauge fixing condition $\mathrm{d}^* A = 0$ in the linear case of Hodge theory is that this is the choice which minimizes the L^2 norm of A over the family of representatives $A - \mathrm{d}\chi$. The proof is immediate from the definition of d^*. We take up the same idea to generalize to the non-linear case; at the same time we can shift from considering connection matrices (i.e. comparing a connection with the product structure) to a more invariant point of view in which we compare pairs of connections.

Suppose A_0 is a connection on a unitary bundle $E \to X$ over a Riemannian manifold X, and consider the gauge equivalence class of another connection A on E:

$$\mathscr{H} = \{u.A \mid u \in \mathscr{G}\} \subset \mathscr{A}.$$

We say that a point B in $\mathscr{H}$ is in *Coulomb gauge* relative to A_0 if

$$d^*_{A_0}(B - A_0) = 0. \tag{2.3.2}$$

(Here $d^*_{A_0}$ is the operator defined in (2.1.24)). This is the Euler–Lagrange equation for the functional $B \to \|B - A_0\|^2$ on the equivalence class $\mathscr{H}$. Indeed if we consider a one-parameter family of gauge transformations $\exp(t\chi)$, where $\chi \in \Omega^0_X(\mathfrak{g}_E)$ has compact support, then

$$\begin{aligned}\frac{\mathrm{d}}{\mathrm{d}t}\|\exp(t\chi)B - A_0\|^2|_{t=0} &= \frac{\mathrm{d}}{\mathrm{d}t}\|(B - t\mathrm{d}_B\chi) - A_0\|^2|_{t=0}\\ &= -\langle\chi, \mathrm{d}^*_B(B - A_0)\rangle.\end{aligned}$$

Here we have used (2.1.20) for the derivative of the action of the gauge group at B.

Notice that this Coulomb gauge condition is *symmetric* in A_0, B. If we put $a = B - A_0$, then for $b \in \Omega^1_X(\mathfrak{g}_E)$,

$$\mathrm{d}^*_B b = d^*_{A_0} b + \{a, b\}, \tag{2.3.3}$$

where the bilinear form $\{,\}$ is the tensor product of the bracket on $\mathfrak{g}_E$ and the Riemannian metric on Ω^1_X. So $\{a, a\} = 0$ and $\mathrm{d}^*_B a = \mathrm{d}^*_{A_0} a$. More conceptually, this symmetry follows from the variational derivation and the fact that $\|A - B\|^2$ is preserved by the gauge group $\mathscr{G}$ acting simultaneously on A, B.

Now, as a special case of this, consider the trivial bundle and let θ denote the product connection. Another connection A is in Coulomb gauge relative to θ if it satisfies $\mathrm{d}^*A = 0$. Here, of course, we are regarding a connection on the trivial bundle (with a fixed trivialization) as a connection 1-form. Shifting point of view slightly, we say that a trivialization τ of a bundle represents a connection in Coulomb gauge if its connection matrix A^τ satisfies $\mathrm{d}^*A^\tau = 0$. It follows from our discussion that this is the choice of trivialization for which the L^2 norm of the curvature matrix is extremal. It is reasonable to hope that, as in the Abelian, linear case, this choice of gauge will actually minimize the L^2 norm and will yield the desired optimal, small, connection matrix.

2.3.2 Application of the implicit function theorem

We will now prove a simple result on the existence of these Coulomb gauge representatives, working over a compact base manifold. This serves as a preliminary to the more refined analysis in the next sections, and the result will be taken up again in Chapter 4.

Let X be a compact Riemannian four-manifold and A be a connection on a unitary bundle E over X.

Proposition (2.3.4). *There is a constant $c(A)$ such that if B is another connection on E and if $a = B - A$ satisfies*

$$\|\nabla_A \nabla_A a\|^2 + \|a\|^2 < c(A),$$

then there is a gauge transformation u such that $u(B)$ is in Coulomb gauge relative to A.

The proof is a simple application of the implicit function theorem, based on the linear model considered above. We have

$$\begin{aligned} u(A + a) &= A + a - (\mathrm{d}_{A+a} u)u^{-1} \\ &= A + uau^{-1} - (\mathrm{d}_A u)u^{-1}. \end{aligned}$$

So the equation to be solved, for u in $\mathscr{G}$, is

$$\mathrm{d}_A^*((\mathrm{d}_A u)u^{-1} - uau^{-1}) = 0. \tag{2.3.5}$$

We write $u = \exp(\chi) = e^\chi$ for a section χ of $\mathfrak{g}_E$, and set

$$G(\chi, a) = \mathrm{d}_A^*((\mathrm{d}_A e^\chi)e^{-\chi} - e^\chi a e^{-\chi}).$$

To apply the implicit function theorem we wish to work in Sobolev spaces, so we extend the domain of G to sections χ in the Sobolev space L_3^2 and bundle-valued 1-forms a in L_2^2. The map G extends to define a smooth map on these Banach spaces, with $G(\chi, a)$ in L_1^2, since the L_3^2 sections χ are continuous (see the Appendix). More precisely, the image of G is contained in the L_1^2 closure of the image of d_A^*. The derivative of G at $\chi = 0, a = 0$ is

$$DG(\xi, b) = \mathrm{d}_A^* \mathrm{d}_A \xi - \mathrm{d}_A^* b. \tag{2.3.6}$$

The implicit function theorem gives a small solution χ to the equation $G(\chi, a) = 0$, for all small enough a, provided that the partial derivative $\xi \mapsto \mathrm{d}_A^* \mathrm{d}_A \xi$ maps onto the image space $\operatorname{im} \mathrm{d}_A^*$. But this surjectivity follows from the Fredholm alternative for the coupled Laplace operator $\mathrm{d}_A^* \mathrm{d}_A$ just as in the linear case considered above. The kernel of this Laplace operator is the kernel of d_A, and its image is the image of d_A^*.

The square root of $\|\nabla_A \nabla_A a\|^2 + \|a\|^2$ is an admissible norm on the Sobolev space L_2^2. So for a small enough constant $c(A)$ we can solve the equation, with $u = \exp(\chi)$, except that χ is at the outset only in L_3^2. The fact that u is actually smooth (when a is) can be obtained by a straightforward elliptic regularity argument. We 'bootstrap', using the equation written in the form:

$$\mathrm{d}_A^* \mathrm{d}_A u = (\mathrm{d}_A u u^{-1}, \mathrm{d}_A u) + u \mathrm{d}_A^* a u^{-1} + (\mathrm{d}_A u, a) + (ua, u^{-1} \mathrm{d}_A u),$$

where $(\ ,\ )$ in this case denotes the contraction on the one-form components only, defined by the metric. If u is continuous and lies in the Sobolev space L^2_l, then the right hand side lies in L^2_{l-1} and so u is in L^2_{l+1}, by elliptic regularity for the second-order elliptic operator $d_A^* d_A$.

2.3.3 *Uhlenbeck's theorem*

We now come to the theorem of K. Uhlenbeck on the existence of local Coulomb gauges, which will provide the essential analytical input for most of the results described in this book.

Let B^4 be the unit ball in $\mathbb{R}^4$ and $m: \mathbb{R}^4 \to S^4$ be the standard stereographic map, a conformal diffeomorphism from $\mathbb{R}^4$ to S^4 minus a point. This maps the unit ball B^4 in $\mathbb{R}^4$ to a hemisphere in S^4. We fix as standard metric on B^4 the pullback by m of the round metric on S^4. This is conformal to the flat metric, so when we look at the L^2 norm on 2-forms, and in particular of curvature tensors, it does not matter which metric we use. On the other hand the d^* operators defined by the two metrics are different. This choice of metric is merely a convenience, which we adopt in order to work on compact manifolds. We could obtain just the same results with the standard flat metric, as indeed is done in the original proof given by Uhlenbeck. Similarly, when we come to discuss the ASD solutions we assume for simplicity that these are defined relative to the standard metric but the results adapt, in an obvious way, to Riemannian metrics which are close to this one.

The setting for this section is that we work with $U(n)$ connections, for fixed n, on the trivial bundle over the four-ball B^4, *i.e.* with connection matrices. As a point of notation, a connection matrix over B^4 is supposed to be smooth on the open ball, unless explicitly stated otherwise: by a connection over $\bar{B}^4$ we mean one which is smooth up to the boundary. We denote by A_r the radial component $\sum (x_i/r) A_i$ of the connection matrix, defined on the punctured ball.

The main result is this:

Theorem (2.3.7). *There are constant* $\varepsilon_1, M > 0$ *such that any connection* A *on the trivial bundle over* $\bar{B}^4$ with $\| F_A \|_{L^2} < \varepsilon_1$ *is gauge equivalent to a connection* $\tilde{A}$ *over* B^4 *with*

(i) $$d^* \tilde{A} = 0,$$

(ii) $$\lim_{|x| \to 1} \tilde{A}_r = 0,$$

(iii) $$\| \tilde{A} \|_{L^2_1} \leq M \| F_{\tilde{A}} \|_{L^2}.$$

Moreover, for suitable constants ε_1, M, *the connection* $\tilde{A}$ *is uniquely determined by these properties, up to the transformation* $\tilde{A} \to u_0 \tilde{A} u_0^{-1}$ *for a constant* u_0 *in* $U(n)$.

We should emphasize that connections are being considered here as connection matrices. The Sobolev norm $\|\tilde{A}\|_{L_1^2}$ is the square root of

$$\int_{B^4} |\nabla \tilde{A}|^2 + |\tilde{A}|^2 \, d\mu,$$

and the reader should recall that there is a Sobolev inequality

$$\|\tilde{A}\|_{L^4} \leq C . \|\tilde{A}\|_{L_1^2}.$$

The precise meaning we attach to the 'boundary condition' (ii) is that for $\rho < 1$ we consider $A_r(\rho, \sigma)$ as a function on S^3, and we require that this function tends to 0 as a distribution on S^3, as $r \to 1$.

This result is very satisfactory. It says that if we stay within the regime of small curvature, measured in L^2, the Coulomb gauge condition can always be satisfied, and yields a small connection matrix, measured in L_1^2. Moreover (although this will not be important for us), the solution enjoys the same uniqueness properties as the elementary radial gauge, provided we impose the boundary condition (ii).

The theorem above will be of most use to us when combined with an auxiliary result for ASD solutions.

Theorem (2.3.8). *There is a constant $\varepsilon_2 > 0$ such that if $\tilde{A}$ is any ASD connection on the trivial bundle over B^4 which satisfies the Coulomb gauge condition $d^*\tilde{A} = 0$ and $\|\tilde{A}\|_{L^4} \leq \varepsilon_2$, then for any interior domain $D \Subset B^4$ and any $l \geq 1$ we have*

$$\|\tilde{A}\|_{L_l^2(D)} \leq M_{l,D} \|F_A\|_{L^2(B^4)},$$

for a constant $M_{l,D}$ depending only on l and D.

Thus, in the small curvature regime the ASD equations in Coulomb gauge enjoy the usual properties of elliptic equations, and the L^2 norm of the curvature controls all derivatives of the connection.

We can combine (2.3.7) and (2.3.8). Put $\varepsilon = \min(\varepsilon_1, \varepsilon_2/CM)$, where C is the Sobolev constant. Then any ASD connection over the unit ball with $\|F\| \leq \varepsilon$ can be represented in a Coulomb gauge in which all derivatives of the connection matrix, over any interior domain D, are bounded in L^2. Using the Sobolev and Ascoli–Arzela theorems we obtain:

Corollary (2.3.9). *For any sequence of ASD connections A_α over $\bar{B}^4$ with $\|F(A_\alpha)\|_{L^2} \leq \varepsilon$ there is a subsequence α' and gauge equivalent connections $\tilde{A}_{\alpha'}$ which converge in C^∞ on the open ball.*

Moreover the same holds true, for a suitable ε, for connections which are ASD with respect to a Riemannian metric on the ball which is close to the Euclidean one. This extension will be clear from our proof.

2.3.4 *Rearrangement argument; the key lemma*

We will give two proofs of Theorem (2.3.7). The first, which extends through Sections 2.3.4 to 2.3.9, is more or less the same as the original proof of Uhlenbeck. It depends upon three basic analytical points which will reappear in slightly different forms in many places in this book. One of these is the use of the implicit function theorem to solve the Coulomb gauge equation, which we have already seen in Section 2.3.2. (The second proof, which we give more briefly in Section 2.3.10, will use the implicit function theorem in a rather different way.) Another idea, which will also appear in the second proof, is an elementary manipulation of the information coming from Sobolev inequalities, which is characteristic of conformally invariant problems.

In this section we will develop this idea in the proof of the key lemma below. A feature of our approach (by no means essential) is that we prefer to carry out the analytical arguments working over compact manifolds. Thus we shall begin by considering connections over the round four-sphere and postpone to Section 2.3.9 below the adaptation of our results to the four-ball.

Lemma (2.3.10). *Let B be a connection on the trivial bundle over S^4 in Coulomb gauge relative to the product connection* (*i.e. with* $\mathrm{d}^*B = 0$). *There are constants $N, \eta > 0$ such that if $\|B\|_{L^4} < \eta$ then $\|B\|_{L^2_1} \leq N\|F_B\|_{L^2}$.*

Proof. Since $H^1(S^4) = 0$, the basic elliptic estimate for the operator $\mathrm{d}^* + \mathrm{d}$ on 1-forms gives a bound of the form

$$\|B\|_{L^2_1} \leq c_1\|\mathrm{d}B\|.$$

Now $F_B = \mathrm{d}B + B \wedge B$, and the L^2 norm of the quadratic $B \wedge B$ term can plainly be bounded by the square of the L^4 norm of B. Using the Sobolev embedding theorem

$$\|f\|_{L^4} \leq \text{const.}\|f\|_{L^2_1},$$

we get

$$\|B \wedge B\| \leq c_2\|B\|_{L^4}\|B\|_{L^2_1}, \quad \text{say.}$$

So

$$\|B\|_{L^2_1} \leq c_1\|F_B\| + c_1c_2\|B\|_{L^4}\|B\|_{L^2_1}.$$

If $\|B\|_{L^4} < (1/2c_1c_2)$, say, we can re-arrange this as

$$\|B\|_{L^2_1}\{1 - c_1c_2\|B\|_{L^4}\} \leq c_1\|F_B\|,$$

to get $\|B\|_{L^2_1} \leq 2c_1\|F_B\|$. So the required result holds with $N = 2c_1$ and $\eta = (1/2c_1c_2)$.

2.3.5 *Estimating higher derivatives: proof of (2.3.8)*

The point to note about the proof of Lemma (2.3.10) above is that it operates on the borderline of the Sobolev inequalities. The given non-linear dependence of the curvature on the connection creates a watershed in the range of available Sobolev spaces. For the weaker Sobolev norms it is not possible to pass from bounds on the curvature to corresponding bounds on the connection, since the quadratic term $B \wedge B$ cannot be controlled by the leading term $\mathrm{d}B$, even in the presence of the Coulomb condition. The lemma deals with the borderline case when we can barely obtain some control—we go from information about the L^4 norm of B to the stronger L_1^2 norm, given L^2 control of the curvature. More to the point, we go from an absolute to a curvature-dependent bound. When we move to stronger norms, involving more derivatives or larger exponents, the estimates become more straightforward. With this in mind, we shall now estimate higher derivatives of B using the gauge-invariant iterated covariant derivatives. These estimates will play an auxiliary role in the proof; we include them so that we can always work with smooth connections.

For a connection B, and $l \geq 1$ put:

$$Q_l(B) = \| F_B \|_{L^\infty} + \sum_{i=1}^{l} \| \nabla_B^{(i)} F_B \|.$$

Here $\nabla_B^{(i)}$ denotes the iterated covariant derivative $\nabla_B \ldots \nabla_B$.

Lemma (2.3.11). *There is a constant $\eta' > 0$ such that if the connection matrix B of Lemma* (2.3.10) *has $\| B \|_{L^2} < \eta'$ then for each $l \geq 1$, a bound,*

$$\| B \|_{L^2_{l+1}} \leq f_l(Q_l(B)),$$

holds, for a universal continuous function f_l, independent of B, *with $f_l(0) = 0$.*

The point of this formulation is, of course, that the functions f_l could be computed explicitly, but we are too lazy to do so.

The proof of this lemma is an exercise in the 'bootstrapping' technique, using elliptic theory and Sobolev and Hölders inequalities. In particular we make heavy use of the multiplication properties for Sobolev spaces, see the Appendix. The proof is by induction on l. We can divide the argument into two parts: a 'stable' range $l \geq 3$ which can be dealt with cleanly, and the first values $l = 1, 2$ which are similar to the proof of (2.3.10) above.

For the stable range, suppose inductively that we have obtained a bound on the L_l^2 norm of B in terms of $Q_{l-1}(B)$, and $l \geq 3$. Multiplication by B gives a bounded map from L_j^2 to L_j^2, for $j \leq l$. It follows that

$$\| \nabla^{(j)} F \|_{L^2} \leq \text{ const. } \| \nabla_B^{(j)} F \|_{L^2},$$

for a constant depending only on the L_l^2 norm of B, and all $j \leq l$. Here ∇ denotes the covariant derivative defined by the product structure, so $\nabla_B = \nabla + B$, and we write F for the tensor F_B. Thus $Q_l(B)$ controls the L_l^2 norm of $F = \mathrm{d}B + B \wedge B$. Now, using the multiplication property again, we have for $l \geq 3$:

$$\| B \wedge B \|_{L_l^2} \leq \text{ const. } \| B \|_{L_l^2}^2.$$

So $Q_l(B)$ controls the L_l^2 norm of $\mathrm{d}B$ (since, by the inductive assumption, it controls the L_l^2 norm of B). Finally we have the elliptic inequality

$$\|B\|_{L_{l+1}^2} \leq \text{const.}\, \|\mathrm{d}B\|_{L_l^2},$$

since $\mathrm{d}^*B = 0$, and this gives us L_{l+1}^2 control of B in terms of $Q_l(B)$, completing the induction, assuming that $l \geq 3$. Notice that for this part of the argument we do not invoke the absolute bound involving η'.

We now go back to consider the first case $l = 1$. Our input is the L_1^2 control of B given by (2.3.10), if we choose $\eta' \leq \eta$. We write $\nabla F = \nabla_B F - [B, F]$, and now we use the L^∞ norm of F, appearing in the definition of $Q_1(B)$. Combined with the L^4 bound on B, this gives us an L^4 bound, and hence L^2 bound on $[B, F]$. So $Q_1(B)$ controls the L_1^2 norm of F. But

$$\|B \wedge B\|_{L_1^2} \leq \|B\|_{L_1^4} \|B\|_{L^4} \leq \|B\|_{L_2^2} \|B\|_{L^4},$$

so we have:

$$\|B\|_{L_2^2} \leq \text{const.}\, \|\mathrm{d}B\|_{L_1^2} \leq \text{const.}\, \|B\|_{L_2^2} \|B\|_{L^4} + \text{const.}\, \|F\|_{L_1^2}.$$

Thus, if $\|B\|_{L^4}$ is small enough, we can apply the same rearrangement argument as in (2.3.10) to obtain an L_2^2 bound on B in terms of $Q_1(B)$. The argument for the next stage, getting an L_3^2 bound on B, is similar and is left as an exercise.

It is convenient to give a proof of Theorem (2.3.8) at this point. We are given a connection matrix $\tilde{A}$ over B^4 which satisfies both the Coulomb gauge condition and the ASD equation. We can write the two together in the form:

$$\delta \tilde{A} + (\tilde{A} \wedge \tilde{A})^+ = 0, \tag{2.3.12}$$

where δ is the elliptic operator $\mathrm{d}^* + \mathrm{d}^+$. For convenience we will work over the compact manifold S^4, so we regard B^4 as being contained in S^4 using the stereographic map m, and for $D \Subset B^4$ we let ψ be a cut-off function supported in B^4 and equal to 1 on D. The connection matrix $\alpha = \psi \tilde{A}$ can be viewed as being defined over all of S^4 (extending by 0). Equation (2.3.12) gives

$$\begin{aligned} \delta(\alpha) = \delta(\psi \tilde{A}) &= \psi \delta(\tilde{A}) + (\mathrm{d}\psi\, \tilde{A})^+ \\ &= -\psi(\tilde{A} \wedge \tilde{A}^+) + (\mathrm{d}\psi\, \tilde{A})^+. \\ &= -(\tilde{A} \wedge \alpha)^+ + (\mathrm{d}\psi \wedge \tilde{A})^+. \end{aligned}$$

We now employ just the same kind of estimate that we used above, but with the operator δ in place of $\mathrm{d}^* + \mathrm{d}$. Again δ has kernel zero on $\Omega_{S^4}^1$ (by (1.1.19)), so

$$\|\alpha\|_{L_2^2} < \|\delta\alpha\|_{L_1^2}.$$

On the other hand, substituting into the equation, we obtain

$$\|\delta\alpha\|_{L_1^2} \leq \|\psi \tilde{A} \wedge \tilde{A}\|_{L_1^2} + \|\mathrm{d}\psi \wedge \tilde{A}\|_{L_1^2}.$$

The last term is bounded by a multiple of $\| A \|_{L_1^2}$, with a constant depending only on ψ. The subtlety enters in the other term on the right hand side. We have

$$\nabla(\psi \tilde{A} \otimes \tilde{A}) = \{\nabla(\psi \tilde{A})\} \otimes \tilde{A} + \psi \tilde{A} \otimes \nabla \tilde{A},$$

and

$$\psi \tilde{A} \otimes \nabla \tilde{A} = \tilde{A} \otimes \{\psi \nabla \tilde{A}\} = \tilde{A} \otimes \nabla(\psi \tilde{A}) - \tilde{A} \otimes \nabla\psi \otimes \tilde{A}.$$

Here we are working with the tensor product as the universal bilinear operation, to avoid introducing special notation. We have then

$$\begin{aligned}\|\nabla(\psi \tilde{A} \otimes \tilde{A})\|_{L^2} &= \|\{\nabla(\psi \tilde{A})\} \otimes \tilde{A} + \tilde{A} \otimes \nabla(\psi \tilde{A}) - \tilde{A} \otimes \nabla\psi \otimes \tilde{A}\|_{L^2} \\ &\leq \|\nabla(\psi \tilde{A}) \otimes \tilde{A}\|_{L^2} + \|\tilde{A} \otimes \nabla(\psi \tilde{A})\|_{L^2} + \|\tilde{A} \otimes \nabla\psi \otimes \tilde{A}\|_{L^2} \\ &\leq \text{const.}\{\|\nabla(\psi \tilde{A})\|_{L^4} \|\tilde{A}\|_{L^4} + \|\tilde{A}\|_{L^4}^2\}.\end{aligned}$$

Contracting the tensor product to the self-dual part of the wedge product, and writing α for $\psi \tilde{A}$, we get

$$\begin{aligned}\|\alpha\|_{L_2^2} &\leq \text{const.}\ \|\delta\alpha\|_{L_1^2} \\ &\leq \text{const.}\ (\|\tilde{A}\|_{L_1^2} \|\alpha\|_{L_2^2} + \|\tilde{A}\|_{L_1^2}^2 + \|\tilde{A}\|_{L_1^2}).\end{aligned}$$

If the L_1^2 norm of $\tilde{A}$ is sufficiently small, we can rearrange this to get a bound on $\|\alpha\|_{L_2^2}$. Then $\alpha = \tilde{A}$ on the domain where $\psi = 1$, so we have gone from an L_1^2 bound over B to an L_2^2 bound over a smaller domain. We can now iterate this argument, much as in (2.3.11), to estimate all higher derivatives on successively smaller domains, all of which can be chosen to contain D.

2.3.6 *Method of continuity*

We now proceed to the second main step in the proof of (2.3.7), working still over the four-sphere. Lemma (2.3.10) might seem at first to be of little help since we assume that we have precisely what we are trying to construct—a small connection matrix in Coulomb gauge. It is here that the subtlety in Uhlenbeck's method enters. We prove the following proposition about one-parameter families of connections, from which we will be able to deduce (2.3.7) quickly in Section 2.3.9. By a one-parameter family we mean a continuous family of smooth connections, i.e. the connection matrices are defined over $S^4 \times \mathbb{R}$, they are smooth in the S^4 variable, and all partial derivatives are continuous in both variables.

Proposition (2.3.13). *There is a constant $\zeta > 0$ such that if B'_t ($t \in [0, 1]$) is a one-parameter family of connections on the trivial bundle over S^4 with $\|F'_{B_t}\| < \zeta$ for all t, and with B'_0 the product connection, then for each t there* *exists a gauge transformation u_t such that $u_t(B'_t) = B_t$ satisfies*

(i) $$d^* B_t = 0, \text{ and}$$

(ii) $$\| B_t \|_{L_1^2} < 2N \| F_{B_t} \|,$$

where N is the constant of (2.3.10).

It is also true that the gauge transformations u_t, and hence the transformed connections B_t, vary continuously with t, although we do not need this.

To prove this proposition we use the 'continuity method'. Let S be the set of points in $[0, 1]$ for which such a u_t exists (with a constant ζ to be chosen in the course of the proof). We show that S is closed and open in $[0, 1]$. It must then be the whole interval since it certainly contains the point $t = 0$. The proofs of these two properties of S are given in the next two sections.

2.3.7 *Closedness: connections control gauge transformations*

To prove that S is closed for a suitable choice of constants, we combine (2.3.10) and (2.3.11) with the third basic analytical point: an elementary observation about the action of the gauge transformations on connections. This observation will be used at a number of other places in this book, so we will spell it out here in detail.

Suppose that for $i \geq 1$ A_i, B_i are unitary connections on the trivial bundle over S^4 which are gauge equivalent, so $B_i = u_i(A_i)$, say. Suppose also that the A_i and B_i converge in C^∞, as $i \to \infty$, to limiting connections A_∞, B_∞. Then we claim that B_∞ is also gauge equivalent to A_∞. This is a simple consequence of the formula:

$$B_i = u_i A_i u_i^{-1} - du_i u_i^{-1},$$

for the action of the gauge transformations, together with the *compactness* of the structure group $U(n)$. For we can write the formula:

$$du_i = u_i A_i - B_i u_i. \tag{2.3.14}$$

Since the sequences B_i and A_i converge, all the multiple derivatives of these connections are bounded, independent of i. It follows then from the formula above that all derivatives of the u_i are bounded. For, inductively, if u_i is bounded in C^r then so also is $u_i A_i - B_i u_i$, and hence u_i is bounded in C^{r+1}. The induction is started with $r = 0$ where the assertion follows automatically from the compactness of $U(n)$.

We can now apply the Ascoli–Arzela theorem to deduce that there is a subsequence, which we may as well suppose is the full sequence, which converges in C^∞ to a limit u_∞ as $i \to \infty$. The gauge relation (2.3.14) is obviously preserved in the limit, so u_∞ gives a gauge transformation from A_∞ to B_∞.

Three remarks about this simple argument are in order. First, the result is false for connections with non-compact structure groups. Second, it is quite unnecessary to suppose that the connections are defined on the trivial bundle: the proof adapts immediately to sequences of connections on any unitary bundle over any manifold. Third, if A_i and B_i are sequences, all of whose derivatives are bounded then, after taking subsequences, we can always suppose that they are convergent. (In the case of connections on a non-trivial bundle the derivatives of the connections are interpreted as the derivatives of their connection matrices in a fixed system of local trivializations.) To sum up then we have:

Proposition (2.3.15). *If A_i, B_i are C^∞-bounded sequences of connections on a unitary bundle over a manifold X, and if A_i is gauge equivalent to B_i for each i, then there are subsequences converging to limiting connections A_∞, B_∞, and A_∞ is gauge equivalent to B_∞.*

We can now get down to the proof that the set S is closed. We choose ζ so that $2CN\zeta$ is less than the constants η and η' of (2.3.10), (2.3.11), where C is the Sobolev constant. Then if t is in S we have $\| B_t \|_{L^4} \leq C \| B_t \|_{L^2_1} \leq 2NC \| F(B_t) \|_{L^2} \leq 2NC\zeta$, so we can apply (2.3.10) and (2.3.11). The first gives

$$\| B_t \|_{L^2_1} \leq N \| F(B_t) \|_{L^2};$$

that is, we have gone from the open condition $\| B_t \|_{L^2_1} < 2N \| F(B_t) \|_{L^2}$ to a stronger, closed, condition. From (2.3.11) we obtain bounds on all the derivatives of B_t, since by gauge invariance of the covariant derivatives of the curvature we certainly have bounds $\| \nabla^{(j)}_{B_t} F(B_t) \|_{L^\infty} \leq K_j$ say, for all t in S and some (possibly very large) constants K_j.

Now suppose that t_i is a sequence in S converging to a limit s in $[0, 1]$. We can apply proposition (2.3.15) to the pairs B_{t_i}, B'_{t_i} to deduce that, after taking a subsequence, the B_{t_i}, converge in C^∞ to a limit B_s, gauge equivalent to B'_s. The conditions of (2.3.13) are preserved in the limit, since we actually have $\| B_s \|_{L^2_1} \leq N \| F(B_s) \|_{L^2}$. So s is in S and we see that the set is closed.

2.3.8 Openness—the implicit function theorem

The proof that S is open is a variant of the proof of (2.3.4) in Section 2.3.2. We apply the implicit function theorem to the gauge fixing equation:

$$d^*(u_t(B'_t)) = d^*(u_t B'_t u_t^{-1} - du_t u_t^{-1}) = 0.$$

Let t_0 be a point in S. We may as well suppose that $B_{t_0} = B'_{t_0}$, which we will just write B. Put

$$B'_{t_0+\delta} = B + b_\delta$$

and seek a solution $u_{t+\delta}$ to the gauge-fixing equation in the form:

$$u_{t_0+\delta} = \exp(\chi_\delta).$$

The equation to be solved is then $H(\chi_\delta, b_\delta) = 0$ where

$$H(\chi, b) = \mathrm{d}^*(e^\chi(B + b)e^{-\chi} - \mathrm{d}(e^\chi)e^{-\chi}).$$

Recall that the image of d* consists of functions with integral zero. For any $l \geq 3$, H defines a smooth map,

$$H: E_l \times F_{l-1} \longrightarrow E_{l-2},$$

where E_l is the Banach space of L_l^2 Lie-algebra valued functions χ with integral zero, and F_{l-1} is the space of L_{l-1}^2 Lie-algebra valued 1-forms. The implicit function theorem asserts that if the partial derivative

$$(D_1 H)_0: E_k \longrightarrow E_{k-2}$$

is surjective, then for small b in F_{k-1} there is a small solution χ to $H(\chi, b) = 0$. If this is so, it follows that the set S contains an interval about t_0. (The bootstrapping argument of Section 2.3.2 goes through without change to show that any L_k^2 solution is smooth.)

Now $D_1 H$, the linearization of the differential operator, is given by

$$(D_1 H)_0 \chi = \mathrm{d}^* \mathrm{d}_B \chi.$$

To show that this is surjective, for suitably small B, we appeal to the Fredholm alternative. If it were not surjective there would be a non-zero smooth η such that

$$\langle \mathrm{d}^* \mathrm{d}_B \chi, \eta \rangle = 0 \quad \text{for all } \chi.$$

Put $\chi = \eta$ and integrate by parts (i.e. use the property of the formal adjoint) to get

$$0 = \langle \mathrm{d}_B \eta, \mathrm{d}\eta \rangle = \| \mathrm{d}\eta \|^2 + \langle [b, \eta], \mathrm{d}\eta \rangle.$$

Now $\| \mathrm{d}\eta \| \geq \text{const.} \| \eta \|_{L_1^2}$, since $\int \eta = 0$, whereas:

$$\begin{aligned} |\langle [b, \eta], \mathrm{d}\eta \rangle| &\leq \text{const.} \| \mathrm{d}\eta \| . \| [B, \eta] \| \\ &\leq \text{const.} \| \mathrm{d}\eta \| . \| B \|_{L^4} . \| \eta \|_{L^4} \\ &\leq \text{const.} \| \mathrm{d}\eta \|_{L^2}^2 \| B \|_{L_1^2}, \end{aligned}$$

where in the last line we have used the Sobolev inequality and the fact that η has integral zero. So if the operator is not surjective we have

$$\| \mathrm{d}\eta \|^2 \leq \text{const.} \| \mathrm{d}\eta \|^2 \| B \|_{L_1^2},$$

for a universal constant. We can cancel the $\| \mathrm{d}\eta \|^2$ term to deduce a lower bound on the L_1^2 norm of B. Conversely if we choose ζ small then we can make the L_1^2 norm of B_t small, for any t in S. So we deduce that if ζ is small the set S is open.

2.3.9 *Completion of proof*

To deduce Theorem (2.3.7) from (2.3.13) we use two simple devices. One is fundamental in this approach to the proof, and one is an auxiliary step which enables us to transfer from the ball to the sphere.

The key observation is that there is a canonical path from any connection A on the trivial bundle over the ball to the product connection. This uses the dilations of $\mathbb{R}^4$, much as in our proof of the integrability theorem in Section 2.2.2. We let $\delta_t: \mathbb{R}^4 \to \mathbb{R}^4$ be the map $\delta_t(x) = tx$ and, for t in $[0, 1]$, let A_t be the connection matrix $\delta_t^*(A)$ over B^4. Clearly A_0 is 0 and A_1 is A. By the conformal invariance of the L^2 norm of curvature in four dimensions we have

$$\int_{B^4} |F(A_t)|^2 \, d\mu = \int_{|x| \le t} |F(A)|^2 \, d\mu \le \int_{B^4} |F(A)|^2 \, d\mu, \tag{2.3.16}$$

where, in the first step, we use the obvious equivalence between the operations of the dilation δ_t in a fixed metric, and the change of metric by a factor t^{-1} with a fixed connection. So the L^2 norm of the curvature on the path A_t is controlled by that of A.

We now proceed to the auxiliary construction, transferring to the four-sphere. Recall that we have identified the four-ball with a hemisphere in S^4. Let $r: S^4 \to S^4$ be the reflection map, equal to the identity on the equatorial three-sphere and interchanging the two hemispheres, and let $p: S^4 \to \bar{B}^4$ be the 'projection' map, equal to the identity on B^4 and to r on the complementary hemisphere. Unfortunately p is not differentiable on the equator. However, it is a Lipschitz map—almost everywhere differentiable with bounded derivative. For a connection matrix α over B^4 the pull-back $\beta = p^*(\alpha)$ (the 'double' of α) makes good sense as an L^∞ 1-form on S^4. Similarly, the curvature $F_\beta = p^*(F_\alpha)$ is an L^∞ 2-form over S^4. Plainly

$$\int_{S^4} |F_\beta|^2 \, d\mu = 2 \int_{B^4} |F_\alpha| \, d\mu. \tag{2.3.17}$$

Now if we ignore the lack of differentiability of p for the moment, we can deduce the main assertions of (2.3.7) from (2.3.13), using the two constructions above. Given a connection matrix A over $\bar{B}^4$, with curvature small in L^2, we join it to 0 using the path A_t as above, and we do not increase this L^2 norm. Let B'_t be $p^*(A_t)$, so B_t is a path of connections over S^4. The L^2 norm of the curvature is increased by only a factor of $2^{1/2}$. We can suppose then that this L^2 norm is less than the constant ζ of Proposition (2.3.13), and conclude that there are connections B_t, gauge equivalent to B'_t, in Coulomb gauge over S^4. Restricting B'_1 to the four-ball we get the desired connection matrix $\tilde{A}$, gauge equivalent to A and satisfying the Coulomb condition. Proposition (2.3.13) gives also an L_1^2 bound on $\tilde{A}$, as stated in Theorem (2.3.7 (iii)).

Three points remain to be cleared up. First, we must get around the fact that the connections $p^*(A_t)$ are not smooth. Second, we must establish the boundary condition (ii) in Theorem (2.3.7), and third we need to establish the uniqueness of the Coulomb gauge representatives, subject to the given conditions. The last two points will not actually be used later in this book, so we will only sketch the argument. Similarly, for our applications we can be content to obtain the Coulomb gauge representative over a subdomain in B^4, and for this the first point is more easily dealt with.

The cleanest way to handle the first point is to develop the relevant parts of the theory for a class of connection matrices large enough to include the $p^*(A_t)$. This is not difficult: for example, we could work with connection matrices A in L^∞ with $\mathrm{d}A$ in L^∞. We shall use a more elementary approach relying on smooth approximations. Clearly we can choose a family of smooth maps $p_\varepsilon: S^4 \to \bar{B}^4$, converging uniformly to p as $\varepsilon \to 0$, with ∇p_ε uniformly bounded and with p_ε equal to p outside the ε-neighbourhood of the equatorial three-sphere. Then for any smooth α over the ball, the $p_\varepsilon^*(\alpha)$ are smooth and

$$\int_{S^4} |F(p_\varepsilon^*(\alpha)|^2 \,\mathrm{d}\mu \leq 2 \int_{B^4} |F(\alpha)|^2 \,\mathrm{d}\mu + \text{const.}\, \varepsilon \, \| F(\alpha) \|_{L^\infty}.$$

So, given A, we can choose ε small enough for Proposition (2.3.13) to apply to the connections $p_\varepsilon^*(A_t)$, provided that the L^2 norm of $F(A)$ is less than $(1/2^{1/2})\zeta$. We then get Coulomb gauge representatives, $\tilde{B}^{(\varepsilon)}$ say, for $p_\varepsilon^*(A)$, and restricting back to the four-ball we have Coulomb gauge representatives $A^{(\varepsilon)}$ for A over arbitrarily large domains in B^4. The point is that the ε we choose depends on A, but this does not affect the universal bounds obtained on the Coulomb solutions.

To go further and obtain the Coulomb representative over the whole ball we let ε tend to zero. We can apply (2.3.11) and (2.3.15) to $\tilde{B}^{(\varepsilon)}$ over any domain $D' \Subset S^4 \backslash S^3$ to obtain a C^∞-convergent subsequence and a limit $\tilde{B}$. Taking an increasing sequence of domains we get a Coulomb representative $\tilde{A}$ for A over the ball. The universal bound on the L_1^2 norm is preserved in the limit. Also, the linear Coulomb condition $\mathrm{d}^*\tilde{B} = 0$ holds, in a distributional sense. Finally we consider the boundary condition (ii) and uniqueness. The point of our passage to the sphere is to avoid discussing a boundary value problem over the ball (as in Uhlenbeck's original proof). However, a standard symmetry argument shows that the two set-ups are equivalent. First, for uniqueness, if B is a connection matrix over S^4 with $\mathrm{d}^*B = 0$, and u is a gauge transformation such that $\mathrm{d}^*(u(B))$ is also 0, then if the L_1^2 norms of B and $u(B)$ are sufficiently small we must have that u is a constant. This is left as an exercise in the techniques we have used in this chapter. Suppose then that a connection matrix B' over S^4 satisfies the symmetry condition: $r^*(B') = B'$. Suppose that B is gauge equivalent to B', satisfies $\mathrm{d}^*B = 0$, and has small L_1^2 norm. Then the uniqueness assertion above implies that we also have $r^*(B) = B$. Now we can choose our smoothings p_ε so that $p_\varepsilon r = p_\varepsilon$; then

$p_\varepsilon^*(A)$ satisfies the symmetry condition, and so also must the Coulomb representative $\tilde{B}^{(\varepsilon)}$. But if a smooth one-form β satisfies $r^*(\beta) = \beta$, then the normal component of β must vanish on the equator. So the approximations to A satisfy the boundary condition (ii). We have to check that, in the precise form stated after (2.3.7), this boundary condition is preserved in the limiting procedure. This is an easy consequence of the symmetry condition on $\tilde{B}$ and the distributional equation $d^*\tilde{B} = 0$. Conversely, if A_2 is another connection matrix over the four-ball satisfying the conditions of (2.3.7), one sees that $\tilde{B}_2 = r^*(A_2)$ satisfies the equation $d^*\tilde{B}_2 = 0$, as a distribution on S^4. Examination of the proof of the uniqueness result mentioned above shows that it also holds for small L^4 connection matrices which satisfy the Coulomb condition in the weak sense, and this gives the uniqueness over the four-ball.

2.3.10 Alternative approach

We sketch an alternative, more differential-geometric proof of Theorem (2.3.7). In this proof the partial differential equation solved is for a connection rather than a gauge transformation. Although the argument is quite different from that above, the basic analytical input—the implicit function theorem and the balancing of non-linear terms using the Sobolev embedding of L_1^2 in L^4—is much the same. The new ingredient is the use of the positivity of the curvature of the metric on the round four-sphere, through Weitzenböck formulae.

If A is a unitary connection over a Riemannian manifold X, we can define operators

$$\Delta_A = d_A^* d_A + d_A d_A^* \quad \text{on} \quad \Omega_X^q(E),$$

mimicking the ordinary Laplacians of Hodge theory. On the other hand the covariant derivative $\nabla_A : \Omega_X^q(E) \to \Omega_X^q(E) \otimes \Omega^1$ and its formal adjoint ∇_A^* combine to give the 'trace Laplacian' $\nabla_A^* \nabla_A$. The two operators differ by algebraic terms involving the curvature F_A and the Riemannian curvature R_X of X. Symbolically:

$$\Delta_A \alpha = \nabla_A^* \nabla_A \alpha + \{R_X, \alpha\} + \{F_A, \alpha\}. \tag{2.3.18}$$

We do not need the precise formulae (Weitzenböck formulae) here. The important point is that if X is the four-sphere, with constant sectional curvature equal to 1, then for $q = 1, 2$ the contribution from the base curvature R_X is strictly positive. In fact R_X acts as multiplication by 3 on 1-forms and by 4 on 2-forms. As we shall see in a moment these formulae give us estimates involving d_A, d_A^* which depend on A only through the norm of F_A. The other useful point we want to note has to do with the Sobolev theorem. For any connection A over X we define norms $\|\alpha\|_{(A,1)}$ on $\Omega_X^q(E)$ by

$$\|\alpha\|_{(A,1)}^2 = \int_X |\nabla_A \alpha|^2 + |\alpha|^2 \, d\mu. \tag{2.3.19}$$

If X is four-dimensional, the Sobolev theorem gives $\|\alpha\|_{L^4} \leq C\|\alpha\|_{(A,1)}$, and the point to note is that the constant C can be taken to be *independent of* A. This is because we have the pointwise inequality (the 'Kato inequality'),

$$|\nabla_A \alpha| \leq \nabla(|\alpha|) \qquad (2.3.20)$$

at all points where α does not vanish. If α has no zeros we have immediately then

$$\|\alpha\|_{L^4} = \|\{|\alpha|\}\|_{L^4} < C\|\{|\alpha|\}\|_{L^2_1} \leq C\|\alpha\|_{(A,1)}, \qquad (2.3.21)$$

where C is the constant in the ordinary Sobolev theorem for functions. The general case, where α has zeros, now follows by an approximation argument. (For example we can increase the rank of E until the generic section has no zeros.)

With these remarks out of the way, we can embark on our proof. Given a connection A over S^4 we write, for a in $\Omega^1(\mathfrak{g}_E)$ and $B = A + a$,

$$P(a) = \mathrm{d}_B \mathrm{d}_B^* a + \mathrm{d}_B^* F_B. \qquad (2.3.22)$$

We will show that if $\|F_A\|$ is small there is a small solution a to the second order equation $P(a) = 0$.

Lemma (2.3.23). *If the base space X is compact, then for any a and $B = A + a$*

$$\mathrm{d}_B \mathrm{d}_B^* a + \mathrm{d}_B^* F_B = 0$$

if and only if $\mathrm{d}_B^* a = 0$ *and* $\mathrm{d}_B^* F_B = 0$.

Proof. One direction is trivial. In the other direction, if $\mathrm{d}_B \mathrm{d}_B^* a + \mathrm{d}_B^* F_B = 0$, take the L^2 inner product with $\mathrm{d}_B^* F_B$ to get

$$\begin{aligned} \|\mathrm{d}_B^* F_B\|^2 &= -\langle \mathrm{d}_B^* F_B, \mathrm{d}_B \mathrm{d}_B^* a\rangle \\ &= -\langle F_B, \mathrm{d}_B \mathrm{d}_B(\mathrm{d}_B^* a)\rangle \\ &= -\langle F_B, [F_B, \mathrm{d}_B^* a]\rangle = 0. \end{aligned}$$

Here we have used the defining property of the curvature on $\mathfrak{g}_E$ and in the last line the algebraic observation that $(\theta, [\theta, \phi])$ vanishes pointwise for any $\mathfrak{g}_E$-valued function ϕ and two-form θ. So our equation implies that both $\mathrm{d}_B^* F_B$ and $\mathrm{d}_B \mathrm{d}_B^* a$ are zero, and taking the inner product of the latter with a we deduce that $\mathrm{d}_B^* a$ vanishes.

Thus our single equation expresses simultaneously the gauge-fixing condition of $B = A + a$ relative to A and the *Yang–Mills equation* for B: $\mathrm{d}_B^* F_B = 0$. Notice that this is a non-linear version of the reduction, in Hodge theory, of the pair of equations $\mathrm{d}\alpha = \mathrm{d}^*\alpha = 0$ to a single second-order equation $\Delta\alpha = 0$.

Lemma (2.3.24). *There is a constant $N > 0$ such that if B is a unitary connection over S^4 with $\mathrm{d}_B^* F_B = 0$ and $\|F_B\| < N$ then B is flat: $F_B = 0$.*

Proof. The Yang–Mills equation and the Bianchi identity imply that $\Delta_B F_B = 0$. We use the Weitzenböck formula on $\Omega^2(\mathfrak{g}_E)$ to write this as

$$\nabla_B^* \nabla_B F_B + 4F_B + \{F_B, F_B\} = 0.$$

Taking the L^2 inner product with F_B we get:

$$\| \nabla_B F_B \|^2 + 4 \| F_B \|^2 = -\langle \{F_B, F_B\}, F_B \rangle.$$

So $\| F_B \|^2_{(1,B)} \leq \text{const.} \| F_B \|^3_{L^3}$. Now $\| F_B \|^3_{L^3} \leq \| F_B \|^2_{L^4} \| F_B \|$ by Cauchy–Schwarz, and we can play the same game as before, trading L^4 for L_1^2 norms and using the discussion above. We get then

$$\| F_B \|^2_{(1,B)} \leq \text{const.} \| F_B \|^2_{(1,B)} \| F_B \|;$$

so if $\| F_B \|$ is sufficiently small, B must be flat.

It suffices now to prove that there is a constant M such that for any A with $\| F_A \|$ sufficiently small there is a solution a to $P(a) = 0$ with

$$\| a \|_{(1,A)} \leq M \| F_A \|.$$

For then, expanding the curvature formula

$$F_B = F_A + \mathrm{d}_A a + a \wedge a,$$

we get

$$\begin{aligned} \| F_B \| &\leq \| F_A \| + \| \nabla_A a \| + \text{const.} \| a \|^2_{L^4} \\ &\leq \| F_A \| + \| a \|_{(1,A)} + \text{const.} \| a \|^2_{(1,A)}. \end{aligned}$$

So there is a constant δ such that if $\| F_A \| < \delta$ then B is flat. But S^4 is simply connected, so B is gauge equivalent to the product connection by some gauge transformation u. Then $\tilde{A} = u(A)$ is the desired connection matrix, in Coulomb gauge relative to the product connection. Finally $\| \tilde{A} \|_{(1)} = \| a \|_{(1,B)}$ can be estimated in terms of $\| a \|_{(1,A)}$ (using the Sobolev inequality once more) and so in terms of $\| F_A \|$. To get Theorem (2.3.7), for connections over B^4, we use the doubling construction discussed above.

To solve the equation $P(a) = 0$ we use the method of continuity and the other Weitzenböck formula. For $t \in [0, 1]$ we seek a small solution a_t of the equation $P(a_t) = t\mathrm{d}_A^* F_A$. When $t = 1$ there is a trivial solution $a_1 = 0$. To bring the Weitzenböck formula into play suppose that for some t we have a solution $a = a_t$ and expand the equation, writing algebraic terms symbolically:

$$(\mathrm{d}_A + a)(\mathrm{d}_A^* a) + (\mathrm{d}_A^* + a)(F_A + \mathrm{d}_A a + a \wedge a) = t\mathrm{d}_A^* F_A.$$

Rearranging we get

$$\mathrm{d}_A \mathrm{d}_A^* a + \mathrm{d}_A^* \mathrm{d}_A a + \{a, F_A\} + \{a, \mathrm{d}_A a\} + \{a, a, a\} = \mathrm{d}_A^*((t-1)F_A + \{a, a\}).$$

The first two terms give Δ_A and so, using the Weitzenböck formula, we have

$$\begin{aligned} \nabla_A^* \nabla_A a + 3a = &\{a, F_A\} + \{a, \mathrm{d}_A a\} + \{a, a, a\} \\ &+ \mathrm{d}_A^* \{(t-1)F_A + a \wedge a\}. \end{aligned}$$

Now take the L^2 inner product with a to obtain

$$\|a\|^2_{(1,A)} \le \text{const.}\,(\|F_A\|^2 + \|\nabla_A a\|)^4\|a\|_{L^4} + \|a\|_{L^4} + \|F_A\|.\|\nabla_A a\|.$$

So

$$\|a\|^2_{(1,A)} \le \text{const.}\,(\|F_A\| + \|a\|^2_{(1,A)})(\|a\|^2_{(1,A)} + \|a\|_{(1,A)}).$$

In the way that is now familiar, we see then that there are constants ε, k, θ such that if $\|F_A\| < \varepsilon$ and $\|a\|_{(1,A)} < \theta$ then $\|a\|_{(1,A)} < k\|F_A\| \le \frac{1}{2}\theta$. It follows that, if $\|F_A\| < \varepsilon$, the set of t for which there is a solution to our equation with $\|a_t\| < \theta$ is closed. To prove openness we use the implicit function theorem. We have to show that if $F(A)$ is small in L^2 and a is small in $\|\ \|_{(1,A)}$, then, with $B = A + a$, the linearized operator,

$$T(h) = \Delta_B h + h d_B^* a + d_B^*(ah) + \{h, F_B\},$$

is invertible. This is a good exercise, using the Fredholm alternative and the Weitzenböck formula on 1-forms.

Notes

Section 2.1

The material in this section is all standard. Good references are Chern (1979), Spivak (1979), Kobayashi and Nomizu (1963), Milnor and Stasheff (1974), Griffiths and Harris (1978) and Wells (1980).

Section 2.1.1

As far as possible we adopt notation which is becoming standard. The reader should be warned that different conventions are used in the literature for the basic definitions and notation for connections and curvature. For example, many authors work with a convention in which the curvature in a local trivialization is given by $dA - A \wedge A$.

Section 2.1.2

For the "highbrow" proof of the Bianchi identity to which we allude, in the analogous and more complicated setting of Riemannian geometry, see Kazdan (1981).

Section 2.1.3

The first systematic treatment of ASD Yang–Mills connections in the framework of Riemannian geometry is given by Atiyah *et al.* (1978*b*). Other good sources are Booss and Bleeker (1985), Bourguignon and Lawson (1982) and Parker (1982).

Section 2.1.4

The general theory of Chern–Weil theory of characteristic classes and curvature is treated in the appendix of Chern (1979), together with the formulae generalizing (2.1.27) (transgression formulae). The classification of $SO(3)$ bundles over four-manifolds is due to Dold and Whitney (1959); see also Freed and Uhlenbeck (1984).

Section 2.1.5

The main result (2.1.53) is originally due to Koszul and Malgrange (1958). The link with the ASD equations has been used by many authors, in different contexts. It forms the basis of the Penrose–Ward 'twistor' description (Atiyah *et al.* 1978*b*). In the physics literature it goes back to Yang (1977).

Section 2.2.1

Theorem (2.2.1) is the fundamental theorem of differential geometry and the proof is very standard. For the general Frobenius theorem see, for example, Spivak (1979) and Warner (1983).

Section 2.2.2

The Koszul–Malgrange integrability theorem for holomorphic bundles is often deduced from the Newlander–Nirenberg theorem for integrable almost complex structures (Atiyah *et al.* 1978*b*). For different proofs of the latter see Newlander and Nirenberg (1957), Hörmander (1973) and Folland and Kohn (1972). However, the bundle case is really a good deal simpler. In either case one can give rather easier proofs if one assumes that the objects are real analytic. One can then complexify and reduce to the Frobenius theorem. On the other hand, the results hold true for structures which are not C^∞; see Atiyah and Bott (1982) and Buchdahl (1988).

Section 2.3.1

Our discussion of radial gauges follows Uhlenbeck (1982*a*). They are the analogues in Yang–Mills theory of exponential co-ordinates in Riemannian geometry. Conversely the analogue of the Coulomb gauge fixing in the Riemannian case is provided by 'harmonic coordinates'; see Jost and Karcher (1982) and Greene and Wu (1988).

Section 2.3.3

The original proof of Uhlenbeck's Theorem is in Uhlenbeck (1982*b*). Our proof is the same except for two technicalities: we avoid boundary value problems and the use of elliptic theory for L^p spaces.

Section 2.3.4

The same kind of 'rearrangement' argument is basic to the analysis of harmonic maps from surfaces (Sacks and Uhlenbeck 1981) and the Yamabe problem (Trudinger 1968; Schoen 1984).

Section 2.3.10

The alternative proof of the gauge fixing theorem is based on a 'gap' theorem for Yang–Mills solutions proved by Min-Oo (1982). For Weitzenböck formulae see Freed and

Uhlenbeck (1984) and Bourguignon (1981). There is another approach which can be used to obtain the main consequences of (2.3.7) and (2.3.8) for ASD (or general Yang–Mills) connections. One shows that if A is Yang–Mills the function $f = |F_A|$ satisfies a differential inequality $\Delta f \le f + f^2$. Multiply this inequality by $\psi^2 f$, where ψ is a cut-off function supported in the ball, and integrate by parts. If the L^2 norm of f is sufficiently small one can rearrange, using the Sobolev inequality, to obtain an L^4 bound on ψf (compare (6.2.19)). Iterating, one gets an L^∞ bound on an interior domain; see Uhlenbeck (1982*a*) and Gilbarg and Trudinger (1983, Chapter 8). Then consider successively the higher derivatives $|\nabla_A^{(l)} F_A|$. These satisfy similar differential inequalities and one deduces uniform bounds on all covariant derivatives of the curvature. One then has bounds on all derivatives of the connection matrix in a radial gauge; see Taubes (1988).

3

THE FOURIER TRANSFORM AND ADHM CONSTRUCTION

We have not yet met any interesting examples of solutions to the anti-self-dual (ASD) equations. In the present chapter we will remedy this and describe geometric constructions for ASD connections over two special four-manifolds: the flat four-torus and round (conformally flat) four-sphere. For connections over the four-torus we define a 'Fourier transform' interchanging ASD solutions on different bundles. These results are of a somewhat theoretical interest since they do not immediately yield any explicit solutions; but we will then go on to apply the same ideas to derive the 'ADHM construction' for connections over S^4 (which we regard as the conformal compactification of $\mathbb{R}^4$). This gives a complete and concrete description of all solutions in terms of certain matrix data, and the results will provide a prototype for the general theory of Chapter 4.

We will base our discussion throughout this chapter on the geometry of spinors and the link between holomorphic bundles and the ASD condition. As we have seen in Chapter 2, the anti-self-dual forms on a Hermitian surface are of type (1, 1), and ASD connections define holomorphic structures. In this chapter we will use a simple converse. We consider $\mathbb{R}^4$ with its standard metric. The compatible complex structures are parametrized by $SO(4)/U(2) = S^2$, the sphere in Λ^+ (cf. Section 1.1.7). Each such complex structure I defines a space $\Lambda_I^{1,1} \supset \Lambda^-$. It is easy to check that the intersection of the $\Lambda_I^{1,1}$, as I ranges over S^2, is precisely Λ^-. So a connection over $\mathbb{R}^4$ is ASD if and only if it defines a holomorphic structure for each complex structure I on the base. From another point of view, with a fixed complex structure we can study the separate equations (2.1.60) and (2.1.61), but the two together have additional Euclidean symmetry: any one of the three equations (2.1.26) can be incorporated into a commutator equation like (2.1.60). This is also the starting point for the 'twistor' description of ASD solutions; see the notes at the end of the chapter.

3.1 General theory

3.1.1 Spinors and the Dirac equation

Spin structures on four-manifolds have already been discussed in Chapter 1. We will now set up the theory in more detail to give a firm base for the constructions of this chapter. Let S be a two-dimensional complex vector space with a Hermitian metric $\langle , \rangle$ and a compatible complex symplectic form—a preferred basis element λ of $\Lambda^2_{\mathbb{C}} S^*$ with $|\lambda| = 2$. Using this data we define an anti-linear map $J: S \to S$ by

$$\langle x, Jy \rangle = \lambda(x, y). \tag{3.1.1}$$

Then $J^2 = -1$ and, together with the complex structure, J makes S into a one-dimensional quaternionic vector space. However, we shall, for the most part, avoid the use of quaternions here. Notice that we can recover λ from the metric and J, and this gives us the natural isomorphism between the groups $SU(2)$ and $Sp(1)$ (the group of unit quaternions). Each is the symmetry group of S with its given algebraic structure. Now let S^+, S^- be a pair of such complex vector spaces and consider the space $\mathrm{Hom}_J(S^+, S^-)$ of complex linear maps which intertwine the J actions (i.e. maps linear over the quaternions). This is a four-dimensional real vector space, a real form of the complex vector space $\mathrm{Hom}(S^+, S^-) = \mathrm{Hom}_J(S^+, S^-) \otimes \mathbb{C}$; it also carries a standard Euclidean metric, normalized so that the unit vectors in $\mathrm{Hom}_J(S^+, S^-)$ give exactly the maps from S^+ to S^- which preserve both the metrics and the symplectic forms. In the opposite direction, given a four-dimensional Euclidean space V we say that a spin structure on V is a pair of complex vector spaces S^+, S^- as above and an isomorphism $\gamma: V \to \mathrm{Hom}_J(S^+, S^-)$ compatible with the Euclidean metrics. Concretely, in standard bases, we can take

$$\gamma(e_0) = \begin{pmatrix} 1 & 0 \\ 0 & 1 \end{pmatrix} \qquad \gamma(e_1) = \begin{pmatrix} i & 0 \\ 0 & -i \end{pmatrix}$$

$$\gamma(e_2) = \begin{pmatrix} 0 & -1 \\ 1 & 0 \end{pmatrix} \qquad \gamma(e_3) = \begin{pmatrix} 0 & i \\ i & 0 \end{pmatrix}.$$

The symmetry group of the pair (S^+, S^-) is $SU(2)^+ \times SU(2)^-$ which is connected, so γ fixes an orientation of V. One can verify easily enough that this notion does indeed correspond to the description of the double cover $\mathrm{Spin}(4)$ of $SO(4)$ as $SU(2) \times SU(2)$ (or $Sp(1) \times Sp(1)$). Given two spin structures (S^+, S^-, γ), (S_0^+, S_0^-, γ_0) on V, there are exactly two isomorphisms (ϕ^+, ϕ^-), $(-\phi^+, -\phi^-): (S^+, S^-) \to (S_0^+, S_0^-)$ between the spin spaces interchanging γ and γ_0. For the purposes of local differential-geometric calculations we can regard the spinors as canonically attached to the Euclidean geometry. Finally observe that, using the symplectic form to identify S^+ with its dual, we can equally well describe V as a real form of $S^+ \otimes S^-$, which was the description of Chapter 1.

Given a spin structure and a vector e in V we let $\gamma^*(e): S^- \to S^+$ be the adjoint of $\gamma(e): S^+ \to S^-$ defined by the Hermitian metrics. So for e, e' in V, the composite $\gamma^*(e)\gamma(e')$ is an endomorphism of S^+. One easily sees that

$$\gamma^*(e)\,\gamma(e) = 1 \quad \text{if } e \text{ is a unit vector}$$

$$\gamma^*(e)\,\gamma(e') + \gamma^*(e')\,\gamma(e) = 0 \quad \text{if } e \text{ and } e' \text{ are orthogonal.} \tag{3.1.2}$$

This means that $\Lambda^2(V)$ acts on S^+ by

$$(e \wedge e')s = -\gamma^*(e)\gamma(e')s,$$

where e and e' are orthogonal. Moreover one checks that Λ^- acts trivially here and we get a natural isomorphism

$$\rho : \Lambda^+ \longrightarrow \mathfrak{su}(S^+), \tag{3.1.3}$$

where the right-hand side denotes the trace-free, skew-adjoint endomorphisms. (This matching of Λ^+ and S^+ gives a way to fix our orientation conventions.) Of course, ρ is just a manifestation of the local isomorphism between $SU(2)$ and $SO(3)$.

With these algebraic preliminaries complete we can turn to the *Dirac operator* over $\mathbb{R}^4$. If we have a spin structure, we write $\Gamma(S^+)$, $\Gamma(S^-)$ for the spaces of sections of the trivial bundles over V with fibres S^+, S^-; that is, the spinor-valued functions. (For simplicity we will denote these bundles also by S^+, S^-.) The Dirac operator $\mathrm{D}:\Gamma(S^+) \to \Gamma(S^-)$ is defined in terms of an orthonormal basis e_i by

$$\mathrm{D}s = \sum_i \gamma(e_i)\frac{\partial s}{\partial x_i}. \tag{3.1.4}$$

More invariantly, this is the contraction of the full derivative,

$$\nabla s \in \Gamma(S^+ \otimes V^*)$$

by the *Clifford multiplication* $c: S^+ \otimes V^* \to S^-$. The operator $D^*: \Gamma(S^-) \to \Gamma(S^+)$ is the formal adjoint of D, given explicitly by

$$D^*s = -\sum \gamma^*(e_i)\frac{\partial s}{\partial x_i}.$$

Of course there is a complete symmetry between S^+, S^-; and D and $-D^*$ are interchanged by reversing the orientation on V.

Now all of this extends readily enough to a general Riemannian four-manifold. A spin structure gives spinor bundles S^+, S^- whose fibres are related algebraically to the tangent spaces of the manifold just as above; the Levi–Civita connection induces connections on the spin bundles and we define Dirac operators D, D^* by the same procedure. More relevant to our present discussion, the definitions extend to give Dirac operators coupled to a connection A on an auxiliary complex vector bundle E. We obtain operators:

$$\begin{aligned} D_A &: \Gamma(E \otimes_{\mathbb{C}} S^+) \longrightarrow \Gamma(E \otimes_{\mathbb{C}} S^-) \\ D_A^* &: \Gamma(E \otimes_{\mathbb{C}} S^-) \longrightarrow \Gamma(E \otimes_{\mathbb{C}} S^+). \end{aligned} \tag{3.1.5}$$

When the base space is $\mathbb{R}^4$ these can be written concretely in terms of the components $\nabla_i = (\partial/\partial x_i) + A_i$ of ∇_A:

$$D_A s = \sum \gamma(e_i)\nabla_i s, \quad D_A^* s = -\sum \gamma^*(e_i)\nabla_i s.$$

Note that the $\gamma(e_i)$ and the ∇_i act on different factors in the tensor product $E \otimes S^+$.

The next lemma provides another example of a Weitzenböck formula, cf. (2.3.18).

Lemma (3.1.6). *Let A be a unitary connection on a bundle E over $\mathbb{R}^4$. For any section s of $E \otimes S^+$ we have*

$$D_A^* D_A s = \nabla_A^* \nabla_A s - F_A^+ \cdot s,$$

where F_A^+ acts on $E \otimes S^+$ through the homomorphism ρ defined above, and $\nabla_A^* \nabla_A$ is the trace Laplacian given in coordinates by $-\sum_i \nabla_i \nabla_i$.

Proof. We have

$$\begin{aligned} D_A^* D_A s &= -\sum_{i,j} \gamma^*(e_i)\gamma(e_j)\nabla_i \nabla_j s \\ &= -\left\{\sum_i \gamma^*(e_i)\gamma(e_i)\nabla_i \nabla_i s\right\} - \left\{\sum_{i \neq j} \gamma^*(e_i)\gamma(e_j)\nabla_i \nabla_j s\right\}. \end{aligned}$$

Using (3.1.2) we write this as

$$\left(-\sum_i \nabla_i \nabla_i s\right) - \sum_{i<j} \gamma^*(e_i)\gamma(e_j)\,(\nabla_i \nabla_j s - \nabla_j \nabla_i s).$$

The first term is $\nabla_A^* \nabla_A s$ and the second is

$$-\Sigma \rho(e_i \wedge e_j) F_{ij} s = -\rho(F^+)s.$$

Thus the ASD condition is exactly equivalent to the absence of curvature terms in this Weitzenböck formula. (In the calculation above we have used the fact that the base manifold is flat. The general formula is: $D_A^* D_A = \nabla_A^* \nabla_A - F_A^+ + \frac{1}{4}S$, where S is the Riemannian scalar curvature of the base space.)

3.1.2 *Spinors and complex structures*

Now suppose that U is a two-dimensional complex vector space. For any u in U we consider the sequence of maps:

$$0 \longrightarrow \mathbb{C} \xrightarrow{\delta_u} U \xrightarrow{\delta_u} \Lambda^2_{\mathbb{C}} U \longrightarrow 0, \tag{3.1.7}$$

where δ_u is the wedge product with u. For non-zero u this is an exact sequence. If we fix a determinant element $\theta \in \Lambda^2 U$ we can identify $\Lambda^2 U$ with $\mathbb{C}$. This exact sequence in elementary linear algebra has two well-known counterparts, one in algebraic geometry and one in differential analysis. For the first we let $\mathscr{A}^0$, $\mathscr{A}^1$, $\mathscr{A}^2$ be the spaces of holomorphic functions on U

with values in, respectively, $\mathbb{C}$, U, $\Lambda^2 U$ and define maps:

$$\mathscr{A}^0 \xrightarrow{\delta} \mathscr{A}^1 \xrightarrow{\delta} \mathscr{A}^2 \tag{3.1.8}$$

by $(\delta\tau)(u) = \delta_u(\tau(u))$. This is the *Koszul complex* over U, giving a resolution of the ideal sheaf of the origin. That is, we have an exact sequence:

$$0 \longrightarrow \mathscr{A}^0 \xrightarrow{\delta} \mathscr{A}^1 \xrightarrow{\delta} \mathscr{A}^2 \xrightarrow{\text{ev}} \Lambda^2 U \longrightarrow 0, \tag{3.1.9}$$

where ev is evaluation at $0 \in U$. Again, if we fix a basis element θ, we can identify $\mathscr{A}^2$ with the ordinary functions $\mathscr{A}^0$. In complex coordinates z_1, z_2 on U we can write the $\boldsymbol{\delta}$-complex explicitly as:

$$f \longrightarrow \begin{pmatrix} z_1 f \\ z_2 f \end{pmatrix}; \quad \begin{pmatrix} g_1 \\ g_2 \end{pmatrix} \longrightarrow (z_2 g_1 - z_1 g_2). \tag{3.1.10}$$

We can of course consider the same Koszul multiplication on smooth functions $\mathscr{A}^i_\infty$. We still get a complex but exactness fails.

The second counterpart of (3.1.7) is the Dolbeault complex:

$$0 \longrightarrow \Omega^{0,0} \xrightarrow{\bar{\partial}} \Omega^{0,1} \xrightarrow{\bar{\partial}} \Omega^{0,2} \longrightarrow 0, \tag{3.1.11}$$

where $\Omega^{0,p} = \Omega_U^{0,p}$ is the space of smooth forms of type $(0, p)$ over U, thought of as a complex manifold. Indeed, the 'symbol sequence' of the $\bar{\partial}$ complex is a sequence of maps:

$$\begin{aligned} \sigma(\bar{\partial}_p) &\in U \otimes_{\mathbb{C}} \operatorname{Hom}(\Lambda^p_{\mathbb{C}} \bar{U}^*, \Lambda^{p+1}_{\mathbb{C}} \bar{U}^*) \\ &\subset U \otimes_{\mathbb{R}} \operatorname{Hom}(\Lambda^p_{\mathbb{C}} \bar{U}^*, \Lambda^{p+1}_{\mathbb{C}} \bar{U}^*). \end{aligned}$$

These are just the maps $\boldsymbol{\delta}$ for $\bar{U}^*$ in place of U. Otherwise stated, we write down the $\bar{\partial}$ complex by replacing the operators 'multiplication by z_i' in the Koszul complex with $\partial/\partial \bar{z}_i$. We recall that the $\bar{\partial}$-Poincaré lemma asserts that the cohomology of the complex (3.1.11) is just the space of holomorphic functions on U, in dimension 0.

To get back to spinors, we now suppose that U has a Hermitian metric and, for simplicity, we also fix a determinant form θ of norm 2. So in standard coordinates z_1, z_2 we could take $\theta = \mathrm{d}z_1 \mathrm{d}z_2$. We use θ to trivialize the lines $\Lambda^2 U$, $\Lambda^2 \bar{U}$, $\Lambda^2 U^*$, $\Lambda^2 \bar{U}^*$ without further special notation.

Now the exact sequence (3.1.7) is of course defined without reference to any metric. But, using the metric, we can define

$$\delta_u^*: \mathbb{C} = \Lambda^2 U \longrightarrow U,$$

adjoint to δ_u. Now let $S^+ = \mathbb{C} \oplus \mathbb{C}$ be the vector space formed from the sum of the outer terms in the basic exact sequence, let $S^- = U$ and for u in U let $\gamma_u: S^+ \to S^-$ be the linear map $\gamma_u = \delta_u + \delta_u^*$. On S^+ we have a metric and determinant form given by the canonical basis elements, which we denote $\langle 1 \rangle$

and $\langle\theta\rangle$; and on $S^- = U$ we have by hypothesis a metric and determinant form. It is easy then to verify that γ, as defined above, does indeed set up a correspondence between U and $\mathrm{Hom}_J(S^+, S^-)$. In sum we have the following algebraic fact:

Proposition (3.1.12). *If U is a two-dimensional complex Hermitian vector space with a determinant form, there is a canonical spin structure on U (regarded as a Euclidean four-space) with $S^+ = \mathbb{C} \oplus \mathbb{C}$, $S^- = U$ and $\gamma_u = \delta_u + \delta_u^*$.*

We should note that while θ has norm 2 in the standard metric on $\Lambda^2 U$, we define the metric on S^+ so that $\langle\theta\rangle$ has norm 1.

This lemma has counterparts for the Koszul and $\bar\partial$ complexes which will lie at the heart of our constructions in this chapter. Let $\gamma: \Gamma(S^+) \to \Gamma(S^-)$ be the multiplication map:

$$\{\gamma(s)\}(x) = \sum_i x_i \gamma(e_i) s(x). \tag{3.1.13}$$

This is a Euclidean-invariant definition (or more precisely a Spin(4)-invariant one). If the underlying Euclidean space is a Hermitian space U as above, we can identify $\Gamma(S^+)$ with $\mathscr{A}^0_\infty \oplus \mathscr{A}^2_\infty = \mathscr{A}^0_\infty \oplus \mathscr{A}^0_\infty$ and $\Gamma(S^-)$ with $\mathscr{A}^1_\infty$. Then γ is identified with $\boldsymbol{\delta} \oplus \boldsymbol{\delta}^*$. Similarly, using the metric to identify U with $\bar{U}^*$, we interpret $\Omega^{0,1}$ as $\Gamma(S^-)$, and $\Omega^{0,0} \oplus \Omega^{0,2}$ as $\Gamma(S^+)$. The Dirac operator D then becomes

$$D = 2\{\bar\partial - \bar\partial^*\}: \Omega^{0,0} \oplus \Omega^{0,2} \longrightarrow \Omega^{0,1}. \tag{3.1.14}$$

(The change in sign comes from the minus sign introduced in the formal adjoint. A point to beware of here is that we are using the norms on the forms obtained under our isomorphism with the spinors. These do *not* agree with the standard norms used in complex geometry, so the formal adjoint operator is not precisely the standard one. With these latter norms the formula would be $D = 2\bar\partial - \bar\partial^*$. Of course, there is some latitude in our definition of γ in terms of δ, and the numerical factors are rather arbitrary.) The other Dirac operator is the adjoint $D^* = 2(\bar\partial^* - \bar\partial): \Omega^{0,1} \to \Omega^{0,0} \oplus \Omega^{0,2}$. Similarly, for coupled operators, we have, in the presence of a complex structure on $\mathbb{R}^4$, identifications:

$$D_A = 2(\bar\partial_A - \bar\partial_A^*), \quad D_A^* = 2(\bar\partial_A^* - \bar\partial_A). \tag{3.1.15}$$

More generally, the Dirac operator over any complex surface with a Kähler metric (or indeed Kähler manifold of any dimension) can be related to a Dolbeault complex. A spin structure on such a manifold is defined by a choice of a square root of the canonical line bundle on X, i.e. a line bundle $K^{1/2}$ such that $K^{1/2} \otimes K^{1/2} = \Lambda^2 T^*X$. The spinors are identified with the $(0, p)$-forms with values in $K^{1/2}$, and the Dirac operator with the operator $2(\bar\partial - \bar\partial^*)$, defined on such twisted forms. In fact on a four-dimensional Euclidean space with spin structure, the two-sphere of complex structures

$S(\Lambda_\cdot^+)$ can be described as the projective space $\mathbb{P}(S^+)$, and a choice of complex structure corresponds exactly to a choice of splitting of S^+ into orthogonal subspaces. In this chapter we are working over flat manifolds where the tangent bundle is trivial and the twisting by $K^{1/2}$ is not important. So we have chosen to fix a trivialization of $\Lambda^2 U$ and avoid introducing square roots. It will be clear at the end of our discussion that the geometry is completely independent of this choice.

3.1.3 Connections and projections

We will now introduce the last piece of general theory which will underlie the constructions of this chapter. Again the emphasis is on the interplay between differential geometry (here, unitary connections) and complex geometry (holomorphic bundles).

Let X be a smooth manifold and K and L be complex vector spaces, which we take to be finite dimensional for the moment. Let $R: X \to \mathrm{Hom}(K, L)$ be a smooth map. So R is a family of linear maps R_x parametrized by X, or equivalently a bundle map,

$$R: \underline{K} \longrightarrow \underline{L}.$$

If R_x is surjective for all x, the kernels form a vector sub-bundle E of the trivial bundle $\underline{K}$ over X, with $E_x = \mathrm{Ker}(R_x)$. Now $\underline{K}$ has the flat product connection ∇. Suppose we are given a smooth bundle projection $\pi\colon \underline{K} \to E$, left-inverse to the inclusion map i. Then we get an induced connection A on E with covariant derivative:

$$\nabla_A = \pi \nabla i. \tag{3.1.16}$$

Of course, if K has a Hermitian metric we can always define π to be the orthogonal projection to E, and A is then a unitary connection.

Turning to holomorphic bundles, suppose that X is now a complex manifold, that K_0, K_1, K_2 are (finite-dimensional) vector spaces and that we have holomorphic bundle maps:

$$\underline{K}_0 \underset{\alpha}{\longrightarrow} \underline{K}_1 \underset{\beta}{\longrightarrow} \underline{K}_2 \tag{3.1.17}$$

with $\beta\alpha = 0$. So we have a family of chain complexes,

$$K_0 \xrightarrow{\alpha_x} K_1 \xrightarrow{\beta_x} K_2, \tag{3.1.18}$$

varying holomorphically with x. Suppose that α is injective and β is surjective. There is then a family of 'cohomology spaces',

$$\mathscr{E}_x = \mathrm{Ker}\, \beta_x / \mathrm{Im}\, \alpha_x$$

which define a holomorphic bundle $\mathscr{E}$. A holomorphic sequence like (3.1.17) is called a 'monad'. One can, of course, extend the theory to sequences in which the original three bundles are non-trivial. The existence of a natural holomorphic structure on the cohomology bundle $\mathscr{E}$ is a completely standard fact but it is instructive to recall the proof. The holomorphic structure can be defined by saying that a local section s of $\mathscr{E}$ is holomorphic if it has a lift to a holomorphic section s' of $\ker \beta \subset \underline{K}_1$. We have to show that there is a local holomorphic section through each point in $\mathscr{E}$. Let x_0 be in X and k_1 be in $\mathrm{Ker}(\beta_x)$. We can choose a right inverse $P: K_2 \to K_1$ for β_{x_0} and seek a holomorphic section of $\mathrm{Ker}\,\beta$ of the form $k_1 + j(x)$, where $j(x_0) = 0$. Put $\beta_x = \beta_{x_0} + \eta_x$. The condition that the section lie in $\mathrm{Ker}\,\beta$ is satisfied if

$$(1 + P\eta_x)j_x = -P\eta_x(k_1).$$

But when x is close to x_0, $P\eta_x$ is small and we can invert $(1 + P\eta_x)$ to find a unique solution j_x to this last equation. Moreover j_x varies holomorphically with x, since η_x does. Using this construction we find a set of local holomorphic sections of $\mathscr{E}$ which form a basis for the fibres near x_0. A similar argument then shows that any other holomorphic section is a holomorphic linear combination of these, so $\mathscr{E}$ is indeed a holomorphic bundle.

The point we wish to bring out in the discussion above is that the holomorphic structure on the cohomology bundle is a more-or-less formal consequence of the existence of *splittings* P_{x_0} for the exact sequences (3.1.18). Hence there are straightforward generalizations of the theory to *split* exact sequences in which the K_i are infinite-dimensional vector spaces.

We now bring these ideas together by supposing that the vector spaces K_i above have Hermitian metrics. Then the fibres $\mathscr{E}_x$ can be identified with the orthogonal complements of $\mathrm{Im}\,\alpha_x$ in $\mathrm{Ker}\,\beta_x$—that is, with $\mathrm{Ker}(R_x)$ where

$$R = \alpha^* + \beta : \underline{K}_1 \longrightarrow \underline{K}_0 \oplus \underline{K}_2. \tag{3.1.19}$$

Then we get a unitary connection, induced by orthogonal projection, on the bundle $\mathscr{E}$, thought of as the kernel of R.

Lemma (3.1.20). *The unitary connection on $\mathscr{E}$ is compatible with the holomorphic structure defined above.*

Proof. The orthogonal projection π from $\underline{K}_1$ to $\mathrm{Ker}(\alpha^*) = (\mathrm{Im}\,\alpha)^{\perp}$ is

$$\pi(k) = k - \alpha(\alpha^*\alpha)^{-1}\alpha^* k.$$

Now α depends holomorphically on x, so $\bar{\partial}\pi$ is contained in $\mathrm{Im}\,\theta$, and $\pi\bar{\partial}\pi = 0$. To any local holomorphic section s' of $\mathrm{Ker}\,\phi$ we associate the section s'' of $\mathrm{Ker}\,R$ which is represented by the projection $s'' = \pi(s')$ in $(\mathrm{Im}\,\theta)^{\perp}$. Then

$$\pi(\bar{\partial}s'') = \pi((\bar{\partial}\pi)s' + \bar{\partial}s') = \pi\bar{\partial}\pi s' = 0.$$

But $\pi\,\bar{\partial}$ on $\mathrm{Ker}\,R$ is precisely the $\bar{\partial}$ operator of the connection A on $\mathscr{E}$, so the two definitions of local holomorphic sections agree.

3.2 The Fourier transform for ASD connections over the four-torus

3.2.1 Definitions

In this section we study connections over a flat Riemannian four-torus $T = V/\Lambda$, where V is an oriented four-dimensional Euclidean space and Λ is a maximal lattice in V. We begin with flat $U(1)$ connections—that is, flat complex Hermitian line bundles. These can be described explicitly in two ways:

(i) Let $\chi : \Lambda \to U(1)$ be a character. We let Λ act on the trivial bundle $\underline{\mathbb{C}}$ over V by:

$$n(x, \sigma) = (x + n, \chi(n)\sigma).$$

Here $n \in \Lambda$, $x \in V$ and $\sigma \in \mathbb{C}$. The action preserves the horizontal foliation in $\underline{\mathbb{C}}$ (defined by the product structure) and this descends to a flat connection on the quotient bundle over T. We can write χ as $\chi(n) = e^{i\langle \xi, n \rangle}$, where ξ is an element of the dual space $V^* = \mathrm{Hom}(V, \mathbb{R})$. Two elements of the dual space give the same character if they differ by $2\pi\Lambda^*$, where $\Lambda^* \subset V^*$ is the *dual lattice* of linear forms which take integer values on Λ.

(ii) Any ξ in V^* can be regarded as a one-form with 'constant coefficients' on T. We define a flat connection on the trivial line bundle over T by the connection form $-i\xi$. If ξ is in $2\pi\Lambda^*$, the $U(1)$-valued function $e^{i\langle \xi, \ \rangle}$ on V descends to T and gives a gauge transformation taking this connection to the product structure. More generally the parallel transport of this connection around the loop in T associated with a lattice point n is $e^{i\langle \xi, n \rangle}$.

The description of the parallel transport shows that these constructions match up. From either point of view we see that the gauge equivalence classes of flat line bundles over T are parametrized by a dual torus

$$T^* = V^*/2\pi\Lambda^*, \tag{3.2.1}$$

and this is of course a special case of Proposition (2.2.3). We will write L_ξ for the flat line bundle over T corresponding to a point ξ in V^*. To simplify notation we will not always distinguish carefully between points of T^* and their representatives in V^*.

We will now define the Fourier transform of an ASD connection A on a Hermitian vector bundle E over T. It is convenient to begin by introducing a special definition:

Definition (3.2.2). *The connection A is WFF (without flat factors) if there is no splitting $E = E' \oplus L_\xi$ compatible with A, for any flat line bundle L_ξ.*

Now for each ξ the 'twisted' bundle $E \otimes L_\xi$ has an induced ASD connection which we denote by A_ξ. This connection gives in turn coupled Dirac operators which we denote

$$\begin{aligned} D_\xi &: \Gamma(E \otimes L_\xi \otimes S^+) \longrightarrow \Gamma(E \otimes L_\xi \otimes S^-) \\ D_\xi^* &: \Gamma(E \otimes L_\xi \otimes S^-) \longrightarrow \Gamma(E \otimes L_\xi \otimes S^+). \end{aligned} \tag{3.2.3}$$

If we think of L_ξ as the trivial line bundle with connection matrix $i\xi$, then these Dirac operators can be regarded as acting on the fixed spaces $\Gamma(E \otimes S^+)$, $\Gamma(E \otimes S^-)$ and we can then write

$$D_\xi = D_A + i\gamma(\xi), \quad D_\xi^* = D_A^* - i\gamma^*(\xi), \tag{3.2.4}$$

where $\gamma : V \to \mathrm{Hom}_J(S^+, S^-)$ is the map defining the spin structure, and we identify V with V^* using the Euclidean metric. Now for each ξ the Weitzenböck formula (3.1.6) gives

$$D_\xi^* D_\xi = \nabla_{A_\xi}^* \nabla_{A_\xi}. \tag{3.2.5}$$

It follows that if A is WFF then D_ξ has kernel zero for all ξ. For if $D_\xi s = 0$ then

$$0 = \langle D_\xi^* D_\xi s, s \rangle = \langle \nabla_{A_\xi}^* \nabla_{A_\xi} s, s \rangle = \| \nabla_{A_\xi} s \|^2,$$

so s is a covariant constant section of $E \otimes L_\xi$ and, if it is not zero, it yields a splitting $E = E' \oplus L_\xi^*$. In turn, by the general Fredholm alternative, if A is WFF then D_ξ^* is surjective for all ξ.

We are now in the position considered in the first construction of Section 3.1.3, with a family of surjective operators D_ξ^* parametrized by $\xi \in V^*$. The only difference is that the spaces are now infinite dimensional. However this causes no real difficulties. Standard results on families of elliptic operators with smoothly varying coefficients show that the kernels of the D_ξ^* (which all have the same dimension—given by the Fredholm index) form the fibres of a smooth vector bundle $\hat{E}$ over V^*, with

$$\hat{E}_\xi = \mathrm{Ker}(D_\xi^*). \tag{3.2.6}$$

In addition, the L^2 Hermitian metric on $\Gamma(E \otimes S^-)$ defines a metric and unitary connection $\hat{A}$ on $\hat{E}$, via the projection formula (3.1.16). Here we are working over V^*; however, the gauge transformations which identify the line bundles L_ξ for parameters which differ by $2\pi\Lambda^*$ give a similar identification of the fibres of $\hat{E}$, and this respects $\hat{A}$. So the bundle and connection descend to a pair, which we also call $\hat{E}$, $\hat{A}$, over T^*.

Definition (3.2.7). *If A is a WFF, ASD connection on a bundle E over T the Fourier transform of A is the connection $\hat{A}$ on the bundle of Dirac operator kernels $\hat{E}$ over T^*, defined by L^2 projection.*

We could avoid going through the covering space V^* in the construction by extending (3.1.16) to sub-bundles of a general ambient bundle with connection.

Theorem (3.2.8). *The Fourier transform $\hat{A}$ is an ASD connection, with respect to the flat Riemannian metric on T^* induced from V^*.*

To prove this theorem we go through the analogous construction in the holomorphic category. Fix a compatible complex structure on V, and for clarity call the resulting complex vector space U. Then T becomes a complex torus covered by U, and we know that ASD connections over T yield holomorphic bundles. Similarly we get a complex structure on $V^* = \mathrm{Hom}(V, \mathbb{R})$ by identifying V^* with $\bar{U}^*$, so T^* is another complex torus. The flat line bundles L_ξ become holomorphic line bundles and, for this choice of complex structure, they form a *holomorphic* family in the parameter ξ. Explicitly, the $\bar{\partial}$ operator of L_ξ, regarded as the trivial C^∞ bundle, is given by

$$\bar{\partial}_\xi = \bar{\partial} + \delta_\xi, \tag{3.2.9}$$

which depends holomorphically on ξ. Now suppose $\mathscr{E}$ is a holomorphic bundle over T. In analogy with (3.2.3) we write $\bar{\partial}_\xi$ for the $\bar{\partial}$ operators on $\mathscr{E} \otimes L_\xi$. We say that the bundle $\mathscr{E}$ is WFF if

$$H^0(\mathscr{E} \otimes L_\xi) = H^2(\mathscr{E} \otimes L_\xi) = 0, \tag{3.2.10}$$

for all ξ (cf. (2.1.52)).

We then have an infinite-dimensional version of the second construction in Section 3.1.3. The $\bar{\partial}$ complexes for $\mathscr{E} \otimes L_\xi$ give a holomorphically varying family of complexes,

$$K_0 \xrightarrow{\bar{\partial}_\xi} K_1 \xrightarrow{\bar{\partial}_\xi} K_2 \tag{3.2.11}$$

with $K_i = \Omega_T^{0,i}(\mathscr{E})$. Once again, the same definitions go through in this infinite-dimensional case to give, if $\mathscr{E}$ is WFF, a holomorphic bundle $\hat{E}$ over T^* with fibres:

$$\hat{\mathscr{E}}_\xi = H^1(T; \mathscr{E} \otimes L_\xi). \tag{3.2.12}$$

As we pointed out in Section 3.1.3, the theory goes through easily for infinite-dimensional complexes with suitable splittings; here these splittings are provided by the Hodge theory. Following Mukai (1981) we call $\hat{\mathscr{E}}$ the *Fourier transform* of $\mathscr{E}$.

Hodge theory also gives the link between these two constructions. Once metrics are fixed, the $\bar{\partial}$-cohomology classes have canonical harmonic representatives. If A is a compatible unitary connection on $\mathscr{E}$, so that $\bar{\partial}_{\mathscr{E}} = \bar{\partial}_A$, we have

$$\begin{aligned} H^1(T; \mathscr{E} \otimes L_\xi) &= \mathrm{Ker}\, \bar{\partial}_\xi^* \cap \mathrm{Ker}\, \bar{\partial}_\xi \\ &= \mathrm{Ker}(\bar{\partial}_\xi^* \oplus (-\bar{\partial}_\xi)) \\ &= \mathrm{Ker}\, D_\xi^*. \end{aligned} \tag{3.2.13}$$

Here we interpret the (0, 1)-forms as spinors, as in Section 3.1.2. Similarly $H^0(T; \mathscr{E} \otimes L_\xi) \oplus H^2(T; \mathscr{E} \otimes L_\xi)$ is identified with $\mathrm{Ker}\, D_\xi$.

So in particular if A is a WFF, ASD connection and $\mathscr{E}$ the associated holomorphic bundle, we see that $\mathscr{E}$ is also WFF, and we have a natural isomorphism of the fibres of $\hat{E}$ and $\hat{\mathscr{E}}$. This is, of course, precisely in line with our discussion of the relation between the two constructions in Section 3.1.3, with the formal adjoint of the $\bar{\partial}$ operator in place of the adjoint α^* (and an irrelevant change of sign). Once again the same argument applies in the infinite-dimensional setting to show that, under this isomorphism, the connection $\hat{A}$ is compatible with the holomorphic structure on $\hat{\mathscr{E}}$. In particular, the curvature of $\hat{A}$ has type (1, 1). But by varying the complex structure on V we get all compatible complex structures on V^*, so appealing to the basic fact noted at the beginning of this chapter, we see that $\hat{A}$ is an ASD connection.

3.2.2 *The inversion theorem*

As Riemannian manifolds, the flat tori T, T^* are, in general, quite distinct. They are however in a symmetrical dual relation with one another. The torus T^* parametrizes the flat line bundles over T and T parametrizes the flat line bundles over T^*. A point x in V yields a character

$$\chi_x : 2\pi\Lambda^* \longrightarrow U(1), \quad \chi_x(2\pi v) = e^{2\pi i\langle v, x\rangle}$$

and the character is trivial if and only if $x \in \Lambda$. So we get flat line bundles $\tilde{L}_x$ over T^* with parallel transport χ_x. The whole picture is neatly summarized by the 'Poincaré bundle' over the product $T \times T^*$. Let us write T_ξ for the 'slice' $T \times \{\xi\}$ in the product, and T^*_x for $\{x\} \times T^*$.

Lemma (3.2.14). *There is a line bundle $\mathbb{P}$ over $T \times T^*$, with a unitary connection, such that the restriction of $\mathbb{P}$ to each T_ξ is isomorphic (as a line bundle with connection) to L_ξ and the restriction to each T^*_x is isomorphic to $\tilde{L}^*_x = \tilde{L}_{-x}$.*

Notice the asymmetry in sign here, together with the factor of 2π above. These are completely analogous to the asymmetries in the ordinary Fourier transform.

Proof. We begin with the covering space $T \times V^*$. To make the argument clearer we suppose that Λ is the standard lattice $\mathbb{Z}^4$ in a coordinate system x_i on V; so Λ^* is also the standard lattice in the dual coordinates ξ_i on V^*. Over $T \times V^*$ we consider the connection one-form: $\mathbb{A} = i\,(\sum \xi_i \,\mathrm{d}x_i)$ on the trivial bundle $\underline{\mathbb{C}}$. Then we lift the action of $2\pi\Lambda^*$ on $T \times V^*$ to $\underline{\mathbb{C}}$ by

$$2\pi v(x, \xi, \sigma) = (x, \xi + v, e^{-2\pi i\langle v, x\rangle}\sigma).$$

This action preserves the connection $\mathbb{A}$. We define $\mathbb{P}$ to be the quotient bundle $\underline{\mathbb{C}}/2\pi\Lambda^*$ over $T \times T^*$, with the connection induced by $\mathbb{A}$.

Consider the situation on a slice T_ξ. The connection form $\mathbb{A}$ agrees with that defining L_ξ by approach (i) and the quotient by $2\pi\Lambda^*$ has no effect. On the other hand, on the covering $\{x\} \times V^*$ of a slice T_x^* the connection form $\mathbb{A}$ vanishes and the action of $2\pi\Lambda^*$ is the inverse of that used to define $\tilde{L}_x$ by approach (ii). (Notice that the connection on $\mathbb{P}$ is *not* flat; it has curvature $\mathrm{d}\mathbb{A} = i\sum \mathrm{d}\xi_i \mathrm{d}x_i$.)

In line with this symmetry, we define the inverse transform $\check{B}$, on a bundle $\check{F}$, of a WFF, ASD connection B on F over T^* by the same procedure as in Section 2.2.1. So the fibre $\check{F}_x$ is the kernel of the operator D_x^* on $F \otimes \tilde{L}_x$. The main result of this section 3.2 is:

Theorem (3.2.15). *If A is a WFF, ASD connection on E over T then $\hat{A}$ is* WFF *and there is a natural isomorphism $\omega: \check{\hat{E}} \to E$ with $\omega^*(A) = (\hat{A})\check{}$.*

Of course we also have the symmetrical formula $(\check{B})\hat{} = B$ and we see that the Fourier transform gives a one-to-one correspondence between WFF, ASD connections over T and T^*. The analogy with the ordinary Fourier transform for functions hardly needs to be pointed out. This is a non-trivial correspondence, even when T and T^* are chosen to be isometric. In general the topological type of the bundles E, $\hat{E}$ are different. The index theorem for families, which we will discuss in Chapter 5, gives the formulae for the Chern classes:

$$\begin{aligned} \operatorname{rank}(\hat{E}) &= c_2(E) - c_1(E)^2 \\ c_1(\hat{E}) &= \sigma(c_1(E)) \\ c_2(\hat{E}) &= \operatorname{rank}(E) + c_1(E)^2, \end{aligned} \tag{3.2.16}$$

where $\sigma: H^2(T) \to H^2(T^*)$ is the composite of the isomorphism $H^2(T) \to H_2(T^*)$, valid for tori in any dimension, with the Poincaré duality isomorphism $H_2(T^*) \to H^2(T^*)$.

To prove the inversion theorem (3.2.15) we will again go via the analogue in the holomorphic category. We fix a complex structure I on T, and so on T^*, and define the inverse transform $\check{\mathscr{F}}$ of a WFF holomorphic bundle $\mathscr{F}$ over T^* in the obvious way:

$$\check{\mathscr{F}}_x = H^1(T^*; \mathscr{F} \otimes \tilde{L}_x).$$

Then we will prove the holomorphic inversion theorem, due to Mukai:

Theorem (3.2.17). *If $\mathscr{E}$ is a WFF holomorphic bundle over the torus T, relative to the complex structure I, then $\hat{\mathscr{E}}$ is WFF and there is a natural holomorphic isomorphism $\omega_I: \check{\hat{\mathscr{E}}} \to \mathscr{E}$.*

We will now concentrate on the proof of this inversion theorem for holomorphic bundles, returning to connections in Section 3.2.5.

Observe first that the Poincaré bundle $\mathbb{P}$ has a holomorphic structure, compatible with its connection. This is equivalent to the fact used above that the L_ξ vary holomorphically with ξ. We will begin the proof of (3.2.17) by constructing a natural isomorphism of the fibres of $\check{\hat{\mathscr{E}}}$ and $\mathscr{E}$ over the point 0 in T. To do this we consider the holomorphic bundle:

$$\mathscr{G} = p_1^*(\mathscr{E}) \otimes \mathbb{P} \tag{3.2.18}$$

over $T \times T^*$, where p_1 is the projection to the first factor. We will obtain our isomorphism by considering the cohomology of $\mathscr{G}$ in two different ways. This could be done using direct image sheaves and the Leray spectral sequences of the two projection maps, but we will set out the argument from first principles, using the pair of spectral sequences associated with a double complex.

3.2.3 *Double complexes and spectral sequences*

A double complex $(\mathscr{C}^{**}, \delta_1, \delta_2)$ is a collection of abelian groups $\mathscr{C}^{p,q}$, labelled by positive integers p, q, and homomorphisms

$$\delta_1 : \mathscr{C}^{p,q} \longrightarrow \mathscr{C}^{p+1,q}, \quad \delta_2 : \mathscr{C}^{p,q} \longrightarrow \mathscr{C}^{p,q+1},$$

such that the combinations δ_1^2, δ_2^2 and $\delta_1\delta_2 + \delta_2\delta_1$ are all zero. We can form the cohomology of the total complex:

$$\delta_1 + \delta_2 : C^n \longrightarrow C^{n+1},$$

where $C^n = \sum_{p+q=n} \mathscr{C}^{p,q}$. On the other hand we have the 'cohomology of the rows':

$$E_1^{p,q} = H^p(\mathscr{C}^{*,q}, \delta_1).$$

The two cohomologies are related by a spectral sequence—a sequence of groups $E_r^{p,q}$ $(r \geq 1)$ with differentials

$$d_r : E_r^{p,q} \longrightarrow E_r^{p+1-r,q+r}$$

such that $d_r^2 = 0$. When $r = 1$ we have the cohomology of the rows above; in general the $E_{r+1}^{p,q}$ are given by the cohomology of d_r on the $E_r^{p,q}$. For large r the $E_r^{p,q}$ with $p + q = n$ are the quotients in a filtration of the cohomology of the total complex $H^n(C^*, \delta_1 + \delta_2)$. The differentials d_r are induced by δ_1 and δ_2. For example, d_1 is the 'vertical' map induced on the δ_1 cohomology classes by δ_2. The set up is symmetrical in the two indices, so we have another spectral sequence $\tilde{E}_r^{p,q}$ beginning with the 'cohomology of the columns':

$$\tilde{E}_1^{p,q} = H^q(\mathscr{C}^{p,*}, \delta_2)$$

and converging to another filtration of the total cohomology. (In fact the spectral sequences that we encounter below will be very simple, and it will be easy to see explicitly how this general theory works out for them.)

The double complex we want is that formed by the $(0, n)$-forms over $T \times T^*$ with values in the holomorphic bundle $\mathscr{G}$. For $0 \leq n \leq 4$ we put

$$\Omega^{0,n}_{T \times T^*}(\mathscr{G}) = \bigoplus_{p+q=n} \mathscr{C}^{p,q}, \tag{3.2.19}$$

where the extra grading comes from the two factors in $T \times T^*$. So $\mathscr{C}^{p,q}$ is spanned by forms of the shape:

$$s(z, \zeta)\, d\bar{z}_I\, d\bar{\zeta}_J; \quad |I| = p, |J| = q,$$

where z_i are local complex coordinates on T and ζ_i on T^*. We define differentials, which we denote $\bar{\partial}_1, \bar{\partial}_2$, using the $\bar{\partial}$ operators in the T and T^* factors respectively; so (up to a sign) $\bar{\partial}_1 + \bar{\partial}_2$ is the $\bar{\partial}$ operator on $\mathscr{G}$. Thus the total cohomology is the Dolbeault cohomology of $\mathscr{G}$.

We have therefore two spectral sequences E_*^{**} and $\tilde{E}_*^{**}$ converging to the cohomology of $\mathscr{G}$. The main result these yield is:

Proposition (3.2.20). (*i*) *For the complex* $(\mathscr{C}^{**}, \bar{\partial}_1, \bar{\partial}_2)$ *defined above, the* $E_2^{p,q}$ groups are:

q			
	0	$H^2(T^*, \hat{\mathscr{E}})$	0
	0	$H^1(T^*, \hat{\mathscr{E}})$	0
	0	$H^0(T^*, \hat{\mathscr{E}})$	0
			$p.$

(ii) *The cohomology groups* $H^i(C^*, \bar{\partial}_1 + \bar{\partial}_2)$ *are zero for* $i \neq 2$ *and there is a natural isomorphism between the two-dimensional cohomology group and the fibre* $\mathscr{E}_0$.

Proof of (3.2.20 (i)). This is quite straightforward. The essential point can be expressed by considering, as in Section 3.1.3, a general family,

$$K_0 \xrightarrow{\alpha_x} K_1 \xrightarrow{\beta_x} K_2$$

of complexes depending smoothly on a space X. Again, suppose that the α_x are all injective and the β_x are surjective, so that we get a cohomology bundle H. Now consider the spaces $\mathscr{K}_0, \mathscr{K}_1, \mathscr{K}_2$ of C^∞ functions on X with values in K_0, K_1, K_2, respectively, and the induced complex $\mathscr{K}_0 \to \mathscr{K}_1 \to \mathscr{K}_2$. In the finite-dimensional situation one sees immediately that the cohomology of this function space complex is 0 in dimensions zero and two and in dimension one is the space of smooth sections of H. In brief, the operations of taking cohomology and of taking smooth sections commute. The same is true in an infinite-dimensional version where we take a smooth family of Dolbeault complexes over T parametrized by T^* (or more precisely V^*). One needs a smooth family of splittings of the Dolbeault complexes, which is provided by general elliptic (Hodge) theory.

Now consider the $E_1^{p,0}$ groups of our complex. We can regard the elements of $\mathscr{C}^{p,0}$ as forms over $T \times V^*$ which are invariant under the Λ^* action used to form the Poincaré bundle. In turn we can regard them as Λ^*-invariant functions on V^* with values in $\Omega_T^{0,p}(\mathscr{E})$. The bottom $\bar{\partial}_1$-complex is then identified with the Λ^*-invariant part of the function space complex defined, as in the previous paragraph, by the family of operators $\bar{\partial}_\xi$ over T. So the only cohomology is in dimension one, where we get the Λ^*-invariant sections of the bundle $\hat{\mathscr{E}}$ over V^*, i.e. the sections of $\hat{\mathscr{E}}$ over T^*. Similarly the $E_1^{p,q}$ group vanishes if $p = 0$ or 2, and if $p = 1$ we get the $(0, q)$-forms on T^* with values in $\hat{\mathscr{E}}$. Finally to get to the $E_2^{p,q}$ terms we have to take the cohomology of the differential on the $E_1^{p,q}$ induced by $\bar{\partial}_2$. But this is, almost tautologically, the ordinary $\bar{\partial}$-complex for $\hat{\mathscr{E}}$. So the $E_2^{p,q}$ groups give the Dolbeault cohomology of $\hat{\mathscr{E}}$, as asserted by Proposition (3.2.20 (i)). To sum up: the E spectral sequence begins by taking the horizontal cohomology of the complex, and this corresponds geometrically to taking the cohomology of $\mathscr{G}$ on the 'horizontal' fibres T_ξ in the product. See Fig. 8.

Proof of (3.2.20 (ii)). This is more involved since we now encounter 'jumping' in the cohomology of $\mathscr{G}$ along the vertical fibres T_x^* in the product. The basic ingredient in the proof is the following lemma, which is the analogue of the distributional equation,

$$\int e^{i\xi x}\, dx = 2\pi\, \delta(\xi)$$

in ordinary Fourier theory.

Lemma (3.2.21). *The cohomology groups $H^i(T^*; \tilde{L}_x)$ vanish if x is not zero in T, and for $x = 0$ they are naturally isomorphic to $\Lambda^i(U)$.*

Proof. Represent $\tilde{L}_x$ by the trivial bundle with $\bar{\partial}$ operator $\bar{\partial}_x = \bar{\partial} + \delta_x$. This has constant coefficients, so we can decompose the whole Dolbeault complex into Fourier components. For each n in Λ write e_n for the trigonometric function

$$e_n(\xi) = e^{i\langle n, \xi\rangle} \tag{3.2.22}$$

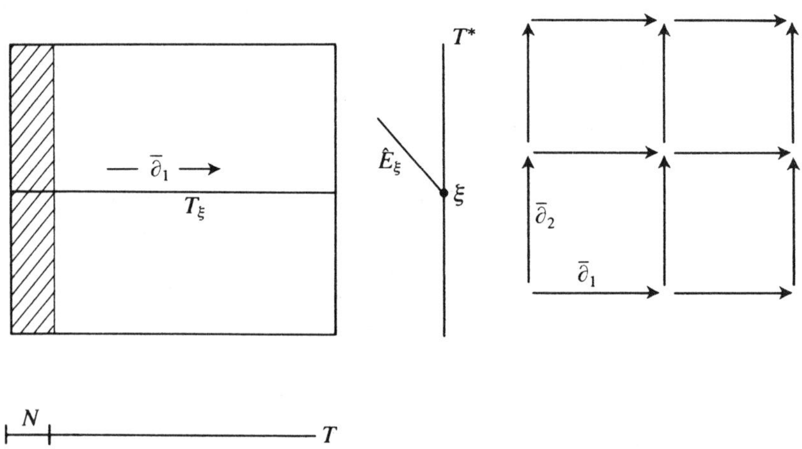

Fig. 8

on T^*. In terms of complex coordinates ζ_1, ζ_2 on T^*, e_n is

$$e_n(\zeta_1, \zeta_2) = \exp(i\,\mathrm{Re}\{n_1\zeta_1 + n_2\zeta_2\}),$$

where n_1, n_2 are the complex coordinates of the lattice point $n \in \Lambda$ in a dual complex coordinate system z_1, z_2 on U. The forms on T^* are spanned by the e_n, $e_n \otimes \mathrm{d}\bar{\zeta}_1$, $e_n \otimes \mathrm{d}\bar{\zeta}_2$ and $e_n \otimes \mathrm{d}\bar{\zeta}_1\mathrm{d}\bar{\zeta}_2$. We have

$$\bar{\partial}_x(e_n) = (n_1 + z_1)(e_n \otimes \mathrm{d}\bar{\zeta}_1) + (n_2 + z_2)(e_n \otimes \mathrm{d}\bar{\zeta}_2)$$

$$\bar{\partial}_x(e_n \otimes \mathrm{d}\bar{\zeta}_1) = -(n_2 + z_2)(e_n \otimes \mathrm{d}\bar{\zeta}_1\mathrm{d}\bar{\zeta}_2)$$

$$\bar{\partial}_x(e_n \otimes \mathrm{d}\bar{\zeta}_2) = (n_1 + z_1)(e_n \otimes \mathrm{d}\bar{\zeta}_1\mathrm{d}\bar{\zeta}_2).$$

Here z_1, z_2 are the coordinates of the point x in T (more precisely, of a representative in U). So we can write the $\bar{\partial}_x$ complex as a direct sum $\bigoplus(D_{n,x}, \delta_{n,x})$ of finite dimensional complexes, where $(D_{n,x}, \delta_{n,x})$, $n \in \Lambda$, is a copy of the δ_{n+x}-complex we began with in Section 3.1.3. So all the $D_{n,x}$ are acyclic if x is not in Λ, and if x is in Λ there is just one term $D_{-x,x}$ contributing to the cohomology. It follows that the cohomology of $\tilde{L}_x$ is zero for non-zero x. For the trivial bundle $\tilde{L}_0$, the Dolbeault cohomology arises from the term $D_{0,0}$, with zero differential, which yields the constant forms. These are invariantly identified with $\Lambda^i(U)$, since $U = \overline{(\bar{U}^*)^*}$.

Note (3.2.22). In place of the direct sum of the finite dimensional complexes above, we should really have the space of rapidly decaying sequences—the Fourier series of smooth functions; but this does not affect the argument.

We now return to the proof of (3.2.20) (ii). Observe first that the restriction of the bundle $\mathscr{G}$ to a slice T^*_x is isomorphic to the tensor product of the vector space $\mathscr{E}_x$ with the line bundle $\tilde{L}_{-x}$. So if x is non-zero the cohomology along the slice vanishes by (3.2.21). If this property held for *all* x the argument used in (i) above would show that the $\tilde{E}_1^{p,q}$ were all zero. In fact we see that the $\tilde{E}_1^{p,q}$ depend only on arbitrarily small neighbourhoods of T^* in $T \times T^*$. To make this precise define another double complex (germ $\mathscr{C}$) to consist of forms defined on arbitrary neighbourhoods of T^*_0 in $T \times T^*$, and identify forms if they agree on any such neighbourhood. Then we have a restriction map from $\mathscr{C}$ to (germ $\mathscr{C}$), commuting with the differentials. This induces maps on the total cohomology and all the terms in the spectral sequences. Then the argument we used in the proof of part (i) shows that the restriction map induces an isomorphism from $\tilde{E}_1^{p,q}$ to (germ $\tilde{E})_1^{p,q}$. Since the restriction maps commute with all the differentials it follows that restriction also gives an isomorphism between the total cohomologies of $\mathscr{C}$ and (germ $\mathscr{C}$). So to prove (ii) we can work with (germ $\mathscr{C}$).

To study the situation near T^*_0 we use the Fourier decomposition introduced in the proof of (3.2.20). Fix a trivialization of $\mathscr{E}$ over a small neighbourhood N of 0 in T, which we identify with a neighbourhood of 0 in V. The double complex for the tube $N \times T^*$ can be written, taking Fourier components in the T^* variable, as a sum of double complexes $\mathscr{D}_n = \Omega_N^{0,p} \otimes D_n \otimes \mathscr{E}_0$.

The differential $\bar{\partial}_2$ goes over to Koszul multiplication by $(x + n)$ on $\mathscr{D}_n$, and $\bar{\partial}_1$ to the ordinary $\bar{\partial}$ operator acting on the $\Omega_N^{0,p}$. The fixed vector space $\mathscr{E}_0$ plays only a passive role. To find the total cohomology of (germ $\mathscr{C}$) we use this decomposition and the *other* sequence (germ $E)_r^{p,q}$, taking $\bar{\partial}_1$-cohomology first. By the Poincaré lemma the groups (germ $E)_r^{p,q}$ vanish for $p > 0$, and for $p = 0$ we get the sum over Λ of the germs of *holomorphic* maps from N to $D_n \otimes \mathscr{E}_0$. But this is just a sum of Koszul complexes with the origin translated by $-n$. So when we proceed to the (germ $E)_2^{p,q}$ term by taking the $\bar{\partial}_2$-cohomology, the only contribution is from D_0, and this yields

$$(\text{germ } E)_2^{p,q} = \begin{array}{|ccc} \mathscr{E}_0 & 0 & 0 \\ 0 & 0 & 0 \\ 0 & 0 & 0. \\ \hline \end{array}$$

So the only non-vanishing cohomology group of the total complex is $H^2(C^*, \bar{\partial}_1 + \bar{\partial}_2) = H^2((\text{germ } C)^*)$ which is isomorphic to $\mathscr{E}_0$. It is easy to check that the isomorphism is independent of the local trivialization of $\mathscr{E}$ used in the argument, so the proof of (3.2.20) is complete.

Note (3.2.23). In line with Note (3.2.22), one should really use here a version of the $\bar{\partial}$-Poincaré lemma incorporating bounds on the solution (over an interior domain), in order to stay within the space of rapidly decaying sequences. This is not difficult: in fact the ordinary proof gives the required bounds.

Combining the results of (3.2.20) we see that $H^0(\hat{\mathscr{E}})$ and $H^2(\hat{\mathscr{E}})$ are both zero, and that there is a natural isomorphism

$$\omega_I : H^1(T^*; \hat{\mathscr{E}}) \longrightarrow \mathscr{E}_0. \tag{3.2.24}$$

This map can be described explicitly as follows: a class in $H^1(\hat{\mathscr{E}})$ is represented by an element α of $\mathscr{C}^{1,1}$ with $\bar{\partial}_1(\alpha) = 0$ and $\bar{\partial}_2(\alpha) = 0$ modulo the image of $\bar{\partial}_1$. Choose an element β of $\mathscr{C}^{0,2}$ with $\bar{\partial}_1(\beta) = -\bar{\partial}_2(\alpha)$, then restrict β to T_0^* to get a form:

$$r_0(\beta) = \beta|_{T_0^*} \in \Omega_{T^*}^{0,2} \otimes \mathscr{E}_0.$$

Here we have used the isomorphism between $\mathbb{P}|_{T^*}$ and the trivial holomorphic line bundle over T^*. Finally take the cohomology class of $r_0(\beta)$ in $H^2(T^*, \tilde{L}) \otimes \mathscr{E}_0$, which is isomorphic to $\mathscr{E}_0$ (using the fixed determinant form θ to trivialize $\Lambda^2 U$). Explicitly,

$$\omega_I([\alpha]) = \int_{T^*} r_0(\beta) \wedge \theta. \tag{3.2.25}$$

3.2.4 Proof of the inversion theorem for holomorphic bundles

We can now proceed to the proof of the inversion theorem (3.2.17). For y in T let $t_y: T \to T$ be the translation map $t_y(x) = y + x$.

Lemma (3.2.26). *For each y in T, a holomorphic bundle $\mathscr{E}$ over T is WFF if and only if $t_y(\mathscr{E})$ is, and in that case there is a natural isomorphism between $\{t_y^*(\mathscr{E})\}^{\wedge}$ and $\hat{\mathscr{E}} \otimes \tilde{L}_y$.*

Proof. The Poincaré bundle $\mathbb{P}$ is characterized as the unique holomorphic bundle on the product space whose restriction to each T_ξ is isomorphic to L_ξ and whose restriction to T_0^* is trivial. This leads to the formula:

$$t_y^*(\mathbb{P}) \otimes p_2^*(\tilde{L}_y) = \mathbb{P}.$$

Now

$$H^i((p_1^*(t_y(\mathscr{E})) \otimes \mathbb{P})|_{T_\xi}) = H^i(\{t_y^*(p_1^*(\mathscr{E}) \otimes \mathbb{P}) \otimes p_2^*(\tilde{L}_y)\}|_{T_\xi})$$

and this space is naturally isomorphic to $H^i(\{p_1^*(\mathscr{E}) \otimes \mathbb{P}\}|_{T_\xi}) \otimes \{\tilde{L}_y\}_\xi$. With $i = 0, 2$ this shows that $t_y^*(\mathscr{E})$ is WFF if and only if $\mathscr{E}$ is; and with $i = 1$ we get natural isomorphisms between the fibres of $(t_y^*(\mathscr{E}))^{\wedge}$ and $\hat{\mathscr{E}} \otimes \tilde{L}_y$ at ξ. These vary holomorphically with ξ to give the required isomorphism of holomorphic bundles.

Now if $\mathscr{E}$ is a WFF bundle over T and x is a point in T, we can compose the isomorphism ω_I of (3.2.25), using $t_y^*(\mathscr{E})$ in place of $\mathscr{E}$, with the tautological isomorphism between $\mathscr{E}_y$ and $(t_y^*(\mathscr{E}))_0$. Combined with (3.2.26) this gives an isomorphism between

$$H^1(\hat{\mathscr{E}} \otimes \tilde{L}_y)) = H^1((t_y^*(\mathscr{E}))^{\wedge})$$

and $\mathscr{E}_y$. Similarly we see that the other cohomology groups of $\hat{\mathscr{E}} \otimes \tilde{L}_y$ vanish. So $\hat{\mathscr{E}}$ is WFF and we have natural isomorphisms between the fibres of $\mathscr{E}$ and $\hat{\mathscr{E}}^{\vee}$. Again, it is really clear from the naturality of the constructions that these fibre isomorphisms fit together to define a holomorphic bundle isomorphism, which we again call ω_I, from $\hat{\mathscr{E}}^{\vee}$ to $\mathscr{E}$. We leave the reader to grapple with the notation required to write out a formal proof of this fact, (but see Section 3.3.5). This completes the proof of (3.2.17).

3.2.5 The inversion theorem for ASD connections

We will now derive the inversion theorem for connections (3.2.15) by translating the constructions above into the language of spinors and Dirac operators. Let A be a WFF, ASD connection on $E \to T$; fix a complex structure I on T, and let $\mathscr{E}$ be the holomorphic bundle defined by A. We have already seen that the connection $\hat{A}$ over T^* defines the holomorphic structure $\hat{\mathscr{E}}$. Using the relation between $\bar{\partial}$-cohomology and harmonic spinors, over T^*, we deduce first from (3.2.17) that $\hat{A}$ is WFF. We also have a natural identification of the bundles $\hat{\mathscr{E}}^{\vee}$ and $\hat{E}^{\vee}$ over T, so our isomorphism of

(3.2.17) goes over to a bundle isomorphism $\omega_I : \hat{E}^{\vee} \to E$. This certainly takes the $\bar{\partial}$ operator defined by A to that defined by $\hat{A}^{\vee}$.

If we can show that ω_I is *independent* of the complex structure I then we can see that it gives a bundle map ω taking A to $\hat{A}^{\vee}$, thus proving (3.2.15). (For it is obvious that if the $\bar{\partial}$ operators defined by two connections are equal for all choices of complex structure I then the connections are the same. Indeed, we only need consider I and the conjugate structure $-I$.) So the proof is completed by the following proposition.

Proposition (3.2.27). *The bundle map $\omega_I : \hat{E}^{\vee} \to E$ is independent of the complex structure I on T.*

Proof. It is clearly sufficient to consider the fibres over 0 in T. We consider then the bundle $\mathscr{G} = p_1^*(E) \otimes \mathbb{P}$ over $T \times T^*$ and identify the term $\mathscr{C}^{1,1}$ in the double complex with bundle-valued 'double spinors': sections of $\mathscr{G} \otimes S_T^- \otimes S_{T^*}^-$. The differential $\bar{\partial}_1$ on $\mathscr{C}$ is given by the family of operators $\bar{\partial}_\xi = \bar{\partial}_{A_\xi}$ over T, and we also have adjoint operators $\bar{\partial}_1^* : \mathscr{C}^{p,q} \to \mathscr{C}^{p-1,q}$ defined via the slices. Similarly we have an adjoint $\bar{\partial}_2^*$.

Now an element on $\hat{E}_0^{\vee}$ is, under this interpretation of the spinors, a 'doubly harmonic' element of $\mathscr{C}^{1,1}$, an element α satisfying:

$$\bar{\partial}_1 \alpha = 0, \quad \bar{\partial}_1^* \alpha = 0 \tag{3.2.28}$$

$$\bar{\partial}_2 \alpha = 0 \bmod \operatorname{Im} \bar{\partial}_1, \quad \bar{\partial}_2^* \alpha = 0 \bmod \operatorname{Im} \bar{\partial}_1^*. \tag{3.2.29}$$

The first conditions in each line are just those considered in the holomorphic theory above and the other two follow by reversing the complex structure, which essentially interchanges $\bar{\partial}$ and $\bar{\partial}^*$. To find $\omega_I(\alpha)$ we have to find β in $\mathscr{C}^{0,2}$ with $\bar{\partial}_1 \beta = \bar{\partial}_2 \alpha$. A solution is provided by the Hodge theory:

$$\beta = G_1^{(I)}(\bar{\partial}_1^* \bar{\partial}_2 \alpha),$$

where $G_1^{(I)}$ is the Green's operator $(\bar{\partial}_1^* \bar{\partial}_1)^{-1}$, defined by looking along the 'horizontal' slices T_ξ. Then we have to evaluate the restriction of β in H^2 along T_0^*. Now, using our trivialization of $\Lambda^2 U$, we can interpret $\mathscr{C}^{0,2}$ as the space of sections of $\mathscr{G}$ over $T \times T^*$, and the evaluation (3.2.25) of $r_0(\beta)$ is then given by plain integration of an E_0-valued function over T^*, with respect to the Riemannian volume element. Moreover, by the Weitzenböck formula (3.1.6) and the relation (3.1.14), we have

$$\bar{\partial}_1^* \bar{\partial}_1 = \tfrac{1}{4} \nabla_1^* \nabla_1.$$

Here, of course, ∇_1 is the covariant derivative in the horizontal direction on $\mathscr{G}$. So, with this interpretation of $\mathscr{C}^{0,2}$, the map $G_1^{(I)}$ is independent of the complex structure I—it is the Green's operator in the T variable,

$$G_1^{(I)} = 4(\nabla_1^* \nabla_1)^{-1} = 4 G_1,$$

say, acting on sections of $\mathscr{G}$.

Finally, use the Riemannian metric to identify the tangent spaces of T and T^* and hence S_T^- with $S_{T^*}^-$. Then the wedge product on S^- becomes an algebraic operator:

$$\kappa : \Gamma(\mathscr{G} \otimes S_T^- \otimes S_{T^*}^-) \longrightarrow \Gamma(\mathscr{G}). \tag{3.2.30}$$

Under the above identifications this becomes a map

$$\kappa : \mathscr{C}^{1,1} \longrightarrow \mathscr{C}^{0,2}.$$

Lemma (3.2.31). *The commutator $[\bar{\partial}_1^*, \bar{\partial}_2]$ is equal to κ on $\mathscr{C}^{1,1}$.*

Proof. Locally we can represent $\bar{\partial}_2$ as the $\bar{\partial}$ operator in the ξ variable and $\bar{\partial}_1^*$ as the operator $\bar{\partial}_A^* + \delta_\xi^*$, acting on $\mathscr{G}$-valued forms. The coefficients of $\bar{\partial}_A^*$ are independent of ξ, so this term commutes with $\bar{\partial}_2$, and

$$[\bar{\partial}_1^*, \bar{\partial}_2] = [\delta_\xi^*, \bar{\partial}_2]. \tag{3.2.32}$$

In standard coordinates (z_1, z_2) on T and (ζ_1, ζ_2) on T^* we have

$$\delta_\xi^*\left(\sum a_\lambda \, \mathrm{d}\bar{z}_\lambda\right) = \sum \bar{\zeta}_\lambda a_\lambda,$$

$$\bar{\partial}_2 = \frac{\partial}{\partial \bar{\zeta}_1} \cdot \mathrm{d}\bar{\zeta}_1 + \frac{\partial}{\partial \bar{\zeta}_2} \cdot \mathrm{d}\bar{\zeta}_2.$$

So, for an element $\sum f_{\lambda\mu} \, \mathrm{d}\bar{z}_\lambda \, \mathrm{d}\bar{\zeta}_\mu$ of $\mathscr{C}^{1,1}$ we have

$$\delta_\xi^* \bar{\partial}_2\left(\sum f_{\lambda\mu} \, \mathrm{d}\bar{z}_\lambda \, \mathrm{d}\bar{\zeta}_\mu\right) = \sum \bar{\zeta}_\lambda \frac{\partial f_{\lambda\mu}}{\partial \bar{\zeta}_\nu} \, \mathrm{d}\bar{\zeta}_\nu \, \mathrm{d}\bar{\zeta}_\mu,$$

and
$$\bar{\partial}_2 \delta_\xi^*\left(\sum f_{\lambda\mu} \, \mathrm{d}\bar{z}_\lambda \, \mathrm{d}\bar{\zeta}_\mu\right) = \sum \frac{\partial}{\partial \bar{\zeta}_\nu} (\bar{\zeta}_\lambda f_{\lambda\mu}) \, \mathrm{d}\bar{\zeta}_\nu \, \mathrm{d}\bar{\zeta}_\mu.$$

The commutator maps $\sum f_{\lambda\mu} \, \mathrm{d}\bar{z}_\lambda \mathrm{d}\bar{\zeta}_\mu$ to $(f_{12} - f_{21}) \, \mathrm{d}\bar{\zeta}_1 \, \mathrm{d}\bar{\zeta}_2$. The form $\mathrm{d}\bar{\zeta}_1 \, \mathrm{d}\bar{\zeta}_2$ is the standard basis element, and this is indeed the operator κ, written in complex coordinates.

Now, returning to Proposition (3.2.27), in this case $\bar{\partial}^* \alpha = 0$ so the lemma gives:

$$\bar{\partial}_1^* \bar{\partial}_2 \alpha = \kappa(\alpha).$$

Then we have, for a section α of $p_1^*(E) \otimes \mathbb{P} \otimes S^- \otimes S^-$ giving an element of $\hat{\check{E}}_0$, the formula:

$$\omega_I(\alpha) = 4 \int_{T_0^*} G_1(\kappa(\alpha)) \, \mathrm{d}\mu. \tag{3.2.33}$$

This formula gives a completely Euclidean-invariant description of ω_I and so completes the proof of (3.2.27) and the inversion theorem.

3.3 The ADHM description of instantons

We will now apply the same ideas to obtain the description due to Atiyah *et al.* (1978*a*) (ADHM) of ASD connections over S^4. These connections are often called 'instantons'. We will think of S^4 as the conformal compactification of $\mathbb{R}^4$, in the standard way. The geometry of the construction is clearer on $\mathbb{R}^4$, but it is useful to pass to the sphere to obtain various analytical facts. In Section 3.3.1 we will state the main result, whose proof will extend through to the end of Section 3.3.

3.3.1 The correspondence

Let A be an ASD connection over $\mathbb{R}^4$ with finite 'energy':

$$\int_{\mathbb{R}^4} |F_A|^2 \, d\mu < \infty .$$

Thanks to the conformal invariance of both the ASD equation and the energy integral, we can regard A as an ASD connection with finite energy over $S^4 \setminus \{\infty\}$. According to the removable singularities theorem of Uhlenbeck, which we shall discuss in the next chapter (Section 4.4), this connection can be extended smoothly over S^4. In particular there is an integer invariant:

$$\kappa(A) = \frac{1}{8\pi^2} \int_{\mathbb{R}^4} |F_A|^2 \, d\mu. \tag{3.3.1}$$

It also makes sense to talk about the fibre, E_∞, of the bundle over the point at infinity and about trivializations of E in a neighbourhood of infinity. Taken back to $\mathbb{R}^4$ such a trivialization leads to a connection matrix A over the complement of a large ball with

$$|\nabla^{(l)} A| = O(|x|^{-3-l}) \quad (l \geq 0). \tag{3.3.2}$$

In particular the curvature of A on $\mathbb{R}^4$ is $O(|x|^{-4})$. Whenever we discuss the asymptotic behaviour of sections of E as $|x| \to \infty$, we assume we are working in such a trivialization.

Recall that on $\mathbb{R}^4$ the ASD equations, expressed in terms of covariant derivatives ∇_i, take the form:

$$[\nabla_1, \nabla_2] + [\nabla_3, \nabla_4] = 0$$

$$[\nabla_1, \nabla_3] + [\nabla_4, \nabla_2] = 0$$

$$[\nabla_1, \nabla_4] + [\nabla_2, \nabla_3] = 0.$$

Throughout this section we will often express our constructions in terms of a coordinate system. Naturally, everything will be fully invariant; when we want to adopt an invariant viewpoint we will write V for the Euclidean space $\mathbb{R}^4$ as in Sections 3.1 and 3.2. We will denote by I, J, K the standard basis for Λ^+, used in the above form of the ASD equations.

The ADHM construction gives a correspondence between ASD solutions, for group $SU(n)$, and certain systems of finite-dimensional algebraic data, indexed by an integer k. The data comprise:

Data (3.3.3).

(i) *A k-dimensional complex vector space $\mathscr{H}$ with a Hermitian metric.*

(ii) *An n-dimensional complex vector space E_∞, with Hermitian metric and determinant form* (i.e. *symmetry group* $SU(n)$).

(iii) *Self-adjoint linear maps $T_i: \mathscr{H} \to \mathscr{H}$ $(i = 1, 2, 3, 4)$, or, invariantly, a linear map $T \in V^* \otimes \mathrm{Hom}(\mathscr{H}, \mathscr{H})$.*

(iv) *A linear map $P: E_\infty \to \mathscr{H} \otimes S^+$, where S^+ is the two-dimensional positive spin space of V, as in Section* 3.1.

(We will often denote such a system of data by (T, P).)

Given such a system and a point x in $\mathbb{R}^4$ we define a linear map,

$$R_x: \mathscr{H} \otimes S^- \oplus E_\infty \longrightarrow \mathscr{H} \otimes S^+, \tag{3.3.4}$$

by

$$R_x = \left(\sum_{i=1}^{4} (T_i - x_i 1) \otimes \gamma(e_i)^* \right) \oplus P,$$

where $\gamma(e_i)^*: S^- \to S^+$ are the adjoints of the maps defining the spin structure.

The product PP^* lies in $\mathrm{End}(\mathscr{H} \otimes S^+) = \mathrm{End}(\mathscr{H}) \otimes \mathrm{End}(S^+)$. The space $\mathrm{End}(S^+)$ contains a direct summand $\mathfrak{su}(S^+)$, which is isomorphic to Λ^+ by the map ρ of (3.1.3). So there is a component of PP^*, which we denote by $(PP^*)_{\Lambda^+}$, in $\mathrm{End}(\mathscr{H}) \otimes \Lambda^+$. In terms of our coordinate system, this has individual components:

$$PP^*_I, PP^*_J, PP^*_K \in \mathrm{End}(\mathscr{H}). \tag{3.3.5}$$

Definition (3.3.6). *A system of 'ADHM data', for group $SU(n)$ and index k, is a system $(\mathscr{H}, E_\infty, T, P)$ as above which satisfies:*

(i) (*The ADHM equations*):

$$[T_1, T_2] + [T_3, T_4] = PP^*_I$$

$$[T_1, T_3] + [T_4, T_2] = PP^*_J$$

$$[T_1, T_4] + [T_2, T_3] = PP^*_K.$$

(ii) (*The non-degeneracy conditions*):

for each x in $\mathbb{R}^4$ the map R_x is surjective.

Note that the ADHM equations could be written in more invariant form

$$[T, T]^+ = PP^*_{\Lambda^+}.$$

The non-degeneracy condition (ii) of (3.3.6) means that the kernels of the linear maps R_x define a sub-bundle E of the trivial bundle with fibre $\mathscr{H} \otimes S^- \oplus E_\infty$. Using the metric on this space we get a connection $A(T, P)$ on E through the projection construction of Section 2.1.3.

There is an obvious notion of equivalence for systems of ADHM data. Otherwise stated, we can fix model spaces $\mathscr{H} = \mathbb{C}^k$, $E_\infty = \mathbb{C}^n$ so that the maps T_i, P become matrices; then the system (T, P) is equivalent to the system (T', P') if

$$T'_i = vT_iv^{-1}, \quad P' = vPu^{-1}, \tag{3.3.7}$$

for v in $U(k)$ and u in $SU(n)$. It is clear that equivalent sets of ADHM data give gauge equivalent connections. The main result we will prove is:

Theorem (3.3.8). *The assignment $(T, P) \to A(T, P)$ sets up a one-to-one correspondence between (a) equivalence classes of ADHM data, for group $SU(n)$ and index k, and (b) gauge equivalence classes of finite energy, ASD $SU(n)$-connections A over $\mathbb{R}^4$ with $\kappa(A) = k$.*

3.3.2 Formal aspects

Our proof of Theorem (3.3.8) will have the same formal structure as that of the inversion theorem for the torus, with systems of ADHM data taking the role of ASD connections over the dual torus. To develop this analogy, we consider the ADHM conditions in the presence of a complex structure. (As before we write U for the base space when it is endowed with a complex structure. We also adopt an auxiliary trivialization of $\Lambda^2 U$.) Note first that our use of I to denote a basis element for Λ^+ and a complex structure is quite consistent, since we have seen that a complex structure precisely corresponds to a choice of unit vector in Λ^+. Given this choice, we introduce new variables:

$$\tau_1 = T_1 + iT_2, \quad \tau_2 = T_3 + iT_4. \tag{3.3.9}$$

Also, we have seen that the complex structure decomposes S^+ into two pieces. We can then write P as

$$P = \pi^* \otimes \langle 1 \rangle + \sigma \otimes \langle \theta \rangle, \tag{3.3.10}$$

where $\pi : \mathscr{H} \to E_\infty$, $\sigma : E_\infty \to \mathscr{H}$. In terms of these new variables the ADHM equations take the form:

$$[\tau_1, \tau_2] + \sigma\pi = 0 \tag{3.3.11}$$

$$[\tau_1, \tau_1^*] + [\tau_2, \tau_2^*] + \sigma\sigma^* - \pi^*\pi = 0. \tag{3.3.12}$$

Of course, this way of writing the equations is analogous to the form of the ASD equations in (2.1.60) and (2.1.61). So we think of the matrices τ_λ, σ, π as analogous to Cauchy–Riemann operators. In the same vein, we construct the analogue of the Dolbeault complex. This is given by maps:

$$\mathscr{H} \xrightarrow{\alpha} \mathscr{H} \otimes U \oplus E_\infty \xrightarrow{\beta} \mathscr{H}, \tag{3.3.13}$$

with α, β given, in standard complex coordinates on $U = \mathbb{C}^2$, by

$$\alpha = \begin{pmatrix} \tau_1 \\ \tau_2 \\ \pi \end{pmatrix} \qquad \beta = (-\tau_2 \quad \tau_1 \quad \sigma). \tag{3.3.14}$$

The composite $\beta\alpha$ is the endomorphism $[\tau_1, \tau_2] + \sigma\pi$ of $\mathscr{H}$, so equation (3.3.11) is equivalent to the condition that (3.3.13) define a complex. We can replace τ_λ by $\tau_\lambda - z_\lambda$, where z_λ are coordinates on $U = \mathbb{C}^2$. This does not affect the condition (3.3.11), so we get a family of complexes:

$$\mathscr{H} \xrightarrow[\alpha_x]{} \mathscr{H} \otimes U \oplus E_\infty \xrightarrow[\beta_x]{} \mathscr{H}, \tag{3.3.15}$$

parametrized by $x = (z_1, z_2)$ in U. Thus we have a holomorphic bundle $\mathscr{E}$ over $\mathbb{C}^2$ described by a monad of a particular kind. On the other hand we have an identification:

$$R_x = \alpha_x^* \oplus \beta_x, \tag{3.3.16}$$

on $\mathscr{H} \otimes U \oplus E_\infty$, interpreted as $\mathscr{H} \otimes S^- \oplus E_\infty$ in the familiar way. This is just the identification of γ_u with $\delta_u^* \oplus \delta_u$ from (3.1.12). It follows that, if the original data satisfy the non-degeneracy condition (ii) of (3.3.6), the maps α_x are injective, and the β_x are surjective. So we arrive at the situation considered in Section 3.1.3. On the other hand we have an interpretation of the ADHM equations:

Proposition (3.3.17). *A system of data (T, P) satisfies the ADHM equations* (3.3.6 (i)) *if and only if, for each choice of complex structure on V, the maps α, β defined by* (3.3.9), (3.3.10) and (3.3.14) *satisfy* $\beta\alpha = 0$.

With these formal aspects in place we will proceed to carry through the scheme of proof used for the inversion theorem. The principal new feature is that we now have a non-compact base manifold $\mathbb{R}^4$, so we cannot appeal to standard Hodge theory to relate harmonic spinors to Dolbeault cohomology. This means we need to develop some additional analytic tools, which is the task of Section 3.3.3. One part of Theorem (3.3.8) is, however, immediately accessible:

Proposition (3.3.18). *For any system of ADHM data (T, P), of index k, the connection $A(T, P)$ is ASD, of finite energy, and $\kappa(A(T, P)) = k$.*

Proof. The fact that $A(T, P)$ is ASD follows immediately from (3.1.20) and (3.3.17), just as in the analogous theorem (3.2.8). To obtain the other two assertions we show directly that the connection has a smooth extension over S^4. This brings in the important point that, while we have developed the

ADHM construction in a form in which symmetry under the Euclidean group is manifest, the construction in fact has symmetry under the group of conformal transformations of S^4. The point at infinity need not be singled out in any way. To see this one must introduce quaternions. If we identify $\mathbb{R}^4$ with the quaternion algebra $\mathbb{H}$, then we can regard both spin spaces S^+, S^- also as copies of the quaternions, and the action γ^* as left quaternion multiplication. Then we can write R_x as:

$$(T-x)\oplus P:\mathbb{H}^k\oplus\mathbb{C}^n \longrightarrow \mathbb{H}^k.$$

Now regard S^4 as the quaternionic projective space $\mathbb{P}(\mathbb{H}^2)$, so a point of S^4 is given by homogeneous coordinates (X_0, X_1). The embedding of $\mathbb{R}^4$ corresponds to the affine patch $\{(1, x)\}$. For any (X_0, X_1), consider the map:

$$R(X_0, X_1) = (X_0T - X_1 1)\oplus X_0P:\mathbb{H}^k\oplus\mathbb{C}^n \longrightarrow \mathbb{H}^k.$$

The kernel of this is unaffected by replacing X_0, X_1 by qX_0, qX_1 for non-zero quaternions q. So the kernels descend to give a smooth vector bundle over S^4, which is canonically identified with the bundle E over $\mathbb{R}^4 \subset S^4$. Note that the fibre of this bundle over the point at infinity is indeed the space $\mathbb{C}^n = E_\infty$. (Viewed over the sphere we can consider any family of maps of the form $X_0L_0 + X_1L_1:\mathbb{C}^{2k+n}\to\mathbb{H}^k$. The choice of the 'base point' at infinity singles out preferred bases in the two spaces. The tie up with the conformal geometry of S^4 comes through the isomorphism between the conformal group and $PGL_2(\mathbb{H})$.)

Now the connection on the kernels of the $R(X_0, X_1)$, induced by projection, descends to S^4 and agrees with $A(T, P)$ over $\mathbb{R}^4$. So this connection certainly has finite energy. To see that $\kappa(A(T, P)) = k$ we have to calculate the Chern class of the bundle E over S^4. This is the kernel of a bundle map from the trivial bundle $\underline{\mathbb{C}}^{n+2k}$ to the direct sum of k copies of the quaternionic 'tautological' bundle U over $\mathbb{HP}^1$. So $\underline{\mathbb{C}}^{n+2k} = E\oplus U^k$ and $c_2(E) = -kc_2(U)$, by the Whitney formula; and it is easy to see that, with standard conventions, $c_2(U) = -1$.

We note in passing that it is possible to extend to other compactifications. In particular, if we fix a complex structure on $\mathbb{R}^4$ to get a natural complex compactification $\mathbb{CP}^2$, then the connection extends smoothly to $\mathbb{CP}^2$. This follows by pulling back under the obvious smooth map from $\mathbb{CP}^2$ to S^4. In particular then, the holomorphic bundle $\mathcal{E}$ defined by the connection extends to $\mathbb{CP}^2$. On the other hand one can see the extension of $\mathcal{E}$ more directly: the monad (3.3.15) over $\mathbb{C}^2$ extends to a monad over $\mathbb{CP}^2$ of the shape

$$\mathcal{O}(-1)\otimes\mathcal{H} \longrightarrow \mathcal{H}\otimes U\oplus E_\infty \longrightarrow \mathcal{O}(1)\otimes\mathcal{H}.$$

3.3.3 Conformally invariant operators

Our derivation of the ADHM construction relies on the properties of two differential operators: the Dirac operator D_A and Laplacian $\nabla_A^*\nabla_A$, on sections of $E \otimes S^-$ and E respectively, over $\mathbb{R}^4$. These both admit conformally invariant interpretations. Consider first the Laplacian. On any Riemannian manifold (M, g) of dimension d, the 'conformal Laplacian' is the operator (acting on the functions on M):

$$L_g = \nabla^*\nabla + \frac{(d-2)}{4(d-1)} S, \tag{3.3.19}$$

where S is the *scalar curvature* of (M, g). Let $g' = e^{2f} g$ be a conformally equivalent metric. Then we have the formula:

$$L_{g'}(\phi) = e^{(d-2)f/2} L_g(e^{-(d-2)f/2}\phi). \tag{3.3.20}$$

Equally, if we have a fixed auxiliary connection A on a bundle over M, the coupled operator,

$$\nabla_A^*\nabla_A + \frac{(d-2)}{4(d-1)} S,$$

acting on sections of the bundle, is conformally invariant in the same sense. Now the scalar curvature of $\mathbb{R}^4$ is, of course, zero while that of the round four-sphere is 12. Thus we can pass from the operator $\nabla_A^*\nabla_A$ over $\mathbb{R}^4$ to $L = \nabla_A^{(*)}\nabla_A + 2$ over S^4 by invoking suitable powers of the stereographic dilation factor of the map. (Here $\nabla^{(*)}$ denotes the adjoint defined by the metric on S^4.)

Now L is a strictly positive, self-adjoint operator, so is invertible, by the Fredholm alternative. The inverse is a singular integral operator with a kernel $k(x, y)$, having a singularity of order $d(x, y)^{-2}$ along the diagonal. (Recall that the Newton potential function in $\mathbb{R}^n$ is proportional to r^{2-n}, $n > 2$.) We can now transform this back to $\mathbb{R}^4$ to obtain:

Proposition (3.3.21).

(i) *There is an integral operator G_A on sections of E over $\mathbb{R}^4$ with kernel $k_A(x, y)$,*

$$(G_A s)(x) = \int_{\mathbb{R}^4} k_A(x, y) s(y)\, \mathrm{d}\mu_y,$$

such that for any compactly supported section s, we have $\nabla_A^*\nabla_A(G_A s) = s$.

(ii) *For each e in the fibre at infinity E_∞, there is a section s_e of E over $\mathbb{R}^4$ with $\nabla_A^*\nabla_A(s_e) = 0$ and*

$$s_e(x) = e + O(1/|x|^2), \quad \nabla s_e = O(1/|x|) \quad \text{as } |x| \longrightarrow \infty.$$

The kernel k_A is obtained from the kernel on S^4 with finite values of the arguments. The harmonic sections s_e are obtained by transforming the function $k(x, \infty)$ on S^4 back to $\mathbb{R}^4$. They have the following uniqueness property:

Lemma (3.3.22). *Let s be any section of E over $\mathbb{R}^4$ with $\nabla_A^*\nabla_A s = 0$, such that $s(x)$ is bounded and ∇s is $O(1/|x|)$ as $x \to \infty$. Then $s = s_e$ for some e in E_∞.*

Proof. We use the conformal invariance to transform the problem to S^4. Choose a local coordinate y about ∞; then in this setting, we are given a solution s of $Ls = 0$ over the punctured space, with $|s| = O(|y|^{-2})$ as $y \to 0$, and $|\nabla s| = O(|y|^{-3})$. We want to show that a distributional equation

$$Ls = \delta_e$$

holds, where δ_e is an evaluation or delta distribution at 0. The regularity of the Laplace operator on distributions then shows that s is a component of the Green's function.

Let ϕ be a standard cut-off function (vanishing near zero) and for small $r > 0$ set $\phi_r(y) = \phi(|y|/r)$. Then $L(\phi_r s)$ is supported in a small ball of radius $O(r)$ and has L^∞ norm $O(r^{-4})$. So if σ is any smooth test function,

$$I(r) = \int_{S^4} (\sigma, L(\phi_r s))$$

is bounded by a constant times $|\sigma(0)|$ as $r \to 0$. But, integrating by parts,

$$I(r) = \int (L\sigma, \phi_r s);$$

so $I(r)$ converges to $T(\sigma)$ as $r \to 0$, where T is the distribution Ls. So

$$|T\sigma| \leq \text{const.}\ |\sigma(0)|.$$

Now choose a family of smooth sections σ_i of E over S^4 whose values at infinity form a basis for the fibre E_∞. We can write any other section η as $(\sum c_i\sigma_i) + \sigma$, where σ vanishes at the point at infinity, and $c_i \in \mathbb{C}$. Then the inequality above gives

$$T(\eta) = \sum c_i(T\sigma_i);$$

so T is indeed a delta distribution at the point at infinity.

We turn now to Dirac operators. Let (X, g) be a Riemannian four-manifold with spin bundles S^+, S^-. If $g' = e^{2f}g$ is a conformally related metric, we can define a new spin structure by keeping the same bundles S^+, S^-, with the same Hermitian metrics, but changing the structure map $\gamma: TX \to \operatorname{Hom}(S^+, S^-)$ to $\gamma' = e^f\gamma$. We then have new Dirac operators D', D'^*, which satisfy the formulae:

$$D's = e^{-f/2}D(e^{3f/2}s), \quad D'^*s = e^{-f/2}D^*(e^{3f/2}s). \tag{3.3.23}$$

The verification of (3.3.20) and (3.3.23) make good exercises. The result extends in the obvious way to coupled Dirac operators.

Now on the compact manifold S^4 we know that D_A and D_A^* have finite-dimensional kernels. In fact the kernel of D_A is zero, although we shall not use this directly, and the important space for us is the kernel of D_A^* on S^4, which we denote by $\mathscr{H}_A$. Using (3.3.23) to transform to $\mathbb{R}^4$, we can represent an element of $\mathscr{H}_A$ as an E-valued spinor field ψ over $\mathbb{R}^4$, with

$$|\psi(x)| = O(|x|^{-3}), \tag{3.2.24}$$

satisfying the differential equation $D_A^*\psi = 0$. Conversely, arguing as in the proof of (3.3.22), one can show that these are the only solutions of $D_A^*\psi = 0$ which are $O(|x|^{-1})$. Alternatively we can assume ψ is in L^2. So, seen from the point of view of $\mathbb{R}^4$, $\mathscr{H}_A$ consists of the 'L^2 A-harmonic spinors'. On the other hand, returning to S^4, we have an evaluation map at the point at ∞:

$$(\text{ev}): \mathscr{H}_A \longrightarrow [S^-]_\infty \otimes E_\infty. \tag{3.2.25}$$

But there is a natural orientation-reversing isometry between the tangent spaces of S^4 at 0 and ∞, so we identify $[S^-]_\infty$ with the positive spin bundle S^+ of $\mathbb{R}^4$. Then (ev) maps to $S^+ \otimes E_\infty$. Clearly $(\text{ev})(\psi)$ will give the leading term in the behaviour of $\psi(x)$ as $|x| \to \infty$. For ρ in S^+ we write l_ρ for the singular section of S^-:

$$l_\rho(x) = |x|^{-4}\{\gamma(x)^*\rho\}, \tag{3.3.26}$$

which decays as $|x|^{-3}$ at infinity. It is easy to verify that l_ρ satisfies the Dirac equation over $\mathbb{R}^4\setminus\{0\}$. The asymptotic behaviour of a general L^2 solution ψ is given by

$$\psi(x) = l_{\text{ev}(\psi)}(x) + O(|x|^{-4}), \quad \text{as } |x| \longrightarrow \infty. \tag{3.3.27}$$

Starting with the unitary ASD solution A over $\mathbb{R}^4$ we now have three of the four objects which will turn out to provide the corresponding ADHM data. These are:

(3.3.28).
(i) *The space $\mathscr{H} = \mathscr{H}_A$ of L^2 harmonic spinors, with the Hermitian metric induced from $L^2(\mathbb{R}^4)$, but normalized by a factor $4\pi^2$, so*

$$\langle\psi, \phi\rangle = \frac{1}{4\pi^2}\int_{\mathbb{R}^4} (\psi, \phi)\,\mathrm{d}\mu.$$

(ii) *The fibre at infinity E_∞, with its given metric.*
(iii) *The complex linear map*

$$P_A: E_\infty \longrightarrow \mathscr{H}_A \otimes S^+$$

derived from the adjoint of the evaluation map $(\text{ev})^*: E_\infty \otimes S^+ \to \mathscr{H}_A$ *using the skew isomorphism between S^+ and its dual.*

We complete the set with the definition of the T_i. These are given by the multiplication action of the coordinate functions x_i on the spinors. We define T_i by the requirement that, for ψ, ϕ in $\mathscr{H}_A$:

$$(T_i\psi, \phi) = \int_{\mathbb{R}^4} (x_i\psi, \phi)\,\mathrm{d}\mu. \tag{3.3.29}$$

Clearly the T_i are self-adjoint. The definition can be thought of as the composition of the multiplication $\psi \to x_i\psi$ with L^2 projection to $\mathscr{H}_A$; the reason for expressing it in the form (3.3.29) is that $x_i\psi$ will not in general be in L^2.

The main content of the ADHM theorem (3.3.8) is:

Proposition (3.3.30). *For any finite energy ASD connection A on a bundle E over $\mathbb{R}^4$, the data $(\mathscr{H}, E_\infty, T_i, P)$ defined in* (3.3.28) *and* (3.3.29) *is a system of ADHM data, and there are natural isomorphisms,*

$$\omega_x : \mathrm{Ker}\, R_x \longrightarrow E_x,$$

giving a bundle map ω with $\omega^*(A) = A(T, P)$.

This will be proved in Section 3.3.4 and 3.3.5.

3.3.4 The double complex

We continue to suppose that A is a finite-energy ASD connection on a bundle E over $\mathbb{R}^4$, and we fix a compatible complex structure I on the base space. This gives us a holomorphic structure on E through the coupled operators $\bar{\partial}_A$. Now, as it stands, this holomorphic structure contains no real information. Indeed it is a general fact that holomorphic bundles over $\mathbb{C}^d$ are all holomorphically trivial. The additional information we exploit is the preferred trivialization of E, outside a compact set, obtained from S^4, as explained in Section 3.3.1. Thus we can talk about sections with given asymptotic behaviour on $\mathbb{R}^4$. We could formalize this notion, within the framework of complex geometry, by introducing a class of holomorphic bundles 'trivialized at infinity'. One way to define these is to introduce a complex compactification S of $\mathbb{C}^2$; for example we could take $S = S^2 \times S^2$ or $\mathbb{CP}^2$. Then we consider holomorphic bundles on S which are trivial on the curve at infinity. These play the role of holomorphic bundles over tori in the Fourier transform. (At the end of the proof of (3.3.18) we observed that the holomorphic bundles in our construction do extend to $\mathbb{CP}^2$.) However, for the proof of (3.3.8) there is no real need to introduce this extra class, and we will proceed more directly.

If s is a section of E which is $O(|x|^{-m})$ as $|x| \to \infty$, the natural growth conditions on the higher derivatives are: $|\nabla^{(l)}s| = O(|x|^{-(m+l)})$. These hold, for example, on products $|x|^{-s}\Pi(x)$, with Π a polynomial. For brevity we shall write these natural growth conditions as $s = O'(|x|^{-m})$. We now define a double complex with groups $\mathscr{A}^{p,q}$ which are the subspaces of

$$\Omega^{0,p}(E) \otimes \Lambda^q(U^*),$$

defined by the growth conditions:

$$\alpha = O'(|x|^{-(p-q+2)}) \quad \text{for } \alpha \in \mathscr{A}^{p,q} \tag{3.3.31}$$

For the differentials in the complex we take the Dolbeault $\bar{\partial}_A$ complex, tensored with the fixed vector spaces $\Lambda^q(U^*)$, in the 'horizontal' directions; and in the 'vertical' directions we take the Koszul multiplication defined using the $\Lambda^*(U^*)$ term. Neglecting the growth condition, this is just the mixed Koszul/Dolbeault complex $\mathscr{D}_0$ considered in the proof of (3.2.20). In a convenient shorthand our double complex $\mathscr{A}^{p,q}$ can be written:

$$\begin{array}{ccccc}
O'(r^0) & \xrightarrow{\bar{\partial}} & O'(r^{-1}) & \xrightarrow{\bar{\partial}} & O'(r^{-2}) \\
\delta \uparrow & & \delta \uparrow & & \delta \uparrow \\
O'(r^{-1}) & \xrightarrow{\bar{\partial}} & O'(r^{-2}) & \xrightarrow{\bar{\partial}} & O'(r^{-3}) \\
\delta \uparrow & & \delta \uparrow & & \delta \uparrow \\
O'(r^{-2}) & \xrightarrow{\bar{\partial}} & O'(r^{-3}) & \xrightarrow{\bar{\partial}} & O'(r^{-4}).
\end{array} \tag{3.3.33}$$

We now look at the cohomology of this double complex in two different ways, using our spectral sequences. In the $\tilde{E}$ sequence we first take the cohomology of the columns, using the Koszul maps. It is easy to check that these are exact on forms supported away from 0; this is just a matter of checking the growth conditions. We conclude in the same way as before that restriction to terms about 0 induces an isomorphism on the $\tilde{E}_1$ term, and hence on the total cohomology. Thus we obtain:

Proposition (3.3.34). *The total cohomology of the complex $\mathscr{A}^{p,q}$ is isomorphic to $\mathscr{E}_0$ in dimension two, and otherwise zero.*

Thus we can use the other spectral sequence to obtain an alternative description of this fibre. In this spectral sequence we begin by taking the $\bar{\partial}$ cohomology of the rows. Thus we have to understand the interaction between the growth conditions and the $\bar{\partial}$ complex. To do this we use the Green's operator G_A of Section 3.3.3. We must first digress to consider the effect of this operator on the growth conditions.

Consider first a general situation:

Lemma (3.3.35). *Let k and f be functions on $\mathbb{R}^d$ with f continuous, k locally integrable and $|f(x)| = O(|x|^{-n})$, $|k(x)| = O(|x|^{-m})$ as $x \to \infty$, where $n + m > d$. Let g be the convolution*

$$g(x) = \int_{\mathbb{R}^d} k(x-y) f(y)\, \mathrm{d}\mu_y;$$

then $|g(x)| = O(|x|^{-n}) + O(|x|^{-m}) + O(|x|^{d-(n+m)})$, as $x \to \infty$.

This is a straightforward estimate obtained by splitting the domain of integration into three regions—balls of radius $\frac{1}{2}|x|$ about x and 0, and their complement. We leave details to the reader.

In our case the dimension d is four and we wish to study the kernel k_A representing the Green's operator. This is not precisely translation invariant, but it is asymptotically so at infinity, so the bound of (3.3.35) applies, with $n = 2$. If s is $O(|x|^{-m})$ for $m > 2$, we can define $G_A(s)$ using the kernel, and it is easy to see (by considering cut-off functions) that $\nabla_A^* \nabla_A G_A(s) = s$. Then (3.3.35) gives:

$$G_A(s)(x) = O(|x|^{-m}) + O(|x|^{-2}) + O(|x|^{2-m}).$$

As for the higher derivatives we have:

Lemma (3.3.36). *If f is a function which is $O(|x|^{-1})$ and Δf is $O'(|x|^{-(l+2)})$ then f is $O'(|x|^{-1})$. (The notation O' was defined above.)*

This follows from the basic elliptic estimates for the Laplacian. Given a point x we consider the restriction of f to a ball of radius $|x|$ centred on x, and rescale this ball to a standard size. Again we leave details to the reader. Similarly one can check that the argument applies with the operator Δ_A in place of Δ.

We now go on to study the cohomology groups, beginning with the bottom row. By the Weitzenböck formula, a holomorphic section s of E satisfies

$$\nabla_A^* \nabla_A s = 4 \bar{\partial}_A^* \bar{\partial}_A s = 0.$$

So if s is also in $\mathscr{A}^{0,0}$, decaying as r^{-2} at ∞, it must be identically zero by (3.3.22). We conclude that $E_1^{0,0} = 0$.

Now let γ be an element of $\mathscr{A}^{2,0}$, decaying as $|x|^{-4}$. The integral defining $G_A(\gamma)$ converges, and $G_A(\gamma)$ is $O(|x|^{-4}) + O(|x|^{-2}) + O(|x|^{-2}) = O(|x|^{-2})$. Since $\nabla_A^* \nabla_A G_A \gamma = \gamma$, we can apply (3.3.36) to see that $G_A \gamma$ is $O'(|x|^{-2})$. Thus the 1-form $\beta = 4 \bar{\partial}_A^* G_A(\gamma)$ is in $\mathscr{A}^{0,2}$ and, by the Weitzenböck formula, satisfies the equation

$$\bar{\partial}_A \beta = 4 \bar{\partial}_A \bar{\partial}_A^* G_A(\gamma) = \gamma.$$

Thus the cohomology group $E_1^{2,0}$ is zero.

We encounter non-trivial cohomology in the middle term. We will show that $E_1^{1,0}$ is naturally isomorphic to the space of harmonic spinors $\mathscr{H}_A$. Let ψ be an element of $\mathscr{H}_A$, interpreted in the familiar way as a (0, 1)-form. Then $\bar{\partial}_A \psi = 0$ and ψ is $O'(|x|^{-3})$, so it lies in $\mathscr{A}^{1,0}$ and defines a cohomology class in $E_1^{1,0}$. Thus we have a linear map:

$$h : \mathscr{H}_A \longrightarrow E^{1,0}.$$

To see that h is injective, suppose $h(\psi)$ is zero so that there is a α in $\mathscr{A}^{0,0}$ with $\bar{\partial}_A(\alpha) = \psi$. Then $\bar{\partial}_A^* \psi = 0$ implies that $\nabla_A^* \nabla_A \alpha = 0$, and appealing again to

(3.3.22) we see that α and ψ must vanish. For the surjectivity, suppose that ψ represents a class in $E_1^{1,0}$, so $\bar{\partial}_A\psi = 0$ and $\bar{\partial}_A^*\psi$ is $O'(|x|^{-4})$. Then $G_A(\bar{\partial}_A^*\psi)$ is defined, lies in $\mathscr{A}^{0,0}$ and

$$\bar{\partial}_A^*\psi = 4\bar{\partial}_A^*\bar{\partial}_A G_A(\bar{\partial}_A^*\psi).$$

Then $\psi' = \psi - 4\bar{\partial}_A G_A \bar{\partial}_A^*\psi$ is another representative for the same cohomology class, with $\bar{\partial}_A^*\psi' = 0$; so we see that the class represented by ψ is in the image of h.

Of course, in all of the above steps we are merely verifying that a version of the Hodge theory holds for the Dolbeault complex over the non-compact base manifold $\mathbb{C}^2$, with the given decay conditions. Just the same arguments apply to the second row to show that the groups $E_1^{0,1}$, $E_1^{2,1}$ are zero and $E_1^{1,1}$ is naturally isomorphic to two copies of $\mathscr{H}_A$, i.e. $E_1^{1,1} = \mathscr{H}_A \otimes U^*$.

The top row behaves rather differently. Consider first the group $E_1^{2,2}$. An element γ in $\mathscr{A}^{2,2}$ is $O(|x|^{-2})$, so the integral representing $G_A(\gamma)$ need not converge. However, the operator $P = \bar{\partial}_A^* G_A$ is represented by a kernel which decays as $|x|^{-3}$, so $P(\gamma)$ is defined. One easily sees, using the same arguments as in (3.3.35) and (3.3.36), that $P(\gamma)$ is $O'(|x|^{-1})$ and $\bar{\partial}_A P(\gamma) = \gamma$. So the cohomology group $E_1^{2,2}$ is zero. We can define a map $h\colon \mathscr{H}_A \to E_1^{1,2}$ just as before, but it is not necessarily true that h is injective, and similarly $E_1^{0,2}$ may not be zero. Both of these phenomena occur because of the existence of bounded sections s_e with $\nabla_A^*\nabla_A s_e = 0$, as in (3.3.21). For any e, the one-form $\psi = \bar{\partial}_A(s_e)$ satisfies the Dirac equation $\bar{\partial}_A s_e = \bar{\partial}_A^* s_e = 0$ and is $O'(|x|^{-1})$, so lies in $\mathscr{H}_A$. We have a natural map,

$$\sigma\colon E_\infty \longrightarrow \mathscr{H}_A,$$

defined by $\sigma(e) = \bar{\partial}_A(s_e)$ and the same arguments as before show that $E_1^{0,2}$ is the kernel of σ and the image of σ is the kernel of h.

Finally we claim that h maps onto $E_1^{1,2}$. The argument we used in the other two cases breaks down, since for a form ψ in $\mathscr{A}^{1,2}$ we only know that $\bar{\partial}_A^*\psi$ is $O(|x|^{-2})$, so the integral we would like to use to define $G_A\bar{\partial}_A^*\psi$ need not converge. To get around this difficulty we use the calculation of the cohomology of the total complex. If $\bar{\partial}_A\psi = 0$, then ψ defines a cohomology class in the total complex, in degree 3. We know that this cohomology group is zero so we can write:

$$\psi = \bar{\partial}_A\alpha + \delta\beta,$$

where $\bar{\partial}_A\beta = \delta\gamma$, for some $\alpha \in \mathscr{A}^{0,2}$, $\beta \in \mathscr{A}^{1,1}$, $\gamma \in \mathscr{A}^{0,2}$. We know that γ lies in the image of $\bar{\partial}_A$, so without loss we may suppose that γ is zero (modifying by $(\bar{\partial}_A + \delta)\beta'$, for some β' in $\mathscr{A}^{1,0}$). Then $\bar{\partial}_A\beta = 0$ and we know that $\beta = \bar{\partial}_A\alpha' + \phi$ for some $\alpha' \in \mathscr{A}^{0,1}$ and $\phi \in \mathscr{H}_A$. So, modifying by $(\bar{\partial}_A + \delta)\alpha'$, we may assume that β lies in $\mathscr{H}_A$. We conclude that any class in the cohomology group $E_1^{1,2}$ can be represented by an element of the form $\psi = \delta\beta$ for some β in $\mathscr{H}_A$. But then $\bar{\partial}_A^*\psi$ is $O'(|x|^{-3})$ so we can define $G_A(\bar{\partial}_A^*\psi)$, and the argument we used before shows that the cohomology class of ψ is in the image of h. We conclude then that $E_1^{1,2}$ is isomorphic to $\mathscr{H}_A/\mathrm{Im}\,\sigma$.

To sum up we have:

Proposition (3.3.37). *The* E_1 *diagram associated with the double complex* $\mathscr{A}^{*,*}$ *is*

$$\begin{array}{|ccc}
\mathrm{Ker}\,\sigma \subset E_\infty & \mathscr{H}_A/\mathrm{Im}\,\sigma & 0 \\
 & d_1 \uparrow & \\
0 & \mathscr{H}_A \otimes U & 0 \\
 & d_1 \uparrow & \\
0 & \mathscr{H}_A & 0. \\
\hline
\end{array}$$

We should emphasize that, so far, we have used the metrics, connections and harmonic theory essentially as an auxiliary tool to control the Dolbeault complex at infinity. In particular, the map σ can be defined without reference to harmonic theory. For e in E_∞ we choose any section s asymptotic to e and with $|\bar{\partial}_A s| = O'(|x|^{-2})$. For example we can take a section which, in the chosen trivialization, is equal to e outside a compact set. Then $\bar{\partial}_A s$ is a representative for $\sigma(e)$ in $\mathscr{H}_A$, viewed as a cohomology group.

Let us note in passing that we can define other complexes like $\mathscr{A}^{**}$ by imposing different growth conditions. A simple alternative is to change the requirement to $\alpha = O'(|x|^{-(p-q+3)})$ in the (p, q) term. The effect of this is to replace term $E_1^{0,2}$ by a term $E_1^{2,0}$—in the opposite 'corner' of the diagram. We leave the details as an exercise.

3.3.5 *Contribution from infinity*

To spell out in concrete terms the algebraic description of E_0 which this spectral sequence yields we choose complex coordinates z_1, z_2 on U and write $\mathscr{H}_A \otimes U = \mathscr{H}_A \oplus \mathscr{H}_A$. Then define maps

$$\tau_\lambda : \mathscr{H}_A \longrightarrow \mathscr{H}_A,$$

to be the components of the differential $\mathrm{d}_1 : \mathscr{H}_A \to \mathscr{H}_A \oplus \mathscr{H}_A$. It is easy to check that the other differential $\mathrm{d}_1 : \mathscr{H}_A \oplus \mathscr{H}_A \to \mathscr{H}_A/\mathrm{Im}\,\sigma$ is the reduction mod $\mathrm{Im}\,\sigma$ of

$$\tilde{\mathrm{d}}_1 = (-\tau_2, \tau_1) : \mathscr{H}_A \oplus \mathscr{H}_A \longrightarrow \mathscr{H}_A.$$

Now, following through the spectral sequence, the only remaining differential is:

$$\mathrm{d}_2 : \{\mathrm{Ker}\,\mathrm{d}_1 \subset \mathscr{H}_A\} \longrightarrow \{\mathrm{Ker}\,\sigma \subset E_\infty\}.$$

The spectral sequence then gives an exact sequence

$$0 \longrightarrow E_3^{0,2} \longrightarrow H \longrightarrow E_3^{1,1} \longrightarrow 0, \tag{3.3.38}$$

where H is the cohomology in dimension two of the total complex, which we know to be isomorphic to the fibre E_0. Similarly we deduce that d_2 is a monomorphism and that $\tilde{d}_1 \oplus \sigma$ maps onto $\mathscr{H}_A$, since the other cohomology groups of the total complex are zero.

The differential d_2 is defined explicitly as follows. If $\psi \in \mathscr{A}^{1,0}$ represents an element of the kernel of d_1 we can write:

$$z_\lambda \psi = \bar{\partial}_A u_\lambda$$

for u_λ in $\mathscr{A}^{0,1}$. Then

$$f = z_2 u_1 - z_1 u_2,$$

is a bounded section of E, and $\bar{\partial}_A f = 0$. So $f = s_e$ for some e and we have $d_2([\psi]) = e$. We shall now show that d_2 can be extended to a map

$$\pi: \mathscr{H}_A \longrightarrow E_\infty.$$

Lemma (3.3.39). *If f is a bounded section of E with $|\bar{\partial}_A f| = O'(|x|^{-2})$ then $f(x)$ tends to a limit in E_∞ as $|x| \to \infty$.*

Proof. Put $\nabla_A^* \nabla_A f = g$. Then g is $O'(|x|^{-3})$ and we can define $f' = G_A(g)$, which is $O'(|x|^{-1})$, and $\nabla_A^* \nabla_A f' = g$. So by (3.3.22) applied to the harmonic section $f - f'$, we must have $f = f' + s_e$ for some e, and f tends to e at infinity.

We now define π on $\mathscr{H}_A$ as follows. For ψ in $\mathscr{A}^{1,0}$ with $\bar{\partial}_A \psi = 0$, we write

$$z_\lambda \psi = \psi_\lambda + \bar{\partial}_A u_\lambda,$$

with $\psi_\lambda = O'(|x|^{-3})$, and $u_\lambda = O'(|x|^{-1})$. Then the section $f = z_2 u_1 - z_1 u_2$ is bounded and has

$$\bar{\partial}_A f = z_1 \psi_2 - z_2 \psi_1 = O'(|x|^{-2}).$$

Then we can apply the lemma above to deduce that f tends to a limit e in E_∞. One easily checks that this limit is independent of the choice of representative ψ in a cohomology class, so we can put $\pi([\psi]) = e$. By the discussion above, π equals d_2 on the kernel of d_1. Also, for any class $[\psi]$, the composite $\sigma\pi([\psi])$ is represented, in the notation above, by $\bar{\partial}_A f = z_1 \psi_2 - z_2 \psi_2$. But this is, by definition, just the commutator $[\tau_1, \tau_2]\psi$. So we have

$$[\tau_1, \tau_2] = \sigma\pi: \mathscr{H}_A \longrightarrow \mathscr{H}_A. \tag{3.3.40}$$

Let us temporarily write $a = (\tau_1, \tau_2): \mathscr{H}_A \to \mathscr{H}_A \oplus \mathscr{H}_A$ and $b = (-\tau_2, \tau_1)$: $\mathscr{H}_A \oplus \mathscr{H}_A \to \mathscr{H}_A$. Our exact sequence (3.3.38) can now be put in the form:

$$0 \longrightarrow \frac{\operatorname{Ker} \sigma}{\operatorname{Im} \pi|_{\operatorname{Ker} a}} \longrightarrow E_0 \longrightarrow \frac{b^{-1}(\operatorname{Im} \sigma)}{\operatorname{Im} a} \longrightarrow 0. \tag{3.3.41}$$

Motivated by (3.3.14) we define

$$\alpha = (a, \pi): \mathscr{H}_A \longrightarrow \mathscr{H}_A \oplus \mathscr{H}_A \oplus E_\infty$$

$$\beta = (b, \sigma): \mathscr{H}_A \oplus \mathscr{H}_A \oplus E_\infty \longrightarrow \mathscr{H}_A.$$

Then, as in Section 3.3.2, the equation (3.3.40) implies that $\beta\alpha = 0$. Define W to be the cohomology group

$$W = \frac{\operatorname{Ker}\beta}{\operatorname{Im}\alpha}.$$

Notice that the exactness of (3.3.41) implies that α is injective and β is surjective. Projection to the factor $\mathscr{H}_A \oplus \mathscr{H}_A$ gives a natural exact sequence:

$$0 \longrightarrow \frac{\operatorname{Ker}\sigma}{\operatorname{Im}\pi|_{\operatorname{Ker}a}} \longrightarrow W \longrightarrow \frac{b^{-1}(\operatorname{Im}\sigma)}{\operatorname{Im}a} \longrightarrow 0. \qquad (3.3.42)$$

Our final observation is that there is a natural map,

$$\omega_I: W \longrightarrow E_0.$$

This map is defined as follows: a triple (ψ_1, ψ_2, e) represents an element of Ker β if and only if there is a section f converging to e at infinity such that $\bar{\partial}_A f = z_1\psi_2 - z_2\psi_1$. Then we put

$$\omega_I(\psi_1, \psi_2, e) = f(0) \in E_0. \qquad (3.3.43)$$

(And, as before, this is independent of representatives.) It is now simple to check that the diagram

$$\begin{array}{ccccccccc}
0 & \longrightarrow & \dfrac{\operatorname{Ker}\sigma}{\operatorname{Im}\pi|_{\operatorname{Ker}a}} & \longrightarrow & W & \longrightarrow & \dfrac{b^{-1}(\operatorname{Im}\sigma)}{\operatorname{Im}a} & \longrightarrow & 0 \\
 & & \| & & \downarrow{\scriptstyle \omega_I} & & \| & & \\
0 & \longrightarrow & \dfrac{\operatorname{Ker}\sigma}{\operatorname{Im}\pi|_{\operatorname{Ker}a}} & \longrightarrow & E & \longrightarrow & \dfrac{b^{-1}(\operatorname{Im}\sigma)}{\operatorname{Im}a} & \longrightarrow & 0
\end{array}$$

commutes; hence ω_I is an isomorphism.

We have now achieved our main goal, showing that the fibre E_0 can be represented as the cohomology of the complex (3.3.15) constructed out of the data $(\tau_1, \tau_2, \pi, \sigma)$, This is the analogue of Proposition (3.2.20) in the torus case. Similarly we can discuss the other fibres. A moment's thought shows that translating the origin by (w_1, w_2) leaves the maps π, σ unchanged, but changes τ_λ to $\tau_\lambda + w_\lambda 1$. So if we define, for $x = (z_1, z_2)$ in $\mathbb{C}^2$, maps α_x, β_x by replacing τ_λ by $\tau_\lambda - z_\lambda$, we get holomorphic bundle maps:

$$\underline{\mathscr{H}}_A \xrightarrow[\alpha]{} \underline{\mathscr{H}}_A \oplus \underline{\mathscr{H}}_A \oplus \underline{E}_\infty \xrightarrow[\beta]{} \underline{\mathscr{H}}_A.$$

These give a holomorphic cohomology bundle $\mathscr{W}$ (the analogue of $\hat{E}\check{}$) and we have a natural bundle map $\omega_I: \mathscr{W} \to \mathscr{E}$, which is an isomorphism on the fibres. As for the Fourier transform, it is rather obvious on general grounds that this is an isomorphism of holomorphic bundles, but it is instructive to verify this fact directly as we shall now do.

Let (t_1, t_2) be coordinates on a neighbourhood in $\mathbb{C}^2$ and $s' = s'(t_1, t_2)$ be a local holomorphic section of $\mathscr{W}$. So we can represent s' by a holomorphic section of ker $\boldsymbol{\beta}$ and in turn by a triple $\psi_1(t)$, $\psi_2(t)$, $e(t)$. For each parameter value there is a unique section f_t of $\mathscr{E}$ with

$$\bar{\partial}_A f_t = (z_1 - t_1)\psi_2(t_\lambda) - (z_2 - t_2)\psi_1(t_\lambda).$$

We can regard $f_t(x)$ as a local section $f(x, t)$ of $p_1^*(\mathscr{E})$ over $\mathbb{C}^2 \times \mathbb{C}^2$. The corresponding section $s = \omega_I(s')$ is obtained by restricting f to the diagonal $s(t) = f(t, t)$. Now f is plainly holomorphic in the t variable since it is obtained from the holomorphic data $\psi_\lambda(t)$, $e(t)$. On the other hand, f is *not* holomorphic in the other variable, with respect to the holomorphic structure defined by A, since $\bar{\partial}_A f_t$ is not zero. However, $\bar{\partial}_A f_t$ does vanish at the point $x = t$, and this means that $f(t, t)$ is a holomorphic section of $\mathscr{E}$ as required.

We have now established the analogue of the inversion theorem for the Fourier transform on holomorphic bundles, and it only remains to give this a Euclidean interpretation.

3.3.6 Euclidean interpretation

We wish to reconcile the ADHM data (T, P) defined in (3.3.28) and (3.3.29) with the holomorphic data $(\tau_\lambda, \pi, \sigma)$ appearing in Section 3.3.5. The relation between the two has already been indicated by our notation in Section 3.3.2. The discussion revolves around contributions from infinity introduced by integration by parts. For this we need an extra piece of notation. For spinors $\alpha \in S^+$, $\beta \in S^-$ we can form a cotangent vector

$$\sum(\gamma(e_i)\alpha, \beta)\,\mathrm{d}x_i$$

and applying the $*$ operator on $\mathbb{R}^4$ we get a 3-form, which we denote by $\{\alpha, \beta\}$. Then we immediately have, from the definition of the Dirac operators,

$$((D_A\alpha, \beta) - (\alpha, D_A^*\beta))\,\mathrm{d}\mu = \mathrm{d}\{\alpha, \beta\}. \tag{3.3.44}$$

We now begin with the definition of the T_i. For ψ in $\mathscr{H}_A$ we have, by the definition of τ_λ,

$$z_\lambda\psi = \tau_\lambda(\psi) + \bar{\partial}_A u_\lambda,$$

with $u_\lambda = O'(|x|^{-1})$. For any other element ϕ in $\mathscr{H}_A$ we consider the integral

$$\int_{\mathbb{R}^4} (z_\lambda\psi, \phi)\,\mathrm{d}\mu = \lim_{R\to\infty} \int_{B(R)} (z_\lambda\psi, \phi)\,\mathrm{d}\mu,$$

where $B(R)$ is the R-ball in R^4. Now $\bar{\partial}_A^*\phi = 0$, so we can integrate by parts to get

$$\int_{B(R)} (\bar{\partial}_A u_\lambda, \phi) = \int_{S(R)} \{u_\lambda, \psi\}\,\mathrm{d}\mu_{S(R)},$$

where $S(R)$ is the three-sphere bounding $B(R)$ and $\{,\}$ is defined by interpreting u_λ as a spinor, under our basic isomorphism. Since ψ is $O(R^{-3})$ on $S(R)$ we see that this integral tends to zero as $R \to \infty$, so we have

$$\int_{\mathbb{R}^4} (z_\lambda \psi, \phi)\,\mathrm{d}\mu = \lim \int_{B(R)} (\tau_\lambda \psi, \phi)\,\mathrm{d}\mu = \int_{\mathbb{R}^4} (\tau_\lambda \psi, \phi)\,\mathrm{d}\mu.$$

It follows immediately that, if we write $\tau_1 = T_1 + iT_2, \tau_2 = T_3 + iT_4$, corresponding to $z_1 = x_1 + ix_2$, $z_2 = x_3 + ix_4$, we do indeed have

$$(T_i \psi, \phi) = \int (x_i \psi, \phi)\,\mathrm{d}\mu, \tag{3.3.45}$$

as in (3.3.29).

We now turn to π and σ. Recall we have described the asymptotic behaviour of the harmonic spinors at infinity in terms of the singular fields $l_\rho(x) = |x|^{-4}\gamma(x)\rho$, for ρ in S^+. Up to a constant, these are the fundamental solutions of the (uncoupled) Dirac equation on $\mathbb{R}^4$. Indeed, we have $l_\rho = -D(|x|^{-2}\rho)$, so

$$\begin{aligned} D^* l_\rho &= -D^* D(|x|^{-2}\rho) \\ &= -\Delta(|x|^{-2})\rho = -4\pi^2 \delta(0)\rho, \end{aligned} \tag{3.3.46}$$

since $\Delta(|x|^{-2}) = 4\pi^2 \delta_0$. (Here δ_0 is the delta distribution at 0; the factor $4\pi^2$ is the volume of S^3.) This distributional formula is equivalent to the equation

$$\int_{S(R)} \{l_\rho, v\} = 4\pi^2(\rho, v), \tag{3.3.47}$$

for all R and spinors ρ, v in S^+. To see this directly, we have for a unit vector x,

$$\begin{aligned} \{l_\rho, v\} &= \sum (\gamma^*(x_i e_i)\gamma(x_j e_j)\rho, v)\,\mathrm{d}\mu_{S^3} \\ &= (\rho, v)\,\mathrm{d}\mu_{S^3}, \end{aligned} \tag{3.3.48}$$

so the integrand is actually constant over the sphere.

Now recall that we have an evaluation map $(\mathrm{ev})\colon \mathscr{H}_A \to E_\infty \otimes S^+$. A choice of complex structure gives a canonical basis $\langle 1 \rangle, \langle \theta \rangle$ for S^+, so we can write

$$(\mathrm{ev})(\psi) = (\mathrm{ev})_1(\psi)\langle 1 \rangle + (\mathrm{ev})_\theta(\psi)\langle \theta \rangle; \tag{3.3.49}$$

for $(\mathrm{ev})_1, (\mathrm{ev})_\theta\colon \mathscr{H}_A \to E_\infty$.

Proposition (3.3.50). *The evaluation map is related to the maps σ, π of Section 3.3.5 by*

(i) $$(\mathrm{ev})_1^* = \sigma\colon E_\infty \longrightarrow \mathscr{H}_A,$$

(ii) $$(\mathrm{ev})_\theta = \pi\colon \mathscr{H}_A \longrightarrow E_\infty,$$

where in (i) *the adjoint is formed using the normalized L^2 metric as in* (3.3.28).

Proof.

(i) We have, for ψ in $\mathscr{H}_A$

$$\langle \sigma(e), \psi \rangle = (4\pi^2)^{-1} \int_{\mathbb{R}^4} (\bar{\partial}_A s_e, \psi) \, d\mu$$

$$= \lim_{R \to \infty} \int_{S(R)} \{s_e, \psi\}.$$

To evaluate this limit we may replace s_e by its leading term, which we should now regard as $e \otimes \langle 1 \rangle$, and ψ by $(\mathrm{ev})_1(\psi) l_{\langle 1 \rangle} + (\mathrm{ev})_\theta(\psi) l_{\langle \theta \rangle}$. Then from (3.3.47) we get

$$\lim_{R \to \infty} \int_{S(R)} \{s_e, \psi\} = 4\pi^2 (\mathrm{ev})_1(\psi),$$

since $\langle 1 \rangle$ and $\langle \theta \rangle$ form an orthonormal basis of S^+. So we have shown that $\sigma^* = 4\pi^2 (\mathrm{ev})_1$, as required.

(ii) For ψ in $\mathscr{H}_A$, we defined $\pi(\psi) \in E_0$ as $\lim_{x \to \infty} f(x)$, where $f = z_1 u_2 - z_2 u_1$ and $z_\lambda \psi = \psi_\lambda + \bar{\partial}_A u_\lambda$.

Let us write $\psi = \alpha_1 \, d\bar{z}_1 + \alpha_2 d\bar{z}_2$. Then, writing Δ_A for the Laplacian, we have

$$\Delta_A u_\lambda = \bar{\partial}_A^*(z_\lambda \psi) = \alpha_\lambda,$$

so $u_\lambda = G_A \alpha_\lambda$.

We claim now that u_λ is asymptotic to $-|x|^2 \alpha_\lambda$. Consider first a function $g(x)$ of the form $|x|^{-4} L(x)$, where L is linear on $\mathbb{R}^4$. Then for the ordinary Laplacian Δ we have

$$\Delta(|x|^{-2} L(x)) = \Delta(|x|^{-2}) L(x) - (\nabla |x|^{-2}, \nabla L) + |x|^{-2} (\Delta L).$$

The first and third terms vanish so we get

$$\Delta g = - |x|^{-4} L(x).$$

Now the components of α_λ have leading (i.e. $O(|x|^{-3})$) terms of this form, and in our given trivialization we have

$$\Delta_A u_\lambda = \Delta u_\lambda + O(|x|^{-5}).$$

It follows then that

$$\Delta_A(u_\lambda + |x|^2 \alpha_\lambda) = O(|x|^{-4}),$$

so $u_\lambda = - |x|^2 \alpha_\lambda + O(|x|^{-2})$. Thus $\pi(\psi)$ is the limit of

$$|x|^2 (z_2 \alpha_1 - z_1 \alpha_2) = |x|^2 (\gamma_x^*(\psi), \langle \theta \rangle),$$

where now we have shifted back to spinor notation. To evaluate this limit we may replace ψ by its leading term. On a term of the form l_ρ we get, as in (3.3.48), $|x|^2(\gamma_x^* \gamma_x(\rho), \langle\theta\rangle) = (\rho, \langle\theta\rangle)$, since $\gamma_x^* \gamma_x = |x|^2$. It follows that the limit of f is $\mathrm{ev}_\theta(\psi)$ as required.

The proof of (3.3.8) is now almost complete. For according to (3.3.45) and (3.3.50), with a given complex structure I, the maps $\alpha_x^* \oplus \beta_x$ formed from the complex (3.3.13) can be identified with the maps R_x defined from (T, P) as in (3.3.14) and (3.3.16). So if we let W be the bundle defined by the kernel of the bundle map $\mathbf{R}$, and A' be the induced connection, the isomorphism ω_I of Section 3.3.5 becomes an isomorphism $\omega_I\colon W \to E$ with $\omega_I^*(\bar\partial_A) = \bar\partial_{A'}$. To complete the picture we must show, as in the torus case, that this is independent of the complex structure I. This uses just the same algebra as in (3.2.31). Suppose (ψ_1, ψ_2, e) is in the kernel of $\alpha_0^* \oplus \beta_0$. Then we have:

$$\bar\partial_A^*(z_2\psi_1 - z_1\psi_2) = \kappa(\psi_1, \psi_2),$$

where κ is the contraction map $S^- \otimes S^- \to \mathbb{C}$. The section s of E with $\bar\partial_A s = z_2\psi_1 - z_1\psi_2$ is

$$\begin{aligned} s &= s_e + 4G_A(\bar\partial_A^*(z_2\psi_1 - z_1\psi_2)) \\ &= s_e + G_A\kappa(\psi_1, \psi_2). \end{aligned}$$

This shows that the map ω_I on the fibres over a point x in $\mathbb{R}^4$ is

$$\omega_I(\underline{\psi}, e) = \{s_e + 4G_A\kappa(\underline{\psi})\}(x), \tag{3.3.51}$$

which makes no reference to a complex structure.

One task remains to complete the proof of the ADHM correspondence. We have to show that if we start with ADHM data $(\mathscr{H}, E_\infty, T, P)$ and construct a connection $A = A(T, P)$ then we recover the same matrix data by the construction of (3.3.28) and (3.3.29). Formally this is another instance of the inversion theorem, although we do not have here the complete symmetry present in the torus case. We follow the same pattern of proof, with the operators R_x playing the role of Dirac operators. Given the matrix data, we fix a complex structure I and look at a double complex:

$$\begin{array}{ccccc}
O'(|x|^{-4}) & \xrightarrow{\alpha} & O'(|x|^{-3}) & \xrightarrow{\beta} & O'(|x|^{-2}) \\
\uparrow{\scriptstyle\bar\partial} & & \uparrow{\scriptstyle\bar\partial} & & \uparrow{\scriptstyle\bar\partial} \\
O'(|x|^{-3}) & \xrightarrow{\alpha} & O'(|x|^{-2}) & \xrightarrow{\beta} & O'(|x|^{-1}) \\
\uparrow{\scriptstyle\bar\partial} & & \uparrow{\scriptstyle\bar\partial} & & \uparrow{\scriptstyle\bar\partial} \\
O'(|x|^{-2}) & \xrightarrow{\alpha} & O'(|x|^{-1}) & \xrightarrow{\beta} & O'(|x|^{0}).
\end{array} \tag{3.3.52}$$

Here the entries are the forms with values in the trivial bundles $\underline{\mathscr{H}}$, $\underline{\mathscr{H}} \oplus \underline{\mathscr{H}} \oplus \underline{E}_\infty$ and $\underline{\mathscr{H}}$ respectively, satisfying the stated growth conditions.

The cohomology of the rows yields the E-valued forms, and then taking the vertical cohomology we get the $E_2^{p,q}$ diagram:

$$\begin{array}{|ccc}
0 & 0 & 0 \\
0 & \mathscr{H}_A & 0 \\
0 & 0 & 0 \\
\hline
\end{array}$$

On the other hand, since all the original bundles are trivial, the only cohomology in the columns comes from the bounded holomorphic sections of $\mathscr{H}$. Thus the $\tilde{E}_1^{p,q}$ diagram is

$$\begin{array}{|ccc}
0 & 0 & 0 \\
0 & 0 & 0 \\
0 & 0 & \mathscr{H}. \\
\hline
\end{array}$$

This yields the desired natural isomorphism between $\mathscr{H}$ and $\mathscr{H}_A$. We leave as an informative exercise for the reader the verification that, under this isomorphism, T_i and P do indeed correspond to the multiplication and evaluation maps, so putting the final touches to the proof of the ADHM theorem (3.3.8).

3.4 Explicit examples

3.4.1 The basic instanton

This ADHM construction gives us a supply of explicit examples of ASD solutions. The simplest example is to take $SU(2)$ solutions with Chern class $c_2 = 1$. Then the T_i are real scalars (1×1 matrices), and P is an element of $\mathrm{Hom}(E_\infty, S^+)$. The ADHM conditions (3.3.6) just assert that P has rank 1. Now, under the symmetry group $SU(2)$ of E_∞, such maps are classified by their norm $\lambda = |P| > 0$. The equivalence classes of solutions are thus parametrized by the manifold $\mathbb{R}^4 \times \mathbb{R}^+$. In fact we can naturally identify the $\mathbb{R}^4$ in this parameter space with the original base manifold. Indeed, to any solution of the ADHM equations we can associate a 'centre of mass' in $\mathbb{R}^4$ with coordinates

$$x_i = k^{-1}\,\mathrm{Trace}(T_i). \tag{3.4.1}$$

The translations of the base space act on the ADHM matrices T_i by $T_i \to T_i + w_i 1$ and preserve P, so this centre of gravity map is equivariant with respect to the translations. Similarly the dilations $x \to tx$ of

$\mathbb{R}^4$ act by $T_i \to tT_i$, $P \to tP$. Thus the translations and dilations together act simply transitively on the parameter space $\mathbb{R}^4 \times \mathbb{R}^+$. In this sense there is really only a single solution. This basic model is the famous 'one-instanton' on $\mathbb{R}^4$.

We will now write down explicit connection matrices on $\mathbb{R}^4$ representing the basic instanton. Recall that the bundle which carries the connection is the kernel of

$$\mathbf{R}: \mathscr{H} \otimes S^- \oplus E_\infty \longrightarrow \mathscr{H} \otimes S^+,$$

regarded as a bundle map between trivial vector bundles. Since $\mathscr{H}$ is now one-dimensional and ψ is an isometry, we can identify $\mathscr{H}$ with $\mathbb{C}$ and E_∞ with S^+. Then

$$R_x = -\gamma(x) \oplus 1 : S^- \oplus S^+ \longrightarrow S^+.$$

A unitary framing for the kernel is provided by the sections

$$\sigma_r(x) = \frac{1}{1 + |x|^2}(1 \oplus \gamma(x)) s_r,$$

where s_1, s_2 is an orthonormal basis of S^-. Let the connection matrix in this trivialization be $\sum A_i \mathrm{d}x_i$, so A_i is a matrix with (p, q) entry:

$$\langle \nabla_i \sigma_p, \sigma_q \rangle = \langle \partial \sigma_p / \partial x_i, \sigma_q \rangle.$$

We can calculate this to be:

$$\frac{1}{1 + |x|^2}(-x_i \langle s_p, s_q \rangle + \langle \gamma(e_i) s_p, s_q \rangle).$$

In other words, the evaluation of the connection 1-form on a tangent vector e at a point x in $\mathbb{R}^4$, is the matrix:

$$\frac{1}{1 + |x|^2}(-\langle x, e \rangle 1 + \gamma(x)^* \gamma(x)).$$

Written out in full, the connection form is

$$A = \frac{1}{1 + |x|^2}(\theta_1 \mathbf{i} + \theta_2 \mathbf{j} + \theta_3 \mathbf{k}), \tag{3.4.2}$$

where $\mathbf{i}, \mathbf{j}, \mathbf{k}$ is a standard basis for $\mathfrak{su}(2)$ and

$$\begin{aligned}
\theta_1 &= x_1 \mathrm{d}x_2 - x_2 \mathrm{d}x_1 - x_3 \mathrm{d}x_4 + x_4 \mathrm{d}x_3, \\
\theta_2 &= x_1 \mathrm{d}x_3 - x_3 \mathrm{d}x_1 - x_4 \mathrm{d}x_2 + x_2 \mathrm{d}x_4, \\
\theta_3 &= x_1 \mathrm{d}x_4 - x_4 \mathrm{d}x_1 - x_2 \mathrm{d}x_3 + x_3 \mathrm{d}x_2.
\end{aligned} \tag{3.4.3}$$

From this one can calculate the curvature:

$$F = \left(\frac{1}{1 + |x|^2} \right)^2 (\mathrm{d}\theta_1 \mathbf{i} + \mathrm{d}\theta_2 \mathbf{j} + \mathrm{d}\theta_3 \mathbf{k}).$$

We see that the pointwise norm $|F|$ has a bell-shaped profile and decays like r^{-4}. We can, of course, apply translations and dilations to obtain formulae for the other connections $A_{y,\lambda}$, with centre y and scale λ. For example we have:

$$|F(A_{y,\lambda})| = \frac{\lambda^2}{(\lambda^2 + |x-y|^2)^2}. \tag{3.4.4}$$

As $\lambda \to 0$ the 'energy density' $|F(A_{y,\lambda})|^2$ converges in the sense of distributions (or of measures) to the delta function at y with mass $8\pi^2$.

3.4.2 Completion of the moduli space

Before considering instantons with larger values of k it will be helpful to reformulate slightly the ADHM construction in the case of the structure group $SU(2)$. As in Section 3.1, if E has an $SU(2)$ structure, then both $E \otimes S^+$ and $E \otimes S^-$ have real structures. With respect to these the Dirac operator is real, so the kernel $\mathscr{H}$ of D_A^* has a real form $\mathscr{H}^R$, i.e. $\mathscr{H} = \mathscr{H}^R \otimes_{\mathbb{R}} \mathbb{C}$. The ADHM data then takes the form $(\mathscr{H}^R, E_\infty, T, P)$ where:

Condition (3.4.5). *(a) $\mathscr{H}^R$ is a real k-dimensional inner product space and each T_i is a symmetric endomorphism of $\mathscr{H}^R$. (b) E_∞ is a two-dimensional complex vector space with an $SU(2)$ structure, and $P: E_\infty \to \mathscr{H}^R \otimes_{\mathbb{R}} S^+$ intertwines the J operators.*

Of course we also have the non-degeneracy condition:

Condition (3.4.6). $(\sum(T_i - x_i)\gamma(e_i)) \oplus P$ *is surjective for all* x.

This description is the first instance ($n = 1$) of a recipe which produces instantons with structure group $Sp(n)$. (There is a similar variant of the ADHM construction to produce instantons for the special orthogonal group.) In the 'complex' notation $(\tau_\lambda, \sigma, \pi)$, the 'reality' conditions become that the τ_λ are symmetric and

$$\pi = \varepsilon\sigma^{\mathrm{T}}, \tag{3.4.6}$$

where $\varepsilon = \begin{pmatrix} 0 & 1 \\ -1 & 0 \end{pmatrix}$, and the operators are written as matrices with respect to standard bases. The effect of this on the monad is to make β the transpose of α, with respect to the natural skew form on $\mathscr{H} \oplus \mathscr{H} \oplus E_\infty$ and the symmetric form on $\mathscr{H}$; so the fibre E_x is the quotient of the annihilator of $\operatorname{Im}\alpha$, under the skew form, by $\operatorname{Im}\alpha$ itself.

If we choose an identification of E_∞ with S^-, we can regard P as a map from $\mathbb{R}^4$ to $\mathscr{H}^R$, since $\mathbb{R}^4 = \operatorname{Hom}_J(S^-, S^+)$. We can then write $P = (P_1, P_2, P_3, P_4)$ and the ADHM equations become:

$$[T_1, T_2] + [T_3, T_4] + P_1 P_2^T - P_2 P_1^T + P_3 P_4^T - P_4 P_3^T = 0, \tag{3.4.7}$$

with the other two equations obtained by permuting (2, 3, 4) cyclically. We now want to analyse the solutions of these real algebraic equations.

To obtain an instanton one must add to the closed constraints (3.4.7) the open conditions corresponding to (3.4.6). We define the *moduli space* M_k to be the set of equivalence classes of matrix data satisfying both conditions, under the action of $O(k) \times SU(2)$ (cf. (3.3.7)). By the ADHM theorem M_k is identified with the space of gauge equivalence classes of ASD $SU(2)$ connections with Chern class k. If we remove the open conditions (3.4.6), we obtain a larger moduli space $\bar{M}_k$, a completion of M_k. We will now explain that this completion can be described in a simple way in terms of the moduli spaces M_j for $j \leq k$. This idea, which we here pursue in the framework of matrix algebra, has a natural extension to moduli spaces over arbitrary four-manifolds, as we explain in Chapter 4. The discussion in the remainder of this chapter can be regarded as a motivation for the general theory in Chapter 4.

Lemma (3.4.8). *Suppose that* $(\mathscr{H}^{\mathbf{R}}, E_\infty, T_i, P_i)$ *is a set of matrix data which is a solution of the algebraic conditions* (3.4.7), *but not necessarily of the non-degeneracy condition* (3.4.6). *Then there is an orthogonal decomposition* $\mathscr{H}^{\mathbf{R}} = \mathscr{H}' \oplus \mathscr{H}''$, *with* $T_i = T_i' \oplus T_i''$ *and* $P_i = P_i' \oplus P_i''$ *such that: (a) the endomorphisms* T_i' *commute; (b) The matrix data* $(\mathscr{H}'', E_\infty, T_i'', P_i'')$ *is a non-degenerate solution (i.e. satisfies* (3.4.6) *and* (3.4.7)).

Proof. Formally the proof here follows exactly the discussion of the flat factors for ASD connections over the torus. This was based on the Weitzenböck formula (3.1.6), which compared the Dirac operator with the covariant derivative. In the ADHM construction the role of the Dirac operator D over the dual torus is played by R, and the family of operators D_ξ by the family R_x. The analogue of the twisted covariant derivatives is given by the family of maps:

$$B_x : \mathscr{H} \longrightarrow (\mathscr{H} \otimes \mathbb{R}^4) \oplus (E_\infty \otimes S^+)$$

$$B_x(h) = i(x_r - T_r)h \oplus P^* h.$$

A little manipulation shows that the ADHM equations (3.4.7) are equivalent to the 'Weitzenböck formula':

$$R_x^* R_x = B_x^* B_x \otimes 1, \tag{3.4.9}$$

in $\mathrm{End}(\mathscr{H} \otimes S^+)$. Suppose our solution is degenerate, so R_x has non-zero kernel for some x in $\mathbb{R}^4$. There is then a vector h in $\mathscr{H}$ with $B_x h = 0$, i.e.

$$P^* h = 0, \quad T_i h = x_i h.$$

This gives a decomposition of $\mathscr{H}$ as $\mathscr{H}' \oplus \mathscr{H}''$ with $\mathscr{H}' = \mathbb{R} \cdot h$ and $\mathscr{H}''$ the orthogonal complement, such that condition (*a*) holds. If (*b*) holds we are done, otherwise we continue to decompose $\mathscr{H}''$ until this condition is met.

This result can be equivalently stated as follows. We have an obvious family of solutions to our equations (3.4.7) with $P = 0$ and T_i commuting (hence simultaneously diagonalizable) symmetric matrices: $T_i = \mathrm{diag}(x_i^j)$ say. By regarding the eigenvalues x_i^j as coordinates of points x^j in $\mathbb{R}^4$ we obtain an unordered 'multiset' (or l-tuple) $(x^1, \ldots, x^l)$ of points in $\mathbb{R}^4$—the 'spectrum' of T. Lemma (3.4.8) asserts that the general solution to the algebraic equations (3.4.7) is a direct sum of a diagonal solution of this kind and a non-degenerate solution.

Corollary (3.4.10). *The set* $\bar{M}_k$ *of solutions of the algebraic conditions* (3.4.7), *up to equivalence, is naturally identified with the disjoint union:*

$$M_k \cup \mathbb{R}^4 \times M_{k-1} \cup s^2(\mathbb{R}^4) \times M_{k-2} \cup \ldots \cup s^k(\mathbb{R}^4).$$

(Here $s^l(\mathbb{R}^4)$ denotes the l-fold symmetric product of $\mathbb{R}^4$.)

3.4.3 Coordinates on an open set in the moduli space

The ideas in Lemma (3.4.8) can be extended to give more explicit information about the actual moduli spaces M_k. We start with the fully diagonalizable solutions, parametrized by the symmetric product $s^k(\mathbb{R}^4)$. Fix such a point $(x^1, \ldots, x^k)$, with all the x^j distinct, and put $T_i = \mathrm{diag}(x_i^j)$. These satisfy the equations (3.4.7), with $P = 0$, since they all commute; but the non-degeneracy condition (3.4.6) fails at the points x^j. We shall find nearby solutions $(T_i + t_i, P)$, with t_i and P small, which are non-degenerate, and thus describe an open set in M_k.

Because we have the freedom to change basis in $\mathscr{H}^R$ by an element of $O(k)$ it is necessary to impose a 'gauge fixing' condition to have an effective parametrization of solutions. A suitable procedure to use is the formal analogue of the Coulomb gauge fixing condition of Chapter 2, with respect to the action, not of the gauge group on connections, but of the group $O(k)$ on the matrix data. This condition is

$$[T_0, t_0] + [T_1, t_1] + [T_2, t_2] + [T_3, t_3] = 0. \tag{3.4.11}$$

It is a simple application of the implicit function theorem in finite dimensions to show that every system of matrices $T_i + t_i$, with t_i small, is conjugate by $O(k)$ to a system satisfying (3.4.11), and that this element is unique up to the stabilizer of (T_i) in $O(k)$. (This is the standard result for the orthogonal slice to a group orbit.) The linearization of the ADHM equation (3.4.7) about the solution $(T_i, 0)$ is

$$[T_1, t_2] - [T_2, t_1] + [T_3, t_4] - [T_4, t_3] = 0 \tag{3.4.12}$$

together with the two cyclic permutations of this. It is helpful here to introduce the algebra of quaternions and put

$$\mathbf{T} = T_1 + \mathbf{i}T_2 + \mathbf{j}T_3 + \mathbf{k}T_4, \quad \mathbf{t} = t_1 + \mathbf{i}t_2 + \mathbf{j}t_3 + \mathbf{k}t_4.$$

Then the linearized equations (3.4.12) and the slice condition (3.4.11) are together equivalent to the single equation:

$$[\mathbf{T}, \mathbf{t}] = 0,$$

where the commutator of quaternionic matrices is defined by

$$[qA, rB] = qr[A, B]$$

for q, r in $\mathbb{H}$ and A, B real. What is made clear is that, since the points x^j are distinct, the map $\mathbf{t} \to [\mathbf{T}, \mathbf{t}]$ is an isomorphism from the space of symmetric quaternion matrices with vanishing diagonal elements to the space of skew symmetric quaternion matrices. So by the implicit function theorem we have:

Lemma (3.4.13). *There exists a $\delta > 0$ such that for all P with $|P| < \delta$ there are unique off-diagonal symmetric matrices $t_i = t_i(P)$ with $|t_i| \leq \text{const.}\ |P|^2$, such that $(T_i + t_i, P)$ satisfies the ADHM equations* (3.4.7).

The solutions constructed in this way are not all distinct; the 'gauge fixing' condition (3.4.11) is preserved by the symmetries of $\mathscr{H}^R$ of the form $\text{diag}(\pm 1, \pm 1, \ldots, \pm 1)$. If we now write P as a $k \times 4$ matrix and call its rows $P^1, \ldots, P^k$ then the effect of such a symmetry is to replace P^j by $\pm P^j$. So the effective parameters are k copies of the quotient of the δ-ball in $\mathbb{R}^4$ by ± 1. The non-degeneracy condition is satisfied precisely when each of the P^j is non-zero. We have then a description of an open set in the moduli space M_k. In fact, coming from our present direction it is more natural to consider first a larger space $\tilde{M}_k$, consisting of equivalence classes of triples (E, A, ρ), where A is an ASD connection on an $SU(2)$ bundle E with $c_2(E) = k$, and $\rho: E_\infty \to S^-$ is an $SU(2)$ isomorphism.

Proposition (3.4.14). *For each set of distinct points $x^1, \ldots, x^k$ in $\mathbb{R}^4$ the above construction produces a family of solutions (E, A, ρ) parametrized by*

$$\prod_{j=1}^{k} ((0, \delta) \times SO(3)).$$

As the x^j vary in small, disjoint open sets $U^j \subset \mathbb{R}^4$ we obtain an open set in $\tilde{M}_k$ parametrized by

$$\prod_{j=1}^{k} (U^j \times (0, \delta) \times SO(3)).$$

(Here we identify $(0, \delta) \times SO(3)$ with the quotient of the punctured ε-ball in $\mathbb{R}^4$ by ± 1; i.e. we identify $SO(3)$ with $\mathbb{RP}^3$.)

The moduli space M_k is the quotient of $\tilde{M}_k$ by an action of $SU(2)$ (the change in the trivialization ρ). In fact, since the automorphism -1 of E_∞ extends to an automorphism of E, it is only the quotient $SO(3) = SU(2)/\pm 1$ which acts effectively. In the description of Proposition (3.4.14) this action corresponds to right multiplication on the $SO(3)$ factors.

3.4.4 *Interpretation of the completion*

The completion $\bar{M}_k$ of the k-instanton moduli space has appeared naturally from the ADHM construction by relaxing the non-degeneracy conditions. It has however a direct interpretation in terms of the connections over $\mathbb{R}^4$. This combines with a direct interpretation of the parameters defined in Proposition (3.4.14) on an open set in M_k, the intersection of M_k with a neighbourhood of a point in $s^k(\mathbb{R}^4) \subset \bar{M}_k$.

In the case when $k = 1$, we have already seen that the point x^1 in $\mathbb{R}^4$ represents the centre of the instanton, the centre of mass of the curvature density, while the scale $\lambda = |P|$ represents the spread of the curvature density function. In the larger space $\tilde{M}_k$, the remaining $SO(3)$ parameter represents the framing ρ at infinity. We shall now explain that, in general, the k-instantons parametrized in Proposition (3.4.14) can be viewed as a 'superposition' of k copies of the one-instanton, centred at the points x^j, and with small scales $\lambda^j = |P^j|$. This is quite easy to see from the ADHM construction. The basic lemma we need is:

Lemma (3.4.15). *Let* $\Delta: V_1 \oplus V_2 \to U_1 \oplus U_2$ *be a linear map between Euclidean vector spaces, written as*

$$\Delta = \begin{pmatrix} \Delta_{11} & \Delta_{12} \\ \Delta_{21} & \Delta_{22} \end{pmatrix}$$

with $\Delta_{pq} \in \mathrm{Hom}(V_q, U_p)$. *Suppose that* Δ_{11} *is invertible and* Δ_{22} *is surjective, with right inverse* Δ_{22}^{-1}. *Let* $\lambda \ll 1$ *and suppose that*

$$\|\Delta_{11}^{-1}\Delta_{12}\| \leq \lambda,$$

$$\|\Delta_{22}^{-1}\Delta_{21}\| \leq \text{const.}$$

Then the kernel of Δ *is close to that of* Δ_{22}: *the distance between them (measured by the corresponding projections) is* $O(\lambda)$.

The proof of (3.4.15) is left to the reader. We shall, in addition, need a version of the statement for families of maps, which we also leave the reader to formulate.

We apply this lemma to the maps $R_x = \Gamma_x \oplus P$, say, whose kernel defines E_x. There are two situations to consider. First we consider the ADHM description away from the points x^j. The key point is that, so long as x stays away from the points x^j, say $|x - x^j| > \eta$, the operator Γ_x is invertible and the norm of its inverse is controlled:

$$\|\Gamma_x^{-1}\| \leq \text{const. } \eta^{-1}.$$

Taking $\Delta = \Gamma_x \oplus P$ with $V_1 = H \otimes S^-$, $V_2 = E_\infty$ and $U_2 = 0$, we deduce that the subbundle $E \subset (\mathscr{H} \otimes S^- \oplus E_\infty)$ is close to the constant subbundle E_∞ over this subset, W say, of $\mathbb{R}$, once $\lambda = \max \lambda^j$ is small. In fact as λ tends to zero the subbundle will converge in $C^\infty(W)$. It follows that the induced connection will converge, over W, to the trivial product connection. We have then:

Proposition (3.4.16). *Over the open set $W = \mathbb{R}^4 \setminus \bigcup_j B_\eta(x^j)$ there is a trivialization $E = S^- \times W$ in which the connection matrix, with all its derivatives, tends to zero with λ.*

We now study the situation in a small ball $B_\eta(x^j)$. We assume η is much less than the distance between the centres x^j in $\mathbb{R}^4$. Let $\mathscr{H}_j \subset \mathscr{H}$ be the span of the jth basis vector and put

$$V_1 = (\mathscr{H}^j)^\perp \otimes S^-, \quad V_2 = \mathscr{H}^j \otimes S^- \oplus E_\infty$$

$$U_1 = (\mathscr{H}^j)^\perp \otimes S^+, \quad U_2 = \mathscr{H}^j \otimes S^+.$$

Then for x in $B_\eta(x^j)$ we have

$$\|\Delta_{11}^{-1}\| \leq \text{const. } \delta^{-1},$$

$$\|\Delta_{22}^{-1}\| \leq \text{const. } \lambda^{-1}, \quad \text{(right inverse)}$$

$$\|\Delta_{12}\| \leq \text{const. } \lambda,$$

$$\|\Delta_{21}\| \leq \text{const. } \lambda^2.$$

It follows then from the lemma that on this ball the bundle E is approximated by the kernel of Δ_{22}. But Δ_{22} is the map:

$$\gamma(x^j - x) \oplus P^j : S^- \oplus E_\infty \longrightarrow S^+,$$

which defines the one-instanton with centre x^j and scale λ^j. The situation is slightly more complicated here, since this 'limit' also depends on λ^j. To get a precise statement we rescale the ball $B_\eta(x^j)$ by a factor λ^j. There is no loss in supposing $x^j = 0$, so we consider the bundle E' defined by the ADHM data $(\lambda^j)^{-1}\Gamma$, $(\lambda^j)^{-1}P$, which is transformed into E by the substitution $x \to \lambda_j x$. Transforming the bounds of (3.4.16) we see that E' converges in C^∞ on compact subsets of $\mathbb{R}^4$ to the basic instanton with centre 0 and scale 1, as $\lambda \to 0$. We have then:

Proposition (3.4.17). *Over a ball $B_\eta(x^j)$, the connection approximates the framed one-instanton corresponding to the data (x^j, P^j), the framing being provided by the trivialization of* (3.4.16), *and the approximation holding in the sense that after rescaling by factor $(\lambda^j)^{-1}$ about x^j, the connection converges on compact sets to the one-instanton with scale* 1.

We see then that the addition of the factor $s^k(R^4)$ in the completed moduli space corresponds to adding 'point instantons', where the scale has shrunk to zero. There is a similar interpretation, which can be obtained in just the same way from the ADHM description, of the intermediate pieces $s^l(\mathbb{R}^4) \times M_{k-l}$.

The connections in M_k near such a piece are formed by superimposing l small instantons on a background connection with Chern class $k - l$.

3.4.5 The case of the torus

We have seen that in the ADHM construction degenerate solutions of the ADHM matrix equations correspond to 'ideal instantons' where the curvature density has become concentrated into δ-functions. There is a similar story for the Fourier transform of ASD connections over the four-torus T. In place of the non-degeneracy condition for matrices we have the WFF condition for connections. Let us write

$$\bar{M}_{k,r}(T)$$

for the set of pairs $([A]; (x_1, \ldots, x_l))$ where A is an ASD connection on a bundle E with rank r and $c_2 - c_1^2 = k - l$, and $(x_1, \ldots, x_l)$ is an l-tuple of (unordered) points of T. The bundle and connection can be decomposed as

$$E = L_{\xi_1} \oplus \ldots \oplus L_{\xi_s} \oplus E_0 \quad (s \geq 0),$$

where the induced connection A_0 on E_0 is WFF. We define the Fourier transform of this pair to be the element of $\bar{M}_{r,k}(T^*)$:

$$([L_{x_1} \oplus \ldots \oplus L_{x_l} \oplus \hat{A}_0]; (-\xi_1, \ldots, -\xi_s)),$$

where $\hat{A}_0$ is the ordinary Fourier transform defined in Section 3.2 (and our notation confuses bundles and connections). It then follows immediately from the ordinary inversion theorem that:

Proposition (3.4.18). *The Fourier transform gives a bijection between $\bar{M}_{k,r}(T)$ and $\bar{M}_{r,k}(T^*)$.*

Now in Chapter 4 we will define a topology on $\bar{M}_{k,r}(T)$ (and show that this space is compact). It can be regarded as a stratified space in two different ways: either according to the number s of flat factors or according to the number l of points. The Fourier transform interchanges these stratifications. It can be shown that, with respect to the topology of Chapter 4, the Fourier transform is a homeomorphism between the compactified spaces. Part of the proof can be based on the ideas used above: for example, if the connection A on $E \to T$ is close to the reducible connection on $L_\xi \oplus E_0$ then it can be deduced from Lemma (3.4.16), applied to the Dirac operator, that $\hat{A}$ is close to the ideal connection $(\hat{A}_0; -\xi)$. One also needs to prove the converse: if A is close to an ideal connection (A', x) then the Fourier transform is close to being reducible as $L_x \oplus \hat{A}'$. The proof of this second part fits into the framework of the excision principle for differential operators discussed in Chapter 7; see in particular Section 7.1.5.

Notes

The material in this chapter is fairly self-contained and we do not give references section-by-section. We refer to Hitchin (1974) for general facts about Dirac operators, including the transformation formula under conformal changes, and the Lichnerowicz–Weitzenböck formula. The transformation law for the conformal Laplacian can easily be obtained from the formula for the variation in scalar curvature; see, for example, Aubin (1982).

The ADHM description was first discovered using twistor methods which go back to Ward (1977). In general the twistor space of a four-manifold is the space of compatible complex structures on the tangent spaces, the two-sphere bundle $S(\Lambda^+)$ of Section 1.1.7. The twistor space of S^4 is $\mathbb{CP}^3$ and Ward showed that there is a one-to-one correspondence between ASD connections over open sets in S^4 and certain holomorphic bundles over the corresponding open sets in $\mathbb{CP}^3$. So the problem of describing all instantons is reduced to the description of holomorphic bundles over projective three-space. In this framework the solutions are obtained from monads over $\mathbb{CP}^3$, and then the Ward correspondence is used to pass back to the four-sphere. See Atiyah (1979) and Atiyah *et al.* (1978*a*, *b*) for this part of the theory.

It was soon realized that the ADHM description was in many ways easier to understand in the framework of Euclidean geometry on $\mathbb{R}^4$, rather than conformal geometry on S^4. The description of the ADHM data along the lines we have given in the text appeared in Corrigan and Goddard, 1984. This approach was taken much further by Nahm who outlined a proof of the main result by a direct calculation, and extended the ADHM construction to obtain 'monopoles'—ASD connections on $\mathbb{R}^4$ invariant under translation in one direction; see Jaffe and Taubes (1980), Nahm (1983) and Atiyah and Hitchin (1989).

Meanwhile, in algebraic geometry, the Fourier transform for holomorphic bundles over complex tori (of any dimension) was developed by Mukai (1981), who gave a general inversion theorem in the 'derived category'. The analogous construction for connections over four-tori was carried through, following Nahm's general scheme, by Schenk (1988) and Braam and van Baal (1989). In general, one can hope to have a form of inversion theorem for ASD connections on $\mathbb{R}^4$ invariant under any group of isometries.

The point of view taken in the text lies roughly midway between these different approaches. We give a framework within which one can verify Nahm's direct calculations, but the calculations themselves are incorporated in well-known general facts of complex geometry. We do not use twistors explicitly but our appeal to the different complex structures is more-or-less equivalent to the use of holomorphic bundles on the twistor space. For the inversion theorem over the torus our spectral-sequence proof is essentially identical to Mukai's, translated into more elementary terms, and using the Dolbeault model for cohomology. For the ADHM theorem, our spectral sequence is closely related to the 'Beilinson spectral sequence' associated with a bundle over projective space; see Okonek *et al.* (1980). In fact these latter spectral sequences enter in two ways. On the one hand they yield a monad description of the Ward bundle over $\mathbb{CP}^3$, while on the other hand they yield a description of bundles over $\mathbb{CP}^2$ trivial on a line. Our monad description in Section 3.3.4 is essentially the monad description of such a bundle over $\mathbb{CP}^2$, regarded as a complex compactification of $\mathbb{R}^4$ (Okonek *et al.* 1980, Chapter II.3). The use of decay conditions on $\mathbb{R}^4$ takes the place of twisting by line bundles over $\mathbb{CP}^2$.

We have not found space in the text to discuss two further facets of the theory. The first involves the relation between holomorphic bundles and ASD connections. In both of the cases we consider, if we fix a complex structure on the base an ASD connection is actually determined by the holomorphic bundle which it defines, via its $\bar{\partial}$ operator. In the case of tori this fact is contained in the general theory of Chapter 6. For the base space $\mathbb{R}^4$ one gets a one-to-one correspondence between 'framed' instantons and holomorphic bundles over $\mathbb{CP}^2$ trivialized over the line at infinity. This can be viewed as an extension of the theory of

Chapter 6 to non-compact base manifolds; on the other hand, it can be proved by elementary means using the monad description (Donaldson 1984*a*). We shall see the key step in this proof in Section 6.5.2, where we find part of the ADHM equations appearing as a 'zero moment map' condition relative to the symmetry group $U(k)$. This approach has been extended by King (1989) to self-dual connections over $\mathbb{CP}^2$. There is also a version of the theory for monopoles (Donaldson 1984*b*).

The other topic concerns the natural metrics on the moduli spaces, defined by the L^2 metrics on one-forms. These are hyperkähler metrics, compatible with an action of the quaternions. Again this property can be derived within a general theory of hyperkähler moment maps (Hitchin *et al.* 1987). Suitable classes of ASD solutions give new hyperkähler metrics on some well-known manifolds (Atiyah and Hitchin 1989; Kronheimer 1989, 1990).

One also has the beautiful fact that, in the torus case, the metric on the moduli space is preserved by the Fourier transform (Braam and Van Baal 1989). A version of this statement in the holomorphic situation, involving the holomorphic symplectic form, is given by Mukai (1984). An analogue for the ADHM construction has been obtained by Maciocia (1990). Again, one expects such results to hold for other translation-invariant solutions. For the case of monopoles see the discussion by Atiyah and Hitchin (1989). When considering instantons on $\mathbb{R}^4$ some analytical points need to be checked to see that the L^2 metric is well defined; see Taubes (1983). The special features which arise, due to the non-compactness of $\mathbb{R}^4$, are similar to those appearing in our version of the Hodge theory for the $\bar{\partial}$-complex in Section 3.3.4.

4

YANG–MILLS MODULI SPACES

In this chapter we introduce moduli spaces of ASD Yang–Mills connections and develop some of their basic properties. Let E be a bundle over a compact, oriented, Riemannian four-manifold X. At the level of sets, the moduli space M_E is defined to be the set of gauge equivalence classes of ASD connections on E: that is, we identify connections which are in the same orbit under the action of the bundle automorphism group $\mathcal{G}$. The moduli space has a natural topology, induced by the topology in the space $\mathcal{A}$ of connections. We shall see that the moduli space is, unlike $\mathcal{A}$, a finite-dimensional object. In addition to its intrinsic topology it carries the structure of a real analytic space. In Section 4.3 we shall see that for most purposes it may be assumed that the moduli space is a smooth manifold, except perhaps for some singular points associated to reducible connections. In Section 4.4 we shall return to the techniques used in Chapter 2 and define a natural compactification of the moduli space. For this we give a proof of another important result of Uhlenbeck: the removability of point singularities for finite-energy ASD connections. We begin this chapter by giving a number of explicit examples of moduli spaces, for standard four-manifolds X, which serve to motivate the general theory.

4.1 Examples of moduli spaces

4.1.1 Example (i). One-instantons over S^4

We have met this example in Chapter 3. Let E be an $SU(2)$ bundle over S^4 with $c_2(E) = 1$. The moduli space $M = M_E$, described from the point of view of $\mathbb{R}^4$, is $\mathbb{R}^4 \times \mathbb{R}^+$, the parameters consisting of a 'centre' and 'scale' which are acted on transitively by the translations and dilations of $\mathbb{R}^4$. We refer to Section 3.4.1 for the explicit formulae.

When viewed from the point of view of the four-sphere it is natural to describe the moduli space as the open five-ball B^5. The full group $\mathrm{Conf}(S^4)$ of orientation-preserving conformal transformations of S^4 acts on the moduli space; the connection $A_{0,1}$ is preserved by the subgroup $SO(5)$ of isometries (in fact this can be identified with the standard connection on the spin bundle S^- of S^4). Thus the moduli space is the open five-ball, viewed as the homogeneous space $\mathrm{Conf}(S^4)/SO(5)$—the space of round metrics in the conformal class. In this description the 1-parameter family $A_{0,\lambda}$, $\lambda > 0$, corresponds to a diameter in the five-ball.

For the next three examples we take the base space X to be the complex projective plane with its standard metric (the Fubini–Study metric). Unlike the four-sphere, the projective plane has no orientation-reversing isometry, so from the point of view of ASD connections we get two distinct oriented manifolds $\mathbb{CP}^2$ and $\overline{\mathbb{CP}}^2$. (Of course, ASD connections on $\mathbb{CP}^2$ are the same as self-dual connections on $\overline{\mathbb{CP}}^2$).

4.1.2 *Example (ii). One-instantons over* $\overline{\mathbb{CP}}^2$

The moduli space associated to an $SU(2)$ bundle E over $\overline{\mathbb{CP}}^2$ with $c_2(E) = 1$ is in many ways similar to that in Example (i). The solutions can be described explicitly using monad constructions similar to those in Chapter 3, exploiting the special geometry of the projective plane. In place of the five-dimensional ball we have the open cone over $\overline{\mathbb{CP}}^2$. The group $PU(3)$ of isometries of the projective plane acts in the obvious way, transitively on the 'level sets' in the cone. To describe the solutions it suffices to treat those corresponding to a generator of the cone, determined by a real parameter $t \in [0, 1)$. We take standard coordinates on an affine patch in $\overline{\mathbb{CP}}^2$ centred on the base of this generator (missing out a line at infinity). The corresponding connections can then be given by the connection matrices:

$$J_t = \frac{1}{1 + |x|^2 - t^2} (\theta_1 \mathbf{i} + t\theta_2 \mathbf{j} + t\theta_3 \mathbf{k}).$$

(Compare (3.4.2).)

4.1.3. *Example (iii). ASD SU(2) connections over* $\mathbb{CP}^2$

The projective plane is, of course, a complex algebraic surface and the Fubini–Study metric is a Kähler metric. As we shall explain in Chapter VI, this allows the moduli spaces of ASD connections to be described by algebro-geometric methods. For $SU(2)$ bundles E it turns out that the moduli space is empty if $c_2 = 1$. For $c_2 = 2$ the moduli space can be described as follows. Introduce the 'dual' projective plane $\mathbb{P}^*$; points of $\mathbb{P}^*$ correspond to complex projective lines in the original plane $\mathbb{P} = \mathbb{CP}^2$ and vice versa. The conics in $\mathbb{P}^*$ (defined by homogeneous equations of degree 2) are parametrized by a copy of $\mathbb{CP}^5$ (there are six coefficients in the defining equation). The moduli space M can be identified with the open subset of $\mathbb{CP}^5$ corresponding to the non-singular conics. Thus M is a smooth manifold of real dimension ten.

The non-singular conics in $\mathbb{P}$ and $\mathbb{P}^*$ can certainly be identified by the classical duality construction in projective geometry. As we shall see, however, the use of the dual plane leads to a more natural compactification of the moduli space. Observe here that the complement $\mathbb{CP}^5 \backslash M$, representing the singular conics, can be identified with the symmetric product $s^2(\mathbb{P})$ of unordered pairs of points in $\mathbb{P}$. For a singular conic is a pair of lines in $\mathbb{P}^*$ (in which we include a 'double line') and lines in $\mathbb{P}^*$ are points in $\mathbb{P}$.

We should emphasize that this example differs from the previous two in that we do not have explicit formulae for the connections represented by points in the moduli space. Our explicit description of the moduli space comes from the abstract existence theorem which we prove in Chapter 6. For larger values of c_2 there are non-empty moduli spaces, which can also be described algebro-geometrically, but it becomes harder to give explicit descriptions as c_2 increases.

4.1.4 Example (iv). ASD SO(3) connections over $\mathbb{CP}^2$

In place of the structure group $SU(2)$ we can consider ASD connections on $SO(3)$ bundles over $\mathbb{CP}^2$ with non-zero Stiefel–Whitney class w_2. Since $H^2(\mathbb{CP}^2)$ has rank one, there is just one such choice of w_2, and we get another family of ASD connections on bundles E with $p_1 = -(3 + 4j)$. These can also be attacked by algebro-geometric methods. When $j = 0$ the moduli space is a single point, corresponding to the standard connection on $\Lambda^-_{\mathbb{CP}^2}$. When $j = 1$ the moduli space can be identified with set of unordered pairs of *distinct* points in $\mathbb{CP}^2$. As in Example (iii) there are more complicated moduli spaces for larger values of j.

4.1.5 Example (v). ASD SU(2) connections over $S^2 \times S^2$

We now consider the four-manifold $S^2 \times S^2$, with its standard Riemannian metric. This admits an orientation reversing isometry, so self-dual and anti-self-dual connections are equivalent. On the other hand, $S^2 \times S^2$ is a Kähler manifold. As we mentioned in Chapter 1, it can be thought of as a non-singular quadric surface Q in $\mathbb{CP}^3$. This means that algebraic geometry can again be brought to bear on the description of moduli spaces. As for $\mathbb{CP}^2$, the moduli space of ASD connections on an $SU(2)$ bundle with $c_2 = 1$ is empty. For $c_2 = 2$ it turns out that the moduli space can be identified with a set M of non-singular quadrics in $\mathbb{CP}^2$. This set contains a distinguished point, the quadric Q itself; the remaining points consists of all non-singular quadrics which intersect Q in four lines (one pair from each of the rulings of Q, and we allow a line to be counted with multiplicity two). This moduli space is a singular complex variety, embedded in the ten-dimensional projective space of quadrics. If Q' is an element of M, different from Q, then all the quadrics in the complex line $Q' + tQ$ $(t \in \mathbb{C})$ meet Q in four lines. Thus M is an open subset of a complex projective cone K (i.e. a variety ruled by projective lines through a common vertex). The complement $K \backslash M$ consists of the singular conics which meet Q in four lines. It is easy to see that these singular conics are just unions of planes $P_1 \cup P_2$ with each P_i tangent to Q at a unique point x_i. So, similar to Section 4.1.3, the complement of M in its natural compactification K can be identified with the symmetric product $s^2(S^2 \times S^2)$, the set of unordered pairs x_1, x_2.

4.2 Basic theory

There are two distinct steps in the definition of the moduli space M: find the solutions to the ASD equation, then divide by the action of the gauge group. It is convenient to consider these separately and in the reverse order. Thus we begin by constructing a space $\mathscr{B}$ of all gauge equivalence classes of connections on E, orbits of the gauge group $\mathscr{G}$ in $\mathscr{A}$. Then we describe the moduli space within $\mathscr{B}$. The whole discussion here consists of an exercise in calculus and differential topology, albeit in infinite-dimensional spaces.

4.2.1 The orbit space

To construct the space of equivalence classes of connections it is most convenient, and standard practice, to work in the framework of Sobolev spaces. The Sobolev embedding theorem in four dimensions tells us that for any $l > 2$ the Sobolev space L^2_l consists of continuous functions. It is then easy to define the notion of a locally L^2_l map from a domain in the four-manifold X to the structure group G of the bundle we wish to consider. For example if G is a unitary group we can regard these maps as a subset of the matrix valued functions of class $L^2_{l,\mathrm{loc}}$, the closure of the smooth maps to the unitary group in the L^2_l norm. Inversion and multiplication of these L^2_l, G-valued functions is defined pointwise. We can then define an L^2_l G-bundle to be a bundle given by a system of L^2_l transition functions. Similarly we can define connections on such a bundle, given in local trivializations by L^2_{l-1} connection matrices, and these have curvature in L^2_{l-2}, thanks to the multiplication $L^2_{l-1} \times L^2_{l-1} \to L^2_{l-2}$, $l > 2$. In fact this multiplication property also holds with $l = 2$, and for many purposes it would be most convenient if one could work throughout with L^2_1 connections. Unfortunately, however, it is not really possible to define the notion of an L^2_2 bundle in a very satisfactory way, so we will stay in the range $l > 2$.

Any L^2_l bundle (indeed, any topological bundle) over the smooth four-manifold X admits a compatible smooth structure, and we can alternatively regard the L^2_{l-1} connections as those which differ from a smooth connection by an L^2_{l-1} section of $T^*_X \otimes \mathfrak{g}_E$. Whichever way one proceeds one obtains a modified version of the usual differential-geometric definitions, in which smooth functions are replaced by those in an appropriate Sobolev space. In this chapter we write $\mathscr{A}$ for the space of L^2_{l-1} connections on a bundle E, and $\mathscr{G}$ for the group of L^2_l gauge transformations. The index $l > 2$ is suppressed; as we shall see, the ultimate description of moduli spaces will be completely independent of this choice, which is essentially an artifact of the abstract machinery employed.

We define $\mathscr{B}$ to be the quotient space

$$\mathscr{B} = \mathscr{A}/\mathscr{G}, \tag{4.2.1}$$

with the quotient topology, and we write $[A]$ for the equivalence class of a connection A—a point in $\mathscr{B}$. The L^2 metric on $\mathscr{A}$,

$$\|A - B\| = \left(\int_X |A - B|^2 \, \mathrm{d}\mu \right)^{1/2}, \tag{4.2.2}$$

is preserved by the action of $\mathscr{G}$, so descends to define a 'distance function' on $\mathscr{B}$:

$$\mathrm{d}([A], [B]) = \inf_{g \in \mathscr{G}} \|A - g(B)\|. \tag{4.2.3}$$

Lemma (4.2.4). *d is a metric on $\mathscr{B}$.*

Proof. The only non-trivial point is to show that $\mathrm{d}([A], [B]) = 0$ implies $[A] = [B]$. This is a manifestation of the general property noted in Section 2.3.7. Suppose $\mathrm{d}([A], [B]) = 0$ and let B_α be a sequence in $\mathscr{A}$, all gauge equivalent to B, converging in L^2 to A. We have to show that A and B are gauge equivalent. If $B_\alpha = u_\alpha(B)$ we have

$$\mathrm{d}_B u_\alpha = (B - B_\alpha) u_\alpha.$$

The u_α are uniformly bounded since the structure group G is compact and this relation shows that the first derivatives $\mathrm{d}_B u_\alpha$ are bounded in L^2. So, taking a subsequence, we can suppose the u_α, regarded as sections of the vector bundle End E, converge weakly in L^2_1 and strongly in L^2 to a limit u. Moreover u satisfies the linear equation,

$$\mathrm{d}_B u = (B - A) u.$$

For if ϕ is any smooth test section of End E we have

$$\langle \mathrm{d}_B u, \phi \rangle = \lim \langle \mathrm{d}_B u_\alpha, \phi \rangle = \lim \langle (B - B_\alpha) u_\alpha, \phi \rangle = \langle (B - A) u, \phi \rangle,$$

since $B_\alpha u_\alpha$ converges to Au in L^1. This equation for u is, in a rather trivial way, an overdetermined elliptic equation with L^2_{l-1} coefficients and, by bootstrapping, we see that u is in fact in L^2_l. The proof is completed by showing that u lies in the subset of unitary sections of End E, and this fact is rather obvious on a moment's reflection. Indeed, let K be a closed subset of some $\mathbb{R}^n$ and let $D_K: \mathbb{R}^n \to \mathbb{R}$ be the function which assigns to a point its distance from K. Then $|D_K(x) - D_K(y)| \leq |x - y|$. So if f_i is a sequence of maps into $\mathbb{R}^n$ which converge in L^2 to a limit f, the composites $D_K f_i$ converge in L^2 to $D_K f$. In particular, if the f_i map into K, then the limit f maps into K almost everywhere. We apply this to $K = G$, embedded in the vector space of matrices, and with maps representing the u_α in local trivializations of E.

We deduce in particular from this lemma that $\mathscr{B}$ is Hausdorff in the quotient L^2_{l-1} topology (finer than the L^2 topology). We now move on to study the local structure of $\mathscr{B}$. The key fact here is that $\mathscr{G}$ is an infinite-dimensional Lie group, modelled on a Banach space. This follows from straightforward properties of Sobolev spaces, exploiting the continuity of the elements of $\mathscr{G}$. We can use the exponential map $\exp:\Omega^0(\mathfrak{g}_E) \to \mathscr{G}$, defined pointwise in the fibres of End E, to construct a chart in which a neighbourhood of 1 in $\mathscr{G}$ is identified with the small L_l^2 sections of $\mathfrak{g}_E$. If an element u of $\mathscr{G}$ is close to the identity in L_l^2 it is also close to the identity pointwise and thus lies in an open set on which the pointwise exponential map is invertible. Moreover the 'logarithm' of u is also in L_l^2. (It is at this point that one runs into severe difficulties if one attempts to construct a quotient of the L_1^2 connections by the L_2^2 gauge transformations.) At bottom, what is being used here is the composition property, if $P: \mathbb{R}^n \to \mathbb{R}^m$ is a smooth function and Z is compact then composition on the left with P gives a smooth map from $L_l^2(Z, \mathbb{R}^n)$ to $L_l^2(Z, \mathbb{R}^m)$, provided we are in a range where a Sobolev embedding $L_l^2 \to C^0$ holds.

By similar arguments one sees that the action, $\mathscr{G} \times \mathscr{A} \to \mathscr{A}$, of $\mathscr{G}$ on $\mathscr{A}$ is a smooth map of Banach manifolds. At a point A of $\mathscr{A}$ the derivative of the action in the $\mathscr{G}$ variable is minus the covariant derivative:

$$-\,\mathrm{d}_A : \Omega^0(\mathfrak{g}_E) \longrightarrow \Omega^1(\mathfrak{g}_E).$$

(Here we are suppressing the Sobolev indices.) The description of the quotient is straightforward given one piece of extra data: the existence of topological complements for the kernel and image of this derivative. Such complements are supplied by the formal adjoint operator d_A^* used already in Chapter 2. Elliptic theory gives topological isomorphisms:

$$\Omega^1(\mathfrak{g}_E) = \operatorname{im} \mathrm{d}_A \oplus \ker \mathrm{d}_A^*, \quad \Omega^0(\mathfrak{g}_E) = \ker \mathrm{d}_A \oplus \operatorname{im} \mathrm{d}_A^*. \tag{4.2.5}$$

Of course, these spaces are orthogonal complements in the L^2 inner product, but the point is that the decompositions are compatible with the higher Sobolev structures.

For $A \in \mathscr{A}$ and $\varepsilon > 0$ we set:

$$T_{A,\varepsilon} = \{a \in \Omega^1(\mathfrak{g}_E) | \mathrm{d}_A^* a = 0, \|a\|_{L_{l-1}^2} < \varepsilon\}. \tag{4.2.6}$$

It now follows, purely as a matter of general theory, that a neighbourhood of $[A]$ in $\mathscr{B}$ can be described as a quotient of $T_{A,\varepsilon}$, for small ε. That is, every nearby orbit meets $A + T_{A,\varepsilon}$. Concretely, this amounts to solving the Coulomb gauge fixing condition, relative to A, as we have done in (2.3.4). To get a more precise statement we must pause to discuss the isotropy groups of connections under the action of $\mathscr{G}$.

4.2.2 Reducible connections

In general one says that a connection A on a G-bundle E is reducible if for each point x in X the holonomy maps T_γ of all loops γ based at x lie in some proper subgroup of the automorphism group Aut $E_x \cong G$. If the base space is connected we can, by a standard argument, restrict attention to a single fibre and we obtain a holonomy group $H_A \subset G$, or more precisely a conjugacy class of subgroups. It can be shown that this is a closed Lie subgroup of G. On the other hand we can define the isotropy group Γ_A of A in the gauge group $\mathcal{G}$:

$$\Gamma_A = \{u \in \mathcal{G} \mid u(A) = A\}. \tag{4.2.7}$$

Lemma (4.2.8). *For any connection A over a connected base X, Γ_A is isomorphic to the centralizer of H_A in G.*

Here we regard H_A and Γ_A as subgroups of Aut E_x for some base point x. We leave the proof of the lemma as an exercise. Note in particular that Γ_A always contains the centre $C(G)$ of G.

Now, as a closed subgroup of G, Γ_A is a Lie group. Its elements are the covariant constant sections of the bundle Aut E, and from this it is clear that the Lie algebra of Γ_A is the kernel of the covariant derivative d_A on $\Omega^0_X(\mathfrak{g}_E)$. Thus the isotropy group has positive dimension precisely when there are non-trivial covariant constant sections of $\mathfrak{g}_E$. The group Γ_A acts on $\Omega^1_X(\mathfrak{g}_E)$ and on $T_{A,\varepsilon}$. We have then:

Proposition (4.2.9). *For small ε the projection map from $\mathcal{A}$ to $\mathcal{B}$ induces a homeomorphism h from the quotient $T_{A,\varepsilon}/\Gamma_A$ to a neighbourhood of $[A]$ in $\mathcal{B}$. For a in $T_{A,\varepsilon}$, the isotropy group of a in Γ_A is naturally isomorphic to that of $h(a)$ in $\mathcal{G}$.*

The proof is a straightforward application of the implicit function theorem, using the argument of (2.3.15) to reduce to a quotient by gauge transformations which are close to the identity. We leave the details as an exercise.

Let us write $\mathcal{A}^*$ for the open subset of $\mathcal{A}$ consisting of connections whose isotropy group is minimal—the centre $C(G)$:

$$\mathcal{A}^* = \{A \in \mathcal{A} \mid \Gamma_A = C(G)\}.$$

Let $\mathcal{B}^* \subset \mathcal{B}$ be the quotient of $\mathcal{A}^*$. Proposition (4.2.9) asserts in particular that $\mathcal{B}^*$ is modelled locally on the balls $T_{A,\varepsilon}$ in the Hilbert spaces $\ker \mathrm{d}_A^* \subset L^2_{l-1}(\Omega^1_X(\mathfrak{g}_E))$. It is easy enough to show that these give charts making $\mathcal{B}^*$ into a smooth Hilbert manifold. This description breaks down at point of $\mathcal{B}\setminus\mathcal{B}^*$. However, the structure at these singular points has a familiar general form. We partition $\mathcal{B}$ into a disjoint union of pieces $\mathcal{B}^\Gamma$ labelled by the conjugacy classes of the isotropy groups Γ_A in G. For each connection A we have a decomposition:

$$\mathfrak{g}_E = V \oplus V^\perp, \tag{4.2.10}$$

where V is the set of elements fixed by Γ_A, and $V^\perp$ is the orthogonal complement. (In fact V is just the Lie algebra of the holonomy group.) The locally closed subset $\mathcal{B}^\Gamma$ is itself a Hilbert manifold, modelled on the space of

1-forms $\ker d_A^* \cap \Omega_X^1(V)$. The structure of $\mathscr{B}$ 'normal' to $\mathscr{B}^\Gamma$ is modelled on

$$\frac{\ker d_A^* \cap \Omega_X^1(V^\perp)}{\Gamma_A}.$$

Moreover there is a semicontinuity property: if a point $[A]$ lies in the closure of $\mathscr{B}^\Gamma$ then Γ_A contains Γ (or, more precisely, a representative from this conjugacy class). All of this can be summarized by saying that $\mathscr{B}$ is a *stratified space* with strata the $\mathscr{B}^\Gamma$. The appearance of such stratified spaces as the quotients of manifolds under group actions is quite typical in both finite and infinite dimensional problems and we should emphasize that in all of this we have only used the general, formal properties of the action on $\mathscr{A}$—the existence of Γ_A-invariant complements for the kernel and cokernel of the linearization d_A and the properness of the action.

It is customary to call the open subset $\mathscr{B}^*$ the manifold of *irreducible connections*. This is not strictly accurate in general since, as we have seen, it is the centralizer of the holonomy group that is relevant. For example, a connection on an $SU(n)$ bundle which happens to reduce to the subgroup $SO(n)$—embedded in the standard way—gives rise to a point of $\mathscr{B}^*$ if $n > 2$ since the centralizer is just the centre $\mathbb{Z}/n$ of $SU(n)$. With this said, however, we will in future just refer to irreducible connections, since in the cases of primary interest—connections on $SU(2)$ or $SO(3)$ bundles over a simply-connected manifold X—the two notions coincide. For $SU(2)$ the only possible reductions are to a copy of $S^1 \subset SU(2)$, or to the trivial subgroup (i.e. when A is the product connection). In the first case the reduction corresponds, in the framework of vector bundles, to a decomposition

$$E = L \oplus L^{-1}, \tag{4.2.11}$$

for a complex line bundle L over X. The corresponding decomposition of $\mathfrak{g}_E$ is

$$\mathfrak{g}_E = \underline{\mathbb{R}} \oplus L^{\otimes 2}. \tag{4.2.12}$$

For $SO(3)$ vector bundles in general, we consider decompositions of the form $\underline{\mathbb{R}} \oplus L$. In (4.2.12) the factor $\Omega_X^1(L^{\otimes 2})$ inherits a complex structure from that on $L^{\otimes 2}$, and $\Gamma_A \cong S^1$ acts by the square of the standard action. So in this case the structure of $\mathscr{B}$ normal to the singular stratum $\mathscr{B}^{S^1}$ is that of $\mathbb{C}^\infty/S^1$, a cone on an infinite-dimensional complex projective space.

The structure around the trivial connection is more complicated since three different strata are involved. The local model is a cone over a space which is itself singular. This illustrates the paradox that the simplest connection, the trivial product structure, is the most complicated from the point of view of the orbit space $\mathscr{B}$.

When working with $SU(2)$ or $SO(3)$ connections over simply connected manifolds we thus have a firm hold on all the reductions. Similarly we have a firm hold on the reducible ASD solutions. By (2.2.6) a line bundle L over the

Riemannian four-manifold X admits an ASD connection precisely when $c_1(L)$ is represented by an anti-self-dual 2-form, and if X is simply connected this connection is unique up to gauge equivalence. If we now start with an $SU(2)$ bundle E the reductions correspond to splittings $E \cong L \oplus L^{-1}$, and a necessary and sufficient condition for such an isomorphism is

$$c_2(E) = -c_1(L)^2. \tag{4.2.13}$$

In the $SO(3)$ case we have:

$$p_1(\underline{\mathbb{R}} \oplus L) = c_1(L)^2. \tag{4.2.14}$$

In sum we obtain:

Proposition (4.2.15). *If X is a compact, simply connected, oriented Riemannian four-manifold and E is an $SU(2)$ or $SO(3)$ bundle over X, the gauge equivalence classes of reducible ASD connections on E, with holonomy group S^1, are in one-to-one correspondence with pairs $\pm c$ where c is a non-zero class in $H^2(X; \mathbb{Z})$ with $c^2 = -c_2(E)$ or $p_1(E)$ respectively.*

We obtain pairs $\pm c$ because, in the $SU(2)$ case, there is complete symmetry between L, L^{-1}. Likewise in the $SO(3)$ case we have to choose a generator for the trivial factor $\underline{\mathbb{R}}$. In general, if we have a connection on a G-bundle E whose holonomy reduces to a subgroup H, there are different ways to obtain an H-bundle from E; the choices are parametrized by $N(H)/H$ where $N(H)$ is the normalizer of H in G.

4.2.3 The moduli space

We now turn to the other step in the construction of the moduli space: examining the solutions of the ASD equation $F^+(A) = 0$. Let us dispose of one point straight away. Our set-up in the previous section depended on the choice of a Sobolev space L^2_l. We temporarily denofe the corresponding orbit space by $\mathcal{B}(l)$, so for each $l > 2$ and a fixed bundle E we have a moduli space $M(l) \subset \mathcal{B}(l)$ of L^2_{l-1} ASD connections modulo L^2_l gauge transformations. *A priori* these depend, both as sets and topological spaces, on l; but in fact we have:

Proposition (4.2.16). *The natural inclusion of $M(l+1)$ in $M(l)$ is a homeomorphism.*

The essence of this is the assertion that if A is an ASD connection (on a C^∞ bundle) of class L^2_{l-1}, for $l > 2$, there is an L^2_l gauge transformation u such that $u(A)$ is in L^2_l, or indeed smooth. We know by (4.2.9) that there is an $\varepsilon > 0$ such that any L^2_{l-1} connection B with $\|A - B\|_{L^2_{l-1}} < \varepsilon$ can be gauge transformed into the horizontal (Coulomb gauge) slice through A; i.e. we can find u in L^2_l with

$$d^*_A(u^{-1}(B) - A) = 0.$$

By the symmetry of the Coulomb gauge condition, A is also in Coulomb gauge relative to $u^{-1}(B)$: that is, $d^*_{u^{-1}(B)}(A - u^{-1}(B)) = 0$; and by the invariance of the condition we have, writing $A' = u(A) = B + a$:

$$d_B^* a = 0.$$

Now the smooth connections are dense in the L^2_{l-1} topology, so we can choose B to be smooth. The difference 1-form a also satisfies

$$d_B^+ a + (a \wedge a)^+ = -F_B^+,$$

this being the ASD equation for A'. Thus $(d_B^* \oplus d_B^+)a$ lies in L^2_{l-1} (the curvature of B is smooth and the quadratic term $(a \wedge a)^+$ is in L^2_{l-1} by the multiplication results for Sobolev spaces). So, by the basic regularity results for the linear elliptic operator $d_B^* \oplus d_B^+$, which has smooth coefficients, we see that a is in L^2_l as desired.

This establishes the surjectivity of the inclusion map on the moduli spaces; the proof of injectivity is rather trivial and the fact that it gives a homeomorphism is left as an exercise. (In the next section we will prove a sharper regularity theorem (4.4.13) for the critical exponent $l = 2$.)

Thanks to this Proposition we can unambiguously refer to the moduli space $M = M_E$ with the induced (metrizable) topology and drop the Sobolev notation. We obtain local models for M within the local models for the orbit space $\mathscr{B}$ discussed above. Let A be an ASD connection and define:

$$\psi: T_{A,\varepsilon} \longrightarrow \Omega^+(\mathfrak{g}_E), \quad \psi(a) = F^+(A + a) = d_A^+ a + (a \wedge a)^+. \tag{4.2.17}$$

Let $Z(\psi) \subset T_{A,\varepsilon}$ be the zero set of ψ. The map h of (4.2.9) induces a homeomorphism from the quotient $Z(\psi)/\Gamma_A$ to a neighbourhood of $[A]$ in M.

4.2.4 Fredholm theory

Recall that a bounded linear map

$$L: U \to V$$

between Banach spaces is *Fredholm* if it has finite-dimensional kernel and cokernel. It follows that the kernel and image of L are closed and admit topological complements, so we can write:

$$U = U_0 \oplus F, \quad V = V_0 \oplus G, \tag{4.2.18}$$

where F and G are finite-dimensional and L is a linear isomorphism from U_0 to V_0. The index of L is the difference of the dimensions:

$$\operatorname{ind}(L) = \dim \ker L - \dim \operatorname{Coker} L = \dim F - \dim G.$$

The index is a deformation invariant, unchanged by continuous deformations of L through Fredholm operators (in the operator norm topology).

Many constructions from linear algebra in finite dimensions can be extended to Fredholm operators. If U and V are finite dimensional, the index is just the difference of their dimensions; roughly speaking a Fredholm operator gives a way to make sense of the difference of the dimensions of two infinite-dimensional Banach spaces. We shall see a number of illustrations of this idea later in the book. For the present we wish to develop an analogous description for certain non-linear maps, which we will apply in Section 4.2.5 to describe the Yang–Mills moduli space.

Let N be a connected open neighbourhood of 0 in the Banach space U. A smooth map $\phi: N \to V$ is called Fredholm if for each point x in N the derivative:

$$(D\phi)_x : U \to V$$

is a linear Fredholm operator. In this case the index of $(D\phi)_x$ is independent of x and is referred to as the index of ϕ.

Let ϕ be such a Fredholm map with index r and $\phi(0) = 0$. We wish to study ϕ locally, in an arbitrarily small neighbourhood of 0. So in this section we will regard maps as being equal if they agree on such a neighbourhood. Suppose first that $L = (D\phi)_0$ is surjective, so the index is the dimension of the kernel of L. The implicit function theorem in Banach spaces asserts that there is then a diffeomorphism f from one neighbourhood of 0 in U to another, such that

$$\phi \circ f = L.$$

We will just say that ϕ is *right equivalent* to the map L if they agree under composition on the right with a local diffeomorphism.

Now consider the general case when L is not necessarily surjective. We fix decompositions as in (4.2.18) and let $\phi' : N \to V_0$ be the composite of ϕ with the linear projection from V to V_0. Then the derivative of ϕ' at 0 is surjective by construction, so by applying a diffeomorphism f in a suitably small neighbourhood of 0 we can 'linearize' ϕ'. We obtain:

Proposition (4.2.19). *A Fredholm map ϕ from a neighbourhood of 0 is locally right equivalent to a map of the form*

$$\tilde{\phi} : U_0 \times F \longrightarrow V_0 \times G, \quad \tilde{\phi}(\xi, \eta) = (L(\xi), \alpha(\xi, \eta))$$

where L is a linear isomorphism from U_0 to V_0, F and G are finite-dimensional, $\dim F - \dim G = \operatorname{ind} \phi$, *and the derivative of α vanishes at* 0.

As an immediate corollary we obtain a finite-dimensional model for a neighbourhood of 0 in the zero set $Z(\phi)$. Under a diffeomorphism of U this is taken to the zero set of the finite-dimensional map:

$$f : F \to G, \quad f(y) = \alpha(0, y).$$

This is as far as we can go in describing the zero set of a Fredholm map in any generality. The point is that all the phenomena we encounter are essentially finite-dimensional. The idea used in (4.2.19) of reduction of an infinite-dimensional, non-linear problem to a linear part and a finite-dimensional non-linear part will appear in a number of places in this book, and especially in Chapter 7.

We can extend this discussion to various other topics in infinite-dimensional differential topology. For example we can define Fredholm maps between Banach manifolds. The most important global notion for us will be that of a Fredholm section of a bundle of Banach spaces $\mathscr{E} \to \mathscr{P}$. In the case when $\mathscr{E}$ is a trivial bundle $\mathscr{P} \times V$ such a section is just a map into V. In general the section is defined to be Fredholm if in local trivializations it is represented by Fredholm maps from the base to the fibre.

4.2.5 Local models for the moduli space

Return now to our local description of the ASD moduli space (4.2.17). The map ψ is a smooth Fredholm map with derivative at 0

$$\mathrm{d}_A^+ : \ker \mathrm{d}_A^* \longrightarrow \Omega_X^+(\mathfrak{g}_E).$$

We know that $\mathrm{d}_A^* \oplus \mathrm{d}_A^+ = \delta_A : \Omega^1 \to \Omega^0 \oplus \Omega^+$ is Fredholm, being an elliptic operator, and this immediately implies that the restriction of d_A^+ to the kernel of d_A^* is Fredholm with index

$$\operatorname{ind} \psi = \operatorname{ind} \delta_A + \dim \ker \mathrm{d}_A = \operatorname{ind} \delta_A + \dim \Gamma_A. \tag{4.2.20}$$

The integer $s = \operatorname{ind} \delta_A$ is thus of vital importance to our understanding of the moduli space. General elliptic theory says that it depends only upon the initial topological data—the bundle E and four-manifold X. The Atiyah–Singer index theorem gives the formula, for a general G-bundle E,

$$s = a(G)\kappa(E) - \dim G(1 - b_1(X) + b_+(X)),$$

where $a(G)$ is an integer depending on G. For $SU(2)$ bundles E this takes the precise form:

$$s = 8c_2(E) - 3(1 - b_1(X) + b_+(X)), \tag{4.2.21}$$

and for $SO(3)$ bundles:

$$s = -2p_1(E) - 3(1 - b_1(X) + b_+(X)). \tag{4.2.22}$$

Here $b_1(X)$ is the first Betti number of X and $b_+(X)$ is the 'positive part' of the second Betti number, as in (1.1.1). The number $b_+(X)$ will be for us the key invariant of an oriented four-manifold, its importance stemming from its place in these index formulae. We will give a direct proof of the index formula (4.2.21) for $SU(2)$ bundles (from which the general case can easily be deduced) in Chapter 7.

Applying the decomposition of (4.2.19) to the Fredholm map ψ we get:

Proposition (4.2.23). *If A is an ASD connection over X, a neighbourhood of $[A]$ in M is modelled on a quotient $f^{-1}(0)/\Gamma_A$ where*

$$f\colon \ker \delta_A \longrightarrow \operatorname{coker} d_A^+$$

is a Γ_A-equivariant map.

Here we choose a Γ_A-invariant complement to the image of d_A^+, for example the L^2 orthogonal complement $\ker d_A \subset \Omega_X^+(\mathfrak{g}_E)$.

We shall sometimes refer to the index $s = \operatorname{ind} \delta_A$ as the 'virtual' dimension of the moduli space. This is motivated by the fact that points of the zero set of f which are both regular points for f and which represent free Γ_A orbits form a manifold of this dimension, since

$$s = \dim \ker \delta_A - \dim \operatorname{coker} d_A^+ - \dim \Gamma_A. \tag{4.2.24}$$

We will now put this discussion into the abstract framework of Fredholm differential topology. For simplicity we restrict to the open subset $\mathcal{B}^*$, so we can ignore stabilizers. The free $\mathcal{G}/C(G)$ action on $\mathcal{A}^*$ makes $\mathcal{A}^* \to \mathcal{B}^*$ into a principal bundle. Now $\mathcal{G}/C(G)$ acts linearly on the vector space $\Omega_X^+(\mathfrak{g}_E)$ so we get an associated bundle of Banach spaces:

$$\mathscr{E} = \mathcal{A}^* \times_{\mathcal{G}/C(G)} \Omega_X^+(\mathfrak{g}_E) \longrightarrow \mathcal{B}^*. \tag{4.2.25}$$

(Remember that we are suppressing the Sobolev spaces; the 2-forms in (4.2.25) are really the L^2_{l-2} forms.) The self-dual part of the curvature gives an equivariant map $F^+\colon \mathcal{A}^* \to \Omega_X^+(\mathfrak{g}_E)$ and this translates into a section Ψ of $\mathscr{E}$. This section is Fredholm of index s, and its zero set is the part of the moduli space in $\mathcal{B}^*$.

In the local models above the two equations $d_A^* a=0$, $d_A^+ a+(a \wedge a)^+ = 0$ play different roles. The second is the ASD equation while the first is an auxiliary construction of a local slice through the orbit. Other choices for the slice can certainly be made (and note that this slice depends upon a metric on X rather than just the conformal class). A more invariant description of the linearized picture is furnished by the 'deformation complex':

$$\Omega_X^0(\mathfrak{g}_E) \xrightarrow{d_A} \Omega_X^1(\mathfrak{g}_E) \xrightarrow{d_A^+} \Omega_X^+(\mathfrak{g}_E). \tag{4.2.26}$$

The ASD condition for A precisely asserts that $d_A^+ \circ d_A = 0$ so this does form a complex, and we get three cohomology groups H_A^0, H_A^1, H_A^2. The middle cohomology, the quotient

$$H_A^1 = \frac{\ker d_A^+}{\operatorname{Im} d_A}$$

represents the linearization of the ASD equations modulo $\mathcal{G}$. By the Hodge theory for the complex we have natural isomorphisms,

$$H_A^1 \cong \ker \delta_A, \quad H_A^2 = \operatorname{coker} d_A^+ \cong \ker d_A \subset \Omega_X^+(\mathfrak{g}_E), \tag{4.2.27}$$

while H^0_A is the Lie algebra of Γ_A. Again nothing is special here to the ASD equations: we will get such a complex any time we have an equation invariant under a group.

In these terms our index s is minus the Euler characteristic of the complex,

$$s = \dim H^1_A - \dim H^0_A - \dim H^2_A, \tag{4.2.28}$$

and the local model above is a map

$$f\colon H^1_A \longrightarrow H^2_A.$$

While the map f considered in (4.2.23) is fully determined by the construction, it is really better to think of a whole class of maps, each giving a model for the moduli space. We can take any Γ_A-invariant submanifold S transverse to the orbit through A, and any equivariant vector sub-bundle Ξ over S of the trivial bundle with fibre $\Omega^+_X(\mathfrak{g}_E)$ such that Ξ_A gives a lifting of H^2_A. We then look at the solutions of the part of the ASD equation:

$$F^+(A) = 0 \bmod \Xi_A, \quad \text{for } A \in S. \tag{4.2.29}$$

These form a submanifold Y of S with tangent space H^1_A at A. Choosing coordinates τ to identify Y with H^1_A, and a trivialization σ of Ξ, we get an equivariant local model for the moduli space in terms of a map

$$f = f_{S,\Xi,\tau,\sigma}\colon H^1_A \longrightarrow H^2_A, \quad F^+(\tau(p)) = \sigma(f(p)) \tag{4.2.30}$$

just as above. The point is that the map we get depends upon the choices made, and so is not intrinsic to the situation. The quadratic part of f is intrinsic, it is induced by the wedge product,

$$f(p) = [(p \wedge p)^+] + O(p^3) \in H^2_A. \tag{4.2.31}$$

The intrinsic structure can be encoded neatly as a *sheaf of rings* over M. With any model f, as above, we associate a ring $\mathcal{O}/\mathcal{I}$ where $\mathcal{O}$ is the ring of germs of Γ_A-invariant functions on H^1_A and $\mathcal{I}$ is the pull-back by f of the ideal of invariant functions on H^2_A which vanish at 0. Then this ring is independent, up to canonical isomorphism, of the local model. The rings fit together to define a 'structure sheaf' on M. The space H^1_A can be obtained intrinsically from this sheaf as the 'Zariski tangent space' to the moduli space at $[A]$.

In fact it is easy to see (by complexification) that the maps throughout the discussion can be taken to be real analytic. Thus the moduli space is a 'real analytic space'.

4.2.6 Discussion of examples

We now examine our five examples of moduli spaces (Section 4.1) in the light of this general theory. First it can be shown using a Weitzenböck formula that the H^2_A spaces are zero except in Example (v). Thus the linearized operators d^+_A are surjective and the only singularities come from reductions.

In Example (i) there are of course no reductions ($H^2(S^4)$ is zero) and we see a smooth moduli space whose dimension, five, agrees with our index formula. Similarly in Example (ii) we get a five-dimensional space, and this is in line with the index formula since while $\overline{\mathbb{CP}}^2$ has $b_2 = 1$ the positive part b_+ is zero. But in Example (ii) we get a singular point, the vertex of the cone corresponding to the unique reduction $L \oplus L^{-1}$, where $c_1(L)$ is a generator of $H^2(\mathbb{CP}^2)$. Now our general theory says that a neighbourhood of this singular point is modelled on $H^1_A/\Gamma_A = H^1_A/S^1$. But H^1_A has six real dimensions, by the index formula, and lies wholly in the L^2 part of $\mathfrak{g}_E$ in the splitting $\mathfrak{g}_E = \underline{\mathbb{R}} \oplus L^2$. So we can regard it as $\mathbb{C}^6$ with the standard circle action (more precisely, Γ_A acts with weight 2). Thus the theory gives the local model $\mathbb{C}^3/S^1$, which is indeed an open cone over $\mathbb{CP}^2$. (Of course our general theory makes no predictions about the global structure of the moduli space.) We can see explicitly in the formula for the connection matrices J_t that J_0 is reducible, involving only the basis element $\mathbf{i}$ of $SU(2)$. This is indeed the standard connection on the Hopf line bundle over $\mathbb{CP}^2$.

Turning to Example (iii), we have now changed orientation so $b_+ = 1$ and we have a ten-dimensional space predicted by the index formula. There are no reductions since the intersection form is positive definite. Similarly in Example (iv) the spaces have no reductions and their dimensions, zero and eight, agree with those given by the index formula for $SO(3)$ bundles.

Example (v) is the most complicated. The dimension is ten as expected, but we again have a reducible solution, corresponding to the quadric Q in our description of the moduli space. The position is summarized by the diagram of $H^2(S^2 \times S^2)$ (Fig. 9).

The reduction corresponds to the class $(1, -1)$ in the standard basis, and this is in the ASD subspace by symmetry between the two factors. Now our deformation complex breaks up into two pieces, corresponding to the terms $\mathbb{R}$ and L^2 in $\mathfrak{g}_E$. The trivial factor contributes cohomology $\mathbb{R}$ to H^2_A, a copy of $H^+(S^2 \times S^2)$, but nothing to H^1_A, since $H^1(S^2 \times S^2) = 0$. On the other hand, as we will see in Section 6.4.3, there is no contribution to H^2_A from the L^2 factor. So we get in sum a local model $f^{-1}(0)/S^1$ where $f\colon \mathbb{C}^6 \to \mathbb{R}$. In Chapter 5 we will show that there is a natural decomposition $H^1_A = \mathbb{C}^3 \times \mathbb{C}^3$ in which a suitable representative f has the form

$$f(z_1, z_2) = |z_1|^2 - |z_2|^2. \tag{4.2.32}$$

To identify a neighbourhood in the moduli space we reverse the complex structure on the second $\mathbb{C}^3$ factor, so $e^{i\theta} \in \Gamma_A$ acts as $e^{2i\theta}$ on the first factor and as $e^{-2i\theta}$ on the second. Now the map $(z_1, z_2) \mapsto z_1 \otimes z_2$ induces a homeomorphism between $f^{-1}(0)/S^1$ and the space of 3×3 complex matrices with rank ≤ 1. It is an interesting exercise to match up this description of a neighbourhood of the singular point with the description in terms of quadrics in Section 4.1.

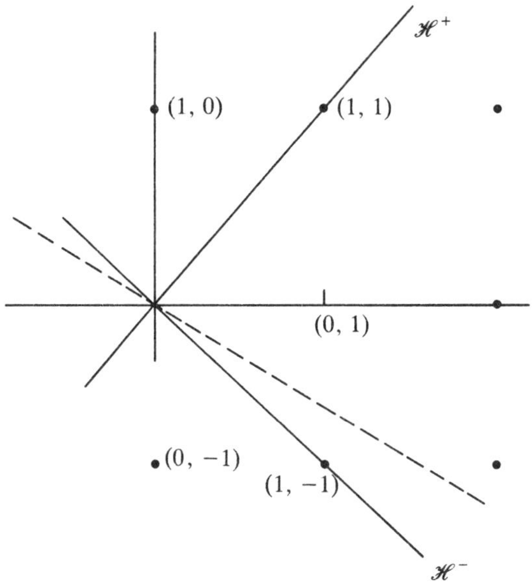

Fig. 9

Notice that this singularity has rather a different nature to that in Example (ii). In the latter case a singular point is present for any metric on the base space, since $b_+ = 0$ and all the classes are represented by purely ASD forms. For $S^2 \times S^2$, by contrast, we can make a small perturbation of the metric under which the reduction in the ASD moduli space disappears. It suffices to take a product metric on round two-spheres with different radii ρ_1, ρ_2, say. Then the ASD subspace is spanned by $(\rho_1^2, -\rho_2^2)$ and avoids the reduction $(1, -1)$. Another interesting exercise is to write down explicit models for the moduli space after such a variation of metric and to see how their topological type changes (cf. Section 4.3.3).

4.3 Transversality

4.3.1 Review of standard theory

We have developed techniques for analysing the local structure of the ASD moduli spaces and tested them against the explicit examples of Section 4.1. In this section we will take the theory further by introducing arguments based on the notion of 'general position'. We have seen that the part of the moduli space M consisting of irreducible connections can be regarded as the zero set of a section Ψ of a bundle $\mathscr{E}$ over the Banach manifold $\mathscr{B}^*$. This depends on a choice of Riemannian metric g on the underlying four-manifold X, so we may write Ψ_g to indicate this dependence. In fact only the conformal class $[g]$ of the metric is relevant, so the abstract picture is that we have a family of equations,

$$\Psi_g([A]) = 0, \tag{4.3.1}$$

for $[A]$ in $\mathscr{B}$, parametrized by the space $\mathscr{C}$ of all conformal structures on X.

Let us then briefly review some standard properties of 'families of equations', beginning in finite dimensions. The simplest situation to consider is a smooth map $F: P \to Q$ between manifolds of dimension p, q respectively. We can regard this map as a family of equations $F(x) = y$ for $x \in P$, parametrized by $y \in Q$. That is, we are looking at the different fibres of the map F. Recall that a point x in P is called a *regular point* for F if the derivative $(DF)_x$ is surjective, and a point y in Q is a *regular value* for F if all the points in the fibre $F^{-1}(y)$ are regular points. If y is a regular value, the implicit function theorem asserts that the fibre $F^{-1}(y)$ is a smooth submanifold of dimension $p - q$ in P. The well-known theorem of Sard affirms that regular values exist in abundance. Recall that a subset of a topological space is of *second category* if it can be written as a countable intersection of open dense sets. By the Baire category theorem, a second category subset of a manifold is everywhere dense. We state the Sard theorem in two parts:

Proposition (4.3.2). *Let $F: P \to Q$ be a smooth map between finite dimensional manifolds.*

(i) *Each point $x \in P$ is contained in a neighbourhood $P' \subset P$ such that the set of regular values of the restriction $F|_{P'}$ is open and dense in Q.*

(ii) *The regular values of F on P form a second category subset of Q.*

For 'most' points y in Q, then, the fibre $F^{-1}(y)$ is a submanifold of the correct dimension $(p - q)$. (For p less than q this is taken to mean that the fibre is empty.) If the map F is proper (e.g. if P is compact) we do not need to introduce the notion of category—the regular values are then open and dense in Q, since if Q' is a compact neighbourhood in Q we can cover $F^{-1}(Q') \subset P$ by a finite number of patches of the form P' as in (4.3.2(i)).

Suppose now that y_0, y_1 are two regular values, so we have two smooth fibres. If the points are sufficiently close together (and the map is, say, proper) these fibres will be diffeomorphic. An extension of the ideas above provides information in the general case when p and q are not close. We assume Q is connected and choose a smooth path $\gamma: [0, 1] \to Q$ between y_0 and y_1. Then we can embed the fibres $F^{-1}(y_0)$, $F^{-1}(y_1)$ in a space:

$$W_\gamma = \{(x, t) \in P \times [0, 1] \mid F(x) = \gamma(t)\}. \tag{4.3.3}$$

As we shall see in a moment, it is always possible to choose a path γ so that W_γ is a $(p - q + 1)$-dimensional manifold-with-boundary, giving a cobordism between the manifolds $F^{-1}(y_0)$, $F^{-1}(y_1)$. The projection map from W_γ to $[0, 1]$ decomposes the cobordism into a one-parameter family of fibres, $F^{-1}(\gamma(t))$. We can think of these as a one-parameter family of spaces interpolating between $F^{-1}(y_0)$ and $F^{-1}(y_1)$, much as we considered in Section 1.2.3. (We could go on to perturb the projection map slightly to make it a Morse function, so that, as in the proof of the h-cobordism theorem, the fibres change by standard surgeries. But this refinement will not be necessary here.)

We can sum up this discussion, for the family of equations $F(x) = y$ parametrized by $y \in Q$, in the slogan: *for generic parameter values the solutions form a manifold of the correct dimension, and any two such solution sets differ by a cobordism within* $P \times [0, 1]$. The same ideas apply to other 'families of equations', depending on parameters, and in particular, as we shall see, to the ASD equations (4.3.1).

A common framework for the 'general position' arguments that we need is provided by the notion of *transversality*. Let $F: P \to Q$ be a smooth map as above, and R be a third manifold. A smooth map $h: R \to Q$ is said to be *transverse* to F if for all pairs (x, r) in $P \times R$ with $F(x) = h(r)$ the tangent space of Q at $F(x)$ is spanned by the images of $(\mathrm{D}F)_x, (\mathrm{D}h)_r$. When this condition holds the set:

$$Z = \{(x, r) \in P \times R \mid F(x) = h(r)\} \tag{4.3.4}$$

is a smooth submanifold (possibly empty) of $P \times R$, with codimension dim Q.

Transversality is a generic property; any map h can be made transverse to F by a small perturbation. If R is compact we can prove this as follows. We consider a family of maps h_s parametrized by an auxiliary manifold S (which we can take to be a ball in a Euclidean space). Precisely, we have a total map:

$$\underline{h}: R \times S \longrightarrow Q \tag{4.3.5}$$

and $h_s(r) = \underline{h}(r, s)$. We suppose that there is a base point $s_0 \in S$ such that $h_{s_0} = h$. Suppose we have constructed a family of this form such that $\underline{h}$ is transverse to F. Then the space,

$$\underline{Z} = \{(x, r, s) \in P \times R \times S \mid h_s(r) = F(x)\}, \tag{4.3.6}$$

is a submanifold of $P \times R \times S$ with a natural projection map $\pi: \underline{Z} \to S$. It is easy to see that the regular values $s \in S$ of π are precisely the parameter values for which h_s is transverse to F. We use Sard's theorem to find a regular value arbitrarily close to s_0, and this gives the desired small, transverse, perturbation of the original map h. The remaining step in the proof of generic transversality is the construction of the transverse family h_s. How best to do this depends on the context. First, suppose that the image space Q is a finite-dimensional vector space U. We can then take S to be a neighbourhood of 0 in U and put

$$h_s(r) = h(r) + s.$$

This clearly has the desired property, since the image of the derivative of h alone is surjective. It may be possible to be more economical; if $V \subset U$ is a linear subspace which generates the cokernel of $(\mathrm{D}F)_x + (\mathrm{D}h)_r$ for all $(x, r) \in Z$ we can use these same variations with S a neighbourhood of 0 in V. In general, cover Q with coordinate balls B_i and find a finite cover of R by open sets $R_n (n = 1, \ldots, N)$ with $h(R_n) \subset \frac{1}{2} B_{i(n)}$. Let R'_n be slightly smaller open sets which still cover R and ψ_n be cut-off functions, supported in R_n and equal to 1 on R'_n. Then take

$$S = \prod_{n=1}^{N} \tfrac{1}{2} B_{i(n)} \tag{4.3.7}$$

and

$$h(r, s_1, \ldots, s_N) = h(t) + \psi(t)s_1 + \ldots + \psi_N(t)s_N.$$

Here the 'addition' of $\psi_n(t)s_n$ is done using the coordinates of $B_{i(n)}$.

(If R is not compact we can still find a transverse perturbation of h, using the argument above on successive compact pieces.)

One application of this theory is the proof of the assertion above on the choice of a path $\gamma: [0, 1] \to Q$ such that W_γ is a submanifold. We take $R = [0, 1]$ and $h = \gamma_0$, for any path γ_0 from y_0 to y_1. Then we find a perturbation γ transverse to F. (Note that we can assume that γ has the same end points, since the map is already transverse there.) Other applications are:

(1) If $K \subset Q$ is a countable, locally-finite union of submanifolds whose codimension exceeds dim R then any map $h: R \to Q$ can be perturbed slightly so that its image does not meet K. In fact the locally finite condition may be dropped, but then one needs a rather longer argument, applying the Baire category theorem in the function space of maps from R to Q.

(2) A section Ψ of a vector bundle $V \to P$ can be perturbed so that it is transverse to the zero section. The zero set is then a smooth submanifold of the base space. To fit this into the framework above we can take F to be the inclusion of the zero section in the total space and h to be the section, regarded as a map from P to V. However, in this situation, if x is a zero of Ψ, we shall usually write $(D\Psi)_x$ for the intrinsic derivative mapping $(TP)_x$ to the fibre V_x. The transversality condition is just that $(D\Psi)_x$ be surjective for all points of the zero set. Since this is a situation we shall want to refer to frequently in this book we introduce the following terminology. A point x in the zero set of a section Φ of a vector bundle V will be called a *regular point* of the zero set if $(\mathrm{D}\Phi)_x$ is surjective. We say that the zero set is *regular* if all its points are regular points.

In the context (2) of vector bundles we can formulate the construction above of a section with a regular zero set in the following way. Given any section Φ we consider an auxiliary space S and a bundle,

$$\underline{V} \longrightarrow P \times S,$$

whose restriction to $P \times \{s_0\}$ is identified with V. In fact we may as well assume that $\underline{V}$ is the pull-back of V to the product. We choose S so that there is a section Φ of $\underline{V}$ which agrees with Φ on $P \times \{s_0\}$, and which has a regular zero set $\underline{Z} \subset P \times S$. Then, as before, we apply Sard's theorem to the projection map from $\underline{Z}$ to S. A regular value s of the projection map yields a perturbation $\Phi_s = \underline{\Phi}|_{P \times \{s\}}$ of Φ, having a regular zero set in P.

4.3.2 The Fredholm case

Transversality theory in finite dimensions does not go over wholesale to the infinite-dimensional setting of smooth maps between Banach manifolds, but to a large extent it does extend to situations where the linear models are Fredholm operators. We begin with the extension, due to Smale, of the Sard theorem. Let $F: \mathscr{P} \to \mathscr{Q}$ be a smooth Fredholm map between paracompact Banach manifolds, and let x be a point of $\mathscr{P}$. We can choose a coordinate patch $\mathscr{P}' \subset \mathscr{P}$ containing x, and a coordinate system so that F is represented, in a neighbourhood of x, by a map:

$$(\xi, \eta) \longmapsto (L(\xi), \alpha(\xi, \eta)),$$

as in (4.2.19), with L a linear isomorphism between Banach spaces and $\alpha: U_0 \times \mathbb{R}^p \to \mathbb{R}^q$. A point (ζ, θ) is a regular value of $F|_{\mathscr{P}'}$ if and only if θ is a regular value for the finite dimensional map,

$$f_\zeta = \alpha|_{L^{-1}(\zeta)}: \mathbb{R}^p \longrightarrow \mathbb{R}^q.$$

It follows easily from the ordinary Sard theorem that the regular values for the restriction of F to a small coordinate patch $\mathscr{P}'$ are open and dense in $\mathscr{Q}$. The Baire category theorem applies equally well to Banach manifolds so, taking a countable cover of $\mathscr{P}$, we obtain the Smale–Sard theorem:

Proposition (4.3.8). *If $F: \mathscr{P} \to \mathscr{Q}$ is a smooth Fredholm map between paracompact Banach manifolds, the regular values of F are of second category, hence everywhere dense in $\mathscr{Q}$.*

If $\mathscr{P}$ is connected then for any such regular value $y \in \mathscr{Q}$ the fibre $F^{-1}(y) \subset \mathscr{P}$ is a smooth submanifold of dimension

$$\dim F^{-1}(y) = \operatorname{ind} F. \tag{4.3.9}$$

Similarly we have a Fredholm transversality theorem:

Proposition (4.3.10). *If $F: \mathscr{P} \to \mathscr{Q}$ is a Fredholm map, as in (4.3.8), and $h: R \to \mathscr{Q}$ is a smooth map from a finite-dimensional manifold R, there is a map $h': R \to \mathscr{Q}$, arbitrarily close to h (in the topology of C^∞ convergence on compact sets) and transverse to F. If h is already transverse to F on a closed subset $G \subset T$ we can take $h' = h$ on G.*

Proof. The proof is much as before. There are two possible approaches. For the first we suppose initially that R is compact. Then the construction we gave for the transverse family h is valid, except that now we must use the more economical, finite-rank, perturbations which suffice to generate the cokernels. Then we can take S again to be finite-dimensional. We obtain a finite-dimensional manifold $\underline{\mathscr{Z}} \subset \mathscr{P} \times R \times S$ and apply the ordinary Sard theorem as before. Then we handle the general case by writing S as a union of compact sets. For the second approach, we work with an infinite set of balls $B_{i(n)}$ and use an infinite-dimensional space S, replacing the product in (4.3.7) by the space of bounded sequences. Then S is a Banach manifold and the projection $\pi : \underline{\mathscr{Z}} \to S$ is Fredholm, with index $\mathrm{ind}(\pi) = \mathrm{ind}(g) + \dim R$. We use (4.3.8) to find a regular value s of π and hence a transverse perturbation h_s.

Our main application of the Fredholm theory will be to sections of vector bundles. Suppose that $\mathscr{V} \to \mathscr{P}$ is a bundle of Banach spaces over a Banach manifold, and Φ is a Fredholm section of $\mathscr{V}$, i.e. represented by Fredholm maps in local trivializations of $\mathscr{V}$. We would like to perturb Φ to find a section with a regular zero set. We cannot now proceed directly to apply (4.3.10) since the hypotheses will not be satisfied if $\mathscr{P}$ is infinite-dimensional. We can however apply the same scheme to analyse perturbations. Following the notation at the end of Section 4.3.2 we consider a bundle $\underline{\mathscr{V}} = \pi_1^*(\mathscr{V}) \to \mathscr{P} \times S$, where S is an auxiliary Banach manifold with base point s_0. Let $\underline{\Phi}$ be a section of $\mathscr{V}$, extending Φ, which is Fredholm in the $\mathscr{P}$ variable. That is, in local trivializations $\underline{\Phi}$ is represented by smooth maps to the fibre whose partial derivatives in the $\mathscr{P}$ factor are Fredholm. For s in S we regard the restriction of Φ to $\mathscr{P} \times \{s\}$ as another section Φ_s of $\mathscr{V}$.

Proposition (4.3.11). *If the zero set $\underline{\mathscr{Z}} \subset \mathscr{P} \times S$ is regular then there is a dense (second category) set of parameters $s \in S$ for which the zero sets of the perturbations Φ_s are regular.*

This follows immediately from (4.3.8), applied to the projection map from $\underline{\mathscr{Z}}$ to S, as before. Notice that, as in our first proof of (4.3.10), if it is possible to choose S to be finite-dimensional then we only need the 'ordinary' Sard theorem. We will return to discuss the construction of such families $\underline{\Phi}$, in the abstract setting, in Section 4.3.6.

4.3.3 *Applications to moduli spaces*

We will now apply this theory to the ASD equations and state our main results. We will defer the proofs of the main assertions, which involve more detailed differential-geometric considerations, to Sections 4.3.4 and 4.3.5. The main results were first proved by Freed and Uhlenbeck and our treatment is not fundamentally different from theirs. Throughout this section we let X be a compact, simply connected, oriented four-manifold. We will use the terminology introduced at the end of Section 4.3.1, so (with a given metric) an irreducible ASD connection A is called *regular* if $H_A^2 = 0$ and we call a moduli space regular if all its irreducible points are regular points. Of course, a regular moduli space of irreducible connections is a smooth manifold of dimension given by the index $s = s(E)$. But the converse is not true; it may happen that the moduli space is homeomorphic to a smooth manifold of the correct dimension, but is not regular. (We will see an example of this in Chapter 10). The regularity condition is equivalent to the condition that as a ringed space the moduli space should be a manifold.

We begin by discussing the natural parameter space in the set-up, the space $\mathscr{C}$ of conformal structures on X. At one point we will want to apply Banach manifold results to this space, so we agree henceforth to work with C^r metrics on X for some fixed large $r (r = 3$ will do). The space $\mathscr{C}$ is the quotient of these metrics by the C^r conformal changes. It is easy to see that $\mathscr{C}$ is naturally a Banach manifold. We can use the construction of Section 1.1.5 to obtain a set of handy charts on $\mathscr{C}$. Given one conformal structure $[g_0] \in \mathscr{C}$ with $\pm$ self-dual subspaces $\Lambda^+; \Lambda^-$, the space $\mathscr{C}$ is naturally identified with the space of C^r maps,

$$m: \Lambda^- \longrightarrow \Lambda^+$$

with $|m_x| < 1$ for all $x \in X$. In particular the tangent space of $\mathscr{C}$ at the given point is naturally identified as:

$$(T\mathscr{C})_{g_0} \cong \mathrm{Hom}(\Lambda^-, \Lambda^+). \tag{4.3.12}$$

We will now consider the abelian reductions in our moduli spaces. We have seen in (2.2.6) that a line bundle $L \to X$ admits an ASD connection if and only if $c_1(L)$ can be represented by an anti-self-dual harmonic form. If we identify $H^2(X; \mathbb{R})$ with the space of harmonic two-forms we have a decomposition $H^2(X; \mathbb{R}) = \mathscr{H}^+ \oplus \mathscr{H}^-$; the condition for an ASD connection is that $c_1(L)$ lies in $\mathscr{H}^-$. If the intersection form of X is negative definite, so $\mathscr{H}^+ = 0$, this is no restriction—any line bundle carries an ASD connection, for any metric on X. If $b^+(X)$ is non-zero on the other hand we see a marked difference—the space $\mathscr{H}^-$ is then a proper subspace of $H^2(X)$ and we would expect that generically it meets the integer lattice $H^2(X; \mathbb{Z}) \subset H^2(X; \mathbb{R})$ only at zero. We introduce some notation. Let Gr be the Grassmann manifold of b^--dimensional subspaces of $H^2(X; \mathbb{R})$ and $U \subset \mathrm{Gr}$ be the open subset of maximal negative subspaces, with respect to the intersection form. So the assignment of the space $\mathscr{H}^-(g)$ of ASD harmonic forms to a conformal class gives a map:

$$P: \mathscr{C} \longrightarrow U. \tag{4.3.13}$$

Now suppose that c is a class in $H^2(X; \mathbb{Z})$ with $c. c < 0$ and define

$$N_c \subset U = \{\mathscr{H}^- | c \in \mathscr{H}^-\}.$$

It is easy to see that N_c is a submanifold of codimension $b^+(X)$ in U. Our main result here is:

Proposition (4.3.14). *The map P is transverse to N_c.*

For any $l > 0$ we let $K_l \subset U$ be the union of the N_c as c runs over all the integer classes with $-l \leq c.c < 0$. One easily sees that K_l is a locally finite union of the manifolds N_c. We can then apply our transversality theory to deduce that any map $h: R \to \mathscr{C}$, with $\dim R < b^+(X)$, can be perturbed slightly to a map h' with $h'(R) \cap P^{-1}(K_l)$ empty. For example we can apply (4.3.10) to the Fredholm inclusions of the components of $P^{-1}(K_l)$, using the local finiteness condition. (In fact, as in application (1) of Section 4.3.1, we could take l to be infinite here, but the simpler result will serve for our applications.)

We obtain then:

Corollary (4.3.15). *If $b^+(X) > 0$ then for any $l > 0$ there is an open dense subset $\mathscr{C}(l) \subset \mathscr{C}$ such that, for $[g] \in \mathscr{C}(l)$, the only reducible g-ASD connection on an $SU(2)$ or $SO(3)$ bundle E over X with $\kappa(E) \leq l$ is the trivial product connection. Moreover if $d < b^+(X)$ and $h: R \to \mathscr{C}$ is a smooth family of metrics parametrized by a d-dimensional manifold R, then there is an arbitrarily small perturbation of h whose image lies in $\mathscr{C}(l)$.*

We now turn our attention to the parts of the moduli spaces representing irreducible connections. We write $M^*(g) = M^*_E(g)$ for the intersection of the moduli space $M(g) = M_E(g)$ defined by the metric g, with the open subset $\mathscr{B}^* \subset \mathscr{B}$. For each conformal class $[g]$ we have a space $\Omega^+_{X,g}(\mathfrak{g}_E)$ of self-dual forms defined by g. This is acted on by the gauge group $\mathscr{G}$, so, as in (4.3.1), we get a quotient bundle $\underline{\mathscr{E}} \to \mathscr{B}^* \times \mathscr{C}$. The self-dual part of the curvature defines a natural section $\underline{\Psi}$ of $\underline{\mathscr{E}}$. So we have exactly the set-up considered in (4.3.11). Let us straight away put this construction into a more manageable form. We fix a reference metric g_0 and identify $\mathscr{C}$ with the maps $m: \Lambda^- \to \Lambda^+$ having operator norm less than 1. Then we can identify the different subspaces Λ^+_g with the fixed space Λ^+, as in Section 1.1.5. This yields a definite isomorphism $\underline{\mathscr{E}} \cong \pi_1^*(\mathscr{E})$, so we can regard our section $\underline{\Psi}$ as a family of sections of $\mathscr{E}$. In these terms the section over $\mathscr{B}^* \times \mathscr{C}$ is represented by:

$$(1 + mm^*)^{-1}(F^+(A) + mF^-(A)),$$

(cf. (1.1.11)). The factor $(1 + mm^*)^{-1}$ is irrelevant for our purposes; it can be absorbed by a change of trivialization, so we may as well write, somewhat imprecisely:

$$\underline{\Psi}(A, m) = F^+(A) + mF^-(A) \in \Omega^+_X(\mathfrak{g}_E). \tag{4.3.16}$$

The main result we need here is due to Freed and Uhlenbeck:

Theorem (4.3.17). *For any $SU(2)$ or $SO(3)$ bundle E over X, the zero set of $\underline{\Psi}$ in $\mathscr{B}^* \times \mathscr{C}$ is regular.*

Thus the natural variations of the ASD equations provided by the conformal structures form a transverse family. Given this result we straight away deduce from (4.3.11):

Corollary (4.3.18). *There is a dense (second category) subset $\mathscr{C}' \subset \mathscr{C}$ of conformal classes, such that for $[g]$ in $\mathscr{C}'$ and for any $SU(2)$ or $SO(3)$ bundle E over X, the moduli space $M_E^*(g)$ is regular (as the zero set of the section Ψ_g determined by $[g]$).*

We shall write $\mathfrak{M}$ for the zero-set of $\underline{\Psi}$ in $\mathscr{B} \times \mathscr{C}$ and $\mathfrak{M}^*$ for the intersection of M with $\mathscr{B}^* \times \mathscr{C}$. Thus Theorem (4.3.17) says that this 'universal moduli space' $\mathfrak{M}^*$ is a Banach manifold, cut out transversely. The individual moduli spaces $M^*(g)$ are the fibres of the projection $\mathfrak{M}^* \to \mathscr{C}$.

We can also consider families of metrics. Let $[g_0]$, $[g_1]$ be two points in the subset $\mathscr{C}'$ of (4.3.18), and $\gamma:[0, 1] \to \mathscr{C}$ be a path between them. We can apply (4.3.10) with $h = \gamma$ and F the projection map π from the universal zero set $\mathfrak{M}^* \subset \mathscr{B}^* \times \mathscr{C}$ to $\mathscr{C}$. We get a new path γ' transverse to π. We may assume the new path has the same end points, since γ is already transverse there. We then obtain, as in Section 4.3.1, a cobordism W_γ between the two moduli spaces $M_E^*(g_0)$, $M_E^*(g_1)$. In our applications we will want to combine this idea with the results above for reducible solutions. Note first that for any l we can choose the path γ' to have the transverse property above for all bundles E with $\kappa(E) \leq l$. Second, if $b^+(X) \geq 2$ we can choose the path to lie in $\mathscr{C}(l)$ (assuming that the given end points lie in this open set). We summarize our results in this case in the following proposition:

Corollary (4.3.19). *Suppose the four-manifold X satisfies the condition $b^+(X) \geq 2$, and fix $l > 0$. Then*

(i) *For a dense (second category) set of metrics g on X the moduli spaces $M_E(g)$ for all $SU(2)$ or $SO(3)$ bundles E with $0 < \kappa(E) \leq l$ contain no irreducible connections and are regular.*

(ii) *Let g_0, g_1 be metrics which satisfy the conditions of (1). Then for a dense set of paths γ from g_0 to g_1 and any bundle E with $0 < \kappa(E) \leq l$ the space:*

$$W_\gamma = \{([A], t) \in \mathscr{B}_E \times [0, 1] \,|\, [A] \in M_E(g_t)\}$$

gives a smooth cobordism between the manifolds $M_E(g_0)$, $M_E(g_1)$, and W_γ lies in $\mathscr{B}^* \times [0, 1]$.

(In (ii) we can replace 'dense' by the stronger condition 'second category'—defined in the topological space of smooth maps from $[0, 1]$ to $\mathscr{C}$, cf. the remarks in application (1) in Section 4.3.1.)

There is only one more point we need to consider for the applications in this book: the behaviour of the moduli space around a reducible connection in the case when $b^+ = 0$. The result here is again due to Freed and Uhlenbeck.

Proposition (4.3.20). *If $b^+(X) = 0$ then for generic metrics $[g]$ and any non-trivial $SU(2)$ or $SO(3)$ bundle E over X, the cohomology groups H_A^2 are zero for all the reducible ASD connections on E, and a neighbourhood in M_E of such a reducible solution is modelled on a cone over $\mathbb{CP}^d$, where:*

$$d = \tfrac{1}{2}(s(E) - 1) = 4\kappa(E) - 2.$$

Here the last part of the statement is just read off from the local model in Section 4.2.5, given that $H_A^2 = 0$. We shall omit the proof of (4.3.20), unlike the other results in this section, since it is similar in style to those of (4.3.14) and (4.3.17) and for our applications a much simpler argument can be used; see Section 4.3.6.

4.3.4 Unique continuation

Our proofs of the results (4.3.14) and (4.3.17) in Section 4.3.5 will depend on the following lemma, which we will also use in Chapter 5.

Lemma (4.3.21). *If A is an irreducible $SU(2)$ or $SO(3)$ ASD connection on a bundle E over a simply connected four-manifold X, then the restriction of A to any non-empty open set in X is also irreducible.*

Suppose on the contrary that A is decomposable on, say, a geodesic ball B_ε. Recall that this means that there is a gauge transformation u defined over B_ε which leaves A invariant (i.e. $d_A u = 0$) but which does not lie in the centre of the gauge group ($\{1, -1\}$ in the $SU(2)$ case). We assume that 2ε is less than the injectivity radius of X, and we will show that u can be extended to the larger ball, with the same centre, $B_{2\varepsilon}$. By a sequence of such extensions u can be extended over all of X—the extension being single-valued since X is simply connected (just as for the analytic continuation of holomorphic functions). This will show that A is reducible on X, contrary to hypothesis. We trivialize E over $B_{2\varepsilon}$ using parallel transport along radial geodesics (cf. Section 2.3.1). Thus we have a connection matrix A in this trivialization with zero radial component, and the gauge transformation u, viewed in this trivialization as a map to the structure group G, satisfies

$$\left\langle d_A u, \frac{\partial}{\partial r} \right\rangle = \frac{\partial u}{\partial r} = 0.$$

So u is a constant over B_ε and can be trivially extended to the larger ball $B_{2\varepsilon}$. We put $A' = u(A) = uAu^{-1}$, so A and A' are two ASD connection matrices, and the condition $d_A u = 0$ on B_ε means that $A = A'$ on this ball. We want to show that A equals A' on the larger ball. What is needed, therefore, is a unique continuation theorem for ASD connections in a radial gauge.

It is convenient to identify the punctured ball $B_{2\varepsilon}\backslash\{0\}$ with the cylinder $(-\infty, \log 2\varepsilon) \times S^3$ under the map $(r, \theta) \mapsto (\log r, \theta)$. If the metric on the ball were flat this would be a conformal equivalence with the standard cylinder metric. In general the conformally equivalent metric on the cylinder has the form:

$$dt^2 + \gamma(t, \theta) d\theta^2,$$

where the second term is shorthand for a metric on S^3 which is close to the round metric. Likewise, we carry the connection forms A, A' over to the cylinder. They have no $\mathrm{d}t$ component, so can be viewed as one-parameter families of connection forms over the three-sphere. The ASD equation takes the form:

$$\frac{\mathrm{d}A}{\mathrm{d}t} = *_t F(A(t)), \tag{4.3.22}$$

where $F(A(t))$ is the curvature of the connection over the 3-sphere and $*_t : \Omega^2 \to \Omega^1$ is the $*$ operator in three dimensions defined by the metric $\gamma(t, \theta)\mathrm{d}\theta^2$. This may be regarded as an ordinary differential equation (ODE) for the path $A(t)$, so the result we require is the unique continuation of solutions to this ODE.

We will apply a result of Agmon and Nirenberg (1967). Let $\mathscr{P}_t : V \to V$ be a smooth one-parameter family of linear differential operators over a compact manifold, each self adjoint with respect to a fixed L^2 norm $\| \ \|$ on V. Suppose that the time derivative of $\mathscr{P}_t$ is controlled by $\mathscr{P}_t$ in that we have a bound,

$$\left| \left\langle \left(\frac{\mathrm{d}\mathscr{P}_t}{\mathrm{d}t} \right) v, v \right\rangle \right| \leq K_1 \, \|\mathscr{P}_t v\| \, \|v\|$$

for v in V. Suppose $w(t)$ is a one-parameter family in V which satisfies a differential inequality:

$$\left\| \frac{\mathrm{d}w}{\mathrm{d}t} - \mathscr{P}_t w \right\| \leq K_2 \|w\|$$

for some constant K_2. Then Agmon and Nirenberg show that if $w(t)$ vanishes for an initial interval it will do so also for all times t. This is proved by establishing a convexity property of $\log(\|w(t)\|)$. To put our problem into a form where this result can be applied we suppose that $A(t)$, $A'(t)$ are two solutions to (4.3.22) which agree for $t \leq \log \varepsilon$. The difference $a(t) = A(t) - A'(t)$ satisfies the equation

$$\frac{\mathrm{d}a}{\mathrm{d}t} = *_t(\mathrm{d}a + [A, a] + [a, A']).$$

So we certainly have

$$\left\| \frac{\mathrm{d}a}{\mathrm{d}t} - *_t \mathrm{d}a \right\| \leq K_2 \|a\|,$$

for some constant K_2 depending on A and A' (since these are smooth). The only difficulty is that the operators $*_t\mathrm{d}$ are not self-adjoint with respect to a fixed L^2 inner product, although they are so with respect to the t-dependent inner product defined by $\gamma(t, \theta)$. This difficulty is, however, remedied by a change of variable as follows. Let $P = *d$ be the 'curl' operator for the round metric and let Γ_t be the multiplication operator on 1-forms given by the matrix

$$\frac{\gamma}{(\det \gamma)^{1/2}}.$$

Then we have:

$$\Gamma_t^2 = *_t(*)^{-1} : \Omega^1 \to \Omega^1.$$

Put $w(t) = \Gamma_t^{-1} a(t)$, and let $\mathscr{P}_t = \Gamma_t P \Gamma_t$. Then

$$\frac{\mathrm{d}w}{\mathrm{d}t} = \Gamma^{-1}\frac{\mathrm{d}a}{\mathrm{d}t} - \Gamma^{-1}\frac{\mathrm{d}\Gamma}{\mathrm{d}t}\Gamma^{-1}a,$$

while, applying Γ^{-1} to the bound above, we have:

$$\left\| \Gamma^{-1}\frac{\mathrm{d}a}{\mathrm{d}t} - \Gamma^{-1} *_t \mathrm{d}a \right\| \leq \text{const. } \|a\|.$$

Combining these we derive a bound on $\mathrm{d}w/\mathrm{d}t - \mathscr{P}_t w$ of the form required. Similarly

$$\left| \left\langle \left(\frac{\mathrm{d}\mathscr{P}_t}{\mathrm{d}t} \right) v, v \right\rangle \right| = 2 \left| \left\langle \mathscr{P}_t v, \Gamma_t^{-1} \left(\frac{\mathrm{d}\Gamma}{\mathrm{d}t} \right) v \right\rangle \right| \leq \text{const. } \|\mathscr{P}_t v\| \, \|v\|,$$

so both of the required estimates hold, and we can derive Lemma (4.3.21) from the result of Agmon and Nirenberg.

We should point out that, in the results of Section 4.3.3, we are using metrics on the base space X which are only C^r. So it is not strictly true that the connections will be smooth in local trivializations; they will only be C^{r+1}. However this makes no difference to the argument.

The proof above applies to ASD connections for the group S^1. We do not have quite the same set-up at the beginning of the argument, since the isotropy group is always S^1 for any S^1-connection. But we can now argue more directly. If A is an ASD S^1-connection which is flat in some ball, then in a radial gauge the connection matrix vanishes over the ball and we deduce that A must be flat everywhere. This is a local argument, so applies to any closed ASD 2-form. Of course, we have just the same results for self-dual forms. We obtain then:

Corollary (4.3.23). *Suppose ω is a closed two-form on X which satisfies $*\omega = \pm\omega$. Then if ω vanishes on a non-empty open set in X it is identically zero.*

4.3.5 Proofs of transversality results

We will now prove the results stated in Section 4.3.3, beginning with (4.3.14).

We first compute the derivative of the map $P: \mathscr{C} \to U \subset \mathrm{Gr}$. At a given reference metric g_0 we identify the tangent space to $\mathscr{C}$ with $\underline{\mathrm{Hom}}(\Lambda^-, \Lambda^+)$. Here we have introduced temporarily the notation $\underline{\mathrm{Hom}}$ to emphasize that we are dealing with bundle maps. In just the same way, using the graph construction, we identify the tangent space to the Grassmannian Gr at $P([g_0])$ with $\mathrm{Hom}(\mathscr{H}^-, \mathscr{H}^+)$. Here $\mathscr{H}^+, \mathscr{H}^-$ are the self-dual and anti-self-dual subspaces of $H^2(X;\mathbb{R})$ determined by g_0. So the derivative of the map P at $[g_0]$ can be viewed as a linear map

$$DP:\underline{\mathrm{Hom}}(\Lambda^-,\Lambda^+) \longrightarrow \mathrm{Hom}(\mathcal{H}^-,\mathcal{H}^+).$$

Lemma (4.3.24). *The derivative of P is*

$$DP(m)(\alpha)=\Pi(m(\alpha)),$$

where $\alpha\in\mathcal{H}^-$, and $\Pi:\Gamma(\Lambda^+)\to\mathcal{H}^+$ is the L^2 projection.

For any m in $\underline{\mathrm{Hom}}(\Lambda^-,\Lambda^+)$ we consider the one-parameter family of conformal classes $[g_t]$ corresponding to tm, for small t in $\mathbb{R}$. Then $P(g_t)$ can be regarded as a linear map from $\mathcal{H}^-$ to $\mathcal{H}^+$. For $\alpha\in\mathcal{H}^-$ we let $\alpha_t=\alpha+P(g_t)\alpha$; so α_t is a closed two-form on X and $*_t\alpha_t=-\alpha_t$, where $*_t$ is the star operator defined by g_t. We differentiate this relation with respect to t at $t=0$ to get

$$\left(\frac{\mathrm{d}}{\mathrm{d}t}*_t\right)(\alpha)+*_0\left(\frac{\mathrm{d}\alpha_t}{\mathrm{d}t}\right)=-\frac{\mathrm{d}\alpha_t}{\mathrm{d}t}.$$

That is,

$$(1+*_0)\left(\frac{\mathrm{d}\alpha}{\mathrm{d}t}\right)\bigg|_{t=0}=2m\alpha,$$

since the derivative of $*_t$ is $2m$ (exercise). But

$$DP(m)(\alpha)=\frac{\mathrm{d}\alpha_t}{\mathrm{d}t}\bigg|_{t=0}\in\mathcal{H}^+,$$

by construction, so $(1+*_0)(DP(m)(\alpha)=2DP(m)(\alpha)+\mathrm{d}\phi$. Since $\mathrm{d}\phi$ is L^2 orthogonal to the subspace $\mathcal{H}^+$, the cohomology class of $\mathrm{d}\alpha_t/\mathrm{d}t$ is represented by $\Pi(m(\alpha))$, as desired.

Now let N_c be the subset of $U\subset \mathrm{Gr}$ determined by an element $c\in H^2(X)$, as in Section 4.3.3. If $\mathcal{H}^-\in N_c$, so $c\in\mathcal{H}^-$, we have an evaluation map

$$e_c:(T\mathrm{Gr})_{\mathcal{H}^-}\cong\mathrm{Hom}(\mathcal{H}^-,\mathcal{H}^+)\longrightarrow\mathcal{H}^+.$$

It is easy to see that the tangent space to N_c is the kernel of e_c, so the normal space can be identified with $\mathcal{H}^+$. So the assertion of (4.3.14), that P is transverse to N_c, amounts to the statement that, if $c\in\mathcal{H}^+(g_0)$, the composite

$$e_c\circ(DP):\underline{\mathrm{Hom}}(\Lambda^-,\Lambda^+)\longrightarrow\mathcal{H}^+$$

is surjective. Let α be the anti-self-dual 2-form on X which represents c. By (4.3.24) the L^2 inner product of $e_c\circ(DP)(m)$ with a class represented by a self-dual form β is $\langle\beta,m(\alpha)\rangle$. Suppose the composite is not surjective, so there is a non-zero form $\beta\in\mathcal{H}^+$ with $\langle\beta,m(\alpha)\rangle=0$ for all m. Then we have

$$\langle\alpha\otimes\beta,m\rangle=0,$$

for all m, and this can only happen if the section $\alpha\otimes\beta$ is identically zero on X. So α must vanish on an open set in X, contradicting (4.3.23).

We now turn to the proof of the Freed–Uhlenbeck theorem (4.3.17). We adopt the notation of Section 4.3.3. Thus if $([A_0],[g_0])$ is a point in the universal zero set $\mathfrak{M}^* \subset \mathscr{B}^* \times \mathscr{C}$ (i.e. A_0 is irreducible and g_0-ASD) the section $\underline{\psi}$ is represented by $F^+(A) + mF^-(A)$. We know that the derivative of $F^+(A)$ at A_0 is given by the operator $d^+_{A_0}$, so the intrinsic derivative

$$D\underline{\Psi}: T\mathscr{B}^* \times \mathscr{C} \longrightarrow \Omega^+_X(\mathfrak{g}_E)$$

is represented by:

$$D\underline{\Psi}(a,m) = d^+_{A_0}a + m(F^+_{A_0}).$$

Suppose, contrary to (4.3.17), that $([A_0],[g_0])$ is not a regular point; so this derivative is not surjective, and we can find a non-zero element $\theta \in \Omega^+_X(\mathfrak{g}_E)$ which is L^2-orthogonal to the image. Thus:

$$\langle \theta, d_{A_0}a\rangle = \langle d_{A_0}\theta, *a\rangle = 0 \quad \text{for all } a \in \Omega^1_X(\mathfrak{g}_E);$$

$$\langle \theta, m(F^-(A_0))\rangle = \langle F^-(A_0).\theta, m\rangle = 0 \quad \text{for all } m.$$

Here $F^-(A_0).\theta$ is the section of $\Lambda^- \otimes \Lambda^+$ obtained by contraction with the metric on $\mathfrak{g}_E$. We deduce then that $d_{A_0}\theta$ and $F^-(A_0).\theta$ are identically zero. The second condition just means that the images of

$$F^-_A : (\Lambda^-)^* \longrightarrow \mathfrak{g}_E, \qquad \theta : (\Lambda^+)^* \longrightarrow \mathfrak{g}_E$$

are pointwise orthogonal in $\mathfrak{g}_E$. Now $\mathfrak{g}_E$ has rank 3 (since we assume $G = SU(2)$ or $SO(3)$) and we deduce then that at each point of X one of F^-_A, θ has rank less than or equal to 1. We need the following observation:

Lemma (4.3.25). *Suppose ϕ is in $\Omega^+_X(\mathfrak{g}_E)$ or $\Omega^-_X(\mathfrak{g}_E)$ and $d_A\phi = 0$. Then on any simply connected open set in X where ϕ has rank 1, the connection A is reducible.*

Locally, in such an open set, we can write $\phi = s \otimes \omega$ where $\omega \in \Omega^+$ or Ω^- and s is a section of $\mathfrak{g}_E$ with $|s| = 1$. Then the equation $d_A\phi = 0$ becomes

$$(d_A s) \wedge \omega + s \otimes d\omega = 0.$$

Now the condition $|s| = 1$ implies that $d_A s$ is pointwise orthogonal to s. So both terms in this expression must vanish and $d_A s \wedge \omega = 0$. But the wedge product with a non-vanishing, purely self-dual or anti-self-dual 2-form ω gives an isomorphism from Ω^1 to Ω^3; so s is covariant constant and the connection is locally reducible. The result holds for simply connected regions by a continuation argument (although any open set would suffice for our argument).

Now we can apply this lemma in the situation discussed above to the two cases $\phi = \theta$ and $\phi = F^-(A) = F(A)$. In the second case we use the Bianchi identity (2.1.21), $d_A F(A) = 0$. There must be a non-trivial open set on which one of these forms has rank 1 exactly, and lemma (4.3.25) gives an open set on which A is reducible. Now invoke (4.3.21) to deduce that A is a reducible connection over all of X, contrary to hypothesis.

4.3.6 Other perturbations

The Freed–Uhlenbeck theorem answers most of our needs in the direction of transversality for the applications in later chapters. There are, however, a number of reasons why one may want to consider also other perturbations of the ASD equations. For example consider the moduli space M_0 when the base manifold X is not simply connected. The moduli space is obtained from representations of the fundamental group and is unaffected by changes in the metric. Or consider moduli spaces for other structure groups G with larger rank, which are not covered by the Freed–Uhlenbeck theorem.

We shall indicate briefly an alternative approach. Returning to the situation described at the end of Section 4.3.2, we consider a Banach vector bundle $\mathscr{V} \to \mathscr{P}$ with a smooth Fredholm section Φ. We wish to construct a perturbation $\Phi + \sigma$, adding a section σ in order to obtain a regular zero set. Comparing this with the finite dimensional problem we encounter three points:

(1) We construct σ using smooth cut-off functions; these do not exist on general Banach manifolds, so we should require our manifold $\mathscr{P}$ to be modelled on a Hilbert space (or, for example, Sobolev spaces L^p_k with p an even integer).

(2) One wants the perturbation to stay within the Fredholm class. This can be achieved by requiring that σ be a 'compact' perturbation in the following sense. We suppose there is another bundle $\mathscr{V}'$ over $\mathscr{P}$ and a bundle map $i: \mathscr{V}' \to \mathscr{V}$ which is a compact inclusion on the fibres; then we ask that $\sigma = i \circ s$, for a smooth section s of $\mathscr{V}$.

(3) Since infinite-dimensional Banach manifolds are not locally compact there is no general reason why transversality in a neighbourhood of a point should be an open condition. Suppose however that $\sigma = i \circ s$ is a compact section as above and a compactness condition of the following kind holds: for all $y_0 \in \mathscr{P}$ and bounded sections s of $\mathscr{V}'$ there is a neighbourhood $\mathscr{P}'$ containing y_0 and an $R > 0$ such that the set

$$\{y \in \mathscr{P}' \mid \Phi(y) \in \mathscr{V}' \quad \text{and} \quad \|(\Phi + s)(y)\|_{\mathscr{V}'} \leq R\}$$

is compact.

Then, as in finite dimensions, the set of sections s such that $\Phi + i \circ s$ is transverse to 0 is open in the C^1 topology.

For our application we take $\mathscr{P}$ to be the Hilbert manifold $\mathscr{B}^*$, formed using connections of class L^2_{l-1} for some $l \geq 3$. For $\mathscr{V}, \mathscr{V}'$ we take the vector bundles ε, $\varepsilon' \to \mathscr{B}^*$ whose fibres consist of the L^2_{l-2} and L^2_{l-1} sections in

$\Omega^+(\mathfrak{g}_E)$ respectively:

$$\varepsilon = \mathscr{A}^* \times_{\mathscr{G}} L^2_{l-2}(\Omega^+(\mathfrak{g}_E)), \quad \varepsilon' = \mathscr{A}^* \times_{\mathscr{G}} L^2_{l-1}(\Omega^+(\mathfrak{g}_E)).$$

As before, Ψ is the canonical section induced by $A \mapsto F^+(A)$. It is easy to verify that for any section s of $\mathscr{V}'$ the properties above hold (the compactness property of (3) follows from the ellipticity of the ($d_A^* \oplus d_A^+$ operators). Then one obtains:

Proposition (4.3.26). *There is a dense (second category) set of perturbations s such that $\Psi + i \circ s$ has a regular zero set.*

Reducible connections can be dealt with similarly. Suppose for example that X is a simply connected manifold with negative-definite intersection form, and A is a reducible solution corresponding to a non-trivial line bundle. Then we have a local model $f^{-1}(0)/\Gamma_A$ where

$$f\colon \mathbb{C}^p \longrightarrow \mathbb{C}^q$$

is an equivariant map. Then we can simply change the ASD equations in the local model by adding a term ρl to f, where l is a generic complex linear map, and ρ is a cut-off function, equal to 1 near the origin.

In some applications these straightforward, but abstractly constructed, perturbations are not adequate because one has to take into account the non-compactness of the moduli spaces, which brings in the moduli spaces for different Chern classes, as we shall see in Section 4.4. One wants to make the perturbations in some more intrinsic way, depending only on the restriction of connections to small regions in X. They can then be defined simultaneously over all the moduli spaces. For example one can make perturbations of this kind using the holonomy of connections around paths.

4.4 Compactification of moduli spaces

Let us return to our five examples and observe that the moduli spaces, while not normally compact, have in each case obvious compactifications suggested by the geometric description. These compactifications involve symmetric products of the underlying four-manifold. In example (i) the natural compactification to take is the closed five-ball, adjoining a copy of the four-sphere itself as a boundary to the moduli space. Our explicit formulae for the ASD solutions explain the geometric meaning of this compactification. We explained in the Euclidean setting of Section 3.4 that the one-parameter family of connections $A_{0,\lambda}$ converges to the flat connection away from 0, and their curvature densities converge to the point mass $8\pi^2\delta_0$, concentrated at the origin. Similarly, viewed on the four-sphere a sequence of connections 'converges' to a point x on the boundary if the curvature densities approach $8\pi^2\delta_x$. We have also seen in Section 3.4 how this can be generalized to 'completions' of the higher-moduli spaces, working again over $\mathbb{R}^4$, and using the ADHM description. Equally one can check that the explicit solutions J_t over $\overline{\mathbb{CP}}^2$ of Example (ii) exhibit much the same behaviour: as $t \to 1$ the curvature densities converge to a point mass at the coordinate origin. Clearly there should be some general theory lying behind these examples, and the development of this theory is our task here.

Section 4.4 is organized as follows. In Section 4.4.1 we define a class of 'ideal ASD connections' and state the main result (essentially due to Uhlenbeck)—the existence of a natural compactification of any ASD moduli space. There are three ingredients in the proof of this result. The main one is the analysis of Section 2.3 of connections with small curvature over a ball. This is extended to a general manifold by a patching argument, discussed in Section 4.4.2, and some elementary properties of the curvature density functions, discussed in Section 4.4.3. The other substantial ingredient is a theorem of Uhlenbeck on the removability of singularities in ASD connections (Theorem (4.4.12)). We postpone the proof of this until the end of the chapter. The proof takes up again the analytical techniques used in Section 2.3.

4.4.1 The compactification

For simplicity we will consider connections on $SU(2)$ bundles here, the extension to general gauge groups being quite straightforward. Thus for an oriented compact Riemannian four-manifold X we have a sequence of moduli spaces M_k, labelled by the Chern class $k \geq 0$. (If X is simply connected, M_0 is a single point representing the product connection.)

Definition (4.4.1). *An ideal ASD connection over X, of Chern class k, is a pair:*

$$([A], (x_1, \dots, x_l)),$$

where $[A]$ is a point in M_{k-l} and $(x_1, \dots, x_l)$ is a multiset of degree l (unordered l-tuple) of points of X. The curvature density of $([A], (x_1, \dots, x_l))$ is the measure:

$$|F(A)|^2 + 8\pi^2 \sum_{r=1}^{l} \delta_{x_r}.$$

So by the Chern–Weil formula (2.1.29) the total mass of the curvature density of an ideal ASD connection is $8\pi^2 k$, where k is the Chern class.

Let A_α, $\alpha \in \mathbb{N}$ be a sequence of connections on $SU(2)$ bundle P_k of Chern class k. We say that the gauge equivalence classes $[A_\alpha]$ *converge weakly* to a limiting ideal ASD connection $([A], (x_1, \dots, x_l))$ if:

Condition (4.4.2).

(i) *The action densities converge as measures, i.e. for any continuous function f on X,*

$$\int_X f|F(A_\alpha)|^2 \, d\mu \longrightarrow \int_X f|F(A)|^2 \, d\mu + 8\pi^2 \sum_{r=1}^{l} f(x_r).$$

(ii) *There are bundle maps* $\rho_\alpha : P_l|_{X \setminus \{x_1, \ldots, x_l\}} \to P_k|_{X \setminus \{x_1, \ldots, x_l\}}$ *such that* $\rho_\alpha^*(A_\alpha)$ *converges (in* C^∞ *on compact subsets of the punctured manifold) to* A.

There is an obvious extension of this definition dealing with the convergence of a sequence of ideal ASD connections. This notion of convergence then endows the set of all ideal ASD connections of fixed Chern class k,

$$IM_k = M_k \cup M_{k-1} \times X \cup M_{k-2} \times s^2(X) \cup \ldots,$$

with a topology. It is not hard to show that this topology is second-countable, Hausdorff and even metrizable. The ordinary moduli space M_k is embedded as an open subset of IM_k. More generally, the induced topology on the different 'strata' $M_{k-l} \times s^l(X)$ is the usual one. We define $\bar{M}_k$ to be the closure of M_k in the space of ideal connections IM_k. This gives a convenient formalism in which to express the main result of this section:

Theorem (4.4.3). *The space* $\bar{M}_k$ *is compact.*

This result follows easily enough from the special case:

Theorem (4.4.4). *Any infinite sequence in* M_k *has a weakly convergent subsequence, with a limit point in* $\bar{M}_k$.

The proof of (4.4.4) takes up the remainder of Section 4.4. We make three remarks before embarking on the proof. First we give the analogue for $SO(3)$ connections. For a fixed value $w \in H^2(X; \mathbb{Z}/2)$ of the Stiefel–Whitney class w_2, we have moduli spaces $M_{k,w}$ on bundles E with $\kappa(E) = -(1/4)p_1(E) = k$. The index k need not be an integer, but any two differ by an integer. The natural compactification of $M_{k,w}$ is now a subset of:

$$M_{k,w} \cup M_{k-1,w} \times X \cup M_{k-2,w} \times s^2(X) \cup \ldots.$$

Second, the reader can check that Theorem (4.4.3) is indeed consistent with all the examples in Section 4.1. Third, it is natural to ask about the structure of the compactified space around the 'points at infinity'. This is a question we will take up in Chapters 7 and 8.

4.4.2 Patching arguments

In Section 2.3 we obtained good control of the ASD connections with small curvature over a ball. We now wish to extend this control to a general manifold using a patching argument. The key point is the conformal invariance of the L^2 norm of curvature in four dimensions. This conformal invariance means that Corollary (2.3.9), stated for connections over the unit ball in Euclidean space, applies *with the same constant* ε to Euclidean balls of arbitrary radius. Further, the result applies, with a small adjustment in the constant, to balls with Riemannian metrics which are close to the Euclidean metric. In particular the results apply to connections over small geodesic balls, of radius r say, in the compact manifold X. For, when rescaled to the standard size, the metric on these balls is within $O(r^2)$ of the Euclidean metric. The first step in the proof of (4.4.4) is a patching argument which takes us from individual small balls to a more global conclusion. In this section we will twice make use of the 'diagonal argument'. Suppose we have a sequence of objects L_α and a countable collection of 'convergence conditions' $C_1, C_2, \ldots$. If we know that for any n and any subsequence $\{\alpha'\} \subset \{\alpha\}$ there is a sub-subsequence $\{\alpha''\} \subset \{\alpha'\}$ which satisfies condition C_n, then we can conclude that there is some subsequence which satisfies *all* the conditions C_n simultaneously. The proof is standard.

We begin with a simple lemma, which contains the crux of the matter. In this section, when we state that a sequence of connections converges, without other qualification, we mean C^∞ convergence over compact subsets.

Lemma (4.4.5). *Suppose that A_α is a sequence of unitary connections on a bundle E over a base manifold Ω (possibly non-compact), and let $\tilde{\Omega} \Subset \Omega$ be an interior domain. Suppose that there are gauge transformations $u_\alpha \in \mathrm{Aut}\, E$ and $\tilde{u}_\alpha \in \mathrm{Aut}\, E|_{\tilde{\Omega}}$ such that $u_\alpha(A_\alpha)$ converges over Ω and $\tilde{u}_\alpha(A_\alpha)$ converges over $\tilde{\Omega}$. Then for any compact set $K \subset \tilde{\Omega}$ we can find a subsequence $\{\alpha'\} \subset \{\alpha\}$ and gauge transformations $w_{\alpha'} \in \mathrm{Aut}\, E$ such that $w_{\alpha'} = \tilde{u}_{\alpha'}$ in a neighbourhood of K and the connections $w_{\alpha'}(A_{\alpha'})$ converge over Ω.*

This follows the line of ideas begun in Section 2.3.7. There is no loss in supposing that the u_α are all the identity, so over $\tilde{\Omega}$ both A_α and $\tilde{u}_\alpha(A_\alpha)$ are convergent sequences of connections. We may suppose, taking a subsequence $\{\alpha'\}$, that the $\tilde{u}_{\alpha'}$ converge over $\tilde{\Omega}$ to a limit $\tilde{u}$. Now, fixing a precompact neighbourhood N of K, we extend $\tilde{u}|_N$ arbitrarily over Ω, to a gauge transformation u^* say. Also, over N we can write

$$\tilde{u}_{\alpha'} = \exp(\xi_{\alpha'})\tilde{u},$$

for sections $\xi_{\alpha'}$ of the bundle $\mathfrak{g}_E$ which converge to 0. Now let ψ be a cut-off function, supported in N and equal to 1 on a neighbourhood of K. We put

$$w_{\alpha'} = \exp(\psi\xi_{\alpha'})u^*.$$

Then the $w_{\alpha'}(A_{\alpha'})$ are convergent on Ω, since the $\xi_{\alpha'}$ converge over the support of ψ. On the other hand $w_{\alpha'} = \tilde{u}_{\alpha'}$ on a neighbourhood of K, as required.

We observe that in this proof we can replace the C^∞ convergence by convergence in suitable Sobolev topologies, for example $L^2_{l-1,\mathrm{loc}}$. We can isolate the property needed for the argument to work. Suppose the topology on connections is defined by a norm $\| \;\|_\nu$ on 1-forms. Then we need to know that any sequence of functions f_α with $\|\mathrm{d}f_\alpha\|_\nu$ bounded has a subsequence converging uniformly on compact subsets. Note, as in Section 4.2.1, that this *fails* for the L^2_1 topology on connections.

There are two useful extensions of this simple result.

Lemma (4.4.6). *Suppose that Ω is exhausted by an increasing sequence of precompact open sets*

$$U_1 \Subset U_2 \Subset \ldots \subset \Omega, \quad \bigcup_{n=1}^{\infty} U_n = \Omega.$$

Suppose A_α is a sequence of connections over Ω and for each n there is a subsequence $\{\alpha'\}$ and gauge transformations $u_{\alpha'} \in \operatorname{Aut} E|_{U_n}$ such that $u_{\alpha'}(A_{\alpha'})$ converges over U_n. Then there is a subsequence, and a sequence of gauge transformations, such that the transformed connections converge over all of Ω.

The proof is an application of the diagonal argument, using Lemma (4.4.5) to choose successive compatible sequences of gauge transformations. We leave details as an exercise.

Lemma (4.4.7). *Suppose Ω is a union of domains $\Omega = \Omega_1 \cup \Omega_2$ and A_α is a sequence of connections on a bundle E over Ω. If there are sequences of gauge transformations $v_\alpha \in \operatorname{Aut} E|_{\Omega_1}$, $w_\alpha \in \operatorname{Aut} E|_{\Omega_2}$ such that $v_\alpha(A_\alpha)$ and $w_\alpha(A_\alpha)$ converge over Ω_1 and Ω_2 respectively, then there is a subsequence $\{\alpha'\}$ and gauge transformations $u_{\alpha'}$ over Ω such that $u_{\alpha'}(A_{\alpha'})$ converges over Ω.*

By Lemma (4.4.6) it suffices to treat a compact subset of Ω, covered by precompact sets $\Omega'_1 \Subset \Omega_1$, $\Omega'_2 \Subset \Omega_2$ say. We apply Lemma (4.4.5) with Ω_1 taking the place of Ω and K a compact neighbourhood of $\Omega'_1 \cap \Omega'_2$ in $\Omega_1 \cap \Omega_2$. After modifying v_α, and taking a subsequence, we may as well suppose that $v_\alpha = w_\alpha$ on $\Omega'_1 \cap \Omega'_2$. Then the two sequences of gauge transformations v_α, w_α glue together to define gauge transformations u_α over the union $\Omega'_1 \cup \Omega'_2$.

We can combine these results into a very simple statement.

Corollary (4.4.8). *Suppose A_α is a sequence of connections on a bundle E over Ω with the following property. For each point x of Ω there is a neighbourhood D of x, a subsequence $\{\alpha'\}$, and gauge transformations $v_{\alpha'}$ defined over D such that $v_{\alpha'}(A_{\alpha'})$ converges over D. Then there is a single subsequence $\{\alpha''\}$ and gauge transformations $u_{\alpha''}$ defined over all of Ω, such that $u_{\alpha''}(A_{\alpha''})$ converges over all of Ω.*

In brief, local and global convergence of connections modulo gauge transformations are equivalent, if we are allowed to take subsequences.

Again, by Lemma (4.4.6), we can restrict attention to a precompact subset of Ω, which we may suppose is a finite union of neighbourhoods $D_1, \ldots, D_m$ of the kind appearing in the hypothesis. We then argue by induction on the number m of balls. If $m = 1$ the assertion is trivial. If we know inductively that, after taking a subsequence and applying gauge transformations, the connections converge over $\Omega_{m-1} = D_1 \cup \ldots \cup D_{m-1}$ we apply Lemma (4.4.5) to the pair Ω_{m-1}, D_m.

In our applications of this Lemma the convergence is obtained from the ASD equation, via Uhlenbeck's theorem.

Proposition (4.4.9). *Let Ω be an oriented Riemannian four-manifold. Suppose A_α is a sequence of ASD unitary connections on a bundle E over Ω with the following property. For each point $x \in \Omega$ there is a geodesic ball D_x, such that for all large enough α,*

$$\int_{D_x} |F(A_\alpha)|^2 \, d\mu \leq \varepsilon^2,$$

where $\varepsilon > 0$ is the constant of (2.3.9). Then there is a subsequence $\{\alpha'\}$ and gauge transformations $u_{\alpha'} \in \operatorname{Aut} E$ such that $u_{\alpha'}(A_{\alpha'})$ converges over Ω.

If the condition holds for a given ball D_x, it also holds for any smaller ball. So we may assume that, when rescaled to standard size, the metric on the ball is arbitrarily close to the Euclidean metric. Then we combine (4.4.8) and (2.3.9).

We will need a slight extension of this result in the case when the limiting connection is actually flat. Suppose Ω is simply connected and for any compact set $K \subset \Omega$,

$$\int_K |F(A_\alpha)|^2 \, d\mu \longrightarrow 0.$$

Then the hypotheses of (4.4.9) are obviously satisfied, and the limiting connection must be flat. For a simply connected base space this implies that the limit is the product connection, so our result asserts that there are connection matrices $A^\tau_{\alpha'}$ which tend to zero, in C^∞ on compact subsets. It is often useful to know more explicitly how small the connection matrices can be made, in terms of the curvature (cf. our discussion in Section 2.1.1). This is specially easy if Ω can be covered in a simple way by balls. Let us say that a domain Ω is *strongly simply connected* if we can find a cover by balls $D_1, \ldots, D_m$ such that for $1 \leq r \leq m$ the intersection $D_r \cap (D_1 \cup \ldots \cup D_{r-1})$ is connected. Here we can take the balls D_r to be any differentiably embedded balls, not necessarily geodesic balls. (This condition is easily seen to imply that Ω is simply connected. The condition is closely related to the existence of a handle decomposition of Ω with no one-handles.)

Proposition (4.4.10). *If Ω is strongly simply connected and $\Omega' \Subset \Omega$ is a pre-compact interior domain, then there are constants $\varepsilon_\Omega, M_{\Omega,\Omega'} > 0$ such that any ASD connection A over Ω with*

$$\int_\Omega |F(A)|^2 \, d\mu \equiv \|F(A)\|^2 < \varepsilon_\Omega^2$$

can be represented over Ω' by a connection matrix A^τ with

$$\int_{\Omega'} |A^\tau|^4 \, d\mu \leq M_{\Omega,\Omega'} \|F(A)\|^4.$$

This gives a general answer to the question posed in Section 2.3 of finding a small connection matrix for an ASD connection with small curvature. We can actually estimate any norm of the derivatives of the connection matrix, with suitable constants, in an interior domain. The version given here, for the L^4 norm, is the one we shall need in Section 4.4.4. The virtue of the L^4 norm on 1-forms is that, like the L^2 norm on 2-forms, it is conformally invariant (in four dimensions). So the constants in (4.4.10) depend only on the conformal structure.

The proof of (4.4.10) is just a matter of following through the argument of (4.4.7), putting in estimates at each stage. For simplicity consider the case when $m = 2$ (for the general case one applies this argument inductively, as above). We use (2.3.7) and (2.3.8) to choose connection matrices A^σ, A^τ over D_1, D_2 respectively, all of whose derivatives (on interior domains) are controlled by the L^2 norm of the curvature. These are intertwined by a transition function u over the intersection. Then

$$\mathrm{d}u = uA^\tau - A^\sigma u,$$

so we get L_l^2 bounds on $\mathrm{d}u$ on interior domains of the intersection. When the curvature is small the A^τ, A^σ are small, so the variation of u is small. Since the intersection is connected, by hypothesis, u is close to a constant u_0. There is no loss in supposing that u_0 is the identity, since we can always conjugate A^σ by u_0 without changing the problem. Then we write $u = \exp(\xi)$ and modify A^σ to

$$A^{\sigma'} = \exp(-\psi\xi)A^\sigma,$$

where ψ is a cut-off function equal to 1 on an interior domain $N \Subset D_1 \cap D_2$, containing $\Omega' \cap D_1 \cap D_2$. So

$$A^\sigma = \exp(-\psi\xi)\exp(\xi)A^\tau,$$

and $A^{\sigma'} = A^\tau$ on N. These connection matrices thus match up and give a connection matrix A^π over all of Ω'. We can estimate the norm of A^π, via the norms of $\mathrm{d}u$ and $\mathrm{d}\xi$, from our estimates on A^σ, A^τ, and hence in terms of the L^2 norm of the curvature.

The attraction of this argument is that the constants ε_Ω, $M_{\Omega,\Omega'}$ can be computed explicitly from the geometry of the cover and the constants in the corresponding local results (2.3.7) and (2.3.8). In fact, the same result is true if one just assumes Ω to be simply connected, or even that $\pi_1(\Omega)$ has no non-trivial representation in the structure group G, but it is then much harder to give explicit constants.

In the discussion above, the ASD condition is only being used in an auxiliary way, to obtain elliptic estimates on the higher derivatives of connection matrices. There are similar results for general connections. In four dimensions, L^2 control of the curvature is not by itself enough, since this gives only L_1^2 control of the connection matrices and we cannot control the variation of the transition functions. However in lower dimensions the theory works well, since we then have a Sobolev embedding $L_2^2 \to C^0$. We note that the proof of Uhlenbeck's theorem applies equally well in lower dimensions. We obtain for example:

Proposition (4.4.11). *Let W be a compact, strongly simply connected manifold of dimension 2 or 3. There are constants η, M such that any connection over W with*

$$\|F(A)\|_{L^2} < \eta$$

can be represented by a connection matrix A^τ with

$$\|A^\tau\|_{L_1^2} \leq M \|F(A)\|_{L^2}.$$

(A corresponding result holds in four dimensions if we are given L^p bounds on the curvature, for some $p > 2$.)

4.4.3 Proof of the compactness theorem

We can now return to our main goal and give a proof of (4.4.4), assuming one extra fact which will be taken up in Section 4.4.4. This is the 'Removable Singularities' theorem of Uhlenbeck. The relevant version for us is:

Theorem (4.4.12). *Let A be a unitary connection over the punctured ball $B^4 \setminus \{0\}$, which is ASD with respect to a smooth metric on B^4. If*

$$\int_{B^4 \setminus \{0\}} |F(A)|^2 < \infty,$$

then there is a smooth ASD connection over B^4 gauge equivalent to A over the punctured ball.

To spell out the precise meaning of the statement: if the connection A in the theorem is a connection on a bundle E over $B^4 \setminus \{0\}$ there is a connection A' on a bundle E' over B^4 and a bundle map $\rho: E \to E'|_{B^4 \setminus \{0\}}$ with $\rho^*(A') = A$.

Given (4.4.12), and (4.4.9) from Section 4.4.2, the proof of (4.4.4) follows easily enough from two pieces of general theory, involving the two interpretations of the curvature density of an ASD connection: as a positive measure on X and as a four-form representing a topological invariant (cf. 2.1.4).

We shall regard the measures as lying in the dual space of $C^0(X)$. Recall first that for any sequence ν_α of positive measures on X with the $\int_X d\nu_\alpha$ bounded, we can find a subsequence $\nu_{\alpha'}$ converging to a limiting measure ν in the sense that for any continuous function f on X,

$$\int_X f \, d\nu_{\alpha'} \longrightarrow \int_X f \, d\nu.$$

This is the property of 'weak-$*$ compactness' of the ball in the dual space. The proof is an easy application of the diagonal argument. We choose a countable sequence of functions f_i whose linear span is dense in $C^0(X)$ (for example smoothings of the characteristic functions of balls). For each i the ν_α integrals of f_i form a bounded sequence of real numbers, and so have a convergent subsequence; the diagonal argument allows us to choose a single subsequence $\{\alpha'\}$ making all these integrals converge simultaneously. The result for the general function follows from the density of the span of the f_i.

The second piece of theory involves the interpretation of the curvature density of an ASD connection A on a bundle E over the closed manifold X as a topological invariant:

$$\int_X |F(A)|^2 = -\int_X \mathrm{Tr}(F(A)^2) = 8\pi^2\kappa(E).$$

The primary role of this in our argument is that it gives a fixed bound on the L^2 norm of the curvature of an ASD connection on a given bundle. We also need an extension to a local version of the formula.

Suppose Z is a compact oriented four-manifold with boundary $\partial Z = W$ and B is a connection over W. Choose any extension of B to a connection A over Z and form the integral $\int_Z \mathrm{Tr}(F(A)^2)$ as above. Modulo $8\pi^2\mathbb{Z}$ this integral depends only on the connection B over W, not on A or Z. To see this one considers two extensions over manifolds Z, Z' and then glues these together to get a connection over the closed manifold $Z \cup_W Z'$. Dividing by $8\pi^2$ we thus get an invariant, the *Chern–Simons invariant* $\tau_W(B) \in \mathbb{R}/\mathbb{Z}$ of the connection over the three-manifold W. This can alternatively be expressed as follows: we choose a trivialization of the bundle over W to represent the connection by a connection matrix, which we also call B. Then:

$$\tau_W(B) = \frac{1}{8\pi^2}\int_W \mathrm{Tr}(\mathrm{d}B \wedge B + \tfrac{2}{3} B \wedge B \wedge B), \quad \mathrm{mod}\ \mathbb{Z}.$$

The proof of this assertion follows easily from the similar formula (2.1.27) in Chapter 2 for the variation of the Chern–Weil form. The only part of this theory we need here is the fact that the gauge invariant quantity $\tau_W(B)$ varies *continuously* with the connection B.

With this background material to hand we can complete the proof of (4.4.4) in short order. Let A_α be a sequence of ASD connections on a bundle E with $c_2(E) = k$ as in the statement of (4.4.4). We show first that there is a finite set $\{x_1, \dots, x_p\}$ in X such that, after taking a subsequence, the punctured manifold $X \setminus \{x_1, \dots, x_p\}$ satisfies the hypotheses of (4.4.9). For this we just choose a subsequence $\{\alpha'\}$ so that the curvature densities $|F(A_{\alpha'})|^2$ converge, as measures, to a limit measure ν. Then

$$\int_X \mathrm{d}\nu = 8\pi^2 k,$$

so there are at most $8\pi^2 k/\varepsilon^2$ points in X which do not lie in a geodesic ball of ν-measure less than ε^2 (otherwise we could take disjoint balls about the x_r which together would have ν-measure more than $8\pi^2 k$). We let these points be $x_1, \ldots, x_p$. Then by (4.4.9) we can take another subsequence $\{\alpha''\} \subset \{\alpha'\}$ and gauge transformations $u_{\alpha''}$ over $X \backslash \{x_1, \ldots, x_p\}$ such that the sequence $u_{\alpha''}(A_{\alpha''})$ converges over this punctured manifold to an ASD connection A on $E|_{X\backslash\{x_1,\ldots,x_p\}}$.

Plainly

$$\int_{X\backslash\{x_1,\ldots,x_p\}} |F(A)|^2 \leq 8\pi^2 k.$$

In particular, the left-hand side is finite. Thus we can invoke Theorem (4.4.12) to deduce that the restriction of A to a punctured ball about any of the x_r extends to a smooth connection over the ball. But this just means that A extends to a connection on a bundle E' over X. Of course E' need not be isomorphic to E; indeed if p is bigger than zero it cannot be since, from the definition of the x_r, we must have strict inequality in the line above—we must 'lose' at least ε^2 units of energy at each point x_r. Similarly, it is easy to see that the limiting measure ν is

$$\nu = |F(A)|^2 + \sum_{r=1}^{p} n_r \delta_{x_r},$$

for some real numbers $n_r \geq \varepsilon^2$. (We point out again that no deep facts about measures are being used in this argument: a direct approach is to choose a countable basis for the topology of X consisting of small geodesic balls D_λ and then arrange, by a diagonal argument, that all the integrals $I(\lambda, \alpha')$ of the $|F(A'_\alpha)|^2$ over the D_λ converge to limits as $\alpha' \to \infty$. Then the x_r are the points which do not lie in any ball D_λ with $I(\lambda, \alpha') < \varepsilon^2$ for all large α'.)

To complete the proof we need only show that each of the coefficients n_r is an integer. We can then define a multiset $(x_1, \ldots, x_l)$, repeating the points according to the multiplicities n_r, and the two properties in the definition of weak convergence are satisfied. This integrality follows from the relative version of the Chern–Weil theory. We choose small disjoint balls Z_r in X centred on the points x_r. Clearly

$$\tau_{\partial Z_r}(A) = \lim \tau_{\partial Z_r}(A''_\alpha) \in \mathbb{R}/\mathbb{Z},$$

since after gauge transformations the connections converge in C^∞ on ∂Z_r. On the other hand, the convergence of the measures gives

$$n_r = \frac{1}{8\pi^2} \lim \int_{Z_i} \operatorname{Tr}(F(A_\alpha)^2) - \operatorname{Tr}(F(A)^2).$$

Using the definition of $\tau_{\partial Z_r}$ in terms of an extension over the ball Z_r we see that $n_r = 0 \bmod \mathbb{Z}$ as required.

4.4.4 *The removable singularities theorem: regularity of L_1^2 solutions*

We will base our proof of (4.4.12) on the gauge fixing theorem (Theorem (2.3.7)) of Chapter 2 and the sharp regularity theorem promised in Section 4.2.3. We begin with the latter. Following our usual practice we set up the problem on the compact manifold S^4.

Proposition (4.4.13). *There is a constant $\zeta > 0$ such that if A is any L_1^2 connection matrix on the trivial bundle over S^4 with:*

(i) $d^*A = 0$
(ii) $F^+(A) \equiv d^+A + (A \wedge A)^+$ *is smooth*
(iii) $\|A\|_{L_1^2} \leq \zeta$

then A is smooth.

Note first that, if we knew that A was in L_3^2 the conclusion would follow from the standard bootstrapping argument, using the ellipticity of $d^* + d^+$, cf. (4.2.16). The point of the result here is precisely that we obtain information on the 'borderline' of the Sobolev inequalities. To achieve this we use much the same 'rearrangement' argument as in (2.3.10). We suppose first that A is smooth and seek to estimate the Sobolev norms of A, using the given equations. Let us write ϕ for $F^+(A)$. First we have

$$\|A\|_{L_1^2} \leq C\|(d^* + d^+)A\|_{L^2} \leq C(\|\phi\|_{L^2} + \|(A \wedge A)^+\|_{L^2},$$

and the last term is estimated, via Sobolev and Hölder inequalities by a multiple of $\|A\|_{L_1^2}^2$. If ζ, and hence the L_1^2-norm of A, is sufficiently small we can rearrange this to get a bound $\|A\|_{L_1^2} \leq \text{const.}\ \|\phi\|_{L^2}$ (as in (2.3.10)). Then for the L_2^2 norm we have

$$\|A\|_{L_2^2} \leq \text{const.}\ (\|\phi\|_{L_1^2} + \|(A \wedge A)^+\|_{L_1^2}).$$

As in (2.3.11) the last term is estimated by $\|A\|_{L^4}\|A\|_{L_2^2}$. Once ζ is small we can rearrange to get a bound on the L_2^2 norm of A. Similarly for the higher norms. Just as in (2.3.11) the picture changes for the L_4^2 norm for which we can use a simpler estimate. The upshot is that we get *a priori* bounds on all the norms of A in terms of the corresponding norms of ϕ, once ζ is small.

The discussion so far may seem perverse since we assume precisely what we want to prove—the smoothness of A. To bring the result to bear we observe that if B is another solution to

$$d^*B = 0, \quad d^+B + (B \wedge B)^+ = \phi$$

with $\|B\|_{L_1^2} \leq \zeta$, then

$$\begin{aligned}(d^* + d^+)(A - B) &= (B \wedge B - A \wedge A)^+ \\ &= ((B - A) \wedge A)^+ + (B \wedge (B - A))^+,\end{aligned}$$

which gives an estimate:

$$\| A - B \|_{L_1^2} \leq \text{const.}(\| A - B \|_{L_1^2} \| A + B \|_{L_1^2}).$$

So, again, when ζ is small, we must have $B = A$. Thus it suffices to show that there is *some* small, smooth, solution to these equations. To do this we use the method of continuity, embedding our equation in the family:

$$\mathrm{d}^* A_t = 0, \quad \mathrm{d}^+ A_t + (A_t \wedge A_t)^+ = t\phi, \tag{4.4.14}$$

for $0 \leq t \leq 1$.

There are constants η_0, C such that if A_t is a solution to (4.4.14) with $\| A_t \|_{L_1^2} < \eta_0$ then $\| A_t \|_{L_1^2} \leq 2C \| \phi \|_{L^2}$. So if the L^2 norm of ϕ is less than $\eta_0/4C$ we have

$$\| A_t \|_{L_1^2} < \eta_0 \Rightarrow \| A_t \|_{L_1^2} \leq \tfrac{1}{2}\eta_0.$$

The condition on the norm of ϕ is arranged by choosing ζ small. It follows then that the open constraint $\| A_t \|_{L_1^2} < \eta_0$ is closed, and by the now-familiar argument we see that the set of times t for which such a small solution A_t exists is closed (cf. Sections 2.3.7 and 2.3.9).

On the other hand, we prove that this set is open using the implicit function theorem. The linearization of the equation (4.4.14) at a solution A_t has the form

$$L(a) \equiv ((\mathrm{d}^* + \mathrm{d}^+) + P)a = \phi,$$

where $P(a) = (A_t \wedge a + a \wedge A_t)^+$. The L_1^2 to L^2 operator norm of P is small when A_t is small in L_1^2. Since $\mathrm{d}^* + \mathrm{d}^+$ is an invertible operator between these spaces, we deduce that L has kernel 0 if ζ is small. Then the Fredholm alternative tells us that L is invertible as a map from L_3^2 to L_2^2. So, by the implicit function theorem, the solution can be continued for a small time interval as an L_3^2 connection matrix. But, by the remark at the beginning of the proof, any L_3^2 solution is smooth, so the proof is complete.

While this regularity result is just what we need for the proof of (4.4.12) we should mention that it leads to a general regularity theorem for L_1^2 solutions of the ASD equations. For this one uses an extension of (2.3.7) to show that any L_1^2 connection can be locally transformed, by an L_2^2 gauge transformation, to satisfy the Coulomb condition.

4.4.5 Cutting off connections

Our strategy of proof of (4.4.12) is similar to that used for a linear equation in (3.3.22). We make a sequence of cut-offs to extend the connection over the singularity, introducing some error term, then examine the behaviour of the error term as the cut-off shrinks down to a point.

Let us write $D(r)$ for the r-ball about the origin in $\mathbb{R}^4$, and for $r < 1$ let $\Omega(r)$ be the complement $B^4 \setminus \overline{D(r)}$. In this section we will prove:

Lemma (4.4.15). *Let A be a connection on a bundle E over $B^4\backslash\{0\}$ which satisfies the hypotheses of* (4.4.12). *Then for all small enough r there is a connection A_r on a bundle E_r over B^4, and a bundle isomorphism*

$$\rho_r : E|_{\Omega(r)} \longrightarrow E_r|_{\Omega(r)}$$

such that:

(i) $\rho_r^*(A_r) = A$ *over* $\Omega(r)$

(ii) $\int_{B^4} |F^+(A_r)|^2 \, d\mu \to 0$, *as* $r \to 0$.

For the proof of (4.4.15) we will make use of a simple 'cutting-off' construction for connections which will appear again in Chapter 7. The general set-up is as follows: we have a connection A on a bundle E over a four-manifold Z and a trivialization τ of the bundle over an open set $\Omega \subset Z$, so A is represented by a connection matrix A^τ over Ω. Let ψ be a smooth function on Z, taking values in $[0, 1]$ and equal to 1 on a neighbourhood of $Z\backslash\Omega$. We define a new connection $A(\tau, \psi)$ on E, equal to A on $Z\backslash\Omega$ and given, in the same trivialization, by the connection matrix ψA^τ on Ω. Clearly these definitions do patch together to yield a connection over all of Z. For brevity we will sometimes just denote this connection by ψA, suppressing the trivialization used over Ω. (Although it should be emphasized that the gauge equivalence class of the connection $A(\tau, \psi)$ does depend on τ.)

The curvature of the connection ψA is

$$F(\psi A) = \psi F(A) + (d\psi)A^\tau + (\psi^2 - \psi)(A^\tau \wedge A^\tau). \qquad (4.4.16)$$

In particular, if A is ASD the self-dual part $F^+(A)$ is supported in Ω and

$$\| F^+(\psi A)\|_{L^2} \le \| d\psi \|_{L^4} \| A^\tau \|_{L^4} + \| A^\tau \|_{L^4}^2. \qquad (4.4.17)$$

We will use this construction most often in the situation where we have a decomposition of a four-manifold X into open sets $X = Z \cup Z'$ and $\Omega = Z \cap Z'$. We suppose ψ is a smooth function over X, vanishing outside Z. Then the connection $A(\tau, \psi)$ has a canonical extension to a connection over X—extending by the product connection (zero connection matrix) outside Z. We will still denote this connection by ψA. Similarly, it may happen that the original connection A was defined over a rather larger subset of X than Z—and we will still write ψA for the connection over X obtained from the restriction of A to Z by the procedure above.

With these general remarks in place we proceed to the proof of (4.4.15). Consider the four-dimensional annulus

$$\mathcal{N} = \{x \in \mathbb{R}^4 \,|\, \tfrac{1}{2} < |x| < 1\},$$

and fix a slightly smaller annulus $\mathcal{N}' \Subset \mathcal{N}$. Then $\mathcal{N}'$ satisfies the 'strongly simply connected' condition of (4.4.10) (it may be covered by two balls meeting in a set which retracts onto a two-sphere). So there are constants $\varepsilon_{\mathcal{N}}$, $M_{\mathcal{N},\mathcal{N}'}$ such that a connection with $\| F \|_{L^2} < \varepsilon_{\mathcal{N}}$ can be represented by a connection matrix A^τ with L^4 norm bounded by $M_{\mathcal{N},\mathcal{N}'}\|F\|_{L^2}$. Now fix a cut-off function ψ as above, equal to 1 on the outer boundary of $\mathcal{N}'$ and vanishing on the inner boundary. The cut-off connection ψA then extends smoothly over the unit ball and we have, combining (4.4.17) with the estimate on the L^4 norm of A^τ:

$$\|F^+(\psi A)\|_{L^2(\mathcal{N}')} \leq C.\|F(A)\|_{L^2(\mathcal{N}')},$$

with a constant C independent of A.

For $r < 1$ let $\mathcal{N}(r)$, $\mathcal{N}'(r)$ be the images of the above annuli under the dilation map $x \mapsto rx$. We can apply the construction equally well to these rescaled annuli, and by the scale invariance the relevant constants will be independent of r. (It is clear that the deviation of the metric on the ball from the Euclidean metric will be irrelevant here.) So now if we have an ASD connection A over the punctured ball $B^4 \setminus \{0\}$ and if $\|F(A)\|_{L^2(\mathcal{N}(r))} < K_{\mathcal{N}}$, we can obtain a new connection, $A_r = \psi_r A$, defined over the whole ball, equal to A outside $D(r)$, and with:

$$\|F^+(\psi_r A)\| \leq C.\|F(A)\|_{L^2(\mathcal{N}(r))}.$$

Now if the curvature of A has finite L^2 norm over the punctured ball, as in the hypotheses of (4.4.12), the L^2 norm of $F^+(\psi_r A)$ tends to 0 with r, and the proof of (4.4.15) is complete.

4.4.6 Completion of proof of removable singularities theorem

To complete the proof we wish to apply (4.4.13), and this requires that we transfer our connections to the four-sphere. (This is an auxiliary step which could be avoided.) We therefore introduce another parameter R, with $r < \frac{1}{2}R < \frac{1}{2}$ and construct another connection $A(r, R)$, modifying A_r by cutting-off in the other direction over the annulus $\mathcal{N}(R)$. Thus, in a suitable gauge over $\mathcal{N}(R)$, we multiply our connection matrices by a cut-off function $(1 - \psi_R)$ vanishing on the outer boundary of $\mathcal{N}(R)$, and equal to 1 on the inner boundary. These connections can then be regarded as connections over $S^4 = \mathbb{R}^4 \cup \{\infty\}$. As in the proof of (4.4.15), the L^2 norm of the curvature of $A(r, R)$ can be made as small as we please by making R small. Thus we can apply (2.3.7) to find trivializations $\tau = \tau(r, R)$ such that the connection matrices

$$A^\tau(r, R)$$

for $A(r, R)$ satisfy the Coulomb condition $d^*A^\tau = 0$. We can also suppose the L^2_1 norms of the $A^\tau(r, R)$ are as small as we please, by fixing R small. In particular we can suppose that $A^\tau(r, R)$ satisfies the condition

$$\|A^\tau\|_{L^2_1} \leq \zeta$$

of (4.4.13). We now fix R, and let r tend to 0. We have a family of connection matrices which are bounded in L^2_1. So, by the weak compactness of the unit ball in a Hilbert space, we can find a sequence $r_i \to 0$ such that the connection matrices $A^\tau(r_i, R)$ converge weakly in L_1^2 to a limit $\tilde{A}^\tau$, which also satisfies the Coulomb condition.

Over any ball $B \Subset S^4 \setminus \{0\}$, we have uniform bounds on the covariant derivatives of the curvature of the $A^\tau(R, r)$ as r tends to 0. This implies, by (2.3.11), that we get a uniform bound on the higher Sobolev norms of the $A^\tau(r, R)$ over B (using again the scale invariance of the L^4 norm). Thus we can suppose that $A^\tau(r_i, R)$ converges in C^∞ to $\tilde{A}^\tau$ over compact subsets of $S^4 \setminus \{0\}$.

The proof of (4.4.12) is now in our hands. We know that $F^+(A^\tau(r, R))$ tends to 0 in L^2 over a neighbourhood of the origin. So $F^+(\tilde{A}^\tau)$ vanishes near 0, and is smooth, since the convergence is in C^∞ away from 0. By construction, the L_1^2 norm of $\tilde{A}^\tau$ is less than or equal to ζ. So by (4.4.13) the connection matrix $\tilde{A}^\tau$ is smooth over all of S^4. As in Section 2.3.7, we can suppose that the bundle trivializations $\tau(r_i, R)$ converge as $i \to \infty$, in C^∞ over compact subsets of $S^4 \setminus \{0\}$ to a limit σ. Now restrict the data to the fixed ball $B(\frac{1}{2}R)$. The connection matrix $\tilde{A}^\tau$ is smooth over the origin. On the other hand, over the punctured ball it is the connection matrix for A in the trivialization σ.

Notes

Section 4.1

These examples are gathered from a number of sources. Example (i) is very well known; see for example Atiyah *et al.* (1978b). For Example (ii) see Buchdahl (1986) and Donaldson (1985*b*). The remaining examples use the correspondence between ASD solutions and stable bundles described in Chapter 6. For examples (iii) and (iv), discussed in the algebro-geometric framework, see Barth (1977) and Okonek *et al.* (1980, Chapter 4). The classification of bundles over $S^2 \times S^2$ used in Example (v) was given by Soberon–Chavez (1985), with an extra technical condition which was removed by Mong (1988).

Sections 4.2.1, 4.2.2 and 4.2.3

This material is standard in Yang–Mills theory; see for example Atiyah *et al.* (1978*b*), Mitter and Viallet (1981) and Parker (1982). A general reference for differential topology in infinite-dimensional spaces is Eells (1966).

Sections 4.2.4 and 4.2.5

The local decomposition of Fredholm maps has been used in many contexts; the application to moduli problems goes back to Kuranishi (1965), in the case of moduli of complex structures. For the standard results on Fredholm operators see, for example, Lang (1969).

Section 4.3.1

For a systematic development of transversality theory we refer to [Hirsch, 1976].

Section 4.3.2

The extension of Sard's theorem to Fredholm maps was given by Smale (1965), together with some applications to differential topology in infinite dimensions and partial differential equations.

Section 4.3.3

The results on moduli spaces for generic metrics were proved by Freed and Uhlenbeck (1984). The discussion of the variation of harmonic forms with the metric is taken from Donaldson (1986).

Section 4.3.4

The result used in the proof of Lemma (4.3.21) is taken from Agmon and Nirenberg (1967). The Corollary (4.3.23), which is also used in the proof of Freed and Uhlenbeck, is usually deduced from the theorem of Aronszajin (1957) for second-order equations.

Section 4.3.6

For discussions of other perturbations of the ASD equations see Donaldson (1983*a*, 1987*b*, 1990*a*) and Furuta (1987).

Section 4.4.1

This compactification of the moduli space was defined by Donaldson (1986), although the idea is essentially implicit in the work of Uhlenbeck. A similar 'weak compactness' theorem was proved by Sedlacek (1982). For a purist the definition of a topology by specifying the convergent subsequences is not very satisfactory in general. However there are no difficulties in this case, since the space is metrizable. A metric on the moduli space M which yields the compactified space $\bar{M}$ as the metric completion is defined by Donaldson (1990*b*).

Section 4.4.2

These patching arguments are basically elementary, and the construction is much the same as that in Uhlenbeck (1982*b*). For the construction of small connection matrices over spheres using radial gauge fixing see Uhlenbeck (1982*a*) and Freed and Uhlenbeck (1984).

Section 4.4.3

There are a number of ways of setting out the proof of the compactness theorem; for an alternative see Freed and Uhlenbeck (1984). For the theory of the Chern–Simons invariant used in our approach, see [Chern, 1979, Appendix].

Section 4.4.4

The removable singularities theorem was first proved by Uhlenbeck (1982*b*). The proof we give here is different and, we hope, simpler. Another approach is to obtain *a priori* bounds on the curvature of a connection over the punctured ball; see for example Freed and Uhlenbeck (1984, Appendix D). Such bounds can be obtained from the results of Section 7.3. For generalizations of the removable singularities theorem see Sibner and Sibner (1988) and the references quoted there.

5

TOPOLOGY AND CONNECTIONS

Let P be a principal G-bundle over a compact connected manifold X, let $\mathscr{A} = \mathscr{A}_{X,P}$ be the space of connections in P, and let $\mathscr{G}$ be the gauge group—the group of bundle automorphisms. The main theme of this chapter is the topology of the orbit space $\mathscr{B} = \mathscr{A}/\mathscr{G}$ introduced in Chapter 4 and of its open subset $\mathscr{B}^* = \mathscr{A}^*/\mathscr{G}$, the space of irreducible connections modulo gauge transformations. Previously we have examined the local structure of the orbit space and seen that $\mathscr{B}^*$ is a Banach manifold, as long as we allow our connection matrices to have entries in a suitable Sobolev space. Now, however, we are interested in the global topology.

Although $\mathscr{A}$ is an affine space, and hence contractible, this is far from being true of $\mathscr{B}$. The non-triviality of the orbit space is a reflection of the impossibility of finding a uniform, global procedure by which to pick out a preferred gauge for each equivalence class of connections; such a procedure would define a section, $s: \mathscr{B} \to \mathscr{A}$, for the quotient map $p: \mathscr{A} \to \mathscr{B}$. In turn, the non-existence of a global gauge-fixing condition can be deduced from the existence of topologically non-trivial families of connections. The notion of a family of connections is central to our discussion, and is introduced in Section 5.1.

The first important result is Proposition (5.1.15), which describes the rational cohomology of $\mathscr{B}^*$ in the case of an $SU(2)$ bundle over a simply-connected four-manifold: as a ring, the cohomology is freely generated, with one two-dimensional generator for each generator of $H_2(X)$ and an extra generator in dimension four. The two-dimensional generators result from a natural map $\mu: H_2(X) \to H^2(\mathscr{B}^*)$, which forms the main subject of Section 5.2. We shall take some time to describe the geometry of this map and to construct explicit cocycle representatives for the classes $\mu(\Sigma)$, from several points of view. This effort is justified by the importance of these constructions in later chapters; as we explain at the beginning of Section 5.2.2, the particular cocycle representatives contain more information than the cohomology classes themselves, and play a significant role in Chapters 8 and 9.

In Section 5.3 we discuss a different route by which the topologies of X and $\mathscr{B}^*$ are related: this is through the notion of a 'concentrated' or 'particle-like' connection, whose curvature is concentrated near a finite collection of points. We have seen in Section 4.3 that such connections can be expected to arise near the boundary of the moduli space of ASD connections. Here we shall examine the topological content of this phenomenon, and its relationship to the Poincaré duality pairing on X.

Finally, in Section 5.4, we prove a result which falls naturally within the framework of this chapter: the orientability of the ASD moduli spaces.

5.1 General theory

5.1.1 Families of connections

In this section, $P \to X$ will be a principal G-bundle over a compact, connected manifold: later we shall restrict ourselves to vector bundles or $SU(2)$ bundles over a four-manifold, but for the moment we can be quite general. As in Chapter 4, we shall allow connection matrices of class L^2_{l-1} and gauge transformations of class L^2_l; for the most part, we are interested only in homotopy-invariant properties which are insensitive to the degree of differentiability, so our particular choice is unimportant. Sometimes, when there may be a doubt about which manifold or bundle is involved, we shall write $\mathscr{B}^*_X$ or $\mathscr{B}^*_{X,P}$ for the orbit space $\mathscr{B}^* = \mathscr{A}^*/\mathscr{G}$. Much of the material of this section is excellently presented elsewhere; some of the original references are listed in the notes at the end of the chapter.

In studying the global topological properties of $\mathscr{B}$ and $\mathscr{B}^*$, some difficulties arise from the fact that the action of $\mathscr{G}$ on $\mathscr{A}$ is not free: even when a connection A is irreducible, the stabilizer $\Gamma_A \subset \mathscr{G}$ may be non-trivial—it coincides with $C(G)$, the centre of the structure group. For this reason, it is convenient to work initially with *framed connections*. If (X, x_0) is a manifold with base-point, a framed connection in a bundle P over X is a pair (A, φ), where A is a connection and φ is an isomorphism of G-spaces, $\varphi: G \to P_{x_0}$. (Such framed connections were used in Section 3.4. Note that for a unitary vector bundle, a framing is equivalent to a choice of orthonormal basis for the fibre E_{x_0} of the associated vector bundle.) The gauge group acts naturally on framed connections, and we write $\tilde{\mathscr{B}}$ for the space of equivalence classes

$$\tilde{\mathscr{B}} = (\mathscr{A} \times \mathrm{Hom}(G, P_{x_0}))/\mathscr{G}. \tag{5.1.1}$$

Another way to think of $\tilde{\mathscr{B}}$ is to regard a framing ϕ as fixed and define $\mathscr{G}_0 \subset \mathscr{G}$ to be its stabilizer, that is

$$\mathscr{G}_0 = \{g \in \mathscr{G} \mid g(x_0) = 1\}.$$

Then $\tilde{\mathscr{B}}$ may be described as $\mathscr{A}/\mathscr{G}_0$. Either way, there is a natural map $\beta: \tilde{\mathscr{B}} \to \mathscr{B}$. In the description (5.1.1), β is the map which forgets the framing; in the second description, β is the quotient map for the remainder of the gauge group,

$$\mathscr{G}/\mathscr{G}_0 \cong \mathrm{Aut}(P_{x_0}) \cong G. \tag{5.1.2}$$

Since the stabilizers $\Gamma_A \subset \mathscr{G}$ consist of covariant-constant gauge transformations, the subgroup $\mathscr{G}_0$ acts freely on $\mathscr{A}$, and $\tilde{\mathscr{B}}$ is therefore a Banach manifold. The fibre $\beta^{-1}([A])$ is isomorphic to G/Γ_A, (where Γ_A is regarded as a subgroup of G via the isomorphism (5.1.2)). In particular, if $\tilde{\mathscr{B}}^* \subset \tilde{\mathscr{B}}$ is the space of framed *irreducible* connections, there is a principal bundle with fibre $G/C(G)$, the *base-point fibration*

$$\beta: \tilde{\mathscr{B}}^* \longrightarrow \mathscr{B}^*. \tag{5.1.3}$$

Our first aim is to describe the homotopy-type of $\tilde{\mathscr{B}}_{X,P}$. This depends only on the homotopy type of X and the bundle P. Indeed, more generally, if $f:(Y, y_0) \to (X, x_0)$ is any smooth map, there is an induced map

$$f^*: \tilde{\mathscr{B}}_{X,P} \longrightarrow \tilde{\mathscr{B}}_{Y, f^*(P)},$$

defined by pulling back connections and framings, and the homotopy class of f^* depends only on the homotopy class of f. This is an important point, for the definition of $\tilde{\mathscr{B}}$ does involve the smooth structure of X, through the notion of connection. The next proposition clarifies the matter by showing how $\tilde{\mathscr{B}}$ can be constructed from X at the level of homotopy, without reference to any finer structure. Recall first that with any topological group G there is an associated *classifying space* BG, which is the base of a G-bundle $EG \to BG$ whose total space EG is contractible. The classifying space is unique up to homotopy equivalence and has the property that for any space Z, the isomorphism classes of G-bundles $P \to Z$ are in one-to-one correspondence with $[Z, BG]$ (the homotopy classes of maps). The correspondence is given by pulling back the bundle EG, so $[f] \mapsto f^*(EG)$. Similarly, if $Y \subset Z$ is a subspace, isomorphism classes of pairs (P, φ) consisting of a bundle $P \to Z$ and a trivialization $\varphi: P|_Y \to Y \times G$ are classified by the homotopy classes of maps of pairs $(Z, Y) \to (BG, *)$, where $* \in BG$ is a base-point.

Proposition (5.1.4). *There is a weak homotopy equivalence*

$$\tilde{\mathscr{B}}_{X,P} \cong \mathrm{Map}^o(X, BG)_P,$$

where Map^o *denotes base-point-preserving maps and* $\mathrm{Map}^o(X, BG)_P$ *denotes the homotopy class corresponding to the bundle* $P \to X$.

Recall that a map $A \to B$ is a weak homotopy equivalence if it gives isomorphisms $\pi_n(A) \to \pi_n(B)$ for all n, or equivalently if the induced map $[T, A] \to [T, B]$ is a bijection whenever T is a compact manifold or cell-complex.

Now the maps $f: T \to \tilde{\mathscr{B}}$ are naturally interpreted in terms of *families of connections*. In general, by a family of connections in a bundle $P \to X$ parametrized by a space T we shall mean a bundle $\underline{P} \to T \times X$ with the property that each 'slice' $P_t = \underline{P}|_{\{t\} \times X}$ is isomorphic to P, together with a connection A_t in P_t for each t, forming a family $\underline{A} = \{A_t\}$. Informally, this is a bundle over $T \times X$ with a connection 'in the X directions'. (If T is just a topological space we must take care that $\underline{P}$ only has a smooth structure in the X directions: it should be given by transition functions whose partial derivatives in the X directions exist and depend continuously on $t \in T$. Similar remarks apply to the connections.) A family of connections is *framed* if an isomorphism is given

$$\underline{\varphi}:\underline{P}|_{T\times\{x_0\}}\to G\times T.$$

Then for each t, the pair (A_t, φ_t) is a framed connection. It is important to realize that the bundle $\underline{P}$ over $T\times X$ need not be isomorphic to $T\times P$.

The proof of Proposition (5.1.4) rests on the existence of a universal family of framed connections, parametrized by $\tilde{\mathscr{B}}$ itself. Let $\pi_2:\mathscr{A}\times X\to X$ be the projection on the second factor and let $\underline{P}\to\mathscr{A}\times X$ be the pull-back $\pi_2^*(P)$—so $\underline{P}=\mathscr{A}\times P$. Then $\underline{P}$ carries a tautological family of connections $\underline{A}$, in which the connection on the slice P_A over $\{A\}\times X$ is $\pi_2^*(A)$. If a framing φ for P at x_0 is chosen, we also obtain a framing $\underline{\varphi}$ for the family. The group $\mathscr{G}_0$ acts freely on $\mathscr{A}\times X$ as well as on $\underline{P}$, and there is therefore a quotient bundle

$$\tilde{\mathbb{P}}\longrightarrow\tilde{\mathscr{B}}\times\underline{X}$$

$$\tilde{\mathbb{P}}=\underline{P}/\mathscr{G}_0. \tag{5.1.5}$$

The family of connections $\underline{A}$ and the framing φ are preserved by $\mathscr{G}_0$, so $\tilde{\mathbb{P}}$ carries an inherited family of framed connections $(\tilde{\mathbb{A}}, \underline{\varphi})$. This is the *universal family* in $P\to X$ parametrized by $\tilde{\mathscr{B}}$.

If a framed family is parametrized by a space T and carried by a bundle $\underline{P}\to T\times X$, there is an associated map $f:T\to\tilde{\mathscr{B}}$ given by

$$f(t)=[A_t,\varphi_t]. \tag{5.1.6}$$

Conversely, given $f:T\to\tilde{\mathscr{B}}$ there is a corresponding pull-back family of connections carried by $(f\times 1)^*(\tilde{\mathbb{P}})$. These two constructions are inverses of one another: if f is determined by (5.1.6), then for each t there is a *unique* isomorphism ψ_t between the framed connections in P_t and $(f\times 1)^*(\tilde{\mathbb{P}})_t$, and as t varies these fit together to form an isomorphism $\psi:\underline{P}\to(f\times 1)^*(\tilde{\mathbb{P}})$ between the two families. (The uniqueness of ψ_t results from the fact that $\mathscr{G}_0$ acts freely on $\mathscr{A}$). Thus:

Lemma (5.1.7). *The maps $f:T\to\tilde{\mathscr{B}}$ are in one-to-one correspondence with framed families of connections on X parametrized by T, and this correspondence is obtained by pulling back from the universal framed family,* $(\tilde{\mathbb{A}}, \tilde{\mathbb{P}}, \underline{\varphi})$.

If f_1 and f_2 are homotopic, the corresponding framed bundles $(\underline{P}_1, \underline{\varphi}_1)$ and $(\underline{P}_2, \underline{\varphi}_2)$ are isomorphic; and conversely, if the families $\underline{A}_1$ and $\underline{A}_2$ are carried by isomorphic framed bundles then, after identifying the two bundles, we can interpolate between the connections with a family $(1-s)\underline{A}_1+s\underline{A}_2$, thus showing that $f_1\sim f_2$. Since every bundle over $T\times X$ carries some family of connections (use a partition of unity), we have:

Lemma (5.1.8). *The homotopy classes $[T,\tilde{\mathscr{B}}]$ parametrize isomorphism classes of pairs $(\underline{P}, \underline{\varphi})$, where*

(i) $\underline{P} \to T \times X$ *is a G-bundle with* $P_t \cong P$ *for all* t.
(ii) $\underline{\varphi}: \underline{P}|_{T \times \{x_0\}} \to T \times G$ *is a trivialization.*

On the other hand the defining property of BG shows that such bundles are classified by homotopy classes of maps of pairs $(T \times X, T \times \{x_0\}) \to (BG, *)$ inducing the bundle P on each slice $\{t\} \times X$. Because T is compact, the exponential law is valid:

$$[T \times X, BG] = [T, \mathrm{Map}(X, BG)],$$

and the end result is a bijection from $[T, \tilde{\mathscr{B}}]$ to $[T, \mathrm{Map}^o(X, BG)_P]$.

Such a bijection is just what is required to establish Proposition (5.1.4). All that is missing is a map $\delta: \tilde{\mathscr{B}} \to \mathrm{Map}^o(X, BG)_P$ by which this bijection is induced. But a suitable δ can be defined by $\delta(b)(x) = \gamma(b, x)$, where $\gamma: (\tilde{\mathscr{B}} \times X, \tilde{\mathscr{B}} \times x_0) \to (BG, *)$ is the classifying map for the bundle $\tilde{\mathbb{P}}$.

The space $\mathscr{B}^*$ does not parametrize a universal family in quite the way that $\tilde{\mathscr{B}}$ does. We do have the following construction however. Let $\mathscr{A}^* \subset \mathscr{A}$ be the space of irreducible connections in $P \to X$ and let $\underline{P} \to \mathscr{A}^* \times X$ be the pull-back bundle $\underline{P} = \mathscr{A}^* \times P$. As before, this carries a tautological family of connections. The gauge group $\mathscr{G}$ acts on this family, but does not act freely on the base $\mathscr{A}^* \times X$ unless $C(G)$ is trivial. Since $C(G)$ acts trivially on the base and non-trivially on the bundle $\underline{P}$, the quotient is not a G-bundle but a bundle whose structure group is the 'adjoint group' $G^{\mathrm{ad}} = G/C(G)$: we define

$$\mathbb{P}^{\mathrm{ad}} \longrightarrow \mathscr{B}^* \times \mathscr{X} \tag{5.1.9}$$

to be the quotient $\mathbb{P}^{\mathrm{ad}} = \underline{P}/\mathscr{G}$. The terminology is not meant to imply the existence of a bundle $\mathbb{P}$ such that $\mathbb{P}^{\mathrm{ad}} = \mathbb{P}/C(G)$. For example, if $G = SU(2)$, then $\mathbb{P}^{\mathrm{ad}}$ is an $SO(3)$ bundle over $\mathscr{B}^* \times X$. It carries a family of connections (without framing) for the $SO(3)$ bundle $P/\{\pm 1\} \to X$ parametrized by $\mathscr{B}^*$, but in general there will be an obstruction to lifting this to an $SU(2)$ family; that is, the second Stiefel–Whitney class $w_2(\mathbb{P}^{\mathrm{ad}})$ may be non-zero. At the Lie algebra level, $SU(2)$ and $SO(3)$ are isomorphic, so the associated adjoint bundle $\mathfrak{g}_{\mathbb{P}}$ is a bundle of Lie algebras with fibre $\mathfrak{su}(2)$, and its pull-back to $\tilde{\mathscr{B}} \times X$ (via the base-point fibration β) is isomorphic to $\mathfrak{g}_{\tilde{\mathbb{P}}}$.

5.1.2 Cohomology

Our next aim is to describe the cohomology of $\tilde{\mathscr{B}}$ and $\mathscr{B}^*$ in the case of an $SU(2)$ bundle over a simply-connected four-manifold. There is a general construction which produces cohomology classes in $\tilde{\mathscr{B}}_{X,P}$, for any G-bundle $P \to X$, using the slant-product pairing

$$/: H^d(\tilde{\mathscr{B}} \times X) \times H_i(X) \longrightarrow H^{d-i}(\tilde{\mathscr{B}}).$$

For each characteristic class c associated with the group G, there is a cohomology class $c(\tilde{\mathbb{P}}) \in H^d(\tilde{\mathscr{B}} \times X)$, where $d = \deg(c)$, so one can define a map

$$\tilde{\mu}_c : H_i(X) \longrightarrow H^{d-i}(\tilde{\mathscr{B}})$$

by

$$\tilde{\mu}_c(\alpha) = c(\tilde{\mathbb{P}})/\alpha.$$

A similar construction produces cohomology classes in $\mathscr{B}^*_{X,P}$ using the bundle $\mathbb{P}^{\text{ad}}$. Thus, given a characteristic class c for the group G^{ad}, there is a map $\mu_c : H_i(X) \to H^{d-i}(\mathscr{B}^*)$ defined by

$$\mu_c(\alpha) = c(\mathbb{P}^{\text{ad}})/\alpha.$$

If T is any $(d-i)$-cycle in $\tilde{\mathscr{B}}$, the class $\tilde{\mu}_c(\alpha)$ can be evaluated on T using the formula

$$\langle \tilde{\mu}_c(\alpha), T \rangle_{\tilde{\mathscr{B}}} = \langle c(\tilde{\mathbb{P}}), T \times \alpha \rangle_{\tilde{\mathscr{B}} \times X}, \tag{5.1.10}$$

which expresses the fact that the slant product is the adjoint of the cross-product homomorphism. The most important instance of this construction, for our applications, is when $G = SU(2)$ and the homology class is two-dimensional:

Definition (5.1.11).

(i) *For an $SU(2)$ bundle $P \to X$, the map $\tilde{\mu} : H_2(X; \mathbb{Z}) \to H^2(\tilde{\mathscr{B}}_{X,P}; \mathbb{Z})$ is given by*

$$\tilde{\mu}(\Sigma) = c_2(\tilde{\mathbb{P}})/[\Sigma].$$

(ii) *The map $\mu : H_2(X; \mathbb{Q}) \to H^2(\mathscr{B}^*_{X,P}; \mathbb{Q})$ is given by*

$$\mu(\Sigma) = -\tfrac{1}{4} p_1(\mathbb{P}^{\text{ad}})/[\Sigma].$$

The second of these definitions is also valid if $G = SO(3)$: in either case, $\mathbb{P}^{\text{ad}}$ is an $SO(3)$ bundle.

In Section 5.2 we shall spend some time in showing how this particular map may be concretely realized. Here though, we shall first discuss the two other non-trivial instances of this construction for the $SU(2)$ case: the maps $\tilde{\mu}_{c_2} : H_i(X) \to H^{4-i}(\tilde{\mathscr{B}})$ for $i = 1$ and 3. Each of these has a straightforward geometrical interpretation.

Let γ be a closed path in X, beginning and ending at x_0 and representing the class $[\gamma] \in H_1(X; \mathbb{Z})$. For each connection A, let $h_\gamma(A)$ denote the holonomy of the connection around the loop. This automorphism of the fibre P_{x_0} depends on the equivalence class of A as a framed connection; so the construction defines a map

$$h_\gamma : \tilde{\mathscr{B}} = (\mathscr{A}/\mathscr{G}_0) \longrightarrow SU(2) \cong S^3.$$

Thus one obtains a cohomology class $h_\gamma^*(\omega) \in H^3(\tilde{\mathscr{B}})$ by pulling back the fundamental class $\omega \in H^3(S^3)$, and the point to be made is that this class coincides with $\tilde{\mu}_{c_2}([\gamma])$. The proof is not difficult, and is left as an exercise;

since the slant-product is the adjoint of the cross-product homomorphism, what has to be shown is that for any three-cycle T in $\tilde{\mathscr{B}}$, we have

$$\langle h_\gamma^*(\omega), T\rangle = \langle c_2(\tilde{\mathbb{P}}), [\gamma] \times T\rangle.$$

Note that the left-hand side is the degree of the map $h_\gamma : T \to S^3$. (Strictly, this equality implies equality of the two classes only over $\mathbb{Q}$; but since $H^3(\tilde{\mathscr{B}}_X; \mathbb{Z})$ is torsion-free in the 'universal' case $X = S^1$, the result is also true over $\mathbb{Z}$.)

Next consider the map $\tilde{\mu}_{c_2} : H_3(X) \to H^1(\tilde{\mathscr{B}})$. If $[Y] \in H_3(X)$ is a class represented by an embedded three-manifold $Y \subset Z$, then for each connection A over X one can calculate the Chern–Simons invariant of $A|_Y$, (see Section 4.4.3). This invariant $\tau_Y(A)$ takes values in S^1, and therefore defines a map $\tau_Y : \tilde{\mathscr{B}} \to S^1$. Again, from the fundamental class of S^1, we obtain by pull-back an element of $H^1(\tilde{\mathscr{B}})$. That this class coincides with $\tilde{\mu}_{c_2}([Y])$ can easily be deduced from the Chern–Weil definition of the second Chern class. Of course, there is no real need here for Y to be an embedded manifold: the Chern–Simons form can equally well be integrated on any C^∞ singular three-cycle.

We return to our main concern, the map $\tilde{\mu}$ defined in (5.1.11). When X is a simply-connected four-manifold, the image of $\tilde{\mu}$ generates all of the rational cohomology of $\tilde{\mathscr{B}}_X$. Precisely, we have the following result.

Proposition (5.1.12). *Let P be an $SU(2)$-bundle over a simply-connected four-manifold X, and let $\Sigma_1, \ldots, \Sigma_b$ be a basis for $H_2(X; \mathbb{Z})$. Then the rational cohomology ring $H^*(\tilde{\mathscr{B}}_{X,P}; \mathbb{Q})$ is a polynomial algebra on the generators $\tilde{\mu}(\Sigma_1), \ldots, \tilde{\mu}(\Sigma_b)$. In particular, $H^{2k}(\tilde{\mathscr{B}}; \mathbb{Q}) \cong s^k(H_2(X; \mathbb{Q}))$.*

To begin the proof, recall from Section 1.2.1 that X has the homotopy-type of a cell-complex, made by attaching a single four-cell to a wedge of two-spheres. So, up to homotopy, X appears in a cofibration

$$\bigvee_1^b S^2 \hookrightarrow X \longrightarrow S^4,$$

in which the two-spheres represent the classes Σ_i. Applying the functor $\mathrm{Map}^o(-, BG)$, using Proposition (5.1.4) and the fact that the mapping functor turns cofibrations into fibrations, we obtain a fibration,

$$\tilde{\mathscr{B}}_{S^4,k} \longrightarrow \tilde{\mathscr{B}}_{X,P} \longrightarrow \prod_1^b \tilde{\mathscr{B}}_{S^2}. \tag{5.1.13}$$

Here $\tilde{\mathscr{B}}_{S^4,k}$ denotes the space of framed connections in the unique $SU(2)$ bundle on S^4 with $c_2 = k = c_2(P)[X]$. (The notation need not mention the bundle on S^2, because any two are isomorphic.) To calculate the cohomology of $\tilde{\mathscr{B}}_{X,P}$ we need to know the cohomology of the fibre and the base in (5.1.13). This information is supplied by the next lemma.

Lemma (5.1.14).

(i) *$H^*(\tilde{\mathscr{B}}_{S^2}; \mathbb{Q})$ is a polynomial algebra on the generator $\tilde{\mu}([S^2])$.*

(ii) *$H^i(\tilde{\mathscr{B}}_{S^4}; \mathbb{Q}) = 0$ for all $i > 0$.*

Proof. According to (5.1.4), the space $\tilde{\mathscr{B}}_{S^n}$ has the weak homotopy type of one component of $\Omega^n(BSU(2))$, (where $\Omega^n(Y)$ stands for $\mathrm{Map}^o(S^n, Y)$, as is usual). Since a bundle on S^n is determined by a transition function on S^{n-1}, there is a homotopy equivalence

$$\Omega^n(BSU(2)) \simeq \Omega^{n-1}(SU(2)) = \Omega^{n-1}S^3.$$

The calculation of the rational cohomology of these spaces is a standard application of the spectral sequence of the path-space fibration. In general, if $F \to P \to B$ is a fibration over a simply-connected base B, there is a spectral sequence $(E_r^{p,q}, d_r)$ whose E_2 term is

$$E_2^{p,q} = H^p(F) \otimes H^q(B)$$

and whose E_∞ terms $E_\infty^{p,q}$ are the quotients for an increasing filtration of the cohomology of the total space, $H^{p+q}(P)$. (The coefficients should be a field; generally we should write $E_2^{p,q} = H^q(B; H^p(F))$.) We apply this Leray–Serre spectral sequence to the fibration

$$\Omega S^3 \longrightarrow PS^3 \longrightarrow S^3,$$

in which PS^3 is the path space

$$PS^3 = \{p:[0,1] \longrightarrow S^3 \,|\, p(0) = \text{North pole}\}$$

and the map $PS^3 \to S^3$ is given by $p \mapsto p(1)$. Since the total space is contractible, we have $E_\infty^{p,q} = 0$ unless $p = q = 0$. The cohomology of the base appears only in dimensions zero and three, so it follows that d_3 is the only non-zero differential and that it is an isomorphism except in dimension zero. It is not hard to deduce that the E_2 and E_3 terms are as shown in

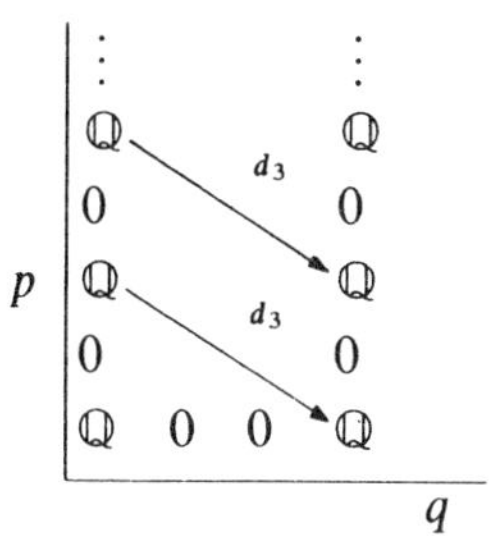

$$E_2^{p,q} = E_3^{p,q}.$$

So $H^p(\Omega S^3; \mathbb{Q})$ is $\mathbb{Q}$ when p is even and 0 otherwise. The ring structure is also easily calculated, using the Leibnitz rule which d_3 obeys: one shows that all powers of the two-dimensional class are non-zero, so that $H^*(\Omega S^3; \mathbb{Q})$ is a polynomial algebra on a two-dimensional generator.

Next one can calculate the cohomology of $\Omega^2 S^3$ using the fibration with base ΩS^3 and total space $P(\Omega S^3)$. Similar reasoning shows the E_2 term to be

$$\begin{array}{|ccccc} \mathbb{Q} & 0 & \mathbb{Q} & 0 & \cdots \\ \mathbb{Q} & 0 & \mathbb{Q} & 0 & \cdots \\ \hline \end{array}$$

which establishes that

$$H^p(\Omega^2 S^3; \mathbb{Q}) = \begin{cases} \mathbb{Q} & p = 0, 1 \\ 0 & p \geq 2. \end{cases}$$

At the next and final stage—the calculation of $H^*(\Omega^3 S^3)$—there is a complication arising from the fact that $\Omega^2 S^3$ is not simply connected:

$$\pi_1(\Omega^2 S^3) = \pi_3(S^3) = \mathbb{Z}.$$

The rational cohomology of this space is the same as that of the circle S^1 and, like S^1, it has a universal covering space $\Omega^2 S^3$ whose cohomology ring is trivial. (This is because the covering transformations act trivially on the cohomology ring.) The kth component $(\Omega^3 S^3)_k$ of the third loop space can be identified with the space of paths in $\Omega^2 S^3$ which begin at the base point and end at its kth translate. The path-space fibration is therefore

$$(\Omega^3 S^3)_k \longrightarrow P(\Omega^2 S^3) \longrightarrow \Omega^2 S^3.$$

From the Leray–Serre spectral sequence, it follows that the rational cohomology ring of $(\Omega^3 S^3)_k$ is trivial also.

The only unproved assertion in the lemma is that $\tilde{\mu}([S^2])$ generates $H^2(\tilde{\mathscr{B}}_{S^2})$. This holds even over $\mathbb{Z}$. By the Hurewicz theorem,

$$\begin{aligned} H_2(\tilde{\mathscr{B}}_{S^2}) &= H_2(\Omega S^3) \\ &= \pi_2(\Omega S^3) = \pi_3(S^3) = \mathbb{Z}. \end{aligned}$$

So it is only necessary to find a two-sphere $T \subset \tilde{\mathscr{B}}_{S^2}$ with $\langle \tilde{\mu}([S^2]), [T] \rangle = 1$. In terms of families of connection (and using (5.1.10)), this means finding an $SU(2)$ bundle $\underline{P} \to S^2 \times S^2$ with $c_2(\underline{P}) = 1$, which of course poses no problem.

In the fibration (5.1.13), the fibre has trivial rational cohomology, by the second part of the lemma, so the cohomology of the total space is isomorphic to the cohomology of the base. By the Künneth theorem then,

$$H^*(\tilde{\mathscr{B}}_X; \mathbb{Q}) \cong \bigotimes_1^b H^*(\tilde{\mathscr{B}}_{S^2}; \mathbb{Q}),$$

which, by the first part, is a polynomial algebra on the generators $\tilde{\mu}(\Sigma_i)$. This completes the proof of Proposition (5.1.12).

This result can be used to calculate also the rational cohomology of $\mathscr{B}^*_{X,P}$, using the base-point fibration β. The class $\mu(\Sigma) \in H^2(\mathscr{B}^*; \mathbb{Q})$ pulls back to the class $\tilde{\mu}(\Sigma)$, but it is not the case that these classes generate the cohomology ring of the base: there is also a four-dimensional class

$$\nu = -\tfrac{1}{4} p_1(\beta) \in H^4(\mathscr{B}^*; \mathbb{Q}),$$

where $p_1(\beta)$ is the Pontryagin class of β regarded as an $SO(3)$ bundle. This definition can be seen as an example of the general construction $\mu_c(\alpha)$ in the case $c = -\frac{1}{4}p_1$ and $\alpha = [x_0] \in H_0(X)$.

Proposition (5.1.15). *Under the same hypotheses as (5.1.5), the rational cohomology ring $H^*(\mathscr{B}^*_{X,P};\mathbb{Q})$ is a polynomial algebra on the four-dimensional generator v and the two-dimensional generators $\mu(\Sigma_i)$:*

$$H^*(\mathscr{B}^*;\mathbb{Q}) = \mathbb{Q}[v, \mu(\Sigma_1), \ldots, \mu(\Sigma_b)].$$

Proof. The total space $\tilde{\mathscr{B}}^*_X$ of the base-point fibration has the same weak homotopy type as $\tilde{\mathscr{B}}_X$. This is true in general as long as $\dim X > 1$, and the reason is that the reducible connections, which make up the complement $\tilde{\mathscr{B}} \backslash \tilde{\mathscr{B}}^*$, have infinite codimension. More precisely, they form a countable union of infinite-codimension submanifolds in the Banach manifold $\tilde{\mathscr{B}}$, so by transversality arguments (see Section 4.2), any map $f: T \to \tilde{\mathscr{B}}$ can be approximated by a map into $\tilde{\mathscr{B}}^*$. Thus (5.1.12) gives the cohomology of the total space in the fibration

$$SO(3) \longrightarrow \tilde{\mathscr{B}}^* \xrightarrow{\ \beta\ } \mathscr{B}^*.$$

The fibre has the rational cohomology of a sphere, so the Leray–Serre spectral sequence can be summarized in a long exact sequence—the Gysin sequence:

$$\ldots \longrightarrow H^{k-4}(\mathscr{B}^*;\mathbb{Q}) \xrightarrow{\ \smile v\ } H^k(\mathscr{B}^*;\mathbb{Q}) \xrightarrow{\ \beta^*\ } H^k(\tilde{\mathscr{B}}^*;\mathbb{Q})$$

$$\xrightarrow{\ \delta\ } H^{k+1-4}(\mathscr{B}^*;\mathbb{Q}) \longrightarrow \ldots$$

Since $\beta^*(\mu(\Sigma)) = \tilde{\mu}(\Sigma)$, Proposition (5.1.12) shows that β^* is surjective. So the connecting homomorphisms δ in the Gysin sequence are zero, and it breaks up into short exact sequences:

$$0 \longrightarrow H^{k-4}(\mathscr{B}^*;\mathbb{Q}) \xrightarrow{\ \smile v\ } H^k(\mathscr{B}^*;\mathbb{Q}) \xrightarrow{\ \beta^*\ } H^k(\tilde{\mathscr{B}}^*;\mathbb{Q}) \longrightarrow 0.$$

It follows that, over $\mathbb{Q}$,

$$H^k(\mathscr{B}^*) = H^k(\tilde{\mathscr{B}}) \oplus (v \smile H^{k-4}(\tilde{\mathscr{B}})) \oplus \ldots.$$

The result now follows from (5.1.12).

For a general manifold X and bundle P, the rational cohomology of $\tilde{\mathscr{B}}_{X,P}$ is the tensor product of a polynomial algebra on some even-dimensional generators and an exterior algebra on some odd-dimensional ones; that is, $H^*(\tilde{\mathscr{B}};\mathbb{Q})$ is generated freely subject only to the commutativity and anti-commutativity relations of the cohomology ring. The generators arise from the same construction: they are the classes $\tilde{\mu}_c(\alpha)$, where α runs through a basis of $H_i(X)$, for $0 < i < \deg(c)$, and c runs through the primitive rational characteristic classes of the structure group. (A class is *primitive* if it is not expressible in terms of characteristic classes of lower degree). The case of $\mathscr{B}^*$ is more complicated in general, but if $\dim X > 1$ and $C(G)$ is trivial, then

$H^*(\mathscr{B}^*; \mathbb{Q})$ is again a free algebra: the generators are the classes $\mu_c(\alpha)$, but now we must allow α also to include the zero-dimensional generator $[x_0]$. A particularly simple example is the case $G = U(1)$ when the base manifold is a Riemann surface S. The space $\tilde{\mathscr{B}}$, $\mathscr{B}$ and $\mathscr{B}^*$ are all the same here. Since there is no integrability condition in complex dimension one, every connection on a line-bundle $L \to S$ gives rise to a holomorphic structure in L (see Section 2.2.2), and this determines a map $\mathscr{B} \to \mathrm{Pic}_k(S)$, where Pic_k denotes the torus consisting of the isomorphism classes of holomorphic line-bundles of degree k. This map is a homotopy equivalence, for its fibre is essentially the space of all hermitian metrics on L. Thus the cohomology of $\mathscr{B}$ is the same as that of the Jacobian; and since the latter is a $2g$-torus, the algebra $H^*(\mathscr{B}; \mathbb{Z})$ is an exterior algebra on $2g$ one-dimensional generators. Identifying these with the classes $\mu_{c_1}(\alpha)$, for $\alpha \in H_1(S)$, amounts to the usual description of the Jacobian as the quotient of $\mathbb{R}^{2g}$ by the lattice of periods. (See also the discussion at the end of Section 2.2.1.)

5.1.3 *K-theory and the index of a family*

We have seen that when a space T parametrizes a family of connections on a manifold X one can define cohomology classes in T by the slant-product construction, reflecting the non-trivial topology in the family. We now wish to describe a parallel constriction with K-theory replacing ordinary cohomology. Recall that the K-theory of a compact space T is the abelian group $K(T)$ with a generator $[E]$ for each complex vector bundle $E \to T$ and a relation $[F] = [E_1] + [E_2]$ whenever $F \cong E_1 \oplus E_2$. Every element of $K(T)$ can be represented as a virtual bundle, i.e. a formal difference of vector bundles $[E_1] - [E_2]$. The same construction can also be made with real vector bundles, and this gives rise to the real K-theory $KO(T)$.

Suppose now that the base is a smooth manifold X, and let $D: \Gamma(V) \to \Gamma(W)$ be an elliptic operator, acting on sections of a vector bundle $V \to X$. Given a bundle $E \to X$ with a connection A, one can form a new operator by coupling D to A; that is, one replaces ordinary partial derivatives by covariant derivatives, to obtain an operator

$$D_A: \Gamma(E \otimes V) \to \Gamma(E \otimes W).$$

We saw this construction for the Dirac operators in Section 3.1. It cannot, in general, be carried out in a *canonical* way, except at the level of the symbols of the operators, as it involves some arbitrary choice of coordinates. Nevertheless, the *index*

$$\mathrm{ind}(D_A) = \dim \mathrm{Ker}(D_A) - \dim \mathrm{Coker}(D_A)$$

is independent of the choices made, and even independent of the connection A, as it is invariant under deformation. We shall use the notation $\mathrm{ind}(D, E)$ to denote this index. The map $\mathrm{ind}(D, -)$ defines a homomorphism $K(X) \to \mathbb{Z}$;

so the index construction plays a role analogous to that of a homology class in ordinary cohomology theory.

The index of a *family* of operators is a construction which plays the rôle of the K-theory slant product. Let $\underline{A} = \{A_t\}$ be a family of connections over X carried by a vector bundle $\underline{E} \to T \times X$, and let D be an operator as before. By the same construction, one then obtains a family of elliptic operators, $D_{\underline{A}} = \{D_{A_t}\}$ parametrized by T. Suppose, for the moment, that the dimensions of $\mathrm{Ker}(D_{A_t})$ and $\mathrm{Coker}(D_{A_t})$ are independent of t. Then as t varies, these two families of vector spaces form locally trivial vector bundles over T. Their formal difference, $[\mathrm{Ker}\, D_{A_t}] - [\mathrm{Coker}\, D_{A_t}]$, defines an element of the K-theory of T, which we shall denote by $\mathrm{ind}(D, \underline{E})$, because it depends, at bottom, only on the bundle $\underline{E}$ on $T \times X$ and the operator D on X. We have already seen an example of this construction in Section 3.2. In describing the Fourier transform of an ASD connection on a four-torus T, we made use of a family of Dirac operators D_ζ^* on T parametrized by $\zeta \in T^*$. Since $\mathrm{Coker}(D_\zeta^*)$ was always zero (by the Weitzenböck formula), the index of the family was the vector bundle with fibre $\mathrm{Ker}(D_\zeta^*)$ over T^*. This was the bundle we called $\hat{E}$ and which carried the 'transform' of the connection. The total bundle on $T^* \times T$ which carried the family of operators was the product $\underline{P} \times E$, where $\underline{P}$ was the Poincaré line-bundle; so at the level of K-theory,

$$[\hat{E}] = \mathrm{ind}(D, [\underline{P} \times E]).$$

In a general family $D_{\underline{A}}$, the kernels and cokernels will 'jump' in dimension at some points or subspaces in T. The index of the family is still well-defined in $K(T)$, by the following construction. First, let us write

$$\mathscr{V}_t = \Gamma(E_t \otimes V) \quad \mathscr{W}_t = \Gamma(E_t \otimes W),$$

so that D_{A_t} defines a linear map $L_t\colon \mathscr{V}_t \to \mathscr{W}_t$, which is Fredholm if we use suitable Sobolev spaces of sections. As t varies, we obtain a homomorphism between vector bundles, $L\colon \mathscr{V} \to \mathscr{W}$ over T. If L were surjective, the index of the family would be well-defined, as the vector bundle $\mathrm{Ker}(L) \subset \mathscr{V}$. In general, as long as T is compact, one can find a trivial vector bundle $\underline{\mathbb{C}}^N = \mathbb{C}^N \times T$ and a homomorphism $\psi\colon \underline{\mathbb{C}}^N \to \mathscr{W}$ such that the sum

$$L \oplus \psi\colon \mathscr{V} \oplus \underline{\mathbb{C}}^N \longrightarrow \mathscr{W}$$

is surjective. The construction of ψ is much the same as the construction of a transverse family of perturbations which was discussed in a general context in Section 4.2. One starts locally: for $t \in T$, put $n_t = \dim \mathrm{Coker}(L_t)$ and choose a map $\psi_t\colon \mathbb{C}^{n_t} \to \mathscr{W}_t$ whose image spans the cokernel of L_t. This will solve the problem in some open neighbourhood U_t, and one can pass to the global solution by taking a finite cover $\{U_{t_i}\}$, using cut-off functions and setting $N = \Sigma\, n_{t_i}$. The index of the family of operators D_{A_t}, or equivalently the index of L, is then defined by

$$\mathrm{ind}(L) = [\mathrm{Ker}(L \oplus \psi)] - [\underline{\mathbb{C}}^N] \in K(T).$$

To see that this is independent of the choice of ψ, suppose we have two such maps $\psi_1 : \underline{\mathbb{C}}^{N_1} \to \mathscr{W}$ and $\psi_2 : \underline{\mathbb{C}}^{N_2} \to \mathscr{W}$. Then the two maps

$$L \oplus \psi_1 \oplus 0 : \mathscr{V} \oplus \underline{\mathbb{C}}^{N_1 + N_2} \longrightarrow \mathscr{W}$$

$$L \oplus 0 \oplus \psi_2 : \mathscr{V} \oplus \underline{\mathbb{C}}^{N_1 + N_2} \longrightarrow \mathscr{W}$$

are both surjective, and they are homotopic through surjective maps, since each is homotopic to $L \oplus \psi_1 \oplus \psi_2$ by a linear homotopy. Their kernels are therefore isomorphic as vector bundles, from which it follows that

$$[\mathrm{Ker}(L \oplus \psi_1)] - [\underline{\mathbb{C}}^{N_1}] = [\mathrm{Ker}(L \oplus \psi_2)] - [\underline{\mathbb{C}}^{N_2}]$$

as required.

This completes the construction of the index of a family as far as we shall need it. What the construction defines is a group homomorphism

$$\mathrm{ind}(D, -) : K(T \times X) \to K(T).$$

This K-theory slant product can be related to the ordinary slant product via the Chern character and the Atiyah–Singer index theorem for families. First, the definition of the Chern class c_i can be extended from vector bundles to virtual bundles by formally manipulating the Whitney product formula. If E is a vector bundle of rank r, it is usual to introduce formal variables $x_1, \ldots, x_n$ such that the Chern classes are the elementary symmetric polynomials:

$$c_i(E) = \sigma_i(x_1, \ldots, x_r) \in H^{2i}(T; \mathbb{Q}).$$

The *Chern character* of E is then defined to be the following symmetric function of these variables:

$$\mathrm{ch}(E) = \sum_{i=1}^{r} e^{x_i}.$$

In terms of the Chern classes, the first few terms of the Chern character are

$$\mathrm{ch}(E) = r + c_1 + \tfrac{1}{2}(c_1^2 - 2c_2) + \cdots.$$

It induces a ring homomorphism $\mathrm{ch} : K(T) \to H^{\mathrm{even}}(T; \mathbb{Q})$, when the multiplication in $K(T)$ is defined by the operation of tensor product. Now let $D : \Gamma(S^+) \to \Gamma(S^-)$ be the Dirac operator on an even-dimensional manifold X, and let $\underline{A} = \{A_t\}$ be a family of connections over X, carried by a bundle $\underline{E} \to T \times X$.

Theorem (5.1.16) (Atiyah–Singer). *The Chern character of the index of the family of Dirac operators $D_A = \{D_{A_t}\}$ is given by*

$$\mathrm{ch}(\mathrm{ind}(D, \underline{E})) = (\mathrm{ch}(\underline{E})\hat{A}(X))/[X].$$

In this formula, the characteristic class $\hat{A}(X)$ is defined by

$$\hat{A}(X) = \prod_{1}^{n} \left(\frac{y_i/2}{\sinh y_i/2} \right),$$

where the y_i are formal variables related to the Pontryagin classes of X by

$$p_k(X) = \sigma_k(y_1^2, \ldots, y_n^2) \in H^{4k}(X; \mathbb{Q}).$$

In the case of a two-manifold Σ or four-manifold X^4, the $\hat{A}$ class is respectively

$$\hat{A}(\Sigma) = 1$$

$$\hat{A}(X^4) = 1 - \tfrac{1}{24} p_1(X).$$

For a manifold with $\hat{A} = 1$, the index theorem says that the Chern character homomorphism intertwines the slant product in cohomology with the index construction in K-theory.

We shall need the construction of the index of a family at a few different points in this book. The first occasion is in Section 5.2, where we shall use the index of the Dirac operator on a two-manifold to give an alternative definition of the map μ. Another application is in Chapter 8 where we need to define certain torsion classes in the cohomology of $\mathscr{B}^*$. To introduce these classes, let X be a spin four-manifold and let $E \to X$ be a vector bundle with structure group $SU(2)$. Let $\tilde{\mathbb{E}}$ be the vector bundle associated with the principal bundle $\tilde{\mathbb{P}} \to \tilde{\mathscr{B}} \times X$, defined at (5.1.5), and let $\tilde{\mathbb{A}}$ be the universal family of $SU(2)$ connections which this bundle carries. Since both E and the spin bundles $S^{\pm}$ have $SU(2)$ structures, given by skew bilinear forms, the product bundles $E \otimes S^{\pm}$ carry symmetric bilinear forms, which determine real subbundles $(E \otimes S^{\pm})_R$, just as in Section 3.4. The Dirac operator preserves these, so by coupling D to $\tilde{\mathbb{A}}$ one obtains a family of *real* operators parametrized by $\tilde{\mathscr{B}}$:

$$(D_{\tilde{\mathbb{A}}})_R = \{D_A : \Gamma(E \otimes S^+)_R \longrightarrow \Gamma(E \otimes S^-)_R\}_{[A] \in \tilde{\mathscr{B}}}.$$

For any compact set $T \subset \tilde{\mathscr{B}}$ there is therefore an element of the real K-theory,

$$\operatorname{ind}(D, \tilde{\mathbb{E}})_R \in KO(T).$$

The technical point here is that the index construction does not define a K-theory element if the parameter space is non-compact: it only defines an element of $\varprojlim K(T)$, as T runs through all compact subsets. Nevertheless the *characteristic classes* of such an element are evidently well-defined. Using the Stiefel–Whitney classes, we make the following definition:

Definition (5.1.17). *The classes $\tilde{u}_i$ are defined when X is spin by*

$$\tilde{u}_i = w_i(\operatorname{ind}(D, \tilde{\mathbb{E}})_R) \in H^i(\tilde{\mathscr{B}}_{X,E}; \mathbb{Z}/2).$$

In Chapter 8 we shall consider whether these classes descend to $\mathscr{B}^*$. A similar definition can be made for any real elliptic operator on X. In Section 5.4 we shall prove the vanishing of the first Stiefel–Whitney class associated with the operator $\delta = (\mathrm{d}^* \oplus \mathrm{d}^+) : \Omega^1 \to \Omega^0 \oplus \Omega^+$.

5.1.4 Links of the reducible connections

The classes ν and $\mu(\Sigma)$ in $H^*(\mathscr{B}^*; \mathbb{Q})$ do not extend from $\mathscr{B}^*$ to $\mathscr{B}$; the obstruction is the non-triviality of these classes on the links of the reducible strata, $\mathscr{B}^\Gamma$, which make up the complement $\mathscr{B} \setminus \mathscr{B}^*$. We want to calculate this obstruction in the cases relevant to our applications. To begin with, consider a general G-bundle, $P \to X$. We shall suppose only that $\dim X > 1$, so that the reducible connections have infinite codimension in $\mathscr{B}$. Let A be a connection in P which is reducible to a subgroup $H \subset G$, and, as in Section 4.2, let $\Gamma \cong C_G(H)$ be the stabilizer of A in $\mathscr{G}$. Write $\Gamma^{\mathrm{ad}} = \Gamma / C(G)$ for the image of Γ in the adjoint group G^{ad}. Finally, let $\mathscr{U}$ be a contractible neighbourhood of $[A]$ in $\mathscr{B}$, and put $\mathscr{U}^* = \mathscr{U} \cap \mathscr{B}^*$. We shall investigate the restriction of the classes $\mu_c(\alpha)$ to $\mathscr{U}^*$, and the first step is to describe the restriction of the universal bundle $\mathbb{P}^{\mathrm{ad}}$ to $\mathscr{U}^* \times X$.

Lemma (5.1.18). *For suitably chosen $\mathscr{U}$, the space $\mathscr{U}^*$ has the weak homotopy type of the classifying space $B\Gamma^{\mathrm{ad}}$.*

Proof. As in Section 4.1, we may take $\mathscr{U} = T/\Gamma = T/\Gamma^{\mathrm{ad}}$, where $T = T_{A,\varepsilon} \subset \mathscr{A}$ is a transverse slice to the $\mathscr{G}$-orbit. The subspace $\mathscr{U}^*$ is then T^*/Γ^{ad}, where $T^* = T \cap \mathscr{A}^*$. The complement $T \setminus T^*$ has infinite codimension, so T^* has the same weak homotopy type as T, i.e. it is weakly contractible. Since Γ^{ad} acts freely, the lemma follows.

The reduction of P to H determines an H-subbundle $Q \subset P$ over X. Since T^* is also a Γ^{ad}-bundle over $\mathscr{U}^*$, their product is a $(\Gamma^{\mathrm{ad}} \times H)$-bundle.

$$T^* \times Q \longrightarrow \mathscr{U}^* \times X. \tag{5.1.19}$$

Let $\varphi : \Gamma \times H \to G$ be the homomorphism which extends the inclusions of Γ and H in G, and let $\varphi^{\mathrm{ad}} : \Gamma^{\mathrm{ad}} \times H \to G^{\mathrm{ad}}$ be the corresponding homomorphism to the adjoint group.

Lemma (5.1.20). *The restriction of $\mathbb{P}^{\mathrm{ad}}$ to $\mathscr{U}^* \times X$ is the G^{ad}-bundle associated with the $(\Gamma^{\mathrm{ad}} \times H)$-bundle (5.1.19) by the homomorphism φ^{ad}.*

Proof. Let $P^{\mathrm{ad}} \to X$ be the bundle $P/C(G)$, and let $p^*(P^{\mathrm{ad}})$ be the pull-back by the map $p : Q \to X$; so $p^*(P^{\mathrm{ad}})$ is a G^{ad}-bundle over Q with a tautological trivialization. The product bundle $T^* \times p^*(P^{\mathrm{ad}})$ is then a G^{ad}-bundle over $T^* \times Q$ (also with a tautological trivialization) and admits an action of $\Gamma^{\mathrm{ad}} \times H$, for which the quotient space is the bundle $\mathbb{P}^{\mathrm{ad}} \to \mathscr{U}^* \times X$. In this action, $\Gamma^{\mathrm{ad}} \times H$ acts on the fibre G^{ad} by the homomorphism φ^{ad}; and from this the lemma follows.

We specialize to the case of an $SU(2)$ vector bundle E which reduces to $L \oplus L^{-1}$. Here $H = \Gamma = S^1$, so by (5.1.18), $\mathscr{U}^*$ has the weak homotopy type of $B(S^1/\{\pm 1\}) = BS^1$. This is something we have already seen in

Section 4.2.2, for the classifying space of S^1 is $\mathbb{C}P^\infty$. Let A be a connection in E which respects the reduction to S^1.

Proposition (5.1.21). *For any $\Sigma \in H_2(X; \mathbb{Z})$, the restriction of $\mu(\Sigma)$ to the copy of $\mathbb{C}P^\infty$ which links the reducible connection A is given by*

$$\mu(\Sigma)|_{\mathbb{C}P^\infty} = -\langle c_1(L), \Sigma \rangle . h,$$

where $h \in H^2(\mathbb{C}P^\infty; \mathbb{Z})$ is the positive generator.

Proof. The homomorphism φ^{ad} is the map $S^1 \times S^1 \to SO(3)$ given by $(z, w) \mapsto zw^2$. If $e_1, e_2 \in H^2(\mathscr{U}^* \times X)$ are the Euler classes of the bundles corresponding to the two S^1 factors, it follows from (5.1.20) that $p_1(\mathbb{P}^{\mathrm{ad}}) = (e_1 + 2e_2)^2$. The two classes are given by

$$e_1 = h \times 1, \quad e_2 = 1 \times c_1(L),$$

so

$$-\tfrac{1}{4} p_1(\mathbb{P}^{\mathrm{ad}}) = -(\tfrac{1}{4} h^2 \times 1 + h \times c_1(L) + 1 \times c_1(L)^2).$$

From the definition (5.1.11 (ii)) we now calculate

$$\begin{aligned}\mu(\Sigma) &= \tfrac{1}{4} p_1(\mathbb{P}^{\mathrm{ad}})/\Sigma \\ &= -(h \times c_1(L))/\Sigma = -\langle c_1(L), \Sigma \rangle . h.\end{aligned}$$

There is a version of this result also for an $SO(3)$ bundle on X which reduces to $K \oplus \mathbb{R}$, where K is a line bundle. The corresponding formula is

$$\mu(\Sigma)|_{\mathbb{C}P^\infty} = -\tfrac{1}{2}\langle c_1(K), \Sigma \rangle . h.$$

Finally one can consider the trivial $SU(2)$ connection θ in the product bundle $E = \mathbb{C}^2 \times X$. The link $\mathscr{U}^*$ has the weak homotopy type of $BSO(3)$, and the restriction of the four-dimensional class ν on $\mathscr{U}^*$ is the four-dimensional generator of the rational cohomology of $BSO(3)$.

5.2 Three geometric constructions

At the beginning of Chapter 1 we focused on three ways by which a two-dimensional cohomology class could be represented: these were, as the first Chern class of a line bundle L, as the Poincaré dual of a codimension two submanifold, or as the class represented by a closed two-form. In this section we shall give geometric descriptions of the classes $\mu(\Sigma)$ on the Banach manifold $\mathscr{B}^*$, following the same three lines. Thus in Section 5.2.1 we construct a line bundle $\mathscr{L}$, the *determinant line bundle* of a family of Dirac operators on Σ, whose first Chern class is $\mu(\Sigma)$. In Section 5.2.2 we discuss the extent to which transversality arguments will allow us to represent $\mu(\Sigma)$ as the dual class to the zero-set of a section of $\mathscr{L}$. And in Section 5.2.3 we describe how a natural connection in the universal bundle $\mathbb{P}^{\mathrm{ad}}$ gives rise to differential forms representing these classes on $\mathscr{B}^*$.

5.2.1 Determinant line bundles

The first Chern class of a virtual bundle $[E_0] - [E_1]$ on a space T is represented by the line bundle

$$\det([E_0] - [E_1]) = \Lambda^{\max} E_0 \otimes (\Lambda^{\max} E_1)^*,$$

where $\Lambda^{\max} V$ denotes the line bundle $\Lambda^{\dim V} V$. Accordingly, if $D_{\underset{\sim}{A}} = \{D_{A_t}\}$ is a family of operators, say

$$D_{A_t}: \Gamma(E_t \otimes V) \to \Gamma(E_t \otimes W),$$

parametrized by T, one defines the *determinant line bundle* of the family to be the line bundle on T whose fibres are

$$\det \operatorname{ind}(D_{A_t}) = \Lambda^{\max}(\operatorname{Ker} D_{A_t}) \otimes (\Lambda^{\max}(\operatorname{Coker} D_{A_t}))^*. \tag{5.2.1}$$

We shall denote the determinant line bundle by $\det \operatorname{ind}(D, \underline{E})$ or $\det \operatorname{ind}(D_{\underset{\sim}{A}})$. As it stands, this definition appears good only when the kernel and cokernel are locally trivial on T, i.e. when their ranks do not jump. It turns out, however, that (5.2.1) *always* defines a line bundle. That is, there exists a continuous line bundle $\mathscr{L} \to T$ whose fibres, $\mathscr{L}_t$, are canonically isomorphic to the family of lines defined by this formula. The construction of $\mathscr{L}$ uses the same device that we described in Section 5.1.3 for the definition of the index of a family. With the same notation as before, let $\psi: \underline{\mathbb{C}}^N \to \mathscr{W}$ be a vector-bundle map such that

$$D_{\underset{\sim}{A}} \oplus \psi: \mathscr{V} \oplus \underline{\mathbb{C}}^N \to \mathscr{W}$$

is surjective, and define

$$\mathscr{L} = (\Lambda^{\max} \operatorname{Ker}(D_{\underset{\sim}{A}} \oplus \psi)) \otimes (\Lambda^N \mathbb{C}^N)^*.$$

The isomorphism between $\mathscr{L}_t$ and (5.2.1) is a consequence of the exact sequence

$$0 \longrightarrow \operatorname{Ker} \psi_t \longrightarrow \operatorname{Ker}(D_{A_t} \oplus \psi_t) \longrightarrow \mathbb{C}^N \longrightarrow \operatorname{Coker} \psi_t \longrightarrow 0$$

combined with the following standard algebraic lemma:

Lemma (5.2.2). *If (V_i, d_i) is a finite exact sequence, there is a canonical isomorphism*

$$\bigotimes_{i \text{ even}} \Lambda^{\max} V_i = \bigotimes_{i \text{ odd}} \Lambda^{\max} V_i.$$

Remarks.

(i) Since the construction of $\mathscr{L}$ is canonical and local, there is no need for T to be compact. One can, for example, construct the line bundle $\det \operatorname{ind}(D, \tilde{\mathbb{E}})$ for the universal family of connections parametrized by $\tilde{\mathscr{B}}$. The determinant line bundle always represents the first Chern class of the index of a family.

(ii) The action of the scalars $\mathbb{C}^*$ on E induces an action on $\mathscr{L}$. In general, the natural action of $\mathbb{C}^*$ on a vector space V induces an action of weight n on $\Lambda^n V$, so we have:

Lemma (5.2.3). *The weight of the scalar action on* $\det \operatorname{ind}(D_{\underline{A}})$ *is equal to the ordinary numerical index*, $\operatorname{ind} D_{A_t}$.

To obtain a line-bundle on $\tilde{\mathscr{B}}$ representing $\tilde{\mu}(\Sigma)$ we construct the determinant line bundle of the universal family of Dirac operators on Σ. Since the group $\mathrm{Spin}(2)$ is the double cover of the circle $SO(2)$, a spin structure on a Riemann surface Σ may be defined to be a choice of a square root of the canonical bundle; that is, a line bundle $K^{1/2}$ with an isomorphism $K^{1/2} \otimes K^{1/2} = K$, where $K = \Lambda^{1,0}$ is the bundle of holomorphic one-forms. Such square roots always exist, because the degree of K is even: the degree is equal to $-\chi(\Sigma)$, where χ is the Euler characteristic. The Dirac operator $D\colon \Gamma(S^+) \to \Gamma(S^-)$ then becomes the usual $\bar{\partial}$ operator twisted by $K^{1/2}$, which we write as

$$\not{\partial}_\Sigma \colon \Omega^{0,0} \otimes K^{1/2} \longrightarrow \Omega^{0,1} \otimes K^{1/2}.$$

(Compare this with Section 3.3.1, where the Dirac operator on a complex surface was given a similar interpretation.) Now let $E \to \Sigma$ be an $SU(2)$-bundle, let $\tilde{\mathscr{B}}_\Sigma$ be the space of framed connections, and let $\tilde{\mathbb{E}} \to \tilde{\mathscr{B}}_\Sigma \times \Sigma$ be the bundle which carries the universal family $\tilde{\mathbb{A}}$. We define $\tilde{\mathscr{L}}_\Sigma \to \tilde{\mathscr{B}}_\Sigma$ as the determinant line bundle for the family of operators $\not{\partial}^*_{\Sigma,\tilde{\mathbb{A}}}$, obtained by coupling $\not{\partial}^*_\Sigma$ (the adjoint of $\not{\partial}_\Sigma$) to the connections A; equivalently,

$$\tilde{\mathscr{L}}_\Sigma = (\det \operatorname{ind}(\not{\partial}_\Sigma, \tilde{\mathbb{E}}))^*. \tag{5.2.4}$$

Proposition (5.2.5). $c_1(\tilde{\mathscr{L}}_\Sigma) = \tilde{\mu}([\Sigma])$.

Proof. This is an application of the index theorem (5.1.16). Since $\hat{A}(\Sigma) = 1$ we have $\operatorname{ch}(\operatorname{ind}(\not{\partial}_\Sigma, \tilde{\mathbb{E}})) = \operatorname{ch}(\tilde{\mathbb{E}})/[\Sigma]$, so

$$\begin{aligned} c_1(\tilde{\mathscr{L}}_\Sigma) &= -c_1(\operatorname{ind}(\not{\partial}^*_\Sigma, \tilde{\mathbb{E}})) = -\operatorname{ch}(\operatorname{ind}(\not{\partial}^*_\Sigma, \tilde{\mathbb{E}}))_2 \\ &= -\operatorname{ch}(\tilde{\mathbb{E}})_4/[\Sigma] = c_2(\tilde{\mathbb{E}})/[\Sigma] \\ &= \tilde{\mu}([\Sigma]). \end{aligned}$$

This proves the result over $\mathbb{Q}$. It holds also over $\mathbb{Z}$, since $H^2(\tilde{\mathscr{B}}_\Sigma; \mathbb{Z})$ has no torsion. In fact the second cohomology is $\mathbb{Z}$; this can be proved first for the case $\Sigma = S^2$ as in Section 5.1.2; the general case then follows from the spectral sequence of the fibration

$$\tilde{\mathscr{B}}_{S^2} \longrightarrow \tilde{\mathscr{B}}_\Sigma \longrightarrow \prod_1^{2g} \tilde{\mathscr{B}}_{S^1} \qquad \text{(cf. (5.1.13))}$$

together with the fact that $\tilde{\mathscr{B}}_{S^1}$ has trivial cohomology below dimension three, as it is simply $SU(2)$.

This shows how the class $\tilde{\mu}(\Sigma) \in H^2(\tilde{\mathscr{B}})$ may be represented by a natural line bundle. The same construction provides a representative for $\mu(\Sigma) \in H^2(\mathscr{B}^*)$, through the following lemma.

Lemma (5.2.6). *The line bundle* $\tilde{\mathscr{L}}_\Sigma$ *descends from* $\tilde{\mathscr{B}}^*_\Sigma$ *to* $\mathscr{B}^*_\Sigma$, *so there is a line bundle* $\mathscr{L}_\Sigma \to \mathscr{B}^*_\Sigma$ *with* $c_1(\mathscr{L}_\Sigma) = \mu([\Sigma])$ *in* $H^2(\mathscr{B}^*_\Sigma; \mathbb{Q})$.

Proof. The point is to show that the centre $\{\pm 1\} \subset SU(2)$ acts trivially on $\tilde{\mathscr{L}}_\Sigma$. If it does, then $SO(3)$ acts freely on the restriction of $\tilde{\mathscr{L}}_\Sigma$ to $\tilde{\mathscr{B}}^*$, and $\mathscr{L}_\Sigma$ can then be defined as the quotient bundle on $\tilde{\mathscr{B}}^*/SO(3) = \mathscr{B}^*$. By (5.2.3), the weight of the action of the scalars on $\tilde{\mathscr{L}}_\Sigma$ is given by the numerical index, $\operatorname{ind}(\not{\partial}_\Sigma, E)$, and this is twice $\operatorname{ind}(\not{\partial}_\Sigma)$, since E is topologically the trivial bundle $\mathbb{C}^2 \times \Sigma$. But the index of $\not{\partial}_\Sigma$ is zero, as one can see either by Serre duality or because $\hat{A}(\Sigma) = 1$. So the scalars, and ± 1 in particular, act trivially.

If X is a four-manifold and $\Sigma \subset X$ an embedded surface then, by restricting connections to Σ, one may define a map

$$r_\Sigma : \tilde{\mathscr{B}}_{X,E} \longrightarrow \tilde{\mathscr{B}}_\Sigma$$

and so obtain a line bundle $r_\Sigma^*(\tilde{\mathscr{L}}_\Sigma)$ representing $\tilde{\mu}([\Sigma]) \in H^2(\tilde{\mathscr{B}}_X)$. Just as in (5.2.6), because the scalars act trivially, there is a quotient bundle, which, with a slight abuse of notation, we shall call $r_\Sigma^*(\mathscr{L}_\Sigma)$. Thus

$$r_\Sigma^*(\mathscr{L}_\Sigma) \stackrel{\text{def}}{=} r_\Sigma^*(\tilde{\mathscr{L}}_\Sigma)/SO(3) \longrightarrow \mathscr{B}_{X,E}^*.$$

The point here is that r_Σ does not carry $\mathscr{B}_X^*$ into $\mathscr{B}_\Sigma^*$, but is only defined on a smaller set, $\mathscr{B}_X^{**}$—the set of connections on X whose restriction to Σ is irreducible. By using the framed connections as a stepping stone, we have essentially shown that the 'honest' pull-back $r_\Sigma^*(\mathscr{L}_\Sigma) \to \mathscr{B}_X^{**}$ extends naturally to $\mathscr{B}_X^*$.

One corollary of (5.2.6) and the argument above is that $\mu(\Sigma) \in H^2(\mathscr{B}_X^*; \mathbb{Q})$ can be represented by an integer class. In fact, since the spectral sequence of the fibration β shows that $\beta^* : H^2(\mathscr{B}^*; \mathbb{Z}) \to H^2(\tilde{\mathscr{B}}; \mathbb{Z})$ is injective, we have

Corollary (5.2.7). *The map μ on rational cohomology arises from a homomorphism $\mu: H_2(X; \mathbb{Z}) \to H^2(\mathscr{B}_X^*; \mathbb{Z})$ defined by $\mu([\Sigma]) = c_1(r_\Sigma^*(\mathscr{L}_\Sigma))$.*

Remark. When using the restriction map r_Σ, one must be careful to choose Sobolev space topologies for which r_Σ is continuous. This does not pose any difficulty; for example, restriction is a continuous map $L_2^2(\mathbb{R}^4) \to L_1^2(\mathbb{R}^2)$, so these topologies are suitable. Ultimately, we are concerned only with the moduli spaces of ASD connections $M \subset \mathscr{B}_X$; and on M the restriction map is defined whatever the topology on $\mathscr{B}_\Sigma$, because of the regularity results of Section 4.2.

The construction of $\mathscr{L}_\Sigma$ gives another interpretation to the calculation (5.1.21). Let A be a reducible connection in an $SU(2)$ bundle $E = L \oplus L^{-1}$ over Σ, and let $\mathscr{U}^* \simeq \mathbb{C}P^\infty$ be the link of $[A]$ in $\mathscr{B}_\Sigma^*$. If $c_1(L) = k$ then by (5.1.21), the bundle $\mathscr{L}_\Sigma$ on $\mathscr{C}P^\infty$ has degree $-k$. To give an alternative proof, let $[A, \varphi] \in \tilde{\mathscr{B}}$ be a framed connection lying over $[A] \in \mathscr{B}$ and consider the action of the stabilizer

$$\Gamma = \left\{ \begin{pmatrix} e^{i\theta} & 0 \\ 0 & e^{-i\theta} \end{pmatrix} \right\}$$

on the fibre of $\tilde{\mathscr{L}}_\Sigma$ at this point. By the Riemann–Roch theorem, the numerical index of $\not{\partial}_\Sigma$ coupled to L is k. Therefore, since

$$\begin{aligned}\tilde{\mathscr{L}}_{\Sigma,[A,\varphi]} &= \det \operatorname{ind}(\not{\partial}^*_\Sigma, E)\\ &= (\det \operatorname{ind}(\not{\partial}_\Sigma, L))^{\otimes -2},\end{aligned}$$

the circle Γ acts with weight $-2k$, by (5.2.3). So the quotient, $\Gamma^{\mathrm{ad}} = \Gamma/\{\pm 1\}$, acts with weight $-k$, and $\mathscr{L}_\Sigma$ is therefore the bundle of degree $-k$ on $B\Gamma^{\mathrm{ad}} = \mathbb{C}P^\infty$.

In particular, if $k = 0$ then Γ acts trivially on the fibre of $\tilde{\mathscr{L}}_\Sigma$ over $[A, \varphi]$ and the quotient bundle $\mathscr{L}_\Sigma$ is therefore defined at this point.

Corollary (5.2.8). *The line bundle $\mathscr{L}_\Sigma \to \mathscr{B}^*_\Sigma$ extends across the singular stratum consisting of reductions of degree zero.*

5.2.2 Codimension-two submanifolds

On a finite-dimensional manifold, the first Chern class of a line bundle $\mathscr{L}$ is dual to the zero-set of a generic smooth section, but when the base is a Banach manifold some care is necessary. Ultimately, we are concerned only with the ASD moduli spaces $M = M_{X,E}$ rather than the Banach manifolds $\mathscr{B}^*_X$ and $\mathscr{B}^*_\Sigma$ themselves, so we might pull back $\mathscr{L}_\Sigma$ to M using the restriction map, and take a transverse section $\bar{s}$ of the pull-back, so obtaining a submanifold $V = \bar{s}^{-1}(0)$ dual to $\mu([\Sigma])$ in M. We wish, however, to have a representative V whose support is in some sense small, so that we have control on the behaviour of V near the ends of M. Our motive will be revealed more clearly in Chapters 8 and 9; what we want to demonstrate eventually is that certain cohomology classes on M have distinguished cocycle representatives which are *compactly supported*. The same result can be reached in more than one way, and just which approach is most convenient may depend on the particular context. The construction we describe here will be quite adequate for the first applications in the later chapters, but in Chapter 10, for example, we will need a slight variant. Some alternative approaches and refinements of the construction will be discussed later, in Section 9.2.2.

The key idea is to ask that the value of the section $\bar{s}$ at a point $[A] \in M$ 'depends' only on the restriction of $[A]$ to Σ. In other words, $\bar{s}$ should have the form $r^*_\Sigma(s)$, where s is a section of $\mathscr{L}_\Sigma$ over $\mathscr{B}^*_\Sigma$. The zero set V then has the form $r_\Sigma^{-1}(V_\Sigma)$, where $V_\Sigma = s^{-1}(0) \subset \mathscr{B}^*_\Sigma$. The advantage of this construction is that it serves for all the $SU(2)$ bundles E at once: although the different bundles are distinguished by their second Chern class on X, their restrictions to Σ are isomorphic, so the same choice of V_Σ will do for all. This will be important later, as it means that, to some extent, the different cocycles in the different moduli spaces will fit together in the compactification $\bar{M}$ (Section 4.3).

As it stands, there is a difficulty with this approach. The restriction of an ASD connection $[A]$ to Σ may be reducible, even when $[A]$ itself is not, so the map $r_\Sigma : M \to \mathscr{B}^*_\Sigma$ may not be defined. This occurs, for example, with the standard inclusion of S^2 in S^4 in the case of the one-instanton moduli space. A satisfactory compromise is to use a tubular neighbourhood $\nu(\Sigma) \supset \Sigma$ rather than the surface Σ itself since, by (4.3.21), an irreducible ASD connection cannot be reducible on an open set. This is the approach we shall adopt here.

So let $\nu(\Sigma)$ be a tubular neighbourhood with smooth boundary in X, and let $\mathscr{A}_{\nu(\Sigma)}$ be the space of L^2_k connections in $E|_{\nu(\Sigma)}$. Just as for closed manifolds, the quotient space $\mathscr{B}_{\nu(\Sigma)} = \mathscr{A}_{\nu(\Sigma)}/\mathscr{G}_{\nu(\Sigma)}$ by the action of the L^2_{k+1} gauge transformations is Hausdorff, and the space of irreducible connections, $\mathscr{B}^*_{\nu(\Sigma)}$, is a manifold modelled on Hilbert spaces. Local coordinates on $\mathscr{B}^*_{\nu(\Sigma)}$ are obtained from the transverse slices to the $\mathscr{G}$-orbits, defined by

$$T_A = \{A + a \,|\, \mathrm{d}_A^* a = 0 \text{ in } \nu(\Sigma) \quad \text{and} \quad (*a)|_{\partial\nu} = 0\}.$$

To show that every orbit near $[A]$ meets T_A, one must solve a Neumann boundary-value problem for the gauge transformation. The solution, using the implicit function theorem, follows (2.3.4).

To avoid introducing to many restriction maps in our notation, we shall simply write $\mathscr{L}_\Sigma \to \mathscr{B}^*_{\nu(\Sigma)}$ for the pull-back to $\mathscr{B}^*_{\nu(\Sigma)}$ of the line-bundle $\mathscr{L}_\Sigma = \det \operatorname{ind}(\bar{\partial}^*_\Sigma)$ on $\mathscr{B}^*_\Sigma$. Strictly, we must use the framed connections $\tilde{\mathscr{B}}_{\nu(\Sigma)}$ as a stepping-stone to define the pull-back, as we did for $\mathscr{B}^*_X$ in the discussion preceding (5.2.7). The next lemma provides a necessary transversality result:

Lemma (5.2.9). *Let $L \to B$ be a line bundle on a Hilbert manifold and let $f: Z \to B$ be a smooth map from a finite-dimensional manifold Z. Then given any section s_0 of L, one can find a section s which is C^∞-close to s_0 and is transverse to f, i.e. the pull-back $f^*(s)$ is transverse to the zero section of $f^*(L) \to Z$.*

Proof. As in many of the arguments in Section 4.3, this comes down to constructing a suitable transverse family of deformations of s_0. Since Hilbert manifolds support smooth bump-functions, one can find a locally finite open cover $\{U_i\}$, a refinement $\{U'_i\}$ (also a cover) with $\bar{U}'_i \subset U_i$, and sections s_i of L such that $U'_i \subset \operatorname{supp}(s_i) \subset U_i$. For every sequence $\mathbf{x} = (x_i)$ in l^∞, let $s_\mathbf{x}$ be the section

$$s_\mathbf{x} = s_0 + \sum x_i s_i.$$

Thus we have a family of perturbations of s_0 parametrized by a Banach space. The derivative of the total map $l^\infty \times B \to L$ is surjective everywhere, so it follows from the usual transversality argument that the set of $\mathbf{x}$ such that $s_\mathbf{x}$ is transverse to f is dense in l^∞: it is the complement of a first-category subset.

Now consider once more the moduli space M_k of ASD $SU(2)$-connections with $c_2 = k$. Recall that for a generic metric on X, the irreducible part of the moduli space, $M^*_k = M_k \cap \mathscr{B}^*$, is a smooth manifold, and that if $b^+(X) > 0$

we can assume that $M_k^* = M_k$ for $k > 0$. By (4.3.21), there is a well-defined restriction map

$$\begin{aligned} r_{\nu(\Sigma)}: M_k^* &\longrightarrow \mathscr{B}^*_{\nu(\Sigma)} \\ [A] &\longrightarrow [A|_{\nu(\Sigma)}]. \end{aligned} \tag{5.2.10}$$

So, by the transversality argument above, there is a smooth section s of $\mathscr{L}_\Sigma$ on $\mathscr{B}^*_{\nu(\Sigma)}$ such that the pull-back to M_k^* has transverse zero-set. We shall denote by $V_\Sigma \subset \mathscr{B}^*_{\nu(\Sigma)}$ the zero-set of s, and with a slight abuse of notation (omitting mention of the restriction map) we shall write $M_k^* \cap V_\Sigma$ for the zero-set of $r^*_{\nu(\Sigma)}(s)$; that is,

$$M_k^* \cap V_\Sigma = \{[A] \in M_k^* | [A|_{\nu(\Sigma)}] \in V_\Sigma\}. \tag{5.2.11}$$

This notation appears often, and it is important to remember that the restriction to $\nu(\Sigma)$ is implied. Similarly, we shall talk of this intersection as being 'transverse' when what is meant is that $r^*_{\nu(\Sigma)}(s)$ vanishes transversely. This condition means that $M_k^* \cap V_\Sigma$ is a codimension-two submanifold of $\mathscr{B}^*|_{\nu(\Sigma)}$ which is dual to the class $\mu([\Sigma])$.

As we have mentioned, because the distinction between the different bundles disappears on restricting to $\nu(\Sigma)$, the restriction maps (5.2.10) are defined simultaneously for all k, and there is no difficulty in extending the argument (5.2.9) to ensure that the intersections $M_k^* \cap V_\Sigma$ are transverse for all k. Finally, there is an obvious extension of this result to the case of more than one surface:

Proposition (5.2.12). *If $\Sigma_1, \ldots, \Sigma_d$ are embedded surfaces with tubular neighbourhoods $\nu(\Sigma_i)$, there are sections $s_i: \mathscr{B}^*_{\nu(\Sigma_i)} \to \mathscr{L}_{\Sigma_i}$ whose zero-sets V_{Σ_i} have the property that all the intersections*

$$M_k^* \cap V_{\Sigma_{i_1}} \cap \ldots \cap V_{\Sigma_{i_r}} \quad (k > 0, i_1 < \ldots < i_r < d)$$

are transverse. The intersection is then a smooth submanifold dual to $\mu([\Sigma_{i_1}]) \smile \ldots \smile \mu([\Sigma_{i_r}])$ in M_k^*.

The following point will be important in our applications. It follows from (5.2.8) that $\mathscr{L}_{\Sigma_i}$ extends across a neighbourhood of the trivial connection $[\Theta] \in \mathscr{B}_{\nu(\Sigma_i)}$. The sections can therefore be chosen so that:

Condition (5.2.13). *s_i extends continuously and is non-zero on a neighbourhood of $[\Theta]$ in $\mathscr{B}_{\nu(\Sigma_i)}$.*

We shall always suppose that the s_i satisfy this condition, which implies, in particular, that the closure of V_{Σ_i} in $\mathscr{B}_{\nu(\Sigma_i)}$ does not contain $[\Theta]$.

Jumping lines. To achieve transversality, it has been convenient to use an arbitrarily chosen section of $\mathscr{L}_\Sigma$. There is, however, a preferred section, canonically determined by the geometry of Σ. This is the 'determinant' of the family of operators $\bar{\partial}_{\Sigma, A}$. It can be defined as the section $\sigma: \mathscr{B}^*_\Sigma \to \mathscr{L}_\Sigma$ which is

'1' when $\not{\partial}_{\Sigma,A}$ is invertible and 0 otherwise. The term '1' has to be interpreted using the definition (5.2.1) of the determinant line, which identifies the fibre of $\mathscr{L}_\Sigma$ with $\mathbb{C}$ canonically whenever the kernel and cokernel are zero; and one must then verify that σ so defined is smooth. There is a simple characterization of the zero-set of σ when $\Sigma = \mathbb{C}P^1$, the Riemann sphere. Given a connection A in an $SU(2)$-bundle $E \to \mathbb{C}P^1$, let $\mathscr{E} = (E, \bar{\partial}_A)$ be the corresponding holomorphic bundle. It can be shown that every holomorphic vector bundle on $\mathbb{C}P^1$ is a direct sum of line bundles, each of the form H^n, where H is the Hopf bundle. So, since $c_1(\mathscr{E}) = 0$, we will have

$$\mathscr{E} = H^n \oplus H^{-n}$$

for some $n \geq 0$. The space of all connections $\mathscr{B}_\Sigma$ now acquires a stratification according to the splitting-type of $\mathscr{E}$—that is, according to the integer n. The case $n = 0$, when $\mathscr{E}$ is holomorphically trivial, is generic. The connections $[A]$ for which $n > 0$ form a subspace $\mathscr{V}_\Sigma \subset \mathscr{B}_\Sigma$ which, at a generic point, is a codimension-two submanifold. The points of $\mathscr{V}_\Sigma$ are characterized by the fact that $\not{\partial}_{\Sigma,A}$ has non-trivial kernel; so $\mathscr{V}_\Sigma \cap \mathscr{B}^*_\Sigma$ is the zero-set of the determinant σ.

As a simple example, consider the holomorphic bundle $\mathscr{E}_t \to \mathbb{C}P^1$ formed using two coordinate patches $U = \mathbb{C}P^1 \backslash \{\infty\}$ and $U' = \mathbb{C}P^1 \backslash \{0\}$ with transition function

$$\varphi_t = \begin{pmatrix} \zeta & t \\ 0 & \zeta^{-1} \end{pmatrix}. \tag{5.2.14}$$

Here ζ is the affine coordinate on U and t is a parameter. When $t \neq 0$, the bundle $\mathscr{E}_t$ is trivial, because φ_t has a factorization as a product of two terms

$$\begin{pmatrix} 0 & t \\ t^{-1} & \zeta^{-1} \end{pmatrix} \begin{pmatrix} -1 & 0 \\ t^{-1}\zeta & 1 \end{pmatrix},$$

which are regular on U' and U respectively. But when $t = 0$, $\mathscr{E}_t$ is isomorphic to $H^{-1} \oplus H$. This is the phenomenon of 'jumping', where the isomorphism class of a holomorphic bundle changes at a special value of the parameter.

Now suppose that a Riemann surface T parametrizes a holomorphic family of $SL(2, \mathbb{C})$-bundles $\mathscr{E}_t$ on $\mathbb{C}P^1$. The generic situation is that there are only finitely many values of t at which jumping occurs, and that in the neighbourhood of each one of these special values, the family $\mathscr{E}_t$ is isomorphic to the family described by (5.2.14). In this case, a straightforward calculation shows that the second Chern class of the total bundle $\mathscr{E}_T \to T \times \mathbb{C}P^1$ is equal to the number of special values. If we choose a compatible family of unitary connections parametrized by T, this equality can be written

$$\langle \mu([\Sigma]), T \rangle = \#(\mathscr{V}_\Sigma \cap T)$$

which expresses the fact that $\mathscr{V}_\Sigma$ is dual to $\mu([\Sigma])$.

5.2.3 *Differential forms*

In de Rham cohomology the slant product is represented by the operation of 'integration along the fibre'. Given $\omega \in \Omega^{p+f}(B \times F)$, with $f = \dim F$, one can define a form

$$\theta = \int_F \omega \in \Omega^p(B)$$

by the formula

$$\theta_b(X_1, \ldots, X_p) = \int_{\{b\} \times F} \iota(\hat{X}_1) \ldots \iota(\hat{X}_p)\omega,$$

where $\hat{X}_i$ is the horizontal lift of $X_i \in T_b B$. This operation on forms induces a homomorphism

$$\int_F : H^{p+f}(B \times F) \longrightarrow H^p(B)$$

which coincides with the map $[\omega] \mapsto [\omega]/[F]$.

To represent the class $\mu([\Sigma])$ in this way one first needs a form on $\mathscr{B}^* \times X$ representing $-\frac{1}{4}p_1(\mathbb{P}^{\mathrm{ad}})$. We shall use the Chern–Weil representative arising from a natural connection in $\mathbb{P}^{\mathrm{ad}}$. We start by reviewing the construction of a connection in a quotient bundle. Let $\hat{E} \to \hat{Y}$ be a vector bundle and suppose a group Γ acts on $\hat{E}$, covering a free action on $\hat{Y}$. Let $E \to Y$ be the quotient bundle, so $E = \hat{E}/\Gamma$, $Y = \hat{Y}/\Gamma$. Suppose also that we are given

(i) a connection $\hat{\nabla}$ in $\hat{E}$ which is invariant under Γ,

(ii) a connection in the Γ-bundle $p: \hat{Y} \to Y$, determined by a horizontal distribution H.

We can then define a connection ∇ in the quotient bundle E by the formula

$$(\nabla_U s)^{\wedge} = \hat{\nabla}_{\hat{U}} \hat{s}, \tag{5.2.15}$$

in which s is a section of E corresponding to an invariant section $\hat{s}: \hat{Y} \to \hat{E}$, and $\hat{U}$ denotes the horizontal vector on $\hat{Y}$ lying over U—horizontal, that is, with respect to H. To compute the curvature, $F(\nabla)$, for this connection, note first that the pull-back $p^*(\nabla)$ differs from $\hat{\nabla}$ by a vertical part:

$$\hat{\nabla} = p^*(\nabla) + B.$$

Here $B \in \Omega^1_{\hat{Y}} \otimes (\operatorname{End} \hat{E})$ is Γ-invariant and vanishes on H, so can be expressed as $B = \Phi \circ \theta$, where $\theta: T\hat{Y} \to \operatorname{Lie}(\Gamma)$ is the connection one-form corresponding to H and $\Phi: \operatorname{Lie}(\Gamma) \to \operatorname{End} \hat{E}$ is a linear map. So we compute

$$\begin{aligned} F(\hat{\nabla}) &= p^*(F(\nabla)) + p^*(\nabla)B + [B \wedge B] \\ &= p^*(F(\nabla)) + p^*(\nabla)\Phi \wedge \theta + \Phi(\mathrm{d}\theta + [\theta \wedge \theta]). \end{aligned}$$

The last term, $\Theta = \mathrm{d}\theta + [\theta \wedge \theta]$, is the curvature of the Γ-bundle p. So finally, if $U, V \in T_y Y$, we have

$$F(\nabla)(U, V) = F(\hat{\nabla})(\hat{U}, \hat{V}) - \Phi \circ (\Theta(U, V)). \tag{5.2.16}$$

We apply this formula in the case $\hat{Y} = \mathscr{A}^* \times X$ and $\Gamma = \mathscr{G}/\{\pm 1\}$. Let $\hat{\underline{E}} = \pi_2^*(E)$ be the $SU(2)$-bundle on $\mathscr{A}^* \times X$ pulled back from X, and let $\hat{\nabla}$ be the connection in $\hat{\underline{E}}$ which is tautological in the X directions and trivial in the $\mathscr{A}^*$ directions. Thus, *at the point* (A, x), we have $\hat{\nabla} = \pi_2^*(\nabla_A)$. Recall from Section 5.1.1 that although there may be no quotient $SU(2)$-bundle $\mathbb{E} \to \mathscr{B}^* \times X$, there is a well-defined $SO(3)$-bundle of Lie algebras, $\mathfrak{g}_{\mathbb{P}}$. This is the quotient of $\mathfrak{g}_{\hat{E}}$ by $\mathscr{G}/\{\pm 1\}$. To define a quotient connection ∇ in $\mathfrak{g}_{\mathbb{P}}$, one needs a horizontal distribution in the principal bundle $\mathscr{A}^* \to \mathscr{B}^*$. This is provided by the family of horizontal slices $T_A \subset \mathscr{A}$ described in Section 4.2. At this point it is necessary to choose a Riemannian metric on X, since the definition of the slice T_A, as the set $\{A + a \mid \mathrm{d}_A^* a = 0\}$, involves the operator d_A^*.

The curvatures of $\hat{\nabla}$ and ∇ have three parts, corresponding to the decomposition

$$\Lambda^2(\mathscr{A}^* \times X) = \Lambda^2(X) \oplus (\Lambda^1(\mathscr{A}^*) \otimes \Lambda^1(X)) \oplus \Lambda^2(\mathscr{A}^*)$$

and the similar decomposition of $\Lambda^2(\mathscr{B}^* \times X)$. For $F(\hat{\nabla})$, these three parts are given at the point (A, x) by the formulae

(i) $$F(\hat{\nabla})(u, v) = F(A)(u, v)$$

(ii) $$F(\hat{\nabla})(a, v) = \langle a, v \rangle$$

(iii) $$F(\hat{\nabla})(a, b) = 0.$$

Here $u, v \in T_x X$ and $a, b \in T_A \mathscr{A}^*$, so the latter are matrix-valued one-forms.

The connection one-form θ on $\mathscr{A}^*$ corresponding to the horizontal distribution T_A is given by

$$\theta_A(a) = -G_A \mathrm{d}_A^* a$$

where $G_A = \Delta_A^{-1}$ is the Green's operator for the Laplacian $\mathrm{d}_A^* \mathrm{d}_A$ on $\Omega^0(\mathfrak{g}_E)$. (One checks that θ_A is zero on T_A and that $\theta_A(-\mathrm{d}_A \xi) = \xi$ for all $\xi \in \mathrm{Lie}(\mathscr{G})$.) The curvature of this connection is expressed by

$$\Theta_A(a, b) = -2G_A\{a, b\}$$

when a, b are horizontal (i.e. $\mathrm{d}_A^* a = \mathrm{d}_A^* b = 0$). Here $\{\,,\}$ is the natural pairing, $\Omega^1(\mathfrak{g}_E) \otimes \Omega^1(\mathfrak{g}_E) \to \Omega^0(\mathfrak{g}_E)$, formed from the metric on X and the Lie bracket on $\mathfrak{g}$. Combining this expression with the formula (5.2.16), we obtain

Proposition (5.2.17). *The three components of the curvature of the connection* ∇ *in the* $SO(3)$*-bundle* $\mathfrak{g}_{\mathbb{P}} \to \mathscr{B}^* \times X$ *are given at the point* $([A], x)$ *by*

(i) $$F(\nabla)(u, v) = F(A)(u, v)$$

(ii) $$F(\mathbb{V})(a, v) = \langle a, v \rangle$$

(iii) $$F(\mathbb{V})(a, b) = -2G_A\{a, b\}|_x.$$

Here $u, v \in T_x X$, $a, b \in \Omega^1(\mathfrak{g}_E)$, and the latter satisfy $\mathrm{d}_A^* a = \mathrm{d}_A^* b = 0$.

There is now an explicit 4-form representing $-\frac{1}{4}p_1(\mathfrak{g}_{\mathbb{P}})$, namely the Chern–Weil form

$$\frac{1}{8\pi^2}\operatorname{Tr}(F(\mathbb{V}) \wedge F(\mathbb{V})).$$

(Here Tr denotes the trace on two-by-two matrices, cf. (2.1.41)). So given $\Sigma \in H_2(X)$, choose a form $\omega \in \Omega^2(X)$ representing its Poincaré dual, and we have

$$\begin{aligned}\mu(\mathrm{PD}[\omega]) &= -\tfrac{1}{4}p_1(\mathfrak{g}_{\mathbb{P}})/(\mathrm{PD}[\omega]) \\ &= -\tfrac{1}{4}(p_1(\mathfrak{g}_{\mathbb{P}}) \smile [\omega])/[X],\end{aligned}$$

which is represented by the 2-form

$$\Omega = \frac{1}{8\pi^2}\int_X \operatorname{Tr}(F(\mathbb{V}) \wedge F(\mathbb{V})) \wedge \omega$$

on $\mathscr{B}^*$. In the product $F \wedge F$, two terms have degree two in the $\mathscr{B}^*$ directions: these are the square of the term (ii) in (5.2.10) and the term (i) $\wedge$ (iii). The final expression for Ω is given by

Proposition (5.2.18). *The 2-form Ω representing $\mu(\mathrm{PD}[\omega])$ on $\mathscr{B}^*$ is given at the point $[A]$ by the formula*

$$\Omega_{[A]}(a, b) = \frac{1}{8\pi^2}\int_X \operatorname{Tr}(a \wedge b) \wedge \omega + \frac{1}{2\pi^2}\int_X \operatorname{Tr}(G_A\{a, b\}F_A) \wedge \omega.$$

Here a, b represent tangent vectors to $\mathscr{B}^*$ and satisfy $\mathrm{d}_A^* a = \mathrm{d}_A^* b = 0$.

The same expression, of course, represents the class $\mu(\mathrm{PD}[\omega])$ on any smooth submanifold of $\mathscr{B}^*$, and in particular on the ASD moduli space M when this is smooth. Here a simplification occurs if ω is a self-dual harmonic form: the second term in (5.2.17) drops out since it involves the product of an anti-self-dual form (the curvature) with a self-dual one. What remains does not involve the Green's operator. It is the integral of a local expression:

$$\Omega(a, b) = \frac{1}{8\pi^2}\int_X \operatorname{Tr}(a \wedge b) \wedge \omega \quad \text{on } M. \tag{5.2.19}$$

Proposition (5.2.17) is valid for any group, and this procedure gives representatives for any rational class $\mu_c(\alpha)$. In the case of $SU(2)$ or $SO(3)$, we mention the 4-form on $\mathscr{B}^*$ representing the class ν (see (5.1.15)).

If $\varphi \in \Omega^4(X)$ has integral 1, so that it represents the Poincaré dual of $[x_0] \in H_0(X)$, the expression is

$$v(a_1, a_2, a_3, a_4) = \text{const.} \sum \varepsilon_{ijkl} \int_X \operatorname{Tr}(G_A\{a_i, a_j\} G_A\{a_k, a_l\})\varphi.$$

5.3 Poincaré duality

5.3.1 Concentrated connections: statement of the result

We saw in Section 3.4 that the moduli space of k-instantons on $\mathbb{R}^4$ contains an open set consisting of connections whose curvature is concentrated near k distinct points, approximating k copies of the standard one-instanton. One of the aims of Chapter 7 will be to investigate how this description generalizes on an arbitrary four-manifold. In this section, however, we shall look at such concentrated connections from a topological point of view in relation to the classes $\mu(\Sigma)$, without bringing in the anti-self-duality equations.

To begin with, we need a working definition of a concentrated connection. It will be simplest to begin by considering connections which are exactly, rather than approximately, trivial outside some small geodesic balls. So let $E \to X$ be an $SU(2)$-bundle with $c_2(E) = k$ on a Riemannian fourmanifold, let $r > 0$ be a number smaller than half the injectivity radius of X, and consider the subset $\mathscr{K}$ of $\mathscr{B}^* \times s^k(X)$ consisting of the pairs $([A], \{x_1, \dots, x_k\})$ satisfying the following conditions:

Conditions (5.3.1)

(i) *the x_i are distinct, and $d(x_i, x_j) > 3r$* if $i \neq j$;

(ii) *the connection A is isomorphic to the product connection Θ on the set $W = X \setminus \cup B_r(x_i)$;*

(iii) *for each i, $\int_{B_r(x_i)} \operatorname{Tr}(F_A \wedge F_A) = 8\pi^2$.*

Here d is the geodesic distance and $B_r(x)$ is the geodesic ball about x. Thus the curvature of A is concentrated near the points x_i; the centres x_i themselves are included as part of the data for convenience. Note that condition (iii) is topological: because of (ii), there is a preferred trivialization, ψ, of E on $\partial B_r(x_i)$, and the condition can be re-phrased as

(iii′) *the relative second Chern class, $c_2(E, \psi) \in H^4(B, \partial B) \cong \mathbb{Z}$, is equal to* 1.

We have a map $p: \mathscr{K} \to s^k(X)$ which forgets the connection and remembers just the centres. On the other hand, forgetting the centres x_i we can regard $\mathscr{K}$ as parametrizing a family of connections and so consider the cohomology classes $\mu(\Sigma) \in H^2(\mathscr{K})$. The main point is that the classes $\mu(\Sigma)$ are pulled back from $s^k(X)$ via p; thus to evaluate these classes on a family of concentrated connections it is only necessary to know the centres, and there is no dependence on the 'internal' parameters of the connections themselves.

To be precise, let $Z \subset s^k(X)$ be the image of p. This is the set of k-tuples satisfying (5.3.1(i)), and is a smooth manifold since the diagonals are excluded. Let $\Sigma \subset X$ be a smoothly embedded surface, and let $\Sigma_k \subset Z$ be the set

$$\Sigma_k = \{\{x_1, \ldots, x_k\} \in Z \mid \text{at least one } x_i \text{ lies on } \Sigma\}.$$

This is a codimension-two subvariety of Z, having normal crossings at the configurations where two or more points lie on Σ. There is a dual class $\mathrm{PD}(\Sigma_k) \in H^2(Z)$.

Proposition (5.3.2). *The pull-back of* $\mathrm{PD}(\Sigma_k)$ *by the map* $p\colon \mathscr{K} \to Z$ *is* $\mu(\Sigma)$:

$$p^*(\mathrm{PD}(\Sigma_k)) = \mu(\Sigma) \in H^2(\mathscr{K}; \mathbb{Z}).$$

To put this into words, suppose T is a two-manifold parametrizing a family of concentrated connections A_t, with centres $x_i(t)$. Then $\langle \mu(\Sigma), T \rangle$ is equal to the number of times that one of the centres $x_i(t)$ crosses the surface Σ, counted with signs, provided that these crossings are transverse, i.e. provided that the element of surface $t \mapsto x_i(t)$ intersects Σ transversely.

This result plays an important rôle in Chapter 8. It takes a particularly simple form in the case $k = 1$:

Corollary (5.3.3). *Let* $E \to X$ *have* $c_2(E) = 1$, *and let* $\tau\colon X \to \mathscr{B}^*_{X,E}$ *be any map with the property that, for all* x, *the connection* $\tau(x)$ *is flat and trivial outside* $B_r(x)$. *Then the composite*

$$H_2(X; \mathbb{Z}) \xrightarrow{\ \mu\ } H^2(\mathscr{B}^*; \mathbb{Z}) \xrightarrow{\ \tau^*\ } H^2(X; \mathbb{Z})$$

is the Poincaré duality isomorphism.

5.3.2 *Proof of a local version*

We need a lemma which involves an elementary patching construction for connections.

Lemma (5.3.4). (*a*) *The map* $p\colon \mathscr{K} \to Z$ *is a Serre fibration: it has the homotopy-lifting property for simplices.* (*b*) *The fibres of* p *are path-connected.*

Proof. (*a*) Consider first the lifting of a path. Let $\gamma\colon [0, 1] \to Z$ be a path, corresponding to a k-tuple of paths in X, say $x_1(t), \ldots, x_k(t)$, and let $([A], \{x_1(0), \ldots, x_k(0)\})$ lie in $p^{-1}(\gamma(0))$. Let $W_t \subset X$ be the complement of the k balls $B_r(x_i(t))$, and let ψ be a trivialization

$$\psi\colon E|_{W_0} \cong \mathbb{C}^2 \times W_0$$

under which A becomes the product connection, i.e. $A^{\psi} = 0$. For each i, choose a point u_i in the annulus $B_{2r}(x_i(0)) \backslash B_r(x_i(0))$, and let φ_i be the restriction of ψ to the fibre at u_i: on the ball $B_{2r}(x_i(0))$ we now have a framed connection (A, φ_i). Define a diffeomorphism

$$h_{t,i}: B_{2r}(x_i(0)) \longrightarrow B_{2r}(x_i(t))$$

by using the exponential maps and parallel transport along the curve $x_i(t)$. Then use $h_{t,i}$ to transfer the connection A from the first ball to the second, so obtaining a connection $(h_{t,i}^{-1})^*(A)$, framed at the point $h_{t,i}(u_i)$. Finally, define a new connection $A(t)$ on the whole of X by the following conditions, which uniquely specify the gauge-equivalence class:

(i) $A(t)$ is isomorphic to the product connection on W_t;
(ii) $A(t)$ is isomorphic to $(h_{t,i}^{-1})^*(A)$ on $B_{2r}(x_i(t))$;
(iii) the framings at the points $h_{t,i}(u_i)$ are all simultaneously compatible with the trivializations resulting from condition (5.3.1 (i)).

The desired lift of γ is now given by the path

$$t \mapsto ([A(t)], \{x_1(t), \ldots, x_k(t)\}).$$

The general case is no more difficult: we only have to regard a homotopy $\gamma: \Delta \times [0, 1] \to Z$ as a family of paths parametrized by the simplex Δ, and the same construction can be used.

Proof. (*b*) Let κ_1 and κ_2 be two points in the same fibre of p, and let A_1, A_2 be the corresponding connections in E. Since both connections are trivial on the set $W = X \backslash \cup B_r(x_i)$, there is a gauge transformation g defined on $E|_W$ such that $g(A_1|_W) = g(A_2|_W)$. Since both connections have relative second Chern class 1 on each ball (condition (5.3.1 (iii$'$)), there is no topological obstruction to extending g to all of E. Once this is done, the path of connections $(1 - t)g(A_1) + tA_2$, for $t \in [0, 1]$, joins the two equivalence classes, while remaining in the same fibre of p.

Corollary (5.3.5). *If $U' \subset U$ are non-empty open sets in Z, then p induces isomorphisms $\pi_n(\mathscr{K}_U, \mathscr{K}_{U'}) \cong \pi_n(U, U')$.*

Here $\mathscr{K}_U$ denotes the inverse image $p^{-1}(U) \subset \mathscr{K}$.

We are going to deduce Proposition (5.3.2) from a 'local' statement about certain relative classes. (It will, in fact, be this relative version which we shall apply in Chapter 8.) The singular set of Σ_k, where two or more points lie on Σ, has codimension four in Z and can therefore be removed without affecting the calculation. So let $\nu(\Sigma)$ be the r-neighbourhood of Σ (which we shall suppose to be tubular) and put

$$Z^1 = \{\{x_1, \ldots, x_k\} \in Z \,|\, \text{at most one } x_i \text{ lies in } \nu(\Sigma)\}.$$

The set Σ_k does not meet the smaller open set $Z^0 \subset Z^1$ defined by

$$Z^0 = \{\{x_1, \ldots, x_k\} | \text{none of the } x_i \text{ lie in } \nu(\Sigma)\}.$$

The complement $Z^1 \setminus \bar{Z}^0$ is a tubular neighbourhood of $\Sigma_k \cap Z^1$, so the dual of Σ_k defines a relative class, $\alpha \in H^2(Z^1, Z^0)$, whose image in $H^2(Z^1)$ is $\mathrm{PD}(\Sigma_k)$. By the Thom isomorphism theorem, $H^2(Z^1, Z^0) \cong \mathbb{Z}$ and α is the generator.

The class $\mu(\Sigma)$ also has a relative version. If $([A], \{x_1, \ldots, x_k\})$ represents a point of $\mathscr{K}_{Z^0}$, then $[A]$ is trivial on Σ; so restriction to Σ defines a map of pairs:

$$(\mathscr{K}_{Z^1}, \mathscr{K}_{Z^0}) \longrightarrow (\mathscr{B}_\Sigma, \{\Theta\}).$$

Because the line-bundle $\mathscr{L}_\Sigma$ extends across Θ (Corollary (5.2.6)), its pull-back to $\mathscr{K}_{Z^1}$ acquires a preferred trivialization, φ, on $\mathscr{K}_{Z^0}$. So there is a relative Chern class

$$\bar{\mu}(\Sigma) = c_1(\mathscr{L}_\Sigma, \varphi) \in H^2(\mathscr{K}_{Z^1}, \mathscr{K}_{Z^0}),$$

whose image in $H^2(\mathscr{K}_{Z^1})$ is $\mu(\Sigma)$. Proposition (5.3.2) now follows from:

Proposition (5.3.6). *Under the map of pairs $p: (\mathscr{K}_{Z^1}, \mathscr{K}_{Z^0}) \to (Z^1, Z^0)$, the pull-back of the class α is $\bar{\mu}(\Sigma)$.*

Proof. By (5.3.5), $p_*: \pi_2(\mathscr{K}_{Z^1}, \mathscr{K}_{Z^0}) \to \pi_2(Z^1, Z^0)$ is an isomorphism; and since $\pi_1(\mathscr{K}_{Z^1}, \mathscr{K}_{Z^0})$ is trivial, the relative Hurewicz theorem implies that p^* is also an isomorphism on the relative second cohomology groups. So $H^2(\mathscr{K}_{Z^1}, \mathscr{K}_{Z^0})$ is generated by $p^*(\alpha)$, and it only remains to determine which multiple of this generator $\bar{\mu}(\Sigma)$ is. By the excision axiom the question now becomes a local one near Σ, and we can therefore get by with calculating a particular case. For example, on $\Sigma \times S^2$, let A be a connection with $c_2 = 1$ which is trivial outside a neighbourhood of (σ_0, s_0). We can arrange that $[A]$ is invariant under the rotations of S^2 which fix s_0, so that there is a well-defined 'translate', $[A(s)]$, centred at (σ_0, s), for each $s \in S^2$. Thus we define a map

$$f: S^2 \longrightarrow \mathscr{K}$$

by

$$f(s) = ([A(s)], \{s\}).$$

We have $\langle p^*(\mathrm{PD}(\Sigma)), f(S^2)\rangle = 1$, because $p \circ f(S^2)$ meets $\Sigma \times \{s_0\}$ once. On the other hand $\langle \mu(\Sigma), f(S^2)\rangle = 1$ also, because the family of connections $A(s)|_{\Sigma \times \{s_0\}}$ is carried by a bundle on $S^2 \times (\Sigma \times \{s_0\})$ with $c_2 = 1$. This completes the proof.

Approximately concentrated connections. Not surprisingly, it makes little difference to the results above if we relax slightly the definition of a concentrated connection. Given any sufficiently small ε, we define a space $\mathscr{K}^+ \supset \mathscr{K}$ by a small modification of Definition (5.3.1). Condition (i) we leave unchanged, but in place of (ii) we require:

(ii′) *On the set W there exists a trivialization $\tau: E|_W \cong \mathbb{C}^2 \times W$ such that $\| A^\tau \|_{L_l^2(W)} < \varepsilon$.*

Here we are using the Sobolev norm L_l^2 for some $l > 2$, but our choice is not important. Note that if τ_1, τ_2 are two trivializations satisfying the conditions of (ii′) then the gauge transformation g by which they differ is nearly constant. In particular, τ_1 and τ_2 are homotopic provided that ε is small enough. Condition (5.3.1 (iii′)) above is therefore unambiguous, and completes the definition of $\mathscr{K}^+$. We also want a still larger space $\mathscr{K}^{++}$, which is defined just as $\mathscr{K}^+$ is, but with $N\varepsilon$ in place of ε. Here N is some large number which will depend on the geometry of X and the chosen radius r (see below), but which will be independent of ε. We shall suppose ε sufficiently small that condition (5.3.1 (iii′)) is unambiguous, even with $N\varepsilon$ in place of ε.

Lemma (5.3.7). *If ε is sufficiently small, then for any compact pair $S' \subset S$ and any map $f_0: (S, S') \to (\mathscr{K}^+, \mathscr{K})$, there exists a homotopy f_t in the larger space $\mathscr{K}^{++}$,*

$$f_t: (S, S') \longrightarrow (\mathscr{K}^{++}, \mathscr{K}),$$

such that $f_1(S) \subset \mathscr{K}$. Furthermore, the homotopy can be chosen to respect p, in that $p \circ f_t = p \circ f_0$.

The first part of this lemma implies that any cohomology class on $\mathscr{K}^{++}$ which is zero on $\mathscr{K}$ is also zero on $\mathscr{K}^+$. The last part of the lemma allows the same statement to be made for the cohomology of $\mathscr{K}_U^+$ or for any pair $(\mathscr{K}_U^+, \mathscr{K}_{U'}^+)$.

Corollary (5.3.8). *The statements of* (5.3.2), (5.3.3) *and* (5.3.6) *continue to hold with $\mathscr{K}^+$ in place of $\mathscr{K}$.*

Proof of (5.3.7) All that is needed is a gauge-invariant procedure by which a connection A as above, which is close to trivial on $W \subset X$, can be deformed so as to be exactly trivial on W. One way is as follows. If ε is small enough, a trivialization τ_c can be found on W such that the connection matrix A^{τ_c} satisfies the Coulomb gauge condition

$$d^* A^{\tau_c} = 0$$

and the boundary condition

$$* A^{\tau_c} = 0 \quad \text{on} \quad \partial W,$$

(cf. (2.3.7)). The solution τ_c is unique, up to an overall constant gauge transformation. By the usual elliptic estimates, the L_l^2 norm of A^{τ_c} is bounded by some fixed multiple of ε. Now let β be a cut-off function on W which is 0 on the neighbourhood of the k spheres which comprise ∂W and is 1 outside the balls of radius, say, $1\frac{1}{4}r$. The connection A can be deformed in a path A_t defined by

$$(A_t)^{\tau_c} = (1 - t\beta)A^{\tau_c}.$$

The final connection A_1 is flat outside the balls of radius $1\frac{1}{4}r$; by using the dilation map in exponential coordinates on X, this connection can then be deformed until its curvature is supported inside the balls $B_r(x_i)$ as required. The whole homotopy can remain inside $\mathscr{K}^{++}$ since we have, for example,

$$\|(A_t)^{\tau_c}\|_{L_l^2} \leq \|(1-t\beta)\|_{L_l^2}\|A^{\tau_c}\|_{L_l^2} \leq \tfrac{1}{2}N\varepsilon, \quad \text{say}.$$

5.4 Orientability of moduli spaces

5.4.1 The orientation bundle Λ

Let $E \to X$ be an $SU(2)$ bundle on a Riemannian four-manifold, and let M be the moduli space of ASD connections. Inside M, let M^s be the smooth open subset consisting of the irreducible connections $[A]$ with $H_A^2 = 0$. In this section we shall prove:

Proposition (5.4.1). *If X is simply connected, the moduli space M^s is orientable.*

Remark. The result holds for any four-manifold, but the proof is considerably simpler under the assumption that X is simply connected.

According to the description given in Section 4.1, the tangent space to the moduli space at a smooth point $[A]$ is the kernel of the elliptic operator

$$\delta_A = (\mathrm{d}_A^* \oplus \mathrm{d}_A^+) : \Omega^1(\mathfrak{g}_E) \longrightarrow \Omega^0(\mathfrak{g}_E) \oplus \Omega_+^2(\mathfrak{g}_E).$$

An orientation of M^s is equivalent to a trivialization of the highest exterior power of the tangent bundle, so we are led to study the real line bundle $\Lambda^{\max}(\operatorname{Ker}\delta_A)$ on M^s. Since the cokernel of δ_A is trivial for $[A]$ in M^s, this line bundle can be identified with the determinant line bundle of δ. The latter extends naturally to all of $\mathscr{B}^*$, as the determinant of δ coupled to the family of $SO(3)$ connections carried by $\mathfrak{g}_{\mathbb{E}}$:

$$\Lambda = \det \operatorname{ind}(\delta, \mathfrak{g}_{\mathbb{E}}). \tag{5.4.2}$$

The orientability of M^s will therefore follow if we can prove:

Proposition (5.4.3). *The orientation bundle $\Lambda \to \mathscr{B}^*$ defined by* (5.4.2) *is topologically trivial.*

Note that this result gives a little more than just the orientability of M^s. Suppose for example that the moduli space consists of n components, each of them orientable. *A priori*, there are 2^n orientations to choose from. Since $\mathscr{B}^*$ is connected however, the triviality of Λ means that there are two preferred choices, differing by an overall change of sign. Later, in Section 7.1.6, we shall see how a canonical trivialization of Λ can be picked out, so as to give a uniquely determined orientation of the moduli space.

5.4.2 Triviality of Λ

First of all it is convenient to use the framed connections $\tilde{\mathscr{B}}$ rather than $\mathscr{B}^*$. A line bundle $\tilde{\Lambda} \to \tilde{\mathscr{B}}$ can then be defined using the framed family:

$$\tilde{\Lambda} = \det \operatorname{ind}(\delta, \mathfrak{g}_{\tilde{\mathbb{E}}}),$$

and on the open set $\tilde{\mathscr{B}}^*$, this coincides with the pull-back of Λ by the base-point fibration β. Since the fibres of β are connected, $\tilde{\Lambda}$ is trivial only if Λ is. The next simplifying step is to stabilize the bundle E. Consider the $SU(3)$ *bundle* $E_+ = E \oplus \mathbb{C}$ and let $\tilde{\mathscr{B}}_+$ be the space of framed $SU(3)$ connections. There is a determinant line bundle $\tilde{\Lambda}_+ \to \tilde{\mathscr{B}}_+$, defined just as $\tilde{\Lambda}$ is, but now using the universal $SU(3)$ family. There is also a natural 'stabilization' map, $s: \tilde{\mathscr{B}} \to \tilde{\mathscr{B}}_+$, defined by $s(A) = A \oplus \theta$, where θ is the rank-one product connection.

Lemma (5.4.4). *The pull-back* $s^*(\tilde{\Lambda}_+)$ *is isomorphic to* $\tilde{\Lambda}$.

Proof. When an $SU(3)$ connection A_+ decomposes as $A \oplus \theta$ there is a corresponding decomposition of the adjoint bundle:

$$\mathfrak{g}_{E_+} = \mathfrak{g}_E \oplus E \oplus \mathbb{R}.$$

The index of any operator coupled to this bundle is a sum of three corresponding terms, so in an obvious notation we have

$$\begin{aligned} s^*(\tilde{\Lambda}_+) &= \det \operatorname{ind}(\delta, \mathfrak{g}_{\tilde{\mathbb{E}}} \oplus \tilde{\mathbb{E}} \oplus \mathbb{R}) \\ &= \tilde{\Lambda}(\mathfrak{g}_{\tilde{\mathbb{E}}}) \otimes \tilde{\Lambda}(\tilde{\mathbb{E}}) \otimes \tilde{\Lambda}(\mathbb{R}). \end{aligned}$$

Since E is complex, the kernel and cokernel of δ coupled to E have canonical orientations, so $\tilde{\Lambda}(\tilde{\mathbb{E}})$ is canonically trivial. The line bundle $\tilde{\Lambda}(\mathbb{R}) \to \tilde{\mathscr{B}}$ is trivial also, for it is a product bundle. Thus $s^*(\tilde{\Lambda}_+)$ is isomorphic to $\tilde{\Lambda}(\mathfrak{g}_{\tilde{\mathbb{E}}})$.

We can iterate this construction, defining a line bundle $\tilde{\Lambda}_{(l)}$ on the space of $SU(l)$ connections $\tilde{\mathscr{B}}_{(l)} = \tilde{\mathscr{B}}_{E \oplus \mathbb{C}^{l-2}}$. By the argument above, $\tilde{\Lambda}$ is trivial if $\tilde{\Lambda}_{(l)}$ is. So the next lemma completes the proof of (5.4.1) and (5.4.3).

Lemma (5.4.5). *For* $l \geq 3$, *the space of framed* $SU(l)$ *connections* $\tilde{\mathscr{B}}_{(l)}$ *is simply connected.*

Proof. We use again the fibration

$$\tilde{\mathscr{B}}_{S^4,k} \longrightarrow \tilde{\mathscr{B}}_{(l)} \longrightarrow \prod_{i=1}^{b} \tilde{\mathscr{B}}_{S^2},$$

where $\tilde{\mathscr{B}}_{S^4,k}$ is the space of $SU(l)$ connections with $c_2 = k$ on S^4. We shall show that both the fibre and base are simply connected. For the base, we have

$$\begin{aligned} \pi_1(\tilde{\mathscr{B}}_{S^2}) &= \pi_1(\mathrm{Map}^o(S^2, BSU(l))) \\ &= [S^3, BSU(l)] \\ &= 0 \end{aligned}$$

since any $SU(l)$ bundle on S^3 is trivial (because a generic section of the associated vector bundle has no zeros and therefore reduces the structure group to $SU(l-1)$ if $l \geq 3$). For the fibre, it is enough to treat the case $k = 0$, since all the components of $\mathrm{Map}^o(S^4, BSU(l))$ have the same homotopy type. We have

$$\begin{aligned}\pi_1(\tilde{\mathscr{B}}_{S^4,0}) &= \pi_1(\mathrm{Map}^o(S^4, BSU(l))) \\ &= [S^5, BSU(l)].\end{aligned}$$

When $l = 2$ there are precisely two $SU(2)$ bundles on S^5, since $\pi_4(SU(2)) = \pi_4(S^3) = \mathbb{Z}_2$. The non-trivial bundle, P, is represented by the principal fibration with total space $SU(3)$,

$$SU(2) \hookrightarrow SU(3) \to S^5,$$

which comes from the natural action of $SU(3)$ on the unit sphere in $\mathbb{C}^3$. This exhibits P as a sub-bundle of the product bundle $SU(3) \times S^5$, which shows that P becomes trivial after stabilization. On the other hand any $SU(l)$ bundle on S^5 can be reduced to $SU(2)$ (and therefore either to the trivial bundle or to P) by choosing $l-2$ generic sections of the associated vector bundle. This shows that every $SU(l)$ bundle on S^5 is trivial, completing the proof.

5.4.3 *SO*(3) *and other structure groups*

The orientability of the moduli space is not special to the case $G = SU(2)$: Proposition (5.4.1) holds also for $SU(l)$, as our proof shows. The case of the unitary groups can be dealt with by a modification of the stabilization trick: one replaces the $U(l)$ bundle E by the $SU(l+1)$ bundle $E_+ = E \oplus \Lambda^l E^*$, corresponding to the homomorphism $\rho: U(l) \to SU(l+1)$,

$$\rho(u) = \begin{pmatrix} u & 0 \\ 0 & \det u^{-1} \end{pmatrix}.$$

The decomposition of the adjoint bundle is then

$$\mathfrak{g}_{E_+} = \mathfrak{g}_E \oplus (\Lambda^l E) \otimes E,$$

and since the second term is complex, the result corresponding to Lemma (5.4.4) still holds.

Consider next an $SO(3)$ bundle $P \to X$. Since X is simply connected, there is an integer class $\alpha \in H^2(X; \mathbb{Z})$ with $\alpha \equiv w_2(P) \pmod 2$, so P lifts to a $U(2)$-bundle $E \to X$ with $c_1(E) = \alpha$. Let $t: \tilde{\mathscr{B}}_E \to \tilde{\mathscr{B}}_P$ be the map associated with the homomorphism $U(2) \to SO(3)$ and let $\tilde{\Lambda}_E$ and $\tilde{\Lambda}_P$ be the orientation bundles. Since the fibres of t are connected, $\tilde{\Lambda}_P$ is trivial if $t^*(\Lambda_P)$ is. On the other hand $t^*(\tilde{\Lambda}_P)$ is isomorphic to $\tilde{\Lambda}_E$ because $\mathfrak{g}_E = \mathfrak{g}_P \oplus \mathbb{R}$. So the case of $SO(3)$ reduces to that of $U(2)$, which has already been treated.

For an arbitrary simple group, the excision axiom discussed in Chapter 7 can be used to reduce the problem to the case of the four-sphere. Lemma

(5.4.5) and its proof then show that the moduli space is orientable whenever $\pi_4(G) = 0$. This condition covers all cases except for the Lie groups locally isomorphic to $Sp(n)$; and since there is no essential difference between $Sp(n)$ and $Sp(n)^{ad}$ as far as the four-sphere is concerned, it remains only to deal with the case of an $Sp(n)$ bundle over S^4. Under the inclusion $Sp(n) \subsetneq Sp(n+1)$, the Lie algebra of the larger group decomposes as

$$\mathfrak{sp}(n+1) = \mathfrak{sp}(n) \oplus V \oplus \mathbb{R}^3,$$

where V is the standard complex representation of dimension $2n$. So the stabilization argument applies, and since the induced map $\pi_4(Sp(n)) \to \pi_4(Sp(n+1))$ is an isomorphism, it is enough to solve the problem for any one value of n. But $Sp(1)$ is isomorphic to $SU(2)$, the group we treated first, so the argument is complete.

Notes

Section 5.1

Good general references for the material of this Chapter are Atiyah and Jones (1978) and Atiyah and Bott (1982).

Section 5.1.1

Another approach is to work with the equivariant cohomology of the space of connections, under the group $\mathcal{G}/Z(G)$, as used by Atiyah and Bott. In fact this is the same as the ordinary cohomology of the space $\mathcal{B}^*$ of irreducible connections modulo equivalence (since the reducible connections have infinite codimension).

Section 5.1.2

The construction of cohomology classes by the slant product procedure has been used by many authors in different contexts, see for example Newstead (1972). The general description of the rational cohomology for the space of connections over any manifold can be proved using the approach of Atiyah and Bott, via the theory of rational homotopy type. For basic material on fibrations, cofibrations and the Serre spectral sequence, we refer to Spanier (1966).

Section 5.1.3

The basic reference for the indices of families is Atiyah and Singer (1971). The parallel between K-theory and ordinary cohomology, in which elliptic operators correspond to homology classes and the index of a family to the slant product goes far beyond the simple examples we consider here; see for example Atiyah (1970) and Douglas (1980).

Section 5.2.1

There is now a large literature on determinant line bundles, much of it motivated by their role in the 'anomalies' of quantum field theory. See for example Atiyah and Singer (1984) and Freed (1986).

Section 5.2.2

These codimension-two submanifolds were used by Donaldson (1986) as convenient representatives for the cohomology classes; the motivation for the idea was the example of jumping lines in algebraic geometry, for which see, for example, Barth (1977). We shall use such codimension-two representatives in this book, although they are not absolutely essential for our arguments, which could all be phrased in more abstract algebro-topological language.

Section 5.2.3

A rather similar calculation is that of the curvature of the orbit space regarded as an infinite-dimensional Riemannian manifold; see Groisser and Parker (1987).

Section 5.3

These results, with rather more complicated proofs, appear in Donaldson (1986).

Section 5.4

The argument here is that of Donaldson (1983*b*). An alternative approach is possible for $SU(2)$ bundles with odd Chern class; see Freed and Uhlenbeck (1984). The proof of orientability without the assumption that the four-manifold be simply connected is given by Donaldson (1987*b*).

6

STABLE HOLOMORPHIC BUNDLES OVER KÄHLER SURFACES

In this chapter we consider the description of ASD moduli spaces over complex Kähler surfaces. This discussion takes as its starting point the relation between the ASD equation and the integrability condition for $\bar{\partial}$ operators which we have seen in Chapter 2. We show that the ASD connections can be identified, using a general existence theorem, with the holomorphic bundles satisfying the algebro-geometric condition of 'stability'. Examples of concrete applications of the theory have been given in Chapter 4; and in Chapters 9 and 10 we will see how these ideas can be used to draw conclusions about the differential topology of complex surfaces. In the present chapter we first introduce the basic definitions and differential-geometric background, and then, in Section 6.2, give a proof of the existence theorem. In the remaining sections we discuss a number of extra topics: the Yang–Mills gradient flow, the comparison between deformation theories of connections and holomorphic bundles, the abstract theory of moment maps and stability, and the metrics on determinant line bundles introduced by Quillen.

6.1 Preliminaries

6.1.1 The stability condition

For simplicity we will restrict attention, for the greater part of this chapter, to rank-two holomorphic vector bundles $\mathscr{E}$, with $\Lambda^2\mathscr{E}$ holomorphically trivial, or in other words to holomorphic $SL(2, \mathbb{C})$ bundles. However the theory applies, with only minor modifications, to more general situations, as we mention briefly in Section 6.1.4.

Let X be a compact complex surface with a Kähler metric. We identify the metric with the corresponding (1, 1)-form ω. For any line bundle L over X we define the *degree* $\deg(L)$ of L to be

$$\deg(L) = \langle c_1(L) \smile [\omega], [X] \rangle \tag{6.1.1}$$

where $[\omega] \in H^2(X; \mathbb{R})$ is the de Rham cohomology class of ω. If X is an algebraic surface and ω is a 'Hodge metric'—compatible with an embedding $X \subset \mathbb{CP}^N$—this definition agrees with the standard notion in algebraic geometry, when L is a holomorphic line bundle $\mathscr{U}$. The basic link between the degree, which is purely topological, and holomorphic geometry stems from the fact that if $\mathscr{U}$ has a non-trivial holomorphic section s then $\deg(\mathscr{U}) \geq 0$, with strict inequality if $\mathscr{U}$ is not the trivial holomorphic bundle. Indeed the zero set Z, say, of s is a positive divisor: $Z = \sum n_i Z_i$ in X, where $n_i > 0$ and Z_i are irreducible complex curves in X. Then, using the fact that Z represents the Poincaré dual of $c_1(\mathscr{U})$, we have

$$\deg(\mathscr{U}) = \sum n_i \int_{Z_i} \omega. \tag{6.1.2}$$

Now the form ω restricts to the Kähler form on each of the complex curves, but this is just the volume form, so the integral of ω over Z_i equals the Riemannian volume of Z_i. Hence each term in (6.1.2) is positive and $\deg(\mathscr{U})$ can only be zero if Z is empty, in which case s defines a trivialization of $\mathscr{U}$.

Definition (6.1.3). *A holomorphic $SL(2, \mathbb{C})$ bundle $\mathscr{E}$ over X is stable (or 'ω-stable') if for each holomorphic line bundle $\mathscr{U}$ over X for which there is a non-trivial holomorphic map $\mathscr{E} \to \mathscr{U}$ we have* $\deg(\mathscr{U}) > 0$.

It is these stable bundles which are, as we shall see, closely related to the ASD solutions. On the other hand we should emphasize that the stability condition is of an algebro-geometric nature. For Hodge metrics ω it agrees with a definition which was introduced by algebraic geometers studying moduli problems. It is also typically quite easy to decide if a bundle defined by some algebro-geometric procedure is stable or not. An equivalent definition is to require that any line bundle which maps non-trivially to $\mathscr{E}$ should have strictly negative degree—for the transpose of a map from $\mathscr{E}$ to $\mathscr{U}$ is a map from $\mathscr{U}^*$ to $\mathscr{E}^*$, and the skew form gives an isomorphism between $\mathscr{E}$ and its dual. Another notion which will enter occasionally is that of a 'semi-stable' bundle: these are defined by relaxing the strict inequality $\deg(\mathscr{U}) > 0$ in (6.1.3) to $\deg(\mathscr{U}) \geq 0$.

Before stating the main results let us recall from Section 2.1.5 the framework for discussing the relationship between holomorphic bundles and unitary connections. First we fix a C^∞ bundle E over X, and a Hermitian metric on E. This amounts to just fixing the topological invariants for the connections we are interested in. Now, within the space $\mathscr{A}$ of unitary connections on E we consider the subspace $\mathscr{A}^{1,1}$ consisting of the connections whose curvature has type (1, 1). As we have explained in Chapter 2, each connection A in $\mathscr{A}^{1,1}$ endows E with the structure of a holomorphic bundle, which we will denote by $\mathscr{E}_A$. Local holomorphic sections of $\mathscr{E}_A$ are the local solutions of the equation $\bar{\partial}_A s = 0$. Recall too that a unitary connection can be recovered from its (0, 1) part $\bar{\partial}_A$ (Lemma (2.1.54)).

Now it will certainly happen that there are connections A_1, A_2 in $\mathscr{A}^{1,1}$ which are not gauge equivalent but which yield isomorphic holomorphic bundles $\mathscr{E}_{A_i}$. The complete picture can be described by introducing another symmetry group, the 'complex gauge group' $\mathscr{G}^c$ of all general linear automorphisms of the complex vector bundle E (covering the identity map on X). This contains as a subgroup the ordinary gauge group $\mathscr{G}$ of automorphisms preserving the Hermitian metric on E, and $\mathscr{G}^c$ can be thought of as the complexification of $\mathscr{G}$. Now the action of $\mathscr{G}$ on $\mathscr{A}$ *extends* to $\mathscr{G}^c$ as follows. For an element g of $\mathscr{G}^c$ we put $\tilde{g} = (g^*)^{-1}$ (so $g = \tilde{g}$ precisely when g lies in the unitary gauge group $\mathscr{G}$). The action of $\mathscr{G}^c$ is given by

$$\bar{\partial}_{g(A)} = g\bar{\partial}_A g^{-1} = \bar{\partial}_A - (\bar{\partial}_A g)g^{-1}$$

$$\partial_{g(A)} = \tilde{g}\partial_A\tilde{g}^{-1} = \partial_A + [(\bar{\partial}_A g)g^{-1}]^*. \tag{6.1.4}$$

When $g = \tilde{g}$ the effect is to conjugate the full derivative ∇_A by g, so the action agrees with the standard one on the subgroup $\mathscr{G}$. The definition can be paraphrased thus: the group $\mathscr{G}^c$ has an obvious action on the $\bar{\partial}$ operators, and we use the fixed metric to identify these operators with the connections, as in (2.1.54).

Now this action of $\mathscr{G}^c$ preserves the subspace $\mathscr{A}^{1,1}$, and it follows immediately from the definitions that holomorphic bundles $\mathscr{E}_{A_1}$, $\mathscr{E}_{A_2}$ are isomorphic if and only if $A_2 = g(A_1)$ for some g in $\mathscr{G}^c$. So the 'moduli set' of equivalence classes of holomorphic bundles of the given topological type can be identified with the quotient:

$$\mathscr{A}^{1,1}/\mathscr{G}^c.$$

The study of unitary connections compatible with a given holomorphic structure is the same as the study of a $\mathscr{G}^c$ orbit in $\mathscr{A}^{1,1}$. (An alternative approach is to fix the holomorphic structure (i.e. $\bar{\partial}$ operator) and vary the Hermitian metric, and this approach is completely equivalent. But we shall stick to the set-up with a fixed metric here.) The stability condition is of course preserved by $\mathscr{G}^c$, so we can speak of the stable $\mathscr{G}^c$ orbits, the orbits of connections A for which $\mathscr{E}_A$ is stable.

Now any ASD connection on E lies in $\mathscr{A}^{1,1}$. Conversely by (2.1.59) a connection A in $\mathscr{A}^{1,1}$ is ASD if and only if $\hat{F}_A = 0$, where we recall that for any connection A we write

$$\hat{F}_A = F_A . \omega \in \Omega^0(\mathfrak{g}_E).$$

This condition is preserved by $\mathscr{G}$ but not by $\mathscr{G}^c$. Our task is to study the equation $\hat{F}_A = 0$ within the different $\mathscr{G}^c$ orbits in $\mathscr{A}^{1,1}$. The main result of this chapter gives a complete solution to this problem, and is stated in the following theorem:

Theorem (6.1.5).

(i) *Any $\mathscr{G}^c$ orbit contains at most one $\mathscr{G}$ orbit of solutions to the equation* $\hat{F}_A = 0$.

(ii) *A $\mathscr{G}^c$ orbit contains a solution to $\hat{F}_A = 0$ if and only if it is either a stable orbit or the orbit of a decomposable holomorphic structure $\mathscr{U} \oplus \mathscr{U}^{-1}$, where* $\deg(\mathscr{U}) = 0$. *In the first case the solution A is an irreducible $SU(2)$ connection* *and in the second it is a reducible solution compatible with the holomorphic splitting.*

In short, combined with (2.1.59), we have:

Corollary (6.1.6). *If E is an $SU(2)$ bundle over a compact Kähler surface X, the moduli space M_E^* of irreducible ASD connections is naturally identified, as a set, with the set of equivalence classes of stable holomorphic $SL(2, \mathbb{C})$ bundles $\mathcal{E}$ which are topologically equivalent to E.*

Notice that the Kähler metric ω enters both into the definition of the ASD equations and into the definition of stability, but in the latter only through the de Rham cohomology class $[\omega]$.

6.1.2 *Analogy with the Fredholm alternative*

As we shall see, the only substantial part of the proof of (6.1.5) is the *existence* of solutions to $\hat{F}_A = 0$ in stable orbits. This fact can be thought of as analogous to the Fredholm alternative for linear equations. Consider for example a non-negative self-adjoint endomorphism T of a Euclidean space V. We have:

either for all y in V there is a unique solution x to the equation $Tx = y$;

or there is a non-trivial solution x to the equation $Tx = 0$.

In our situation we have an alternative, for each $\mathcal{G}^c$ orbit:

either the orbit contains an (essentially unique) solution of the equation $\hat{F}_A = 0$,

or there is a holomorphic line bundle $\mathcal{U}$ over X with $\deg(\mathcal{U}) \leq 0$ and a non-trivial holomorphic map from $\mathcal{E}$ to $\mathcal{U}$; that is, for any A in the orbit there is a solution $s \in \Omega_X^0 (\mathcal{E}_A^* \otimes \mathcal{U})$ to the equation $\bar{\partial}_A s = 0$.

Of course the analogy should not be pushed too far. The point we want to bring out can be seen if we consider the following method of solving the equation $Tx = y$. Starting from any x_0 we solve the ODE for x_t:

$$\frac{\mathrm{d}x_t}{\mathrm{d}t} = y - Tx_t.$$

Then, as the reader can verify:

either x_t converges as $t \to \infty$ to a limit x_∞ with $Tx_\infty = y$,

or $\|x_t\|$ tends to infinity with t, and the normalized vectors:

$$x_t^* = \frac{1}{\|x_t\|} x_t$$

converge to a limit x_∞^*, as $t \to \infty$, and $Tx_\infty^* = 0$.

We will obtain our alternative in a similar fashion by studying the behaviour as $t \to \infty$ of the solutions to an evolution equation. Before introducing this equation however we will dispose of some simple observations in the next section.

6.1.3 *The Weitzenböck formula and some corollaries*

A good deal of information about ASD connections on holomorphic bundles can be garnered from a simple Weitzenböck formula. This can be seen as a generalization of the Weitzenböck formula (3.1.6) for the Dirac operator over flat space used in Chapter 3.

Suppose that A is a unitary connection on a bundle E over a Kähler surface X, so we have operators

$$\bar{\partial}_A : \Omega^0_X(E) \longrightarrow \Omega^{0,1}_X(E)$$
$$\partial_A : \Omega^0_X(E) \longrightarrow \Omega^{1,0}_X(E),$$

making up the full covariant derivative:

$$\nabla_A = \partial_A + \bar{\partial}_A.$$

We can then form three Laplace operators on $\Omega^0_X(E)$, to wit

$$\nabla_A^* \nabla_A, \quad \bar{\partial}_A^* \bar{\partial}_A, \quad \partial_A^* \partial_A.$$

These are related as follows:

Lemma (6.1.7). *For any connection A*

$$\bar{\partial}_A^* \bar{\partial}_A = \tfrac{1}{2} \nabla_A^* \nabla_A + i\hat{F}_A$$
$$\partial_A^* \partial_A = \tfrac{1}{2} \nabla_A^* \nabla_A - i\hat{F}_A$$

on $\Omega^0_X(E)$.

Here $i\hat{F}_A$ acts on $\Omega^0_X(E)$ by the standard algebraic action of $\mathfrak{g}_E$ on E. Of course, there are similar formulae, proved by replacing E with $\mathrm{End}(E)$, for the Laplacians on $\mathfrak{g}_E$, in which $i\hat{F}_A$ acts by the adjoint action. We should emphasize that these formulae are only valid if the metric ω is Kähler.

Proof. To prove the Lemma we introduce the 'Kähler identities':

$$\bar{\partial}_A^* = i[\partial_A, \Lambda], \quad \partial_A^* = -i[\bar{\partial}_A, \Lambda] \tag{6.1.8}$$

on general (p, q) forms $\Omega^{p,q}_X(E)$. Here

$$\Lambda : \Omega^{p,q} \longrightarrow \Omega^{p-1,q-1}$$

is the adjoint of the wedge multiplication by ω, an algebraic operator. Note in particular that for a (1, 1) form ϕ we have

$$\Lambda\phi = \phi \,.\, \omega,$$

so that for any connection A,

$$\hat{F}_A = \Lambda F_A.$$

The special case of the Kähler identities we need is that

$$\bar{\partial}_A^* = i\Lambda\partial_A, \quad \partial_A^* = -i\Lambda\bar{\partial}_A$$

on $\Omega^{0,1}$, $\Omega^{1,0}$ respectively. These are easy to verify directly using the identity

$$\int \alpha \wedge \bar{\alpha} \wedge \omega = i\int |\alpha|^2 \, d\mu,$$

for (0, 1) forms α.

Given these identities we have

$$\begin{aligned}\nabla_A^*\nabla_A &= i\Lambda(\partial_A - \bar{\partial}_A)(\partial_A + \bar{\partial}_A)\\ &= i\Lambda\{\partial_A\bar{\partial}_A - \bar{\partial}_A\partial_A\}.\end{aligned}$$

Now, by the definition of curvature,

$$\partial_A\bar{\partial}_A + \bar{\partial}_A\partial_A = F_A^{1,1},$$

so

$$i\Lambda\{\partial_A\bar{\partial}_A + \bar{\partial}_A\partial_A\} = \hat{F}_A.$$

Hence

$$\begin{aligned}\nabla_A^*\nabla_A &= 2\bar{\partial}_A^*\bar{\partial}_A - 2i\hat{F}_A\\ &= 2\partial_A^*\partial_A + 2i\hat{F}_A\end{aligned}$$

as required.

This Weitzenböck formula immediately gives us a corresponding vanishing theorem. We say:

$$i\hat{F}_A > 0 \quad \text{(respectively } i\hat{F}_A \geq 0\text{)},$$

if $i\hat{F}_A$ is positive (respectively semi-positive) as a self-adjoint endomorphism of E.

Corollary (6.1.9). *If $A \in \mathscr{A}^{1,1}$ is a unitary connection over a compact Kähler surface, defining a holomorphic structure $\mathscr{E}_A$, and if $i\hat{F}_A \geq 0$, then any holomorphic section s of $\mathscr{E}_A$ is covariant constant (i.e. $\bar{\partial}_A s = 0$ implies $\nabla_A s = 0$). If $i\hat{F}_A > 0$ then s is identically zero.*

The proof is the usual integration by parts, as in Section 3.2.1. Of course the result also applies to sections of the bundle of endomorphisms End $\mathscr{E}$. With these vanishing theorems we can quickly deduce the 'subsidiary clauses' in the main theorem (6.1.5). Suppose, as before, that E is a rank two bundle with $\Lambda^2 E$ trivial.

Proposition (6.1.10). *A $\mathscr{G}^c$ orbit in $\mathscr{A}^{1,1}$ contains at most one $\mathscr{G}$ orbit of ASD connections.*

Proof. Suppose that A_1, A_2 are two $\mathscr{G}^c$ equivalent connections. We consider the bundle $\text{Hom}(E, E)$ endowed with the connection, $A_1 * A_2$ say, induced from A_1 on the first factor and A_2 on the second. Then the element g of $\mathscr{G}^c$ intertwining A_1 and A_2 can be regarded as a holomorphic section of $\text{Hom}(E, E)$. But if A_1 and A_2 are ASD so also is $A_1 * A_2$. Hence, by the previous corollary,

$$\bar{\partial}_{A_1 * A_2} g = 0 \Rightarrow \nabla_{A_1 * A_2} g = 0,$$

and g is $A_1 * A_2$-covariant constant. This implies that $h = g^* g$ is a covariant constant relative to the connection induced by A_1, and in turn that

$$u = gh^{-\frac{1}{2}}$$

is also $A_1 * A_2$-covariant constant. Then u is a unitary transformation of E and the equation $\nabla_{A_1 * A_2} u = 0$ means precisely that u gives a gauge transformation from A_1 to A_2.

Proposition (6.1.11). *If a $\mathscr{G}^c$ orbit contains an ASD connection it is either stable or corresponds to a decomposable holomorphic structure $\mathscr{U} \oplus \mathscr{U}^{-1}$ with* $\deg(\mathscr{U}) = 0$. *In the latter case the connection is reducible. Conversely such a decomposable holomorphic structure admits a compatible reducible ASD connection.*

Proof. We consider first the situation for a holomorphic line bundle $\mathscr{U}$. Let α be any unitary connection on $\mathscr{U}$, with curvature F. Then, by the Chern–Weil theory for line bundles of Section 2.1.4,

$$\deg(\mathscr{U}) = \frac{i}{2\pi} \int_X F \wedge \omega = \frac{i}{2\pi} \int_X (\Lambda F)\, d\mu,$$

since $F \wedge \omega = \frac{1}{2}(\Lambda F)\omega^2 = (\Lambda F)\, d\mu$. Now consider a complex gauge transformation $g = \exp(\xi)$, for a real valued function ξ on X. The curvature of $\alpha' = g(\alpha)$ is

$$F_{\alpha'} = F_\alpha + 2i\, \bar{\partial}\partial\xi,$$

so $\hat{F}_{\alpha'} = \hat{F}_\alpha + \Delta\xi$. By the Fredholm alternative for the Laplacian we can find ξ, unique up to a constant, so that $i\hat{F}_{\alpha'}$ is constant over X. The constant value is then fixed by the Chern–Weil theory to be

$$i\hat{F}_{\alpha'} = \frac{2\pi}{\text{Vol}} \deg(\mathscr{U}),$$

where Vol is the volume of X.

Now let $\mathscr{U}$ be a line bundle with $\deg(\mathscr{U}) \leq 0$, and suppose $\mathscr{E}$ admits an ASD connection A. We consider the connection B on $\text{Hom}(\mathscr{E}, \mathscr{U})$ induced by A and the constant curvature connection α'. Then $i\hat{F}_B = (2\pi/\text{Vol}) \deg(\mathscr{U}) \leq 0$, so by (6.1.8) any holomorphic section s of $\text{Hom}(\mathscr{E}, \mathscr{U})$ is B-covariant constant. If s is not the zero section we must have $\deg(\mathcal{U}) = 0$ and s induces a splitting $\mathcal{E} = \mathcal{U} \oplus \mathcal{U}^{-1}$, compatible with the connections. So the connection A is then the reducible connection induced by α'. On the other hand, if A is not reducible there are no non-trivial holomorphic maps from $\mathcal{E}$ to any line bundle with negative degree and we deduce that $\mathcal{E}$ is stable. The last part of the proposition follows immediately from the existence of the above 'constant curvature' connections on line bundles.

To complete this list of simple properties we have:

Proposition (6.1.12). *All the connections in a stable orbit are irreducible.*

This is obvious—an S^1 reduction of the connection gives *a fortiori* a holomorphic decomposition $\mathcal{E}_A = \mathcal{U} \oplus \mathcal{U}^{-1}$; one of the factors has negative degree and gives a projection map contradicting stability.

6.1.4 Generalizations

As we have mentioned before, the theory of this chapter can be developed for general complex vector bundles, and indeed for principal holomorphic G^c-bundles, where G^c is the complexification of a compact group G. One gets a correspondence between certain holomorphic G^c-bundles and ASD G-connections. In another direction the theory can be generalized to holomorphic bundles over general compact Kähler manifolds. We will pause here to consider some of these generalizations for which we will have applications in Chapters 9 and 10.

Suppose first that $\mathcal{E}$ is a rank n holomorphic vector bundle over our Kähler surface X, and that $\deg(\mathcal{E}) = c_1(\mathcal{E}) \smile \omega$ is non-zero. The Chern–Weil representation then shows that $\mathcal{E}$ cannot admit any ASD connection. The correct generalization of the theory is suggested by the case of line bundles, as in (6.1.10) above. We say that a unitary connection A on $\mathcal{E}$ is *Hermitian Yang–Mills* if $i\hat{F}_A$ is a constant multiple of the identity endomorphism. The constant is then fixed by the Chern–Weil theory to be

$$i\hat{F}_A = \frac{2\pi}{\mathrm{Vol}} \frac{\deg(\mathcal{E})}{\mathrm{rank}(\mathcal{E})}.$$

On the other hand we say that $\mathcal{E}$ is *stable* if for any other bundle $\mathcal{V}$ which admits a non-trivial holomorphic map from $\mathcal{E}$ to $\mathcal{V}$, the inequality

$$\frac{\deg(\mathcal{E})}{\mathrm{rank}(\mathcal{E})} < \frac{\deg(\mathcal{V})}{\mathrm{rank}(\mathcal{V})}$$

holds. The main theorem is then just as before: a bundle admits an irreducible Hermitian Yang–Mills connection if and only if it is stable, and the connection is then unique.

Now these Hermitian Yang–Mills connections have structure group $U(n)$. The induced connection on the bundle of projective spaces is an ASD $PU(n)$ connection, where $PU(n) = U(n)$/scalars. Since $PU(2)$ is naturally isomorphic to $SO(3)$ this yields a convenient way to study certain $SO(3)$ connections. Suppose, for simplicity, that X is simply connected and V is an $SO(3)$ bundle over X such that $w_2(V)$ is the reduction of an integral class c which can be represented by a form of type $(1, 1)$. (For example this occurs if $w_2(V) = w_2(X)$.) Then, as in Section 2.1.4, V is associated with a $U(2)$ bundle E. Any ASD connection on V can be lifted uniquely to a Hermitian Yang–Mills connection on E. By applying the theory to these $U(2)$ connections one easily obtains:

Proposition (6.1.13). *If V is an $SO(3)$ bundle over a compact, simply connected, Kähler surface X with $w_2(V)$ the reduction of a $(1, 1)$ class c, there is a natural one-to-one correspondence between the moduli space M_V^* of irreducible ASD connections on V and isomorphism classes of stable holomorphic rank-two bundles $\mathscr{E}$ with $c_1(\mathscr{E}) = c$ and $c_2(\mathscr{E}) = \frac{1}{4}(c^2 - p_1(V))$.*

Next we mention the situation for bundles over a complex curve (compact Riemann surface) C. The degree of a bundle is then defined simply by evaluating the first Chern class on the fundamental two-cycle of the Riemann surface. The definition of stability is the same. If the Riemann surface is endowed with a metric, we define Hermitian Yang–Mills connections to be those with constant central curvature, and the same relation between stability and the existence of these connections holds good. Not surprisingly this theory was developed, by Narasimhan and Seshadri (1965), some years before the higher-dimensional theory considered above. In the simplest case we get a one-to-one correspondence between stable holomorphic $SL(2, \mathbb{C})$ bundles and flat $SU(2)$ connections over C, i.e. conjugacy classes of irreducible representations of $\pi_1(C)$ in $SU(2)$. These are parametrized by a moduli space:

$$\begin{aligned} W_C &\subset \mathrm{Hom}(\pi_1(C), SU(2))/SU(2) \qquad (6.1.14)\\ &= \text{isomorphism classes of stable holomorphic } SL(2, \mathbb{C})\text{-bundles } \mathscr{E} \to C, \end{aligned}$$

which is a complex manifold of complex dimension $3(\mathrm{genus}(C) - 1)$.

The extensions of the theory sketched so far can all be obtained easily by minor modifications of the existence proof for $SU(2)$ connections over Kähler surfaces, which we present in Section 6.2. The theory can be generalized still further to higher-dimensional Kähler manifolds. Here, by contrast, both the definition of stability and the proof of the existence theorem (due to Uhlenbeck and Yau (1986)) are more involved. However, the bulk of the discussion in this chapter can be adapted easily to any dimension. What is really special in the case of complex surfaces is the simple algebraic fact which we have seen in Chapter 2, that in this dimension the twin equations $F^{0,2} = 0$, $F.\omega = 0$ become the single, Riemannian invariant, anti-self-dual equation.

6.2 The existence proof

6.2.1 The gradient flow equation

Our proof of the core of Theorem (6.1.5)—the existence of ASD connections on stable bundles—is based on a natural evolution equation associated to the problem. Recall from Section 2.1 that for any $SU(2)$ connection over a compact four-manifold

$$\int |F_A|^2 \, d\mu = 8\pi^2 k + 2 \int |F_A^+|^2 \, d\mu,$$

and hence that an ASD connection is an absolute minimum of the Yang–Mills functional:

$$J(A) = \int_X |F_A|^2 \, d\mu. \tag{6.2.1}$$

Now

$$J(A + a) = J(A) + 2\langle F_A, d_A a\rangle + O(\|a\|^2),$$

so the first variation of J is represented, in classical notation, by

$$\delta J = 2\langle d_A^* F_A, \delta A\rangle. \tag{6.2.2}$$

That is, the gradient vector of J with respect to the L^2 inner product is $2d_A^* F_A$. We consider the integral curves of this gradient vector field—one-parameter families A_t satisfying the partial differential equation

$$\frac{\partial}{\partial t} A_t = - d_{A_t}^* F_{A_t}. \tag{6.2.3}$$

We can consider this evolution equation over any Riemannian base manifold. In the case of Kähler manifolds we shall see that it fits in very tidily with the holomorphic geometry; in fact one of the interesting things about this approach to the existence theorem is that one gets as a by-product rather precise information about particular solutions to this 'non-linear heat equation'.

Suppose then that X is a compact Kähler surface and that $A \in \mathscr{A}^{1,1}$ has curvature of type (1, 1). The Bianchi identity $d_A F_A = 0$ implies that both $\partial_A F_A$ and $\bar{\partial}_A F_A$ vanish. Then, by the Kähler identities (6.1.8) above, we have

$$d_A^* F_A = i(\bar{\partial}_A - \partial_A)\hat{F}_A. \tag{6.2.4}$$

We see from this, first, that any critical point of J (i.e. a solution of the Yang–Mills equations $d_A^* F_A = 0$) which also lies in $\mathscr{A}^{1,1}$ is either ASD or reducible. For, taking account of bi-degree,

$$d_A^* F_A = 0 \Leftrightarrow \nabla_A \hat{F}_A = 0. \tag{6.2.5}$$

If $\hat{F}_A$ is zero then A is ASD; if not, the eigenspaces of the (covariant constant, skew adjoint) endomorphism $\hat{F}_A$ decompose the bundle in factors, compatible with the connection. Second, we see that on $\mathscr{A}^{1,1}$ the gradient vectors $d_A^* F_A$ lie in the tangent spaces to the $\mathscr{G}^c$ orbits. For the Lie algebra of $\mathscr{G}^c$ is $\Omega^0(\mathrm{End}_0 E) = \Omega^0(\mathfrak{g}_E \otimes \mathbb{C})$ and the derivative of the $\mathscr{G}^c$ action at A is

$$\phi \longrightarrow -\bar{\partial}_A \phi + \partial_A(\phi^*), \tag{6.2.6}$$

for $\phi \in \Omega^0(\mathrm{End}_0 E)$ (cf. (2.1.20)). So the tangent vector $-d_A^* F_A$ represents the infinitesimal action of the element $i\hat{F}_A$ of the Lie algebra of $\mathscr{G}^c$.

This calculation suggests that the integral curves of the gradient field should preserve the $\mathscr{G}^c$ orbits in $\mathscr{A}^{1,1}$. To make this precise some work is needed. For example, we need to show that solutions to the evolution equation exist. We postpone the discussion of this aspect to Section 6.3 and state here the required result.

Proposition (6.2.7). *For A in $\mathscr{A}^{1,1}$ there is a unique solution A_t, $t \in [0, \infty)$, to the equation $\partial A_t/\partial t = -d_{A_t}^* F_{A_t}$ with $A_0 = A$. Moreover there is a smooth one-parameter family g_t in $\mathscr{G}^c$ such that $A_t = g_t(A)$.*

6.2.2 Outline of proof: closure of $\mathscr{G}^c$ orbits

We can now explain the basic idea of our proof. Let Γ be an orbit of $\mathscr{G}^c$ in $\mathscr{A}^{1,1}$, i.e. the set of unitary connections compatible with a given holomorphic structure $\mathscr{E}$. We know that an ASD connection in Γ, if one exists, minimizes J over all of $\mathscr{A}$, so *a fortiori* over Γ. Thus it is reasonable to hope to find this connection as the limit of the gradient flow of J. So pick any connection A_0 in Γ and consider the solution A_t of the evolution equation starting from A_0. Proposition (6.2.7) asserts that this solution exists and that A_t lies in Γ for all t. Now suppose, for the sake of this exposition, that as $t \to \infty$ the A_t converge to a critical point A_∞ of J which also lies in $\mathscr{A}^{1,1}$. Then as we have seen above, A_∞ is either ASD or reducible. On the other hand, A_∞ lies in the closure $\bar{\Gamma}$ of the orbit Γ. So, given this convergence assumption, we would have to show that if Γ is a *stable* orbit then A_∞ actually lies inside Γ. Conversely, for an unstable orbit Γ we expect that the non-existence of an ASD connection can be ascribed to another orbit Γ', containing a critical point and meeting the closure of Γ.

The discussion above, which we will take up again in Section 6.5, shows that the key to understanding the stability condition can be found in the structure of the $\mathscr{G}^c$ orbits and their closures. (One should set this in contrast with the fact, which we have used many times, that the orbits of the ordinary unitary gauge group $\mathscr{G}$ are closed; cf. (2.3.15) and (4.2.4).) To this end we introduce a simple fact about stable bundles which will bring the stability condition to bear in our existence proof.

Proposition (6.2.8). *Suppose $\mathscr{E}$ and $\mathscr{F}$ are rank-two holomorphic bundles over X with $\Lambda^2\mathscr{E}$, $\Lambda^2\mathscr{F}$ both trivial, and that there is a non-zero holomorphic bundle map $\alpha: \mathscr{E} \to \mathscr{F}$. Suppose that $\mathscr{E}$ is stable and that $\mathscr{F}$ is either stable or a sum of line bundles $\mathscr{U} \oplus \mathscr{U}^{-1}$ with $\deg(\mathscr{U}) = 0$. Then α is an isomorphism.*

Proof. The determinant of α can be regarded as a holomorphic function on X, since $\Lambda^2\mathscr{E}$ and $\Lambda^2\mathscr{F}$ are trivial. But X is compact, so the determinant is constant: we have to show that this constant is not zero. Suppose, on the contrary, that $\det \alpha$ is everywhere zero, so α has rank 0 or 1 at each point of X. We claim then that there is a factorization:

$$\begin{array}{ccc} \mathscr{E} & \xrightarrow{\beta} & \mathscr{L} \\ & \underset{\alpha}{\searrow} & \downarrow \gamma \\ & & \mathscr{F}, \end{array}$$

where $\mathscr{L}$ is a holomorphic line bundle over X. This is a standard fact from analytic geometry. The claim is essentially local so we may regard α as a 2×2 matrix of holomorphic functions (α_{ij}), defined relative to bases for $\mathscr{E}$ and $\mathscr{F}$. If the α_{ij} have a common factor f we can replace α by $f^{-1}\alpha$, so we may as well assume that the highest common factor of the entries is 1, and α vanishes on isolated points. Away from these zero points α has rank one and $\mathscr{L}$ has an obvious definition as the image of α. We have to see that this line bundle defined on the punctured manifold has a natural extension over the zero points. We first show that for each zero point x in X there is a neighbourhood N of x such that $\mathscr{L}$ is holomorphically trivial over $N \backslash \{x\}$. Over such a punctured neighbourhood $\mathscr{L}$ is generated by the two sections $\alpha(s_1)$, $\alpha(s_2)$, where s_1, s_2 are basis elements for $\mathscr{E}$. Since α has rank one it is clear that these satisfy a relation:

$$p\alpha(s_1) + q\alpha(s_2) = 0,$$

for holomorphic functions p and q on N; removing common factors we may suppose that p and q are coprime and their common zero set is just the point x. But then we can rewrite the relation in the form:

$$\frac{1}{q}\alpha(s_1) + \frac{1}{p}\alpha(s_2) = 0,$$

so the section σ of $\mathscr{L}$ defined over $N \backslash \{q = 0\}$ by $(1/q)\alpha(s_1)$ and over $N \backslash \{p = 0\}$ by $-(1/p)\alpha(s_2)$ gives a trivialization of $\mathscr{L}$ over $N \backslash \{p = q = 0\} = N \backslash \{0\}$.

Using this trivialization we can immediately extend $\mathscr{L}$, as an abstract line bundle rather than a subbundle of $\mathscr{F}$, over all of X. The extension is unique, up to canonical isomorphism, by the well known theorem of Hartogs which asserts that a holomorphic function on a punctured neighbourhood $N \backslash \{0\}$ automatically extends over the puncture. For the same reason the factorization $\mathscr{E} \to \mathscr{L} \to \mathscr{F}$, which is manifest away from the zeros, extends to the whole of X.

Now, given this factorization, consider the number $d = \deg(\mathscr{L})$. The non-trivial map β and the stability of $\mathscr{E}$ show that $d > 0$. On the other hand the transpose of γ, defined using the trivialization of $\Lambda^2 \mathscr{F}$ to identify $\mathscr{F}$ with $\mathscr{F}^*$, is a non-trivial map from $\mathscr{F}$ to $\mathscr{L}^*$. If $\mathscr{F}$ is stable or a sum of degree zero line bundles this requires that $\deg(\mathscr{L}^*) = -d > 0$ so we have a contradiction. (Note by the way that the same proof works if we assume that $\mathscr{F}$ is semi-stable.)

To fit this result into the picture envisaged above we consider for any pair of connections A_1, A_2 in $\mathscr{A}^{1,1}$ the $\bar{\partial}$ operator $\bar{\partial}_{A_1 * A_2}$ of the connection $A_1 * A_2$ on $\mathrm{Hom}(E, E)$. The connections are in the same $\mathscr{G}^c$ orbit if there is an everywhere *invertible* section g of $\mathrm{Hom}(E, E)$ with $\bar{\partial}_{A_1 * A_2} g = 0$. On the other hand the condition that $\bar{\partial}_{A_1 * A_2}$ has a non-trivial kernel is a closed condition in the variables A_1, A_2 (in the L^2_2 topology, for example). So if A_2 is in the closure of the orbit of A_1, there is still a non-trivial endomorphism g with $\bar{\partial}_{A_1 * A_2} g = 0$; but if A_2 is not actually in the orbit, g cannot be invertible. Proposition (6.2.8) now appears in this abstract picture as the assertion that if Γ_1, Γ_2 are, for example, distinct stable orbits, then the intersections:

$$\Gamma_1 \cap \bar{\Gamma}_2, \bar{\Gamma}_1 \cap \Gamma_2$$

are both *empty*. Now in our discussion at the beginning of this section we assumed that there is a limiting connection A_∞ for the gradient flow which is a critical point of J, so lies in either a stable or reducible orbit. In the former case we would now conclude that in fact A_∞ lies in the original orbit. It would remain to examine the case when A_∞ is reducible, compatible with a splitting $\mathscr{U} \oplus \mathscr{U}^{-1}$ with $\deg \mathscr{U} > 0$. If we could show that the bundle map $g_\infty : \mathscr{E} \to \mathscr{U} \oplus \mathscr{U}^{-1}$ had a non-zero component mapping into $\mathscr{U}^{-1}$ the proof would be complete since this would contradict the stability of $\mathscr{E}$.

We shall not take the discussion any further at this stage. It certainly need *not* be the case that the gradient flow lines converge in $\mathscr{A}^{1,1}$, so the assumptions we have made are not realistic. However, as we shall see, the basic ideas go through if we replace convergence over all of X by a notion of 'weak convergence' similar to that in Chapter 4. But to obtain this convergence we have to delve into the analytical properties of the gradient flow.

6.2.3 *Calculations with the gradient flow equation*

Suppose that A_t is a solution of the gradient flow equation in $\mathscr{A}^{1,1}$ provided by Proposition (6.2.7). We certainly have that $J(t) = J(A_t)$ is decreasing with time, indeed

$$\frac{\mathrm{d}}{\mathrm{d}t} J(t) = -\int_X |\mathrm{d}_A^* F_A|^2 \,\mathrm{d}\mu = -\|\mathrm{d}_A^* F_A\|^2. \tag{6.2.9}$$

(We will often denote A_t by A for tidiness, leaving the t dependence to be understood.) This is, of course, true for the Yang–Mills gradient flow over any

Riemannian base manifold. A stronger property follows from the Kähler condition on X. Define functions e_t on X by

$$e_t = |\hat{F}_{A_t}|^2. \tag{6.2.10}$$

Then we have:

Lemma (6.2.11). *For a solution A_t of the gradient flow equation,*

$$\frac{\partial e_t}{\partial t} = -\Delta e_t - |\mathrm{d}_A^* F_A|^2.$$

Proof. We have

$$\begin{aligned}\frac{\partial}{\partial t}\hat{F}_A &= \Lambda \mathrm{d}_A\left(\frac{\partial A}{\partial t}\right)\\ &= i\Lambda \mathrm{d}_A(\bar{\partial}_A - \partial_A)\hat{F}_A\\ &= i\Lambda(\partial_A\bar{\partial}_A - \bar{\partial}_A\partial_A)\hat{F}_A = -\nabla_A^*\nabla_A\hat{F}_A,\end{aligned}$$

where in the last step we use Lemma (6.1.7). Now

$$\Delta\{|\hat{F}_A|^2\} = 2(\hat{F}_A, \nabla_A^*\nabla_A\hat{F}_A) - |\nabla_A\hat{F}_A|^2$$

and $|\nabla_A\hat{F}_A|^2 = |\mathrm{d}_A^*F_A|^2$. So

$$\frac{\partial}{\partial t}|\hat{F}_A|^2 = -2(\hat{F}_A, \nabla_A^*\nabla_A\hat{F}_A) = -\Delta|\hat{F}_A|^2 - |\mathrm{d}_A^*F_A|^2,$$

as required.

Corollary (6.2.12). $\sup_X e_t$ *is a decreasing function of t.*

This follows from the previous lemma and the maximum principle for the heat operator $(\partial/\partial t) + \Delta$ on X: the maximum value of any function f on $X \times [0, \infty)$ with $((\partial/\partial t) + \Delta)f \leq 0$ is attained at time 0.

It is instructive to note that we do not have this strong pointwise control of the full curvature tensor F_A. This satisfies the equation

$$\frac{\partial}{\partial t}F_A = -\Delta_A F_A,$$

where Δ_A is the Laplacian on forms (cf. Section 2.3.10). By the Weitzenböck formula (2.3.18) we have

$$\Delta_A F_A = \nabla_A^*\nabla_A F_A + \{F_A, F_A\} + \{R_X, F_A\}$$

and the argument above yields:

$$\left(\frac{\partial}{\partial t} + \Delta\right)|F_A|^2 \leq |(\{F_A, F_A\}, F_A) + (\{R_X, F_A\}, F_A)|.$$

Without extra information, the cubic term in F_A on the right-hand side of this inequality prevents us from making any useful deductions from the formula (consider the solutions of the ODE $\mathrm{d}e/\mathrm{d}t = e^{3/2}$ for comparison). Indeed we shall see that $\sup |F_A|$ can indeed blow up as t tends to infinity. From this general point of view the key feature of the Kähler case is that for the $\hat{F}$ component the non-linear terms disappear from the Weitzenböck formula.

Now, given our solution A_t, let us write

$$I(t) = \|\nabla_A \hat{F}_A\|^2 = \|\mathrm{d}_A^* F_A\|^2. \tag{6.2.13}$$

Proposition (6.2.14). *$I(t)$ tends to 0 as t tends to infinity.*

Proof. The time derivative of $\nabla_A \hat{F}_A$ has two contributions, one from the variation of $\hat{F}_A$ and one from the variation of ∇_A:

$$\begin{aligned}\frac{\partial}{\partial t}\nabla_A \hat{F}_A &= \left(\frac{\partial}{\partial t}\nabla_A\right)\hat{F}_A + \nabla_A\left(\frac{\partial}{\partial t}\hat{F}_A\right)\\ &= [i(\bar{\partial}_A - \partial_A)\hat{F}_A, \hat{F}_A] - \nabla_A \nabla_A^* \nabla_A \hat{F}_A.\end{aligned}$$

Taking the inner product with $\nabla_A \hat{F}_A$ and integrating over X we get

$$\begin{aligned}\frac{\mathrm{d}I}{\mathrm{d}t} &= \left\langle \frac{\partial}{\partial t}(\nabla_A \hat{F}_A), \nabla_A \hat{F}_A \right\rangle\\ &= \langle [i(\bar{\partial}_A - \partial_A)\hat{F}_A, \hat{F}_A], (\bar{\partial}_A + \partial_A)\hat{F}_A \rangle - \|\nabla_A^* \nabla_A \hat{F}_A\|^2,\end{aligned}$$

where we have integrated by parts in the second term. We know by Corollary (6.2.12) that $|\hat{F}_{A_t}|$ is bounded. So the first term on the right-hand side is certainly bounded by a multiple of $\|\nabla_A \hat{F}_A\|^2$. For the second term we use the Cauchy–Schwarz inequality:

$$I(t) = \|\nabla_A \hat{F}_A\|^2 = \langle \hat{F}_A, \nabla_A^* \nabla_A \hat{F}_A \rangle \leq \|\hat{F}_A\| \, \|\nabla_A^* \nabla_A \hat{F}_A\|.$$

Using the L^2 bound on $\hat{F}_A$, we get

$$\frac{\mathrm{d}I}{\mathrm{d}t} \leq c_1 I - c_2 I^2,$$

for positive constants c_1 and c_2. (With $c_2 = (\max_t \|\hat{F}_A\|)^{-2}$.) On the other hand $\mathrm{d}J/\mathrm{d}t = -I$, and J is nonnegative, so

$$\int_0^\infty I(t)\,\mathrm{d}t < \infty.$$

The fact that $I(t)$ tends to zero as $t \to \infty$ follows from these two properties by the following elementary calculus argument.

First we can suppose the constants c_1, c_2 in the differential inequality are both 1 (rescale the variables). Then the solution of

$$\frac{\mathrm{d}f}{\mathrm{d}t} = f - f^2$$

with $f(0) = \frac{1}{2}$ is the monotone increasing function:

$$f(t) = \frac{e^t}{1 + e^t}.$$

Given $\varepsilon_1, \varepsilon_2$ with $0 < \varepsilon_1 < \varepsilon_2 < 1$, let

$$K(\varepsilon_1, \varepsilon_2) = \int_{f^{-1}(\varepsilon_1)}^{f^{-1}(\varepsilon_2)} f(t)\,\mathrm{d}t.$$

Now the convergence of $\int I\,\mathrm{d}t$ implies that $\liminf_{t \to \infty} I(t) = 0$. If I does not tend to zero there is an $\varepsilon > 0$ and an increasing sequence $t_n \to \infty$ with $I(t_n) = \varepsilon$, and we can choose this such that there is an interleaved sequence s_n $(t_{n-1} < s_n < t_n)$, with $I(s_n) = \frac{1}{2}\varepsilon$. It suffices to show that

$$\int_{s_n}^{t_n} I(t)\,\mathrm{d}t > K(\tfrac{1}{2}\varepsilon, \varepsilon),$$

for this implies that $\int I$ diverges and we have a contradiction. But if we choose τ so that $f(t_n - \tau) = \varepsilon$, the differential inequality implies that

$$I(t) \geq f(t - \tau) \quad \text{for } t \leq t_n.$$

So $f(s_n - \tau) \leq \frac{1}{2}\varepsilon$ and

$$\int_{s_n}^{t_n} I(t)\,\mathrm{d}t \geq \int_{f^{-1}(\frac{1}{2}\varepsilon)+\tau}^{t_n} I(t)\,\mathrm{d}t \geq \int_{f^{-1}(\frac{1}{2}\varepsilon)+\tau}^{t_n} f(t - \tau)\,\mathrm{d}t = K(\tfrac{1}{2}\varepsilon, \varepsilon).$$

The argument is made clearer in Fig. 10.

6.2.4 Weak convergence of connections

Recall that Uhlenbeck's theorem in Chapter 2 applied to connections over a ball whose curvature had L^2 norm less than ε. If we have any sequence of connections A_α over the compact manifold X whose curvatures satisfy a common L^2 bound, then the argument of Section 4.4.3 applies to give a subsequence $\{\alpha'\}$, a finite set $\{x_1, \ldots, x_p\}$ in X and a cover of $X \setminus \{x_1, \ldots, x_p\}$ by a system of balls D_i such that each connection $A_{\alpha'}$ has curvature with L^2 norm less than ε over each D_i. So we can put the connections in Coulomb gauge over these balls. If we know that in these Coulomb gauges the connection matrices converge in a suitable function space, the patching argument of Section 4.4.2 can be applied to obtain the corresponding convergence over a compact subsets of the punctured manifold. (See the remarks after Lemma (4.4.6).)

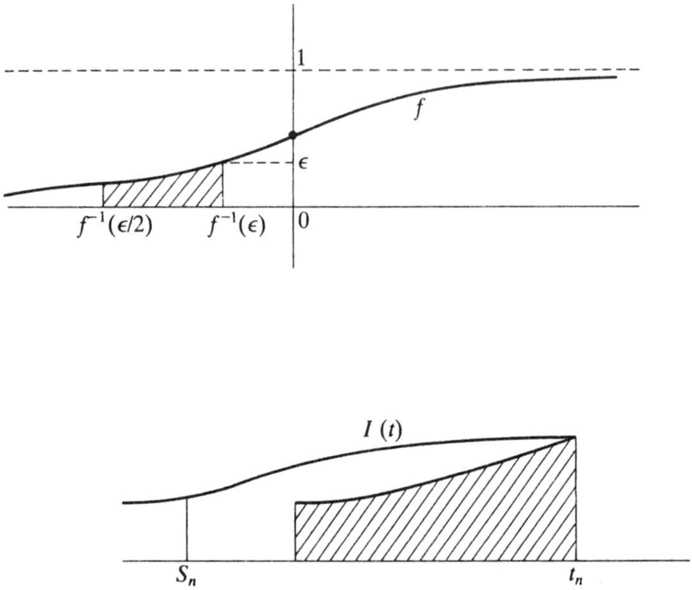

Fig. 10

Now let A_t be a one-parameter family of connections generated by the gradient flow in $\mathscr{A}^{1,1}$ as above. The curvatures $F_t = F_{A_t}$ are bounded in L^2, and we know that:

(i) $F^+_{A_t}$ is uniformly bounded (Corollary (6.2.12));

(ii) $\nabla_{A_t} F^+_{A_t} \to 0$ in L^2 (Proposition (6.2.14)).

It is a good exercise in elliptic estimates to show that, given (i) and (ii) and the Coulomb gauge condition over a small ball provided by Uhlenbeck's theorem, one gets an L^2_2 bound on the connection matrices over interior domains (provided, as usual, that the curvature is small enough in L^2), and that there is a sequence $t_\alpha \to \infty$ such that the connections $A_\alpha = A_{t_\alpha}$ converge over the balls, strongly in L^2_2. The transition functions relating these connection matrices on the overlaps of the balls can then be supposed to converge in L^2_3 and hence in C^0. So the criterion for the application of the patching argument is satisfied, and we get L^2_2 convergence, over compact subsets of the punctured manifold, to a limit A_∞, say.

Now, by property (ii), A_∞ satisfies the equation $\nabla_{A_\infty} \hat{F}_{A_\infty} = 0$, which is elliptic in a Coulomb gauge. Elliptic regularity implies that A_∞ is smooth over $X\backslash\{x_1, \ldots, x_p\}$. Again, this is a good exercise in the techniques used in Chapters 2 and 4. Then, just as in (6.1.10), we have the alternative:

either (*a*) $\hat{F}_{A_\infty} = 0$, so A_∞ is ASD.

or (*b*) $\hat{F}_{A_\alpha}$ is non-zero and A_∞ is a reducible connection on a holomorphic bundle $\mathscr{U} \oplus \mathscr{U}^{-1}$, induced from a constant curvature connection on $\mathscr{U}$, with $\deg(\mathscr{U}) > 0$.

In either case A_∞ extends, in suitable local gauges, over the punctures to give a smooth connection on a bundle E' over X. (The proof in the reducible case is similar to that in Section 4.4, but is easier, since one is working with a linear differential equation.) In sum we have:

Proposition (6.2.15). *There is a sequence $t_\alpha \to \infty$ and a unitary connection A_∞ on a bundle $E' \to X$ satisfying one of the alternatives* (*a*) *and* (*b*), *together with a sequence of unitary bundle maps ρ_α: $E'|_{X\setminus\{x_1,\dots,x_p\}} \to E|_{X\setminus\{x_1,\dots,x_p\}}$ such that $\rho_\alpha^*(A_\alpha)$ converges to A_∞ in L^2_2 on compact subsets.*

Notice that the occurrence of (a) or (b) is independent of the sequence t_α chosen to achieve convergence. It is just a question of the limit of the L^2 norm of $\hat{F}_{A_t}$. Indeed, we have

$$\deg(\mathscr{U})^2 = \frac{\mathrm{Vol}}{8\pi^2} \lim_{t\to\infty} \|\hat{F}_t\|^2, \tag{6.2.16}$$

since the constant curvature connection on $\mathscr{U}$ has $\hat{F} = -2\pi i \deg(\mathscr{U})/\mathrm{Vol}$. In case (*a*) the limiting connection endows E' with a holomorphic structure $\mathscr{E}'$, which by (6.1.10), is either stable or a sum of degree zero line bundles, since it carries an ASD connection.

6.2.5 The limit of the complex gauge transformations

The existence proof is completed by showing that if the original bundle $\mathscr{E}$ is stable then alternative (*a*) occurs, and that $\mathscr{E}'$ is isomorphic to $\mathscr{E}$. To do this we recall that each of the A_α is related to the original connection A by a complex gauge transformation g_α say. Equivalently g_α: $E \to E$ is a bundle isomorphism with $\bar{\partial}_{A*A_\alpha} g_\alpha = 0$. We suppose that $\det(g_\alpha) = 1$ and let

$$v_\alpha = \|g_\alpha\|_{L^2}, \quad f_\alpha = \frac{1}{v_\alpha} g_\alpha.$$

So the f_α are normalized complex linear automorphisms of E, with L^2 norm 1. Also, we write

$$B_\alpha = \rho_\alpha^*(A_\alpha)$$

and

$$h_\alpha = \rho_\alpha^{-1} f_\alpha,$$

so the h_α are bundle maps from E to E' over the punctured manifold, which satisfy

$$\bar{\partial}_{A*B_\alpha} h_\alpha = 0,$$

and the connections B_α converge to A_∞ over the punctured manifold.

The equation $\bar{\partial}_{A*B_\alpha} h_\alpha = 0$ is an elliptic equation for h_α, with A and B_α regarded as fixed. By construction the h_α are bounded in L^2 and an easy bootstrapping argument shows that the h_α are bounded in L^2_3 over compact subsets. Taking a further subsequence if necessary, we can suppose the h_α converge to a limit h, weakly in L^2_3 over compact subsets. This limit satisfies the Cauchy–Riemann equation $\bar{\partial}_{A*A_\infty} h = 0$ and so is a holomorphic bundle map

$$h\colon \mathscr{E}|_{X\setminus\{x_1,\ldots,x_p\}} \longrightarrow \mathscr{E}'|_{X\setminus\{x_1,\ldots,x_p\}}.$$

By Hartogs's theorem h extends over the punctures to a holomorphic map from $\mathscr{E}$ to $\mathscr{E}'$. Our next task is to show that h is not identically zero.

For a complex gauge transformation g of E let us write

$$\tau = \mathrm{Trace}(g^* g),$$

so the L^2 norm of g is the integral of the function τ over X. Now if A is a connection and $g(A) = B$ we have

$$\begin{aligned}\Delta\tau &= \mathrm{Tr}\{\nabla^*\nabla(g^*g)\}\\ &= \mathrm{Tr}\{(\nabla^*\nabla g^*)g + g^*\nabla^*\nabla g - 2(\nabla g^*).(\nabla g)\}\\ &\leq \mathrm{Tr}\{(\nabla^*\nabla g^*)g + g^*(\nabla^*\nabla g)\}.\end{aligned}$$

Here we have written ∇ for the covariant derivative ∇_{A*B} on the endomorphisms. Now $\bar{\partial}_{A*B} g = 0$ implies trivially that

$$\bar{\partial}^*_{A*B}\bar{\partial}_{A*B} g = 0.$$

So, by the Weitzenböck formula, applied to $\mathrm{Hom}(E, E)$,

$$\nabla^*\nabla g = i\{\hat{F}_A g - g\hat{F}_B\}, \tag{6.2.17}$$

and it follows that

$$\Delta\tau \leq i\,\mathrm{Tr}\{\hat{F}_A g^* g - g^*\hat{F}_B g + g^*\hat{F}_B g - g^* g\hat{F}_A\},$$

which gives

$$\Delta\tau \leq 2(\sup|\hat{F}_A| + \sup|\hat{F}_B|)\tau. \tag{6.2.18}$$

We apply this with $g = f_\alpha$ and B the connection A_α. Then by (6.2.12) the curvature terms are bounded and we get

$$\Delta\tau \leq C.\tau,$$

for a constant C independent of α.

Lemma (6.2.19). *If a non-negative function τ on X satisfies $\Delta\tau \leq C\tau$ and $\int_X \tau\, d\mu = 1$ then there is a constant C'' depending on C and X such that for any r-ball $B(r)$ in X,*

$$\int_{B(r)} \tau\, d\mu \leq C'' r^3.$$

In fact we can bound the supremum of f, and so improve this $O(r^3)$ bound to $O(r^4)$, but any positive power suffices for our application.

Proof. Multiply both sides of the inequality by τ and integrate to get

$$\int_X |\nabla\tau|^2 \, \mathrm{d}\mu \leq C \int_X \tau^2 \, \mathrm{d}\mu.$$

So, by the Sobolev inequality in four dimensions,

$$\int_X \tau^4 \, \mathrm{d}\mu \leq C' \left\{ \int_X \tau^2 \right\}^2.$$

Now use Hölders inequality with exponents 3, 3/2 to write

$$\int_X \tau^2 \, \mathrm{d}\mu \leq \left(\int_X \tau \, \mathrm{d}\mu \right)^{2/3} \left(\int_X \tau^4 \, \mathrm{d}\mu \right)^{1/3}.$$

Combining the two inequalities and rearranging we get

$$\int_X \tau^2 \, \mathrm{d}\mu \leq (C')^{1/3} \left(\int_X \tau \, \mathrm{d}\mu \right).$$

Then, substituting back into the first inequality, we see that:

$$\int_X \tau^4 \, \mathrm{d}\mu \leq (C')^3 \int_X \tau \, \mathrm{d}\mu.$$

Now for any r-ball $B(r)$ in X,

$$\int_{B(r)} \tau \, \mathrm{d}\mu \leq \left(\int_{B(r)} \tau^4 \, \mathrm{d}\mu \right)^{1/4} \left(\int_{B(r)} 1 \, \mathrm{d}\mu \right)^{3/4} \leq (C')^3 (\mathrm{Vol}\, B(r))^{3/4},$$

and $\mathrm{Vol}\,(B(r))$ is $O(r^4)$.

This lemma says that for all α the contribution to the L^2 norm of f_α from r-balls around the x_i is $O(r^3)$. So we can choose a compact domain $K \subset X \backslash \{x_1, \ldots, x_p\}$ such that

$$\int_K |h_\alpha|^2 \, \mathrm{d}\mu = \int_K |f_\alpha|^2 \, \mathrm{d}\mu \geq \tfrac{1}{2}, \text{ say,} \quad \text{for all } \alpha,$$

and it follows that the limit h is non-trivial, since the h_α converge strongly in L^2 to h over K.

6.2.6 Completion of existence proof

Let us now take stock of our results. Suppose that the original bundle $\mathscr{E}$ is stable and that the alternative (*a*) holds. Then we have a non-trivial holomorphic map h from $\mathscr{E}$ to the holomorphic bundle $\mathscr{E}'$ defined by the limiting connection A_∞, and $\mathscr{E}'$ is either stable or a sum of zero-degree line bundles. So we can apply (6.2.8) to see that h is actually a holomorphic bundle isomorphism, and hence A_∞ represents the solution to the existence problem. So it remains only to show that for a stable bundle $\mathscr{E}$ the second alternative cannot hold, and this is the business of the present subsection.

We want to prove that if the limit is a sum of line bundles $\mathscr{U} \oplus \mathscr{U}^{-1}$ with $\deg(\mathscr{U}) = d > 0$, then the original bundle $\mathscr{E}$ cannot be stable. We know that there is a non-trivial holomorphic bundle map

$$h = h^+ \oplus h^- : \mathscr{E} \longrightarrow \mathscr{U} \oplus \mathscr{U}^{-1}.$$

It suffices to show that the component h^- mapping to $\mathscr{U}^{-1}$ is non-zero, for this contradicts the stability of $\mathscr{E}$. To make the calculations clearer we will suppose, as is clearly permissible, that the volume of X is normalized to be 2π, so the constant curvature connection on $\mathscr{U}$ has $i\hat{F} = -d$.

We can suppose that $\|g_t\|$ tends to ∞ with t, otherwise the bundle maps g_{t_α} would converge without normalization and their limit would also have determinant 1, hence be an isomorphism from $\mathscr{E}$ to $\mathscr{E}'$. There is thus a sequence t_α of times, $t_\alpha \to \infty$, with

$$\frac{\mathrm{d}}{\mathrm{d}t}\{\|g_t\|^2\}|_{t=t_\alpha} > 0. \tag{6.2.20}$$

We can then extract subsequences, as above, starting from this sequence. As usual we relabel and just call the eventual subsequence t_α, and write g_α for g_{t_α}. Now $\partial g_t/\partial t = i\hat{F}_{A_t} g_t$, so

$$\frac{\mathrm{d}}{\mathrm{d}t}\|g_t\|^2 = 2\int_X \mathrm{Tr}(g_t^* i\hat{F}_A g_t)\,\mathrm{d}\mu. \tag{6.2.21}$$

We choose a compact subset $K = X \setminus \bigcup B(r, x_i)$ over which the connections $\rho_\alpha^*(A_\alpha)$ converge, and such that, as before, the integral of $|h_\alpha|^2$ over K is at least $\frac{1}{2}$. Also the contribution

$$\int_{B(r,x_i)} \mathrm{Tr}(g_\alpha^* i\hat{F}_{A_\alpha} g_\alpha)\,\mathrm{d}\mu$$

is bounded by a multiple of $v_\alpha^2 r^3$, since the $\hat{F}_{A_\alpha}$ are uniformly bounded, $g_\alpha = v_\alpha f_\alpha$ and we can apply (6.2.19) to $\tau = |f_\alpha|^2$. We choose r so small that

$$\int_{X\setminus K} \mathrm{Tr}(g_\alpha^* i\hat{F}_{A_\alpha} g_\alpha)\,\mathrm{d}\mu \le \tfrac{1}{4} d v_\alpha^2 \tag{6.2.22}$$

say. Now over K we write $B_\alpha = \rho_\alpha^*(A_\alpha)$, so the B_α are connections on the C^∞ bundle underlying $\mathscr{U} \oplus \mathscr{U}^{-1}$, converging in L_2^2 to the constant curvature connection A_∞. We have, over K,

$$\mathrm{Tr}(g_\alpha^* i\hat{F}_{A_\alpha} g_\alpha) = v_\alpha^2 \,\mathrm{Tr}(h_\alpha^* i\hat{F}_{B_\alpha} h_\alpha). \tag{6.2.23}$$

We know that $i\hat{F}_{B_\alpha}$ converges to the constant endomorphism of $\mathscr{U} \oplus \mathscr{U}^{-1}$:

$$\Lambda = i\hat{F}_{A_\infty} = \begin{bmatrix} -d & 0 \\ 0 & d \end{bmatrix}.$$

So

$$\int_K \mathrm{Tr}(h_\alpha^* i\hat{F}_{A_\alpha} h_\alpha)\,\mathrm{d}\mu = \int_K \mathrm{Tr}(h_\alpha^* \Lambda h_\alpha)\,\mathrm{d}\mu + \varepsilon_\alpha,$$

where

$$|\varepsilon_\alpha| \le \int_K |h_\alpha|^2 . |i\hat{F}_A - \Lambda|\,\mathrm{d}\mu$$

$$\le \|h_\alpha\|^2_{L^4(K)} \,\|i\hat{F}_{A_\alpha} - \Lambda\|_{L^2(K)}.$$

The h_α are bounded in L^4 and $i\hat{F}_{A_\alpha}$ converges in $L^2(K)$ to Λ, so ε_α tends to zero as α tends to infinity.

On the other hand, if we write

$$h_\alpha = h_\alpha^+ \oplus h_\alpha^- : E \longrightarrow \mathscr{U} \oplus \mathscr{U}^{-1},$$

we have

$$\mathrm{Tr}(h_\alpha^* \Lambda h_\alpha) = d(|h_\alpha^-|^2 - |h_\alpha^+|^2).$$

The proof is now in our hands: we know that

$$0 \le \int_X \mathrm{Tr}(g_\alpha^* i\hat{F}_{A_\alpha} g_\alpha)\,\mathrm{d}\mu = \left\{ \int_{X\setminus K} + \int_K \right\} \mathrm{Tr}(g_\alpha^* i\hat{F}_{A_\alpha} g_\alpha)/\mathrm{d}\mu,$$

so

$$\int_K \mathrm{Tr}(g_\alpha i\hat{F}_\alpha g_\alpha)\,\mathrm{d}\mu = v_\alpha^2 \int_K \mathrm{Tr}(h_\alpha i F_{B_\alpha} h_\alpha)\,\mathrm{d}\mu$$

$$\ge -\int_K \mathrm{Tr}(g_\alpha i\hat{F}_{A_\alpha} g_\alpha) \ge -\tfrac{1}{4}\mathrm{d}v_\alpha^2,$$

by (6.2.7). Hence

$$\int_K \mathrm{Tr}(h_\alpha i\hat{F}_{B_\alpha} h_\alpha) = d(\|h_\alpha^-\|^2_{L^2(K)} - \|h_\alpha^+\|_{L^2(K)}) + \varepsilon_\alpha$$

$$\ge -\tfrac{1}{4}d.$$

On the other hand $\|h\|^2_{L^2(K)} = \|h_\alpha^+\|^2_{L^2(K)} + \|h_\alpha^-\|^2_{L^2(K)} \ge \tfrac{1}{2}$, so

$$\| h_\alpha^- \|_{L^2(K)}^2 \geq \tfrac{1}{4} - (\varepsilon_\alpha/d),$$

and since ε_α tends to zero the limit h^- must indeed be a non-trivial bundle map from $\mathscr{E}$ to $\mathscr{U}^{-1}$.

6.2.7 Semi-stable bundles and compactification of moduli spaces

We have now completed the main task of this chapter, the proof of Theorem (6.1.5). For the remainder of the chapter we discuss a number of additional topics which add colour to the correspondence between stable bundles and ASD connections. We complete Section 6.2 by giving a partial algebro-geometric interpretation of the compactifications of moduli spaces introduced in Chapter 4.

It can certainly happen that one has a family $\mathscr{E}_t$ of holomorphic bundles, parametrized by $t \in \mathbb{C}$ say, such that $\mathscr{E}_t$ is stable for $t \neq 0$, but $\mathscr{E}_0$ is not stable. Our general theorem tells us that $\mathscr{E}_t$ $(t \neq 0)$ corresponds to an ASD connection A_t. It is natural to expect that the behaviour of the family A_t as $t \to 0$, will reflect properties of the holomorphic bundle $\mathscr{E}_0$. This is indeed the case, and the ideas involved are very similar to those used above to study the Yang–Mills gradient flow. While the theory can be developed in some generality, we shall consider here a special but typical case, for which we will have an application in Chapter 10.

We consider the case when $\mathscr{E}_0$ is a semi-stable, but not stable $SL(2, \mathbb{C})$-bundle over a surface X. By definition, this means that there is a destabilizing map from $\mathscr{E}_0$ to a line bundle $\mathscr{U}$ of degree 0. We assume that $\mathscr{U}$ is actually the trivial bundle, so the transpose of this map is a section s of $\mathscr{E}_0$. Let $(x_1, \ldots, x_k) \in s^k(X)$ be the zeros of α, counted with multiplicity (this multiplicity can be defined topologically, in the usual way). Note that s is unique up to scalars, except in the special case when $\mathscr{E}_0$ is trivial, and in any case the multi-set $(x_1, \ldots, x_k)$ is uniquely determined by $\mathscr{E}_0$. The set-up can be conveniently expressed, in a framework which we will develop in Chapter 10, by an exact sequence of sheaves,

$$0 \longrightarrow \mathcal{O} \xrightarrow{\ s\ } \mathscr{E}_0 \xrightarrow{\ s^{\mathrm{T}}\ } \mathscr{I} \longrightarrow 0, \tag{6.2.24}$$

where $\mathscr{I} \subset \mathcal{O}$ is an ideal sheaf, with $\mathcal{O}/\mathscr{I}$ supported on $\{x_1, \ldots, x_k\}$.

Proposition (6.2.25). *Let $\mathscr{E}_t$ be a family of bundles parametrized by $t \in \mathbb{C}$, with $\mathscr{E}_t$ stable for $t \neq 0$, and $\mathscr{E}_0$ semistable, destabilized by a holomorphic section s with zeros $(x_1, \ldots, x_k)$. Let A_t be the ASD connection corresponding to $\mathscr{E}_t$ for non-zero t. Then the family $[A_t]$ converges weakly to the ideal connection $([\theta], x_1, \ldots, x_k)$ as t tends to 0.*

To prove this we choose a continuous family of connections B_t on a C^∞ bundle E, such that B_t defines the holomorphic structure $\mathscr{E}_t$. Our compactness theorem tells us that any sequence $t_\alpha \to 0$ has a subsequence for which the corresponding ASD connections $[A_{t_\alpha}]$ converge weakly. To prove the proposition it suffices to show that for any such convergent sequence the limit is $([\theta], x_1, \ldots, x_k)$. So, switching notation, let A_α be a sequence converging to $([A], y_1, \ldots, y_l)$, where A is an ASD connection on a bundle E'.

Let $\mathscr{E}'$ be the holomorphic structure on E' defined by A. Then, just as in Section 6.2.5, we get a non-trivial holomorphic map $h: \mathscr{E}_0 \to \mathscr{E}'$ over all of X.

We claim first that $\mathscr{E}'$ is the trivial bundle, so A is the product connection. To see this, form the transpose $h^{\mathrm{T}}: \mathscr{E}' \to \mathscr{E}_0$ using the trivializations of Λ^2. If $\mathscr{E}'$ is stable, the composite

$$s^{\mathrm{T}} \circ h^{\mathrm{T}}: \mathscr{E}' \longrightarrow \mathcal{O}$$

must be zero, so h^{T} lifts over $\mathcal{O} \to \mathscr{E}_0$ to give a map from $\mathscr{E}_0$ to $\mathcal{O}$. Again this must be zero if $\mathscr{E}'$ is stable, so h^{T} is zero, which is a contradiction.

The more difficult task is to show that the multiset $(y_1, \ldots, y_l)$ is $(x_1, \ldots, x_k)$. Notice that, due account being taken of multiplicities, we must have $l = k$ for topological reasons. The argument we will use extends this idea. We first consider the situation over $X \setminus \{x_1, \ldots, x_k\}$. On this subset the section s represents a trivial subbundle $\mathcal{O}_S \subset \mathscr{E}_0$. If we choose a C^∞ complement, giving a trivialization of $\mathscr{E}_0$ over K, we express the operator $\bar{\partial}_{B_0}$ in the matrix form:

$$\bar{\partial}_{B_0} = \bar{\partial} + \begin{pmatrix} \phi & \beta \\ 0 & -\phi \end{pmatrix},$$

where $\bar{\partial} + \phi$ is a $\bar{\partial}$ operator on the trivial bundle $\mathcal{O}_S$. Then in this same trivialization we represent $\bar{\partial}_{B_t}$ in the form:

$$\bar{\partial}_{B_t} = \bar{\partial} + \begin{pmatrix} \phi + \varepsilon_1(t) & \beta + \varepsilon_2(t) \\ \varepsilon_3(t) & -(\varphi + \varepsilon_1(t)) \end{pmatrix},$$

where $\varepsilon_i(t)$ tends to zero with t.

We next find complex gauge transformations g_t over X such that $g_t(B_t)$ converges to the trivial connection over $X \setminus \{x_i\}$. To construct these, we first use a complex gauge transformation of the trivial line bundle to reduce to the case when $\phi = 0$. Then we make a further complex gauge transformation of the form

$$\begin{pmatrix} \lambda(t) & 0 \\ 0 & \lambda^{-1}(t) \end{pmatrix}$$

for constants $\lambda(t)$ in $\mathbb{C}$. In our trivialization over $X \setminus \{x_1, \ldots, x_k\}$ these transform $\bar{\partial}_{B_t}$ to

$$\bar{\partial} + \begin{pmatrix} 0 & \lambda(t)^2(\beta + \varepsilon_2(t)) \\ \lambda(t)^{-2}\varepsilon_3(t) & 0 \end{pmatrix}.$$

If we take $\lambda(t) = \|\varepsilon_3(t)\|^{1/4}$, say, then these operators do indeed converge to the standard operator on $\mathcal{O}_X \oplus \mathcal{O}_X$ over $X \setminus \{x_1, \ldots, x_k\}$. (Strictly we should work over a compact subset here, but this is not important.)

We can express this slightly differently as follows: there is trivialization τ of E over $X \setminus \{x_1, \ldots, x_k\}$, and a sequence of connections B'_α on E over X which converge to the product connection defined in this trivialization over $X \setminus \{x_1, \ldots, x_k\}$, but which represent the holomorphic structures $\mathcal{E}_{t_\alpha}$ over X.

Turning now to the ASD connections, we can express our conclusions as follows: there is a trivialization σ of E over $X \setminus \{y_1, \ldots, y_l\}$ and a family of connections A'_{t_α} on E over X which converge to the product connection defined by the trivialization σ over $X \setminus \{y_1, \ldots, y_k\}$, and with

$$|F_{A'_\alpha}|^2 \longrightarrow 8\pi^2 \sum \delta_{y'_i}$$

but which also represent the same holomorphic structures E_{t_α}. Thus there are complex gauge transformations g_α of E over X with $g_\alpha(B'_\alpha) = A'_\alpha$.

Over the 'doubly punctured' manifold $X \setminus \{x_1, \ldots, x_k\} \setminus \{y_1, \ldots, y_k\}$ we can represent g_α by a matrix-valued function, using the trivializations σ, τ, and, as before, we can suppose these converge to a limit g. This limit extends to a holomorphic matrix-valued function over X and hence is a constant.

We now introduce the topological input. With each point x_i we can associate a degree of the trivialization τ over a small sphere about x_i, relative to a trivialization of E which extends over x_i. Here we use the isomorphism:

$$H_3(GL(2, \mathbb{C})) = \mathbb{Z}.$$

It is easy to see from the definition of τ that this is just the multiplicity of the zero of s at this point. Similarly, with each point y_i we associate a number by the degree of σ, and it is easy to see from Chern–Weil theory that this is just the multiplicity of y_i in the multiset $(y_1, \ldots, y_k)$. It follows then that for any point z in $\{x_1, \ldots, x_k, y_1, \ldots, y_k\}$ the difference of the multiplicity of z in the multisets $\{x_1, \ldots, x_k\}$, $\{y_1, \ldots, y_k\}$ is the degree of the map g_α over a small three-sphere S about z (where g_α is viewed as a matrix valued function using the trivializations). On the other hand g_α converges uniformly to the constant matrix g over S. If g is an invertible matrix it is obvious that the degree of g_α is zero for large α; hence the multiplicities agree, and we are finished. In general we use the following lemma:

Lemma (6.2.26). *Let g be a non-zero 2×2 complex matrix and N_r be the intersection of the r-ball about g with the open subset $GL(2, \mathbb{C})$ of invertible matrices. Then $H_3(N_r; \mathbb{Z}) = 0$ for small r; in particular any map $g_\alpha: S^3 \to N_r$ has degree zero.*

This lemma follows from the fact that N_r is the complement of a smooth complex hypersurface in a small ball in $\mathbb{C}^4$, and so is homotopy equivalent to a circle.

Using this lemma we deduce that the degree of g_α is zero for large α, so the multiplicities agree and the proof of (6.2.25) is complete.

6.3 The Yang–Mills gradient equation

6.3.1 Short-time solutions

We will now go back to discuss Proposition (6.2.7) on the existence of solutions to the Yang–Mills gradient equation

$$\frac{\partial A}{\partial t} = -\,\mathrm{d}_A^* F_A \tag{6.3.1}$$

given an initial connection $A_0 \in \mathscr{A}^{1,1}$ over a compact Kähler surface. The discussion falls naturally into parts: first show that solutions exist for a short time, and then go on to obtain estimates which permit continuation to all positive time. The first object can be achieved in two ways, one using the Kähler condition and one in the more general setting of Riemannian geometry. We begin with the latter.

Standard theory gives the short time existence of solutions to *parabolic* equations over compact manifolds with given initial conditions. The problem we have to overcome is that the heat equation (6.3.1) for A_t is not parabolic. For example in the case of an S^1 bundle we get the linear equation:

$$\frac{\partial a}{\partial t} = -\,\mathrm{d}^*\mathrm{d}a,$$

for a one-form a, and this is not parabolic since $\mathrm{d}^*\mathrm{d}$ is not an elliptic operator. This failure of parabolicity occurs for much the same reason as the Yang–Mills equations themselves fail to be elliptic, i.e. due to the presence of the infinite-dimensional gauge symmetry group. The heat equation asks for a one-parameter family of connections A_t; however, we expect that the geometric content of the solution should be contained in the one-parameter family of gauge equivalence classes $[A_t]$ in the quotient space $\mathscr{A}/\mathscr{G}$. Suppose that (B_t, ϕ_t) is a one-parameter family of pairs consisting of a connection B_t and section ϕ_t of $\Omega_X^0(\mathfrak{g}_E)$ which satisfy the coupled equation:

$$\frac{\partial B_t}{\partial t} = -\,\mathrm{d}_B^* F_B + \mathrm{d}_B \phi_t. \tag{6.3.2}$$

Then from the point of view of the quotient space $\mathscr{A}/\mathscr{G}$ the path $[B_t]$ is quite equivalent to a solution of the heat equation (6.3.1), since the two time derivatives differ by a vector along the $\mathscr{G}$-orbit. We get around the lack of parabolicity in the same way as we make the Yang–Mills equations elliptic; by imposing the Coulomb gauge condition and breaking the invariance under the gauge group. We write our connections as $A_t = A_0 + a_t$ and consider the equation for a_t,

$$\frac{\partial A_t}{\partial t} = -(\mathrm{d}_A^* F_A + \mathrm{d}_A \mathrm{d}_A^* a). \tag{6.3.3}$$

(Here we have, as usual, suppressed the subscript t on the right-hand side of the equation. We emphasize that the variable is a_t, which determines A_t.) A solution to this equation gives a solution $(A_0 + a_t, \mathrm{d}_A^* a_t)$ to (6.3.2). On the other hand (6.3.3) is a parabolic equation for a_t: the linearization about $a_t = 0$ is the standard bundle-valued heat equation,

$$\frac{\partial a}{\partial t} = -\Delta_A a.$$

Thus the general theory of non-linear parabolic equations gives a solution to (6.3.3) for a short time interval $[0, \varepsilon)$, with $a_0 = 0$.

We now return to examine the relation between equations (6.3.1) and (6.3.2) in more detail. This is not strictly necessary for our application but it makes an interesting digression. For simplicity we assume that we do not encounter any reducible connections in our discussion—this will certainly be the case in our application to stable bundles.

More generally, suppose we have any path B_t of irreducible connections, parametrized by t in $[0, T)$ (where T may be infinity). We claim that there is a unique one-parameter family of gauge transformations u_t, with $u_0 = 1$ and such that $A_t = u_t(B_t)$ satisfies:

$$\mathrm{d}_A^*\left(\frac{\partial A_t}{\partial t}\right) = 0. \tag{6.3.4}$$

This is just the condition that A_t be the horizontal lift of the path $[B_t]$ in the quotient space, relative to the connection on the principal fibration $\mathscr{A}^* \to \mathscr{A}^*/\mathscr{G}$ given by the Coulomb gauge slices. The connection on the infinite-dimensional bundle along the path $[B_t]$ is represented, in terms of the given lift B_t, by the 'connection matrix':

$$\mathscr{M}_t = G_B \mathrm{d}_B^*\left(\frac{\partial B}{\partial t}\right). \tag{6.3.5}$$

(cf. Section 5.2.3). Thus $\mathscr{M}_t$ lies in $\Omega^0(\mathfrak{g}_E)$—the Lie algebra of the gauge group. The equation to be solved for u_t is then:

$$\frac{\partial u_t}{\partial t} = \mathscr{M}_t u_t. \tag{6.3.6}$$

This is a family of ordinary differential equations (ODEs) in the t variable, parametrized by the compact space X, and standard theory for ODEs gives a unique solution with $u_0 = 1$, smooth in all variables.

To apply this to our problem, suppose we start with a solution (B_t, ϕ_t) to (6.3.2). We find a smooth path $A_t = u_t(B_t)$ as above, with $d_A^*(\partial A/\partial t) = 0$. Then

$$\frac{\partial A}{\partial t} = -\left(d_A^* F_A + d_A\left(u\phi u^{-1} + \frac{\partial u}{\partial t} u^{-1}\right)\right) = -(d_A^* F_A + d_A \psi),$$

say. Now we have $d_A^* d_A^* F_A = \{F_A, F_A\}$, where $\{\,,\}$ denotes the tensor product of the symmetric inner product on two-forms and the skew symmetric bracket on the Lie algebra. So $\{\,,\}$ is skew and $d_A^* d_A^* F_A = 0$. Thus $d_A^* d_A \psi = 0$ and, taking the inner product with ψ, we get $d_A \psi = 0$. So the lift A_t of the path does indeed satisfy the equation (6.3.1).

Similarly, in the Kähler case, if A_t satisfies (6.3.1) and $A_0 \in \mathscr{A}^{1,1}$ we can differentiate to see that A_t lies in $\mathscr{A}^{1,1}$ for all t. Then we can define a one-parameter family of complex gauge transformations g_t by

$$\frac{\partial g}{\partial t} = -i\hat{F}_{A_t} g, \quad g_0 = 1, \tag{6.3.7}$$

and we have $A_t = g_t(A_0)$. This completes the first proof of short time existence.

For the second approach, special to the Kähler case, we work directly with the complex gauge transformations. We can regard equation (6.3.7) as an evolution equation for g_t, with A_t defined to be $g_t(A_0)$. Again this is not a parabolic equation for g_t. But if we put $h_t = g_t^* g_t$, so that h_t is a self-adjoint endomorphism of E, a little calculation shows that

$$g^{-1}\hat{F}_{g(A_0)} g = \hat{F}_{A_0} + \Lambda\bar{\partial}_{A_0}(h^{-1}\partial_{A_0} h), \tag{6.3.8}$$

so that:

$$\frac{\partial h_t}{\partial t} = -2ih(\hat{F}_{A_0} + \Lambda\bar{\partial}_{A_0}(h^{-1}\partial_{A_0} h)). \tag{6.3.9}$$

Now (6.3.9) is a parabolic equation for h_t so a short-time solution exists. Then if we choose any $\tilde{g}_t$ with $\tilde{g}_t^* \tilde{g}_t = h_t$, for example $\tilde{g}_t = h_t^{1/2}$, the connections $B_t = \tilde{g}_t(A_0)$ satisfy (6.3.2) for a suitable ϕ_t. Then we can proceed to find the horizontal lift as before.

6.3.2 *Long-time existence*

There is a standard procedure to follow to attempt to show that the short-time solution of a parabolic evolution equation can be continued for all positive time. One tries to find uniform estimates for all derivatives of a solution α_t defined for t in an interval $[0, T)$ and then to deduce that α_t converges in C^∞ to a limit α_T as t tends to T. Then one can glue on the short-time solution with initial condition α_T to extend the solution to a larger interval. We shall indicate how this can be done for the Yang–Mills flow on (1, 1) connections over a Kähler manifold, appealing to the references cited in the notes for more detailed treatments. In the familiar way, once one has obtained some critical initial estimates, the higher derivatives can be dealt with by a bootstrapping argument. In the Yang–Mills case, over a four-dimensional base manifold X, the crux of the problem is the search for local L^2 estimates on the curvature. More precisely we define, for a one-parameter family of connections A_t, $0 \le t < T$,

$$\delta(r) = \sup_{\substack{x \in X \\ 0 \le t < T}} \int_{B(x,r)} |F_{A_t}|^2 \, d\mu, \tag{6.3.10}$$

where $B(x, r)$ is the r-ball about X.

Proposition (6.3.11). *Let $A_t \in \mathscr{A}^{1,1}$ be a solution to the gradient flow equation over a Kähler surface defined for $0 \le t < T$. Suppose $\delta(r)$ tends to 0 with r; then the solution can be continued to an interval $0 < t < T + \varepsilon$, for some $\varepsilon > 0$.*

We omit the detailed proof of (6.3.11), which follows standard lines. The condition that $\delta(r)$ tends to zero means that there is a fixed cover of X by small balls over which Uhlenbeck's gauge-fixing theorem can be applied to all the A_t. On the other hand, the evolution equation implies that $\nabla_{A_t} \hat{F}_{A_t}$ is bounded in L^2 (cf. Proposition (6.2.14)), so one can apply elliptic estimates in these small balls to obtain an L^2 bound on the covariant derivative of the full curvature tensor $\nabla_{A_t} F_{A_t}$, and hence an L^4 bound on F_{A_t}. Then one can iterate the argument, deriving differential inequalities for the iterated covariant derivatives of the curvature, and deduce that these are all bounded in L^2 over the interval $[0, T)$, and from this point the proof is routine.

We now want to argue that the hypothesis of Proposition (6.3.11) is always fulfilled. Our starting point is the fact that for any solution A_t the component $\hat{F}_{A_t}$ of the curvature is uniformly bounded (Corollary (6.2.12)). If $A_t = g_t(A_0)$ we have then that

$$\left| \frac{\partial g_t}{\partial t} g_t^{-1} \right|$$

is uniformly bounded. By integrating this we get a uniform bound on g_t and g_t^{-1} over $X \times [0, T)$.

To show that $\delta(r)$ tends to zero with r we argue by contradiction. If not, we could find a sequence of times $t_\alpha \to T$ and small balls $B(x_\alpha, r_\alpha)$ with $r_\alpha \to 0$ and a $\delta > 0$ such that:

$$\int_{B(x, r_\alpha)} |F_{A_\alpha}|^2 \, d\mu > \delta. \tag{6.3.12}$$

Now identify the $B(x_\alpha, r_\alpha^{1/2})$ with balls in $\mathbb{C}^2$ by local holomorphic coordinates and then rescale by a factor r_α^{-1}. We get rescaled connections A'_α, say, over large balls $B(0, r_\alpha^{-1/2})$ in $\mathbb{C}^2$. The uniform bound on $\hat{F}_{A_\alpha}$ means that $\hat{F}_{A'_\alpha}$ is $O(r_\alpha^2)$, and similarly the L^2 norm of $\nabla_{A'_\alpha} \hat{F}_{A'_\alpha}$ is $O(r_\alpha)$. So after gauge transformations the rescaled connections converge, in L_2^2 on compact subsets of

$\mathbb{C}^2$, to a finite-energy ASD connection A over $\mathbb{C}^2$, and the condition (6.3.12) implies that A is non-trivial. On the other hand we can suppose, as in Section 6.2.5, that the rescaled versions of the complex gauge transformations g_{t_α} converge to a limit

$$g:\mathbb{C}^2 \longrightarrow GL(2, \mathbb{C}),$$

with g and g^{-1} bounded and such that $A = g(\theta)$, where θ is the trivial product connection over $\mathbb{C}^2$. We then obtain the desired contradiction from the following lemma.

Lemma (6.3.13). *Suppose A is a finite energy ASD connection on the trivial bundle over $\mathbb{C}^2$ which can be written as $g(\theta)$ for a complex gauge transformation $g:\mathbb{C}^2 \to GL(2, \mathbb{C})$, with g and g^{-1} bounded. Then A is a flat connection.*

Proof. We know that A extends to a smooth connection on a bundle E over S^4. Consider the obvious map $s:\mathbb{CP}^2 \to S^4$ which collapses the line at infinity in $\mathbb{CP}^2$ to the point at infinity in S^4, but is the identity on the common open subset $\mathbb{C}^2 = \mathbb{R}^4$. The connections $s^*(A)$ has curvature of type (1, 1) and defines a holomorphic structure $\mathscr{E}$ on the bundle $s^*(E)$ over $\mathbb{CP}^2$. From this point of view g represents a holomorphic trivialization of $\mathscr{E}$ over $\mathbb{C}^2 \subset \mathbb{CP}^2$. Since g and g^{-1} are both bounded this trivialization extends, by the Riemann extension theorem for bounded holomorphic functions, over the line at infinity. In particular the bundles $s^*(E)$ and E are topologically trivial and the ASD connection A must be flat.

(An alternative proof of this lemma is to observe that the function $\tau = \mathrm{Tr}(g^*g)$ is subharmonic (cf. (6.2.18)) and appeal to the Liouville theorem for bounded subharmonic functions.)

6.4 Deformation theory

Theorem (6.1.5) identifies the equivalence classes of irreducible ASD connections and stable holomorphic bundles at the level of sets, but for most purposes one wants information about the structure of the ASD moduli space. In this section we will explain how this structure can be recovered from holomorphic data.

6.4.1 Versal deformations

Recall from Section 4.2 that if A is an ASD connection over a Riemannian four-manifold X, a neighbourhood of $[A]$ in the moduli space has a model

$$f^{-1}(0)/\Gamma_A,$$

where $f: H_A^1 \to H_A^2$ is a smooth map between the cohomology groups of the deformation complex,

$$\Omega_X^0(\mathfrak{g}_E) \xrightarrow{d_A} \Omega_X^1(\mathfrak{g}_E) \xrightarrow{d_A^+} \Omega_X^+(\mathfrak{g}_E), \tag{6.4.1}$$

and the isotropy group Γ_A has Lie algebra H_A^0. Different choices of the map f can be made; the intrinsic structure on the moduli space is encoded in a sheaf of rings, making it a real analytic space.

Now suppose that Z is a compact complex surface and $\mathscr{E}$ is a holomorphic bundle over Z. The same theory can be used to describe the deformations of $\mathscr{E}$ as a holomorphic bundle. We fix a C^∞ bundle E and look at the space of $\bar{\partial}$ operators on E; if we fix an auxiliary metric this can be identified with the space $\mathscr{A}$ of unitary connections. Roughly speaking, we wish to describe a neighbourhood in the quotient space $\mathscr{A}^{1,1}/\mathscr{G}^c$.

The important difference from the ASD case is that the orbits of the symmetry group $\mathscr{G}^c$, unlike those of $\mathscr{G}$, are not in general closed, and the full space $\mathscr{A}^{1,1}/\mathscr{G}^c$ will not be Hausdorff in any useful topology. It is precisely this phenomenon which led algebraic geometers to introduce the notion of stability in the global moduli problem. At the level of local deformations one can avoid these difficulties by means of the notion of a 'versal deformation'. If T is a complex space with base point t_0 we say that a deformation of the holomorphic bundle $\mathscr{E}$ over Z, parametrized by T, is a holomorphic bundle $\mathbb{E}$ over $Z \times T$ which restricts to $\mathscr{E}$ on $Z \times \{t_0\}$. Given a deformation over (T, t_0) and a map $(S, s_0) \to (T, t_0)$ we get, by pull-back, an induced deformation over (S, s_0). We now introduce the corresponding notions at the level of germs, i.e. we regard two spaces as being equivalent if there is an isomorphism between some neighbourhoods of their base points, and maps as being equivalent if they agree in such neighbourhoods. We say that a deformation of $\mathscr{E}$ parametrized by (T, t_0) is *versal* if any other deformation can be induced from it by a map, and that the deformation is *universal* if the map is unique. Throughout the above we can consider parameter spaces T which are arbitrary complex spaces, including singularities and nilpotent elements; we just interpret 'bundles over $T \times Z$' as locally free sheaves.

The theory developed in Chapter 4 can now be used to construct a versal deformation of any holomorphic bundle $\mathscr{E}$. If we identify the tangent space to $\mathscr{A}$ with $\Omega_Z^{0,1}(\text{End } E)$, the derivative of the action of the complex gauge group $\mathscr{G}^c$ at a connection A, with $\bar{\partial}$ operator $\bar{\partial}_A = \bar{\partial}_{\mathscr{E}}$, is

$$\bar{\partial}_{\mathscr{E}}: \Omega_Z^0(\text{End } E) \longrightarrow \Omega_Z^{0,1}(\text{End } E),$$

and similarly the derivative of the map $A \to F^{0,2}(A)$, whose zero set is $\mathscr{A}^{1,1}$, is the $\bar{\partial}_{\mathscr{E}}$ operator on $\Omega^{0,1}$. The analogue of the ASD deformation complex (6.4.1) is the Dolbeault complex

$$\Omega_Z^0(\text{End } E) \xrightarrow{\bar{\partial}_{\mathscr{E}}} \Omega_Z^{0,1}(\text{End } E) \xrightarrow{\bar{\partial}_{\mathscr{E}}} \Omega_Z^{0,2}(\text{End } E), \qquad (6.4.2)$$

with cohomology groups $H^i(\text{End } E)$. The space $H^0(\text{End } E)$ is the Lie algebra of the complex Lie group $\text{Aut } \mathscr{E}$ of automorphisms of $\mathscr{E}$. If we work with bundles having a fixed determinant, for example with $SL(2, \mathbb{C})$ bundles, we can replace $\text{End } E$ with the bundle $\text{End}_0 E$ of trace-free endomorphisms throughout. The main result, essentially due to Kuranishi, can be summarized as follows:

Proposition (6.4.3).

(*i*) *There is a holomorphic map* ψ *from a neighbourhood of* 0 *in* $H^1(\text{End}\,\mathscr{E})$ *to* $H^2(\text{End}_0\,\mathscr{E})$, *with* ψ *and its derivative both vanishing at* 0, *and a versal deformation of* $\mathscr{E}$ *parametrized by* Y *where* Y *is the complex space* $\psi^{-1}(0)$, *with the naturally induced structure sheaf (which may contain nilpotent elements).*

(*ii*) *The two-jet of* ψ *at the origin is given by the combination of cup product and bracket*:

$$H^1(\text{End}\,\mathscr{E}) \otimes H^1(\text{End}\,\mathscr{E}) \longrightarrow H^2(\text{End}_0\,\mathscr{E}).$$

(*iii*) *If* $H^0(\text{End}_0\,\mathscr{E})$ *is zero, so that the group* Aut $\mathscr{E}$ *is equal to the scalars* $\mathbb{C}^*$, *then* Y *is a universal deformation, and a neighbourhood of* $[\mathscr{E}]$ *in the quotient space* $\mathscr{A}^{1,1}/\mathscr{G}^c$ *(in the quotient topology) is homeomorphic to the space underlying* Y. *More generally, if* Aut $\mathscr{E}$ *is a reductive group we can choose* ψ *to be* Aut $\mathscr{E}$ *equivariant, so* Aut $\mathscr{E}$ *acts on* Y *and a neighbourhood in the quotient is modelled on* $Y/\text{Aut}\,\mathscr{E}$ *(which may not be Hausdorff).*

To prove this proposition, in the differential geometric setting, one applies the procedure used for the ASD equations modulo the unitary gauge group to the equation $F_A^{0,2} = 0$, modulo $\mathscr{G}^c$. All we need to know, abstractly, is that the $\bar{\partial}$ complex is split, which follows from Hodge theory. Then we get a map ψ in just the same way that we obtained the map f in the case of ASD connections. We also see, much as before, that the zero set Y is independent, as a ringed space, of the choices made, and that Y, or a quotient of Y, gives a local model for $\mathscr{A}^{1,1}/\mathscr{G}^c$, at least in the case when the automorphism group is reductive (the complexification of a compact group).

The existence of the deformation parametrized by Y is rather obvious if Y is reduced. In general one has to extend the integrability theorem for $\bar{\partial}$ operators on vector bundles. Suppose that $\bar{\partial}_\chi$ is a holomorphic family of $\bar{\partial}$ operators over a polydisc parametrized by $\chi \in \mathbb{C}^\nu$. Let p be a polynomial on $\mathbb{C}^\nu$ and suppose that

$$\bar{\partial}_\chi \bar{\partial}_\chi = p(\chi) G_\chi$$

for a family of operators G_χ; that is, $\bar{\partial}_\chi^2 = 0 \bmod (p)$. Then one has to see that, over a smaller polydisc, we can find a family of complex gauge transformations g_χ such that:

$$g_\chi \bar{\partial}_\chi g_\chi^{-1} = \bar{\partial} + \alpha_\chi$$

where $\alpha_\chi = 0 \bmod (p)$. This additional information can be obtained quite easily from our proof of the integrability theorem, introducing χ as an auxiliary parameter.

Finally, one needs to verify the versal property of the deformation. Again this was done by Kuranishi (1965) in the reduced case, and the addition of nilpotent elements causes no great problems.

6.4.2 Comparison of deformation theories

We will now examine the relation between the deformation theories for ASD connections and holomorphic bundles, so we suppose A is an ASD connection on a unitary bundle E over the Kähler surface X. We begin at the linearized level, with the cohomology groups of the deformation complexes (6.4.1) and (6.4.2). For simplicity we work with $SL(2, \mathbb{C})$ bundles. We use the Hodge theory for each complex to represent the cohomology groups by harmonic elements. Then the algebraic isomorphisms

$$\Omega^{0,1}(\mathrm{End}_0 E) = \Omega^1(\mathfrak{g}_E)$$

$$\Omega^{0,2}(\mathrm{End}_0 E) \oplus \Omega^0(\mathrm{End}_0 E) = \Omega^0(\mathfrak{g}_E) \oplus \Omega^+(\mathfrak{g}_E)$$

together with the Kähler identities give canonical linear isomorphisms:

$$H^1(\mathrm{End}\,\mathscr{E}) = H^1(\mathrm{End}_0\,\mathscr{E}) = H^1_A, \quad H^0(\mathrm{End}_0 E) = H^0_A \otimes \mathbb{C},$$

$$H^2_A = H^2(\mathrm{End}_0\,\mathscr{E}) \oplus H^0_A . \omega.$$

Now we know that $\mathscr{E}_A$ is either stable or a direct sum of line bundles. We begin by considering the first case; in this case H^0 is zero so $\mathscr{E}_A$ has no nontrivial automorphisms and the zero set Z parametrizes a universal deformation of $\mathscr{E}$. We divide the ASD equations up into two parts, in the familiar way. Now for each α in $\Omega^{0,1}(\mathfrak{g}_E)$ we consider the equations for an element $g = 1 + u$ of $\mathscr{G}^c$,

$$\hat{F}(g(A + \alpha - \alpha^*)) = 0, \quad \mathrm{d}^*_A(g(A + \alpha - \alpha^*) - A) = 0.$$

The linearization is $\Delta_A u = -\bar{\partial}^*_A \alpha$.

Since Δ_A maps onto the trace-free endomorphisms, the implicit function theorem gives a solution g_α for all small enough α. Now let H be a fixed subspace of $\Omega^{0,2}(\mathrm{End}_0\,\mathscr{E})$ representing $H^2(\mathrm{End}_0\,\mathscr{E})$, for example the harmonic subspace, and consider the vector bundle Ξ over a neighbourhood of the origin in $\mathrm{Ker}\, \mathrm{d}^*_A$ with fibre

$$\Xi_{\alpha - \alpha^*} = g_\alpha H g_\alpha^{-1} \subset \Omega^{0,2}(\mathrm{End}_0 E).$$

Following the procedure of Section 4.2.5 we construct a model for the ASD moduli space using this bundle over the transversal $\ker d^*_A$, in the form of a map $f: H^1_A \to H^2_A$. On the other hand a model $\psi: H^1_A \to H^2_A$ for the universal deformation is obtained from the H component of $F^{0,2}_A$ on the harmonic subspace $\ker \bar{\partial}^*_A \cap \ker \bar{\partial}_A = \ker \mathrm{d}^*_A \cap \ker \mathrm{d}^+_A$, and a little thought shows that f and ψ are equal. It follows then that the local structure of the ASD moduli space is compatible with that of the universal deformations. In sum we have:

Proposition (6.4.4). *If X is a complex Kähler surface and E an $SU(2)$ bundle over X, the moduli space M^*_E of irreducible ASD connections is a complex* *analytic space and each point in* M_E^* *has a neighbourhood which is the base of the universal deformation of the corresponding stable vector bundle among* $SL(2, \mathbb{C})$ *bundles.*

The situation around reducible solutions is rather more complicated. First, the bundle $\mathscr{E}_A$ is now a sum of line bundles, and $\mathrm{Aut}(\mathscr{E}_A)/\mathbb{C}^*$ is the complexification of $\Gamma_A = S^1$. The group H_A^2 now has components $H_A^2 = H^2(\mathrm{End}_0 E) \oplus \mathbb{R}$. It is still true that, with suitable choices, the Γ_A-equivariant ASD model $f: H_A^1 \to H_A^2$ has an $H^2(\mathrm{End}_0 \mathscr{E})$ component ψ which defines an $\mathrm{Aut}\,\mathscr{E}$-invariant versal deformation Z, but now there is a further component $f_0: H_A^1 \to \mathbb{R}$, of f. So a neighbourhood in the ASD moduli space has the form

$$\{z \in Z \mid f_0(z) = 0\}/\Gamma_A,$$

while a neighbourhood in $\mathscr{A}^{1,1}/\mathscr{G}^c$ has the form $Z/\mathbb{C}^*$. We will see more exactly how the two descriptions are related in the second example of Section 6.4.3.

A final remark which fits in here concerns the orientations of the moduli spaces. In Section 5.4 we have seen that these can be derived from an orientation of the determinant line bundle $\Lambda \to \mathscr{B}$, whose fibres are the tensor products

$$\Lambda_A = \Lambda^{\max} \ker \delta_A \otimes (\Lambda^{\max} \ker \delta_A^*)^*.$$

Now suppose that the base space X is a Kähler surface and A is any unitary connection, not necessarily in $\mathscr{A}^{1,1}$. By the Kähler identities we can identify

$$\delta_A: \Omega_X^1(\mathfrak{g}_E) \longrightarrow \Omega_X^0(\mathfrak{g}_E) \oplus \Omega_X^+(\mathfrak{g}_E)$$

with

$$\bar{\partial}_A^* \oplus \bar{\partial}_A: \Omega_X^{0,1}(\mathrm{End}_0 E) \longrightarrow \Omega_X^0(\mathrm{End}_0 E) \oplus \Omega_X^{0,2}(\mathrm{End}_0 E).$$

Hence the kernel and cokernel are complex vector spaces, and so have canonical orientations. This gives a canonical orientation for the line bundle Λ. If we now deform the metric on X to some generic Riemannian metric for which the moduli spaces are regular, the orientation of Λ can also be deformed in a unique way to give an orientation of these moduli spaces. Of course if the moduli spaces for the original metric were regular, hence complex manifolds, this orientation would agree with the standard orientation of complex manifolds.

6.4.3 Examples

We will illustrate the above ideas by considering two examples.

Example (i)

This is an 'imaginary' example, though we shall encounter something very similar in Chapter 10. Suppose that $\mathscr{E}$ is a stable bundle over a Kähler surface

X, with $\dim H^1(\mathrm{End}_0\,\mathscr{E}) = \dim H^2(\mathrm{End}_0\,\mathscr{E}) = p$. Suppose the universal deformation Z is defined by a map

$$\psi: \mathbb{C}^p \longrightarrow \mathbb{C}^p$$

with an isolated zero at the origin. Thus the topological space underlying Z is a single point, and the corresponding ASD connection A is an isolated point in the moduli space M_E. Now suppose we perturb the metric on X in a one-parameter family g_t. We know from Section 4.2.5 that the moduli space $M_E(g_t)$ is modelled on the zeros of a small deformation,

$$\psi_t: \mathbb{C}^p \longrightarrow \mathbb{C}^p,$$

of ψ. If g_t is generic, ψ_t will have regular zeros, unlike ψ.

We can see then how knowledge of the map ψ, or more invariantly of the structure ring supported on Z, gives additional information about the moduli spaces, for nearby generic metrics. We associate an integer multiplicity $m > 1$ to the original map ψ—the degree of the restriction

$$\frac{\psi}{|\psi|}: S^{2p-1} \longrightarrow S^{2p-1}.$$

Standard arguments tell us then that the isolated zero of ψ splits up into at least m regular zeros of ψ_t, each representing a point of $M_E(g_t)$. Moreover it is easy to see that, counted with the signs given by the canonical orientation, the algebraic sum of these points is precisely m.

Example (ii)

The second example is very concrete. We consider the reducible solution in the moduli space $M_2(S^2 \times S^2)$ described in Sections 4.1.5 and 4.2.6. This corresponds to the decomposable bundle $\mathscr{E} = \mathscr{U} \oplus \mathscr{U}^{-1}$ where, in standard notation,

$$\mathscr{U} = \mathscr{O}(1, -1).$$

Let us see what the deformation theory tells us about the structure of the moduli space near this reducible point. First,

$$H^2(\mathrm{End}_0\,\mathscr{E}) = H^2(\mathscr{O} \oplus \mathscr{U}^2 \oplus \mathscr{U}^{-2}),$$

which vanishes, since $H^2(\mathscr{O}) = H^2(\mathscr{O}(2, -2)) = 0$ (by the Künneth formula for sheaf cohomology). So the obstruction space H^2_A in the ASD deformation theory is made up entirely of the piece $H^0 . \omega$. Our versal deformation space is a neighbourhood of 0 in

$$H^1(\mathrm{End}\,\mathscr{E}) = H^1(\mathscr{O}(2, -2)) \oplus H^1(\mathscr{O}(-2, 2)) = U_1 \times U_2,$$

say, where U_1, U_2 are each three-dimensional complex vector spaces. The quadratic term in the map

$$f_0: U_1 \times U_2 \longrightarrow H^0_A = \mathbb{R},$$

is identified by (4.2.31) as

$$\int_X \mathrm{Tr}(a_1 + a_2) \wedge (a_1 + a_2) \wedge (\omega\gamma),$$

where $a_i \in U_i$ are now viewed as one-forms over $S^2 \times S^2$ with values in $\mathfrak{g}_E$. Here we have written γ for the generator of the S^1 action on $\mathfrak{g}_E$, which we choose to have weight 1 on $\mathcal{O}(2, -2)$. If we write $a_1 = \alpha_1 - \alpha_1^*$ for a bundle valued (0, 1)-form α_1, then α_1 lies in the component $\mathcal{O}(2, -2)$ in End $\mathscr{E}_0$. In matrix notation,

$$a_1 = \begin{pmatrix} 0 & \alpha_1 \\ -\bar{\alpha}_1 & 0 \end{pmatrix}$$

while $\gamma = \begin{pmatrix} 1 & 0 \\ 0 & -1 \end{pmatrix}$. Thus $\mathrm{Tr}(a_1 a_1)(\omega\gamma) = \alpha_1 \wedge \bar{\alpha}_1 \wedge \omega = |\alpha_1|^2$. Similarly if we write $a_2 = \alpha_2 - \alpha_2^*$ then α_2 lies in the other factor $\mathcal{O}(-2, 2)$ and $\mathrm{Tr}(a_2 a_2)(\omega\gamma) = -|\alpha_2|^2$. Thus we have

$$f_0(a_1, a_2) = |a_1|^2 - |a_2|^2 + O(a_i^3).$$

Let us suppose for simplicity that in suitable holomorphic coordinates the function f_0 is given exactly by its quadratic part.

We can now see explicitly how the relation between stability and the ASD solutions works in this local picture. A neighbourhood in the ASD moduli space is given by

$$\{(a_1, a_2) \in U_1 \times U_2 \,|\, |a_1|^2 = |a_2|^2\}/S^1.$$

On the other hand a neighbourhood in the space of isomorphism classes of holomorphic bundles is given by $U_1 \times U_2/\mathbb{C}^*$ where $\mathbb{C}^*$ acts by

$$\lambda(a_1, a_2) = (\lambda a_1, \lambda^{-1} a_2),$$

and S^1 is embedded in $\mathbb{C}^*$ in the standard way. Consider a point (a_1, a_2) in $U_1 \times U_2$ with each of a_i non-zero. We can then clearly find a λ such that $(a_1', a_2') = \lambda(a_1, a_2)$ satisfies $|a_1'|^2 = |a_2'|^2$, and λ is unique up to S^1. These are the points corresponding to stable bundles, which admit ASD connections. The exceptional points of the form $(a_1, 0)$ and $(0, a_2)$ correspond to unstable bundles, in fact just to bundles which can be written as extensions,

$$0 \longrightarrow \mathcal{O}(1, -1) \longrightarrow \mathscr{E} \longrightarrow \mathcal{O}(-1, 1) \longrightarrow 0,$$

$$0 \longrightarrow \mathcal{O}(-1, 1) \longrightarrow \mathscr{E} \longrightarrow \mathcal{O}(1, -1) \longrightarrow 0,$$

respectively. (As we explain in Section 10.3.1, these extensions are indeed parametrized by $U_1 = H^1(\mathcal{O}(2, -2))$, $U_2 = H^1(\mathcal{O}(-2, 2))$.) In this way we can verify our main theorem for bundles close to $\mathscr{U} \oplus \mathscr{U}^{-1}$ by examining the relation between the deformation theories for the two structures.

6.5 Formal aspects

In this section we will describe how the relation between stable holomorphic bundles and ASD connections can be fitted tidily into a rather general formal picture. This picture is not special to complex dimension two: it covers all the generalizations sketched in Section 6.1.4, and in particular the theory of stable bundles and flat unitary connections over Riemann surfaces. At the end of this section we discuss the curvature of the connection defined by Quillen (1985) on the determinant line bundle over the moduli space of stable bundles on a Riemann surface, which plays an important role in the abstract theory.

6.5.1 Symplectic geometry and moment maps

Let (V, Ω) be a symplectic manifold, so Ω is a nondegenerate closed 2-form on V. This 2-form gives an isomorphism between tangent and cotangent vectors,

$$v \longrightarrow i_v(\Omega),$$

where i_v is the contraction operation. We denote the inverse map by $R: T^*V \to TV$. Suppose v is a vector field on V whose associated one-parameter group of diffeomorphisms preserves the symplectic structure, i.e. $L_v\Omega = 0$. The 'homotopy' formula for the Lie derivative on forms,

$$L_v\Omega = (i_v \mathrm{d} + \mathrm{d} i_v)\Omega = \mathrm{d} i_v \Omega,$$

shows that the corresponding one-form $i_v\Omega$ is closed. Now suppose that a group K acts on V, preserving the symplectic form. A *momentum* (or 'moment') map for the action is a map

$$m: V \longrightarrow \mathfrak{k}^*$$

to the dual of the Lie algebra of K, such that

$$\mathrm{d}(\langle m, \xi \rangle) = i_{v(\xi)}\Omega, \tag{6.5.1}$$

for all ξ in the Lie algebra $\mathfrak{k}$. Here $\langle m, \xi \rangle$ is the function on V obtained from m by the pairing between $\mathfrak{k}$ and its dual. This concept generalizes that of the Hamiltonian for one-parameter groups. (The terminology comes from the case when V is the phase space of a mechanical system and K is a group of translations and rotations; the components of the momentum mapping are then the linear and angular momenta in the ordinary sense.) The momentum map is called equivariant if it intertwines the K action on V with the co-adjoint action on $\mathfrak{k}^*$.

From a momentum mapping m we define the *co-momentum* map $m^*: \mathfrak{k} \to C^\infty(V)$ by

$$m^*(\xi) = \langle m, \xi \rangle.$$

If m is equivariant, the co-momentum map is a lifting of the infinitesimal action by a Lie algebra homomorphism:

$$\begin{array}{ccccc} & & \mathfrak{k} & & \\ & & \downarrow m^{*} & \searrow \rho & \\ \mathbb{R} & \longrightarrow & C^{\infty}(V) & \longrightarrow & \mathrm{Vect}(V), \end{array}$$

where $C^{\infty}(V)$ is viewed as a Lie algebra under the Poisson bracket

$$\{f, g\} = \Omega(R\,\mathrm{d}f, R\,\mathrm{d}g),$$

and the homomorphism from $C^{\infty}(V)$ to $\mathrm{Vect}(V)$ maps a function f to $R(\mathrm{d}f)$.

Given an equivariant momentum map one can construct a 'symplectic quotient' of V by the group action. Suppose for simplicity that K acts freely on V (although the theory extends to the case when the stabilizers of points are finite). This means that the momentum map has maximal rank at each point and the zero set $m^{-1}(0) \subset V$ is a smooth submanifold. Since m is equivariant this zero set is preserved by the action of K, and the symplectic quotient U is defined to be:

$$U = m^{-1}(0)/K. \tag{6.5.2}$$

We define a symplectic structure on U, induced from that on V. Let x be a point in V with $m(x) = 0$. We have linear maps

$$\mathfrak{k} \xrightarrow[\rho]{} (TV)_x \xrightarrow[\mathrm{d}m]{} \mathfrak{k},$$

and the kernel of $\mathrm{d}m$ is the annihilator of the image of v, under the skew pairing Ω_x. So Ω passes down to the tangent space of U at $[x]$:

$$(TU)_{[x]} = \mathrm{Ker}\,\mathrm{d}m/\mathrm{Im}\,\rho.$$

It is a simple exercise to check that this defines a closed non-degenerate form on U.

This construction can be generalized in the case when $\mathfrak{k}$ has a non-trivial centre. We can then take the inverse image under m of any vector c in the dual of the centre (under the canonical decomposition of $\mathfrak{k}^*$). For example, take K to be the circle S^1 acting by multiplication on $\mathbb{C}^n$ with the standard constant symplectic form $\Omega = \sum \mathrm{d}x_i\,\mathrm{d}y_i$. The moment map is then

$$m(z_1, \ldots, z_n) = \Sigma |z_i|^2,$$

and, with $c = 1$, the symplectic quotient is the manifold $m^{-1}(1)/S^1 = \mathbb{CP}^n$, with its standard symplectic structure.

The momentum map can be given a geometric interpretation in terms of complex line bundles. Suppose that L is a complex line bundle over V, with a unitary connection whose curvature form is $-2\pi i\Omega$. A moment map for the action of K on V gives a lift of the action of the Lie algebra to L, covering that on V and preserving the $U(1)$ action. To see this we can consider the universal case when $\mathfrak{k}$ is the algebra $C^\infty(V)$ with Poisson bracket. We work on the principal $U(1)$ bundle. Let f be a function on V with corresponding vector field $v = R(df)$. We consider a lift of v to the bundle of the form

$$v^* = \tilde{v} + 2\pi f t, \tag{6.5.3}$$

where $\tilde{v}$ is the horizontal lift defined by the connection and t is the 'vertical' field generating the $U(1)$ action. Let g be another function and

$$w^* = \tilde{w} + 2\pi g t$$

be the corresponding lift. The bracket is then

$$\begin{aligned}[v^*, w^*] &= [v, w] + 2\pi(\nabla_v g - \nabla_w f)t \\ &= \widetilde{[v, w]} + 2\pi(\nabla_v g - \nabla_w f - \Omega(v, w))t,\end{aligned}$$

using the definition of curvature. But

$$\nabla_v g = -\nabla_w f = \Omega(v, w) = \{f, g\},$$

so

$$[v^*, w^*] = \widetilde{[v, w]} + 2\pi\{f, g\}t,$$

and our rule gives a Lie algebra homorphism from $C^\infty(V)$ to the vector fields on the circle bundle covering the action on V. We assume that this can be exponentiated to define an action of K on L.

This construction with line bundles fits in with that of the symplectic quotient above in that it gives an induced line bundle L_U over the symplectic quotient U, with a connection whose curvature is $-2\pi i$ times the induced symplectic form. Sections of L_U can be identified with K-invariant sections of L over $m^{-1}(0)$. In the case when we take the inverse image of a vector in the centre, sections of L_U correspond to sections of L which transform by the appropriate weight. In the example above we get the standard connection on the Hopf line bundle over complex projective space.

6.5.2 *Kähler manifolds*

We now enrich the discussion by supposing that V is a complex Kähler manifold, with Kähler form Ω, and that K acts isometrically on V. We assume that there is an invariant positive-definite inner product on $\mathfrak{k}$, so we can identify $\mathfrak{k}^*$ with $\mathfrak{k}$. We also suppose that there is a complexification K^c of K with Lie algebra $\mathfrak{k}^c = \mathfrak{k} \otimes_{\mathbb{R}} \mathbb{C}$. Then the action of K on V extends, by complexification, to an action of K^c. The elements of K^c respect the complex structure on V, but not in general the metric and symplectic form. In this situation there is an intimate relationship between the symplectic quotient of V by K and the space of orbits of K^c in V and, as we shall explain below, this provides a general setting for our discussion of stable holomorphic bundles and curvature. In that case V and K will be infinite dimensional, but for the moment we consider a situation where V and K are compact.

There are two slightly different ways of understanding the relationship between the moment map and the action of the complexified group, based on the study of the critical points of two real-valued functions. In the first approach we consider the function $\phi: V \to \mathbb{R}$ defined by $\phi(x) = |m(x)|^2$. We calculate the gradient vector field grad ϕ at a point x of V:

$$\begin{aligned}\langle \text{grad } \phi, w \rangle &= 2\langle m_x, \text{d}m_x(w) \rangle \\ &= 2\langle I\rho(m_x), w \rangle.\end{aligned}$$

So

$$\text{grad}_x \phi = 2I\rho(m_x) = 2\rho(Im_x), \tag{6.5.4}$$

where I denotes the complex structure on TV_x and $\mathfrak{k}^c$ and we have extended ρ to the complexified Lie algebra. In particular the gradient flow lines of the function ϕ on V are contained in the orbits of K^c. Let $\Gamma \subset V$ be such an orbit. Then we see that the critical points of the restriction of ϕ to Γ are also critical points for ϕ on V. If x is such a critical point, $\rho(m_x)$ is zero. So if m_x is not itself zero, x has a non-trivial isotropy group under the K action.

Now if V is compact, the descending gradient flow lines converge to the critical set of ϕ. One can then deduce the following alternative: *either* a descending flow line converges to a point of $m^{-1}(0)$ in Γ (an absolute minimum of ϕ) *or* a subsequence in the flow converges to a critical point outside the orbit, but lying in its closure $\bar{\Gamma}$. Moreover, as we shall see in a moment, in the first case the zero of m in Γ is unique, up to the action of K (which acts as a symmetry group of the whole situation).

For the second approach we suppose that the symplectic form corresponds to a line bundle L over V with a unitary connection, as described above. The curvature of the connection has type (1, 1), so L is a holomorphic line bundle. The momentum map gives a lift of the action of K and this extends to the complexification K^c. So we have orbits of K^c in L lying over those in V. Let $\tilde{\Gamma} \subset L$ be such an orbit and consider the function

$$h: \tilde{\Gamma} \longrightarrow \mathbb{R}$$

defined by $h(\gamma) = -\log |\gamma|^2$, where we use the given hermitian metric on the fibres of L. (The minus sign can be removed by replacing the positive bundle L by its dual.) The critical points of h are precisely the points lying over the zeros of the momentum map in Γ. For if γ is a point in L lying over x in V and $\Xi = \xi_1 + i\xi_2$ is a vector in the complexified Lie algebra (with ξ_1, ξ_2 in $\mathfrak{k}$), the action of Ξ on γ is given by the horizontal lift of the action on x plus $2\pi\{i\langle \xi_1, m_x \rangle - \langle \xi_2, m_x \rangle\}\gamma$. The derivative of h in this direction is thus $-4\pi\langle \xi_2, m_x \rangle$, which vanishes for all Ξ if and only if $m_x = 0$.

This gives another way to find zeros of the momentum map in a given orbit Γ, by seeking critical points of h on any lifted orbit $\tilde{\Gamma}$. The uniqueness property mentioned above can be deduced in the present setting from a convexity property of h. If we choose a base point in $\tilde{\Gamma}$ we can pull back h to get a function on the group K^c; this is invariant under K so we have an induced function H on the space $Q = K^c/K$. The geodesics in Q with respect to the natural invariant metric are the translates of images of one-parameter subgroups in K^c of the form $\exp(it\xi)$, for ξ in $\mathfrak{k}$. A short calculation shows that along such a geodesic the second derivative of H, with respect to path length, is given by

$$\frac{\mathrm{d}^2}{\mathrm{d}t^2} H = |\rho(\xi)|^2. \tag{6.5.5}$$

Thus H is convex along geodesics and it follows immediately that it can have at most one critical point (if there is no isotropy group).

The alternative above now translates into the following: *either* H has a unique minimum on Q *or* the minimum is attained 'at infinity'; that is, a minimizing sequence diverges. The two pictures, using the functions ϕ and H, are quite compatible. Indeed the gradient flow lines of H on Q map to the gradient flow lines of ϕ on Γ in V.

In this abstract picture we can divide the orbits of K^c into two classes: 'stable' and 'unstable'. We say that a K^c-orbit is stable if the associated function H on Q is proper. By the discussion above this is equivalent to the two conditions

(i) there is a zero of the moment map in the orbit.

(ii) the points of the orbit have finite stabilizers, under the K action.

We call a point of V stable if it lies in a stable orbit, and write $V_S \subset V$ for the set of stable points. The upshot then is that the *complex quotient* of V_S by K^c can be identified with the *symplectic quotient* U^* of V^* by K where $V^* \subset V$ is the subset of points with no continuous isotropy groups. Notice that, by the second description, U^* is Hausdorff, in the quotient topology. The same is certainly not true, in general, of the full complex quotient V/K^c. In fact U^* is a Kähler manifold (or, more precisely, a Kahler orbifold), with the complex structure induced from the first description and the Kähler two-form induced from the second. If the Kähler form on V is defined by an equivariant positive line bundle L, we get, by the moment map construction, an induced positive line bundle L_U over the quotient. (More precisely, some power of L descends to U^*.)

This completes our outline of the general theory. We now turn to the rich source of examples provided by linear actions. We suppose V is $\mathbb{CP}^n$ with its Fubini–Study metric, so L is the Hopf line bundle. We suppose, as above, that a compact group K and its complexification K^c acts on $\mathbb{CP}^n$. A lift of the action to the line bundle is equivalent to a lift to a linear action on the underlying vector space $\mathbb{C}^{n+1}$; that is, we are considering the action induced by a linear representation of K. What is the meaning of stability in this case?

The non-zero orbits in the tautological bundle L^{-1} can be identified with those in $\mathbb{C}^{n+1}$, and the function h becomes just the logarithm of the Euclidean norm in $\mathbb{C}^{n+1}$. It is clear then that an orbit in $\mathbb{CP}^n$ is stable if and only if it is covered by an orbit $K^c x$ in $\mathbb{C}^{n+1}$ for which the map

$$K^c \longrightarrow \mathbb{C}^{n+1}$$

defined by $g \to g(x)$ is proper. (Equivalently, the K^c orbit is closed and there is only finite isotropy.) Note that this condition is of a purely complex algebraic nature, and makes no mention of the compact group K, metrics or connections. In this setting of linear actions, the stability condition is just that which was discovered by algebraic geometers, working on invariant theory.

In this algebraic situation we consider tensor powers of the holomorphic line bundle L_U over the quotient U^*. Holomorphic sections of these bundles are given by invariant holomorphic sections of $L = \mathcal{O}(1)$ over $\mathbb{CP}^n$, i.e. by invariant polynomials on $\mathbb{C}^{n+1}$. It is a general fact that for large d the invariant polynomials of degree d define an embedding of U^* in projective space, with image a quasi-projective subvariety. This notion is the starting point for the algebraic theory. We say that an orbit $\tilde{\Gamma} \subset \mathbb{C}^{n+1}$ is semi-stable if its closure does not contain the origin, and define an equivalence relation on semi-stable orbits by identifying orbits whose closures intersect. Then the points of the projective variety associated abstractly to the graded ring of K^c-invariant polynomials naturally correspond to equivalence classes of semi-stable orbits, and this variety contains U^* as an open set.

In practice the stable orbits can be identified using the 'Hilbert criterion' which asserts that a point $x \in \mathbb{C}^{n+1}$ defines a stable orbit for the K^c action if and only if the same is true for the action of each one-parameter subgroup, i.e. complex homomorphism $\mathbb{C}^* \to K^c$. In one direction this is trivial; the force of the assertion is that if the map from $K^c \to \mathbb{C}^{n+1}$ is not proper we can find such a 'destabilising subgroup' for which the composite $\mathbb{C}^* \to K^c \to \mathbb{C}^{n+1}$ is not proper. This is quite easy to see analytically, using the compactness of the unit sphere in $\mathfrak{k}^c/\mathfrak{k}$.

We now give four examples of this theory. First we consider the action of $\mathbb{C}^* = K^c$ on $\mathbb{C}^2$ given by the matrices:

$$\begin{pmatrix} \lambda & 0 \\ 0 & \lambda^{-1} \end{pmatrix}, \quad \lambda \in C^*.$$

In $\mathbb{C}^2$ the orbits consist of the 'hyperbolae' $\{xy = c\}$ for non-zero constants c, together with three exceptional orbits, $\{(x, 0) | x \neq 0\}$, $\{(0, y) | y \neq 0\}$ and $\{(0, 0)\}$. These latter three are exactly the unstable orbits. In the projective space $\mathbb{CP}^1$ we have just three orbits in total, one of which is stable. Notice that if we take the topological quotient of $\mathbb{CP}^1$ by $\mathbb{C}^*$ we get a non-Hausdorff topology on the set with three elements. Now, taking the standard metric on $\mathbb{C}^2$ we can restrict to the compact subgroup $K = S^1$; the moment map on $\mathbb{CP}^1$ is represented by $m(x, y) = |x|^2 - |y|^2$ (on the unit sphere in $\mathbb{C}^2$) and the relation between the zeros of m and stability is immediately apparent. (One should compare this example with the second example of Section 6.4.3.)

A more interesting case is the action of $K^c = GL(l, \mathbb{C})$ on pairs (A, v) consisting of an $l \times l$ matrix A and a l-vector v, given by $g(A, v) = (gAg^{-1}, gv)$. It is a simple exercise to verify, using the Hilbert criterion, that the stable points are exactly those for which v is a cyclic vector for A (i.e. the vectors $A^r v$ span $\mathbb{C}^l$). The momentum map is represented by

$$m(A, v) = i([A, A^*] + vv^*) \tag{6.5.6}$$

(restricted to the sphere $|A|^2 + |v|^2 = 1$).

The third example is similar; we simply consider the adjoint action of $GL(l, \mathbb{C})$ on the $l \times l$ matrices. The moment map is

$$m(A) = i[A, A^*].$$

In this case there are no stable orbits, since every matrix has a continuous isotropy group. However, as we have mentioned above, this condition is not essential in the theory. One can work almost equally well with the *closed* orbits in $\mathbb{C}^{n+1}$. For this adjoint action, these are just the orbits of diagonalizable matrices. The general link between zeros of the moment map and 'almost-stability' becomes the assertion that a matrix which commutes with its adjoint is diagonalizable.

For the fourth example we take the action of $GL(k, \mathbb{C})$ on quadruples $(\tau_1, \tau_2, \sigma, \pi)$, where τ_i are $k \times k$ matrices, σ is $n \times k$ and π is $k \times n$. We restrict ourselves to the subvariety defined by the complex equation $[\tau_1, \tau_2] + \sigma\pi = 0$. This is one part of the ADHM equations of Section 3.3.2, defined by a choice of complex structure on $\mathbb{R}^4$. One finds that the stability condition is just the non-degeneracy condition for ADHM data, and moreover that the momentum map is represented by

$$i([\tau_1, \tau_1^*] + [\tau_2, \tau_2^*] + \sigma\sigma^* - \pi^*\pi).$$

So the zeros of the moment map are the systems of ADHM data, and the quotient is the moduli space of framed $SU(n)$ instantons.

6.5.3 Connections over Kähler manifolds

We now return, after our long digression, to connections on holomorphic bundles over Kähler manifolds. At the formal level these furnish an infinite-dimensional example of the general theory above, as we shall now explain.

We begin by considering the space $\mathscr{A}$ of connections on a unitary bundle E over a general compact symplectic base manifold (X, ω), where X has dimension $2n$. This infinite-dimensional space is endowed with the symplectic form:

$$\Omega(a, b) = \frac{1}{8\pi^2}\int_X \operatorname{tr}(a \wedge b) \wedge \omega^{n-1}. \tag{6.5.7}$$

Here $a, b \in \Omega_X^1(\mathfrak{g}_E)$ represent tangent vectors in $\mathscr{A}$ as usual. The gauge group $\mathscr{G}$ of unitary automorphisms of E acts on $\mathscr{A}$, preserving the symplectic form. The Lie algebra of $\mathscr{G}$ is $\Omega_X^0(\mathfrak{g}_E)$ and we use integration over X to identify the dual of the Lie algebra with $\Omega_X^{2n}(\mathfrak{g}_E)$ (or, more precisely, with the appropriate distributions). Then we have:

Proposition (6.5.8). *The map* $m: \mathscr{A} \to \Omega_X^{2n}(\mathfrak{g}_E)$ *given by* $m(A) = (8\pi^2)^{-1} \times F_A \wedge \omega^{n-1}$ *is an equivariant momentum map for the action of* $\mathscr{G}$ *on* $\mathscr{A}$.

Proof. The derivative of $8\pi^2 m$ at a connection A is

$$8\pi^2(\delta m)_A(a) = \mathrm{d}_A a \wedge \omega^{n-1}.$$

(We denote the exterior derivative on $\mathscr{A}$ by δ, for clarity.) The pairing of this with an element ξ of the Lie algebra is

$$\langle(\delta m)_A(a), \xi\rangle = \int_X \mathrm{tr}(\mathrm{d}_A a\omega^{n-1}\xi).$$

Integrating by parts, using the fact that ω is closed, we rewrite this as

$$-\int_X \mathrm{tr}(a\mathrm{d}_A\xi)\,\omega^{n-1};$$

but this is just $8\pi^2\Omega(a, \rho(\xi))$, since the infinitesimal action $\rho(\xi)$ of ξ is $-\mathrm{d}_A\xi$. We have thus verified the momentum condition (6.5.1) and equivariance with respect to the action of $\mathscr{G}$ is clear.

Now suppose that X is a complex Kähler manifold, and ω is the Kähler form. We give $\mathscr{A}$ a complex structure by identifying tangent vectors with their (0, 1) parts (i.e. we represent connections by their $\bar{\partial}$ operators). Then $\mathscr{A}$ becomes a (flat) Kähler manifold and $\mathscr{G}$ acts by isometries. The complexification of the action is just the action of the group $\mathscr{G}^c$ we considered before. To tie up with the theory of holomorphic bundles we consider the subspace $\mathscr{A}^{(1,1)} \subset \mathscr{A}$ of connections having curvature of type (1, 1). This is an infinite-dimensional complex subvariety, and it inherits a Kähler structure from the ambient space. (Actually $\mathscr{A}^{(1,1)}$ may have singularities, and this will lead to the fact that, as we have seen, the moduli space of holomorphic bundles can be singular; but for the present we ignore this aspect.) We have

$$F_A \wedge \omega^{n-1} = \frac{1}{n}(F_A \cdot \omega)\omega^n = \frac{1}{n}\hat{F}_A\omega^n. \tag{6.5.9}$$

So, under the obvious embedding of the Lie algebra of $\mathscr{G}$ in its dual, the momentum map is given, up to a constant, by the component $\hat{F}_A$ of the curvature. So the zeros of the moment map in $\mathscr{A}^{1,1}$ are just the Hermitian Yang–Mills connections (in the case when $c_1 = 0$). On $\mathscr{A}^{1,1}$ the square-norm of the momentum map, $\|\hat{F}_A\|^2$, agrees up to a constant with the Yang–Mills functional $\|F_A\|^2$. (In the case $n = 2$ of complex surfaces this follows from the fact that $\hat{F} = F^+$ on $\mathscr{A}^{1,1}$, and in general it can be proved by a similar manipulation of the Chern–Weil integrand.) So the Yang–Mills flow is just the analogue of the gradient flow of the function ϕ in Section 6.5.2, and (6.2.4) is a special case of (6.5.4). In these terms our main theorem, for the case $n = 2$, should be viewed as the assertion that the definition (6.1.1) of stability is precisely what is required for the identification of the 'symplectic quotient' M_E^* of irreducible ASD connections and the complex quotient $\mathscr{A}_S^{1,1}/\mathscr{G}^c$, just as in the finite dimensional case.

6.5.4 *The curvature of the determinant line bundle*

We can apply the general theory of symplectic quotients to see that there is a natural Kähler structure on the moduli space of stable bundles (ASD connections) over a Kähler surface, provided this moduli space is regular. The Kähler form is that induced from (6.5.7), and the metric is in fact the natural 'L^2-metric', given by the L^2 norm of tangent vectors $a \in \Omega_X^1(\mathfrak{g}_E)$ satisfying the Coulomb condition $d_A^* a = 0$.

There is another aspect of the general theory of symplectic quotients which is interesting in this infinite-dimensional example and which fits into a theme running through this book. This is the realization of the symplectic structure by an equivariant line bundle over the space of connections. We seek a line bundle $\mathscr{L}$ over $\mathscr{A}^{(1,1)}$ acted on by $\mathscr{G}$ and $\mathscr{G}^c$, and a unitary connection on $\mathscr{L}$ having curvature $-2\pi i\Omega$. Now in a sense we already know the answer to this problem. We assume we have such a line bundle and that the centre ± 1 of $\mathscr{G}$ acts trivially. Then we get, as in Section 6.5.1, an induced bundle, which we also call $\mathscr{L}$, over the moduli space M^*, with curvature $-2\pi i$ times the Kähler form. Hence the first Chern class of $\mathscr{L}$ in $H^2(M)$ is the Kähler class $[\Omega]$. However, comparing (6.5.7) with (5.2.19), we see that

$$[\Omega] = \mu(\mathrm{PD}\,\omega) \in H^2(M). \tag{6.5.10}$$

Now if the metric ω is a Hodge metric, corresponding to a line bundle $L \to X$, and so the Poincaré dual PD ω is represented by a surface Σ, then we know that the determinant line bundle det ind $\not{\partial}_\Sigma$ has first Chern class $\mu(\Sigma)$ over M^* (Section 5.2.1). So the natural candidate for the equivariant line bundle generating the symplectic structure is the determinant line bundle of the Dirac operator over a Riemann surface $\Sigma \subset X$, pulled back by the restriction map from $\mathscr{A}_X$ to $\mathscr{A}_\Sigma$. In fact, following this topological route, one can show that the first Chern class of this line bundle in the '$\mathscr{G}$-equivariant' cohomology of $\mathscr{A}$ is represented by the 2-form Ω, and it then follows purely from general theory that this equivariant line bundle has an invariant connection with curvature form $-2\pi i\Omega$. However, it is interesting to construct this connection on the determinant line bundle more explicitly, and this is the task of the present section.

As we have said, all this theory can be carried out over a general compact Kähler manifold Z of any dimension. To obtain an equivariant positive line bundle over $\mathscr{A}^{1,1}$ one needs a corresponding line bundle L over Z, with $c_1(L) = [\omega]$, so we should start with a projective manifold Z. The cleanest construction of the desired bundle $\mathscr{L}$ uses, in place of restriction to a Riemann surface $\Sigma \subset Z$, a line bundle formed by a combination of determinant line bundles of Dirac operators over Z. We shall see this construction in Chapters 7 and 10 below. The uses of these combinations is equivalent topologically to the restriction operation, but is more natural geometrically. However, here for simplicity we will make a rather *ad hoc* construction of the connection, which pushes the real labour down onto a Riemann surface.

Suppose, as above, that Σ is a surface in a four-manifold X which is Poincaré dual to $[\omega]$. Let $\mathscr{L}_\Sigma \to \mathscr{A}_\Sigma$ be the determinant line bundle for Σ and let $\mathscr{L}$ be the pull-back to $\mathscr{A} = \mathscr{A}_X$. Suppose we have constructed a $\mathscr{G}$-invariant connection on the determinant line bundle $\mathscr{L}_\Sigma$ with curvature form:

$$\Omega_\Sigma(a, b) = \frac{1}{8\pi^2}\int_\Sigma \mathrm{Tr}(a \wedge b). \tag{6.5.11}$$

Then we get an induced connection, by pull-back, on $\mathscr{L}$ whose curvature is given by the same formula. Now if we have a 1-form Φ on $\mathscr{A}$ with exterior derivative $\delta\Phi = \Omega - \Omega_\Sigma$, we can modify the pull-back connection by the 1-form Φ to get a new connection which has curvature form Ω. If Φ is $\mathscr{G}$-invariant, the new connection will be invariant, and thus gives a solution to our problem.

There is, however, an obvious 1-form Φ with the desired properties. Let T_Σ be the current associated with Σ, i.e. the map on 2-forms over X:

$$T_\Sigma(\theta) = \int_\Sigma \theta.$$

The condition that Σ be Poincaré dual to ω means that there is a singular 1-form ϕ on X satisfying the distributional equation

$$\omega = T_\Sigma - \mathrm{d}\phi.$$

That is, for any 2-form θ,

$$\int_\Sigma \theta - \int_X \theta \wedge \omega = \int \mathrm{d}\theta \wedge \phi.$$

We now define the 1-form Φ on $\mathscr{A}$ by

$$\Phi_A(a) = \frac{1}{8\pi^2}\int_X \mathrm{Tr}(F_A \wedge a) \wedge \phi, \tag{6.5.12}$$

and we leave to the reader the verification that $\delta\Phi$ is indeed $\Omega_\Sigma - \Omega$.

We have thus reduced the problem to the construction of a connection on the determinant line bundle $\mathscr{L}_\Sigma$ over $\mathscr{A}_\Sigma$. This was done by Quillen (1985). As a first point, in Chapter 5 we worked with the Dirac operator over Σ. However this agrees with the $\bar\partial$ operator after twisting by a square root $K_\Sigma^{1/2}$. So, if we work with $U(n)$ bundles, we may as well consider the $\bar\partial$ operator, for which the notation is slightly simpler. (See also Section 10.1.3.)

We consider then the equivariant line bundle $\mathscr{L}_\Sigma$ over $\mathscr{A}_\Sigma$ with fibres

$$\Lambda^{\max}(\operatorname{Ker}\bar\partial_A)^* \otimes \Lambda^{\max}(\operatorname{Ker}\bar\partial_A^*).$$

The action of $\mathscr{G}^c$ is visible from the equivalent description of $\ker\bar\partial_A^*$ as the cokernel of $\bar\partial_A$, and this description also makes the holomorphic structure of $\mathscr{L}_\Sigma$ apparent, since the operator $\bar\partial_A$ depends holomorphically on A, with the complex structure on $\mathscr{A}$ obtained by the identification with $\Omega_\Sigma^{0,1}(\operatorname{End} E)$. Given this holomorphic structure, a compatible unitary connection is specified by a *hermitian metric* on the fibres of $\mathscr{L}_\Sigma$.

Quillen defines a hermitian metric on the determinant lines as follows. Suppose we are in a situation where the $\bar\partial$ operators are invertible on a dense open set $U \subset \mathscr{A}_\Sigma$. Then, as in Section 5.2, there is a canonical holomorphic section σ of $\mathscr{L}_\Sigma$, vanishing on the 'jumping divisor' $\mathscr{A}_\Sigma \backslash U$ and giving a trivialization of the line bundle over U. So to specify a metric on $\mathscr{L}_\Sigma$ it suffices, by continuity, to give a function

$$D\colon \mathscr{A}_\Sigma \to \mathbb{R}$$

with $D(A) = |\sigma(A)|^2$, which vanishes on the jumping divisor to second order. To do this, Quillen introduces the regularized determinants of the Laplace operators $\bar\partial_A^*\bar\partial_A$, which we denote here by Δ_A. Formally we want to write

$$D(A) = \det \Delta_A = \prod \lambda,$$

where λ runs over the eigenvalues of Δ_A, counted with multiplicity. Of course, this product is wildy divergent, but it can be defined in a formal way using the ζ-function of the operator.

$$\zeta_A(s) = \sum \lambda^{-s}. \tag{6.5.13}$$

This sum converges for $\operatorname{Re}(s)$ large, and so defines a holomorphic function on a half-plane in $\mathbb{C}$. As we shall see, this function can be continued to a meromorphic function on $\mathbb{C}$, holomorphic at 0. One then defines the regularized determinant $D(A)$ as

$$D(A) = \exp(-\zeta_A'(0)). \tag{6.5.14}$$

Note that this is formally correct since if there were only a finite number of eigenvalues we would have

$$\frac{\mathrm{d}}{\mathrm{d}s}\sum \lambda^{-s} = -\sum \lambda^{-s}\log\lambda$$

which takes the value $-\log(\Pi\lambda)$ at $s = 0$.

It is not hard to see that this does indeed define a smooth metric on the determinant line bundle. In general the Quillen metric is defined as follows. For each $c > 0$ we let U_c be the open set of connections A for which c is not an eigenvalue of Δ_A. Over U_c we have a finite-dimensional vector bundle $\mathscr{H}_c^+$ defined by the span of the eigenspaces of Δ_A belonging to eigenvalues $\lambda < c$. Similarly there is a vector bundle $\mathscr{H}_c^-$ defined by the corresponding eigenspaces of the other Laplacian $\bar{\partial}_A \bar{\partial}_A^*$. There is a canonical isomorphism, over U_c, between $\mathscr{L}_\Sigma$ and

$$\Lambda^{\max} \mathscr{H}_c^- \otimes (\Lambda^{\max} \mathscr{H}_c^+)^*,$$

(since $\bar{\partial}_A$ gives an isomorphism between the non-zero eigenspaces). Under this isomorphism we get a metric $|\ |_c$ on $\mathscr{L}_\Sigma$ over U_c using the L^2 metrics on $\mathscr{H}_c^+, \mathscr{H}_c^-$. We then define the true metric to be

$$|\ |^2 = \left(\prod_{\lambda > c} \lambda \right) |\ |_c^2,$$

where the product $\prod_{\lambda > c} \lambda$ is defined by ζ-function regularization as before.

Quillen's result is then:

Theorem (6.5.15) (Quillen). *The curvature of the connection defined by the metric $|\ |$ on the determinant line bundle $\mathscr{L}_\Sigma$ is $-2\pi i \Omega_\Sigma$ where*

$$\Omega_\Sigma(a, b) = \frac{1}{8\pi^2} \int_\Sigma \operatorname{tr}(a \wedge b).$$

6.5.5 *Quillen's calculation*

For simplicity we work in the case when the $\bar{\partial}$ operators are generically invertible. If the metric on a holomorphic line bundle is given in a local holomorphic trivialization by a function h, the curvature of the canonical connection is $\bar{\partial}\partial h$. So in our case the curvature is given by the (1, 1)-form

$$\delta'' \delta' \log D$$

on $\mathscr{A}_\Sigma$. Here we write δ', δ'' for the holomorphic and anti-holomorphic parts of the exterior derivative on the infinite-dimensional space $\mathscr{A}_\Sigma$. Before beginning the calculation we review the meromorphic extension of the ζ-function. To see this one uses a representation in terms of the 'heat kernel' $\exp(-t\Delta)$ (from now on we write Δ for Δ_A). For positive t this is an integral operator given by a smooth kernel $k_t(x, y)$—the fundamental solution of the heat equation:

$$\frac{\partial f}{\partial t} = -\Delta f \quad (t > 0).$$

In the flat model—the ordinary heat equation on $\mathbb{R}^2$—the fundamental solution is given by the well-known formula:

$$\frac{1}{4\pi t}\exp\left(\frac{-|x|^2}{2t}\right). \tag{6.5.16}$$

Let ρ be the injectivity radius of Σ and $k_t^{\#}(x, y)$ be the bundle-valued kernel on the Riemann surface given by:

$$k_t^{\#}(x, y) = \begin{cases} \dfrac{1}{4\pi t}\exp\left(\dfrac{-\mathrm{d}(x, y)^2}{2t}\right)P(x, y) & \text{if } \mathrm{d}(x, y) < \rho \\ 0 & \text{if } d(x, y) \geq \rho, \end{cases} \tag{6.5.17}$$

where $P(x, y): E_x \to E_y$ denotes the parallel transport along the minimal geodesic from x to y, defined by the connection A. It is not hard to see that $k_t(x, y) = k_t^{\#}(x, y) + \varepsilon_t(x, y)$, where ε is bounded as t tends to zero. In particular the trace,

$$\mathrm{Tr}(\exp(-t\Delta)) = \sum e^{-\lambda t} = \int_\Sigma \mathrm{tr}\, k_t(x, x)\mathrm{d}\mu_x, \tag{6.5.18}$$

is

$$\frac{n\,\mathrm{Vol}(\Sigma)}{4\pi t} + O(1)$$

as t tends to 0. This is the first term in an asymptotic expansion

$$\mathrm{Tr}(\exp(-t\Delta)) \sim \sum_{i=-1}^{\infty} a_i t^i, \quad t \to 0. \tag{6.5.19}$$

Returning to the ζ-function, we have for $\lambda > 0$,

$$\lambda^{-s} = \frac{1}{\Gamma(s)}\int_0^\infty e^{-\lambda t}t^{s-1}\,\mathrm{d}t;$$

so, for $\mathrm{Re}(s)$ large,

$$\sum \lambda^{-s} = \frac{1}{\Gamma(s)}\int_0^\infty \mathrm{Tr}(\exp(-t\Delta))t^{s-1}\,\mathrm{d}t. \tag{6.5.20}$$

But, taking the first two terms of the asymptotic expansion (6.5.19),

$$\mathrm{Tr}(\exp(-t\Delta)) = \frac{a_{-1}}{t} + a_0 + \eta(t)$$

say, where $\eta(t)$ is $O(t)$ for small t. Thus

$$\int_0^1 \mathrm{Tr}(\exp(-t\Delta))t^{s-1}\,\mathrm{d}t - \left(\frac{a_{-1}}{s-1} + \frac{a_0}{s}\right)$$

is holomorphic in $\{s|\mathrm{Re}(s) > -1\}$. The remainder

$$\int_1^\infty \mathrm{Tr}(\exp(-t\Delta))t^{s-1}\,\mathrm{d}t$$

is an entire function of s, so we see that

$$\int_0^\infty \mathrm{Tr}(\exp(-t\Delta)t^{s-1}\,\mathrm{d}t$$

has a meromorphic extension to $\{\mathrm{Re}(s) > -1\}$, with simple poles at $s = 0, 1$. Since the Gamma function $\Gamma(s)$ has a simple pole at 0 we see that $\zeta(s)$ is indeed holomorphic at $s = 0$. (One can use the higher terms in the asymptotic expansion to obtain the meromorphic extension of ζ_A over $\mathbb{C}$, but we do not need this.)

We now go on to compute the curvature $\delta''\delta \log D$. It is instructive to begin with a completely formal calculation. If we write $D(A) = \det\Delta = \det(\bar{\partial}_A^*\bar{\partial}_A)$ and differentiate formally, we get

$$\begin{aligned}\delta' \log\det\Delta = \mathrm{Tr}(\Delta^{-1}\delta'\Delta) &= \mathrm{Tr}(\Delta^{-1}\bar{\partial}_A^*(\delta' A))\\ &= \mathrm{Tr}(\bar{\partial}_A^{-1}(\delta' A)).\end{aligned}$$

We might then argue that $\bar{\partial}_A$ varies holomorphically with A, so that $\bar{\partial}_A^{-1}(\delta' A)$ is a holomorphic operator-valued 1-form on the space of connections; hence its trace should also be holomorphic and so $\delta''\delta' \log D = 0$. This is, of course, not correct. The divergence between the true behaviour of the regularized determinant and the formal properties one might expect illustrates the way that the curvature of determinant line bundles appears as an 'anomaly'.

We now proceed with the genuine calculation of the curvature. We have

$$\delta' \mathrm{Tr}(\exp(-t\Delta)) = -t\,\mathrm{Tr}(\exp(-t\Delta)\delta'\Delta), \tag{6.5.21}$$

so for large $\mathrm{Re}(s)$

$$\delta'\zeta_A(s) = \frac{-1}{\Gamma(s)}\int_0^\infty \mathrm{Tr}(\exp(-t\Delta)\,t\delta'\Delta)t^s\,\mathrm{d}t. \tag{6.5.22}$$

Now $\delta'\Delta = \Delta\bar{\partial}_A^{-1}(\delta' A)$, so (6.5.20) gives

$$\begin{aligned}\delta'\zeta_A(s) &= \frac{-1}{\Gamma(s)}\int_0^\infty \mathrm{Tr}(\exp(-t\Delta)\Delta\bar{\partial}_A^{-1}(\delta' A))\,t^s\,\mathrm{d}t\\ &= \frac{-1}{\Gamma(s)}\int_0^\infty -\frac{\mathrm{d}}{\mathrm{d}t}(\mathrm{Tr}(\exp(-t\Delta)\bar{\partial}_A^{-1}(\delta' A)))t^s\,\mathrm{d}t.\end{aligned}$$

(We will not pause to justify the commutation property of the trace used in the step above.) Now integrate by parts to write this as

$$\frac{-s}{\Gamma(s)} \int_0^\infty \mathrm{Tr}(\exp(-t\Delta)\bar{\partial}_A^{-1}(\delta' A))t^{s-1}\,\mathrm{d}t.$$

The function $s/\Gamma(s)$ has a double zero at $s = 0$, so the variation $\delta'\zeta'_A(0)$ is minus the residue of

$$\int \mathrm{Tr}(\exp(-t\Delta)\bar{\partial}_A^{-1}(\delta' A))t^{s-1}\,\mathrm{d}t,$$

which gives

$$\delta' \log D = -\lim_{t\to 0} \mathrm{Tr}(\exp(-t\Delta)\bar{\partial}_A^{-1}(\delta' A)). \tag{6.5.23}$$

If we set, formally, t to be zero we get the previous formula. In sum the ζ-function determinant leads to a regularization of the trace by composing with the smoothing operator $\exp(-t\Delta)$, and then taking a limit as t tends to 0. As we shall now see, the smoothing operator does not depend holomorphically on A, and this leads to the 'anomaly'.

We will now examine the operators in the formula (6.5.23) for $\delta' \log D$ in more detail. The exponential $\exp(-t\Delta)$ is given, as we have already said, by a smooth kernel k_t, which differs from the explicit approximation $k_t^{\#}$ by $O(1)$. The inverse $\bar{\partial}_A^{-1}$ of the bundle-valued $\bar{\partial}$ operator is likewise given by a kernel L, which is singular on the diagonal. The flat model is of course the Cauchy kernel, and in a local holomorphic trivialization for the bundle and local complex coordinates on Σ the kernel has the form:

$$L(w, z) = \left(\frac{1}{2\pi(w - z)} + f(w, z)\right)\mathrm{d}z, \tag{6.5.24}$$

where f is holomorphic across the diagonal. (Note that the differential $\mathrm{d}z$ appears, so that for a (0, 1)-form ϕ the integral $\int_{z\in\Sigma} L(w, z)\,\phi(z)$, which yields $\bar{\partial}_A^{-1}(\phi)$, is intrinsically defined.)

To simplify our notation we will now consider a fixed variation of the connection by $a = \alpha - \alpha^*$, for a bundle valued (0, 1)-form α. Thus the pairing between $\delta' \log D$ and a is represented by the trace of the operator

$$\exp(-t\Delta)\,\bar{\partial}_A^{-1}\,\alpha,$$

where α is now regarded as a multiplication operator on E. This trace is the integral over Σ of the 2-form τ_t given by the expression:

$$\tau_t(z) = \int_\Sigma \mathrm{tr}\,(k_t(z, w)L(w, z)\alpha(z))\,\mathrm{d}\mu_w. \tag{6.5.25}$$

Now $\varepsilon_t = k_t - k_t^{\#}$ is bounded and tends to zero away from the diagonal, as t tends to 0, while $L(z, w)$ is integrable. It follows that

$$\int \operatorname{tr}(\varepsilon_t(z, w)L(w, z)\alpha(z))\, d\mu_w \longrightarrow 0$$

as t tends to 0, uniformly in z. Thus to calculate the limit above we may replace the heat kernel k_t by its explicit approximation $k_t^{\#}$ and the form τ_t by

$$\tau_t^{\#}(z) = \int_\Sigma \operatorname{tr}(k_t^{\#}(z, w)L(w, z)\alpha(z))\, d\mu_w.$$

Note that the limit exists pointwise on Σ, i.e. $\tau_t^{\#}(z)$ converges to a limit $\tau(z)$:

$$\tau(z) = \lim_{t \to 0} \int \operatorname{tr}(k_t^{\#}(z, w)L(w, z)\alpha(z))\, d\mu_w. \tag{6.5.26}$$

The existence of this limit depends on a cancellation mechanism—if we replaced the terms by their pointwise norms the corresponding integral would clearly tend to infinity as $t \to 0$. The situation becomes more transparent if one considers the flat-space model, where L is the Cauchy kernel. Then τ_t vanishes for all t by reasons of symmetry; in general the limit detects the 'constant' term in the expansion (6.5.24) of L, which cannot be determined by local considerations.

We now perform the second differentiation to evaluate $\delta''\delta' \log D$. The argument above has given us the formula:

$$\langle \delta' \log D, \alpha \rangle = \lim_{t \to 0} \int_\Sigma \tau_t^{\#}(z),$$

where

$$\tau_t(z) = \int_\Sigma k_t^{\#}(z, w)L(w, z)\alpha(z)\, d\mu_w.$$

Apart from $k_t^{\#}$, the terms on the right-hand side in this formula depend holomorphically on the connection A (the basis of our erroneous formal calculation above). Thus if we consider another variation $b = \beta - \beta^*$ the pairing of the 2-form $\delta''\delta' \log D$ with (α, β) is

$$\delta''\delta' \log D(\alpha, \beta) = \int_\Sigma \delta''\tau(z),$$

where

$$\delta''\tau(z) = \lim_{t \to 0} \int_\Sigma \operatorname{tr}\{(\delta'' k_t^{\#}(z, w))L(w, z)\alpha(z)\}\, d\mu_w,$$

$\delta'' k_t^{\#}$ being the anti-holomorphic derivative of $k_t^{\#}$ along b. The kernel $k_t^{\#}$ depends on the connection A only through the parallel transport operator $P(x, y) \in \operatorname{Hom}(E_x, E_y)$. We shift for a moment to real variables. The derivative

$\delta P(x, y)$ of the parallel transport with respect to the connection lies in $\mathrm{Hom}(E_x, E_y)$ and vanishes when $x = y$. So there is an intrinsically defined space derivative $\nabla\{\delta P\}$, evaluating at $x = y$, which lies in $(T_X^* \otimes \mathfrak{g}_E)_x$. It is easy to see that this is minus the variation of the connection, evaluated at x. Shifting to complex variables this gives us the formula

$$\nabla'\{\delta'' P\} = \tfrac{1}{2}\beta^*,$$

where ∇' denotes the $(1, 0)$ part of the derivative ∇ on Σ.

We now fix a point z in Σ and evaluate the expression above for $\delta''\tau(z)$. In a local coordinate system w we can take z to be the origin and we write α_0, β_0 for the values of α and β at z. Then the formula above can be written

$$(\delta'' P(0, w))\,\mathrm{d}\bar{z} = \tfrac{1}{2}\beta_0^* w + O(w^2),$$

where the differential $\mathrm{d}\bar{z}$ is the canonical basis element in $\Lambda^{0,1}$ at 0, in the coordinate system. Our formula for $\delta''\tau$ is

$$\delta''\tau(0) = \lim_{t\to 0}\int h_t(w)\operatorname{tr}\{(\delta'' P(0, w))L(w, 0)\alpha_0\}\,\mathrm{d}\mu_w,$$

where $h_t(w)$ is a scalar, essentially the fundamental solution of the scalar heat equation. Thus $h_t(w)$ tends to the delta distribution as t tends to 0. Now the asymptotic formula (6.5.24) for L, and the formula above for $\delta'' P$ along the diagonal, give

$$\delta'' P(0, w)L(w, 0) = \frac{1}{4\pi}\beta_0^* + O(w),$$

and it follows that

$$\delta''\tau(0) = \frac{1}{4\pi}\operatorname{tr}(\beta_0^* \wedge \alpha_0).$$

Thus the pairing between the curvature form $\delta''\delta' \log D$ and (α, β) is

$$\int_\Sigma \delta''\tau(z) = \frac{1}{4\pi}\int_\Sigma \operatorname{tr}(\beta^* \wedge \alpha),$$

and so the curvature form evaluated on (a, b) is $-\dfrac{i}{4\pi}\displaystyle\int \operatorname{tr}(a \wedge b)$, as asserted in (6.5.15).

Notes

Sections 6.1 and 6.2

The relation between stable bundles and Yang–Mills theory goes back to the work of Narasimhan and Seshadri (1965) who dealt with projectively flat unitary connections over Riemann surfaces. Their results were explicitly formulated in terms of connections by Atiyah and Bott (1982), who developed the picture involving the orbits of the complexified gauge group. An analytical proof of the theorem of Narasimhan and Seshadri was given by Donaldson (1983*a*). The extension of the theory to higher dimensions followed conjectures made independently by Hitchin (1980) and Kobayashi (1980). Preliminary results were obtained by Kobayashi and by Lubke (1982, 1983). The existence proof for bundles over algebraic surfaces was given by Donaldson (1985*a*), and the general result for vector bundles over arbitrary compact Kähler manifolds was proved by Uhlenbeck and Yau (1986). Extensions to arbitrary structure groups were obtained by Ramanathan and Subramanian (1988) and to general Hermitian surfaces by Buchdahl (1988). See also the survey article by Margerin (1987).

The proof we give here is similar in outline to that of Uhlenbeck and Yau. The main simplification in four dimensions is that we can appeal directly to Uhlenbeck's gauge fixing result from Chapter 2. In place of the gradient flow equation, Uhlenbeck and Yau use a more direct continuity method. This avoids some technical difficulties, although it is perhaps not so elegant. The gradient flow approach also fits in well with the picture given by Atiyah and Bott of the stratification of the space of connections. In all of these approaches the main idea is to study the limiting behaviour of a family of connections in a G^c orbit. A rather different proof of the general result for bundles over projective manifolds is given by Donaldson (1987*d*), extending some of the techniques of Donaldson (1985*a*). This proof works with *a priori* estimates to control the gradient flow in the stable case, using induction on the dimension of the base manifold, an integral formula to pass down to a general hypersurface and the theorem of Mehta and Ramanathan (1984) mentioned in Chapter 10.

Other developments in this direction consider coupled equations for a connection and a section of some associated bundle; these also have important geometric applications. See Hitchin (1987), Simpson (1989), Corlette (1988) and Bradlow (1990).

Section 6.3.1

The use of non-linear parabolic equations to find solutions of corresponding elliptic equations in differential geometry goes back to Eells and Sampson (1964). For the general theory of parabolic equations on manifolds see Hamilton (1975). One can avoid the gauge fixing procedure used here by recourse to an existence theorem for equations which are 'parabolic modulo a group action'; compare Hamilton (1982) and Deturck (1983). For detailed treatments of the evolution equation used here see Kobayashi (1987) and Jost (1988).

Section 6.3.2

For a slightly different approach to the long-time existence problems see Donaldson (1985*a*), Kobayashi (1987) and Jost (1988).

Section 6.4.1

Deformation theory in holomorphic geometry is a well-developed and large subject. Most references concentrate on deformations of complex manifolds but the results can all be transferred to holomorphic bundles, and many of the proofs are rather simpler in this setting. For techniques based on partial differential equations we refer to Kodaira and Spencer (1958), Kuranishi (1965), and Sunderaraman (1980). Another approach is to construct deformations of the transition functions using power series; see for example Forster (1977) and Palamodov (1976).

Section 6.4.2

The comparison of the deformation theories is given by Donaldson (1987*a*). For more details on orientation questions see Donaldson (1987*b*).

Sections 6.5.1 and 6.5.2

For the relation between the symplectic quotient and stable orbits via the moment map, we refer to Kirwan (1984), Guillemin and Sternberg (1982), and Kempf and Ness (1988). For the theory of stable and semistable points under linear actions see Mumford and Fogarty (1982), Newstead (1978) and Gieseker (1982). There is a transcendental proof of the Hilbert criterion in Birkes (1971).

The fourth of our examples was studied by Donaldson (1984*a*); see the notes on Chapter 3 above. (It is a striking fact that both the ASD equations and the ADHM equations appear as zero moment map conditions—for two quite different symmetry groups. This is another manifestation of the formal similarity between the equations, which we tried to bring out in Chapter 3.)

Section 6.5.3

The discussion here generalizes that of Atiyah and Bott (1982) for connections over Riemann surfaces. See also Donaldson (1985*a*) and Kobayashi (1987). The Kähler metric on the moduli spaces was found by a direct calculation by Itoh (1983).

Section 6.5.4

The basic mathematical reference for connections on determinant line bundles is Quillen (1985). The constructions were generalized substantially by Bismut and Freed (1986). One can obtain a connection on the determinant line bundle over $\mathscr{A}_X$ with the desired curvature more directly by using the description of Section 7.1.4 and the connection defined by Bismut and Freed; see Donaldson (1987*d*). There are deep relations between determinants of $\bar{\partial}$-Laplacians and algebraic geometry; see Bismut et al. (1988).

Section 6.5.5

Here we follow Quillen's paper very closely. For the theory of the asymptotic expansion of the heat kernel see, for example, Gilkey (1984), and for ζ-functions see Ray and Singer (1973).

7

EXCISION AND GLUING

This chapter brings together a number of loosely related topics from analysis, the general context being the description of solutions to differential equations, depending on a parameter, for limiting values of the parameter. We begin by considering the excision principle for the index of linear elliptic operators. As we shall explain below, this principle leads rapidly to the proof of the vital index formula given in Chapter 4 for the virtual dimension of Yang–Mills moduli spaces. We shall show that, as an alternative to the standard proof using pseudo-differential operators, one can prove this principle by introducing a suitable deformation of the differential operator. This discussion sets the scene for the more specialized geometrical topics considered in the rest of the chapter.

In Section 7.1.5 we show that a determinant line bundle can be extended over the compactified moduli spaces introduced in Chapter 4. This involves the asymptotic analysis of the coupled Dirac operators, with respect to the 'distance' to the points at infinity in the moduli space. The main business of the chapter is taken up in Section 7.2 where we describe ASD connections over connected sums. In this case the relevant parameter is the size of the 'neck' in the connected sum. We obtain a rather general description of the moduli space in this situation. This will be applied to prove 'vanishing theorems' in Chapter 9. The same theory also gives the description of neighbourhoods of the points at infinity in the compactified moduli spaces, and this aspect will be taken up in Chapter 8. The last section, Section 7.3, of this chapter contains a proof of a technical decay estimate for ASD connections over a cylinder (or annulus) which is needed to control the solutions over the neck in the connected sum.

7.1 The excision principle for indices

7.1.1 Pseudo-differential operators

Consider the following general situation:

Condition (7.1.1).

(i) *Z is a compact manifold decomposed as a union of open sets $Z = U \cup V$.*

(ii) *$L:\Gamma(\xi) \to \Gamma(\eta)$ and $L':\Gamma(\xi') \to \Gamma(\eta')$ are a pair of elliptic differential operators over Z.*

(iii) *There are bundle isomorphisms $\alpha:\xi|_V \to \xi'|_V$, $\beta:\eta|_V \to \eta'|_V$ such that $L' = \beta^{-1}L\alpha$ over V.*

Informally, L and L' are operators that agree over V. Each of these elliptic operators has an associated index:

$$\text{index}(L) = \dim \ker L - \dim \operatorname{coker} L.$$

Now while the operators agree over V the kernels and cokernels are of course global objects—it does not make sense to talk about the parts of the kernel depending on the restrictions of the operators to U and V. The excision property for indices states that, nevertheless, the index behaves as though we did have such a notion: in brief, the difference of the indices $\text{ind}(L') - \text{ind}(L)$ depends only on the data over U. To be quite precise, suppose that $(Z_1, U_1, V_1, L_1, L'_1, \alpha_1, \beta_1), (Z_2, U_2, V_2, L_2, L'_2, \alpha_2, \beta_2)$ are two sets of data as above. Suppose there is a diffeomorphism $\tau: U_1 \to U_2$ covered by bundle isomorphisms between each of the four pairs of bundles involved. We denote these bundle maps also by τ. Suppose also that τ is compatible with the maps α_i, β_i, in the obvious sense. In brief we can say that the two sets of data are isomorphic over U_1 and U_2. Then the excision property is the assertion:

Proposition (7.1.2). *For any two sets of data, as above, which are isomorphic over open sets U_1, U_2 we have:*

$$\text{ind}(L'_2) - \text{ind}(L_2) = \text{ind}(L'_1) - \text{ind}(L_1).$$

Before proving this we explain its relevance to the derivation of the index formula,

$$\text{ind}(\delta_A) = 8c_2(E) - 3(1 - b^1(X) + b^+(X)), \tag{7.1.3}$$

for the operator $\delta_A = d_A^* + d_A^+$ coupled to the Lie algebra bundle $\mathfrak{g}_E$ associated to an $SU(2)$-bundle E over a compact four-manifold X (cf. (4.2.21)). We know that E is trivial on the complement of a point in X, so we can consider the situation above with $Z = X$, $\xi = \Lambda^1 \otimes \mathfrak{g}_E$, $\eta = (\Lambda^0 \oplus \Lambda^+) \otimes \mathfrak{g}_E$, $L = \delta_A$, for a connection A which is trivial outside a small neighbourhood $U \subset X$ of a point. We then compare with the ordinary δ operator on $\xi' = \Lambda^1 \otimes \mathfrak{g}$, where $\mathfrak{g}$ is the Lie algebra of $SU(2)$. We can choose the base metric and bundle connection in U to have a standard form. Our excision property then tells us that the difference of the indices depends only on $c_2(E)$ not on X. By considering successive changes of $c_2(E)$ by 1 we deduce that there is an index formula of the shape

$$\text{ind}(\delta_A) = R\,.\,c_2(E) + \psi(X),$$

for some constant R and numerical invariant ψ of X. Now taking the trivial bundle and evaluating the index by the Hodge theory of Section 1.1.6, we obtain

$$\ker(\delta) = H^1(X;\mathbb{R}), \quad \ker(\delta^*) = (H^0(X;\mathbb{R}) \oplus H^+(X));$$

thus we see that $\psi(X) = -3(1 - b^1(X) + b^+(X))$. So to derive the complete formula it suffices to know the index in any one other case, where c_2 is not zero. For example we can take the basic one-instanton on S^4 which, as we have seen in Chapter 3, lives in a five-dimensional moduli space. (One has then to see that the cokernel of d_A^+ is 0, to evaluate the index.) Alternatively one might take a reducible connection over $S^2 \times S^2$, and reduce the problem to a model index calculation over the two-sphere (cf. Section 6.4.3).

The excision property was formulated by Atiyah and Singer (1968), and played a vital role in their proof of the general index theorem. We will now review the standard proof of this excision property. This involves the notion of a *pseudo-differential operator*. We recall the definition in outline. A pseudo-differential operator (ψDO) acting on vector valued functions over $\mathbb{R}^n$ is an operator P which can be written in the form

$$(Pf)_x = (2\pi)^{-n/2} \int_{\mathbb{R}^n} p(x, \xi) \hat{f}(\xi) e^{-ix.\xi} \, d\mu_\xi, \tag{7.1.4}$$

where $\hat{f}$ is the Fourier transform of f and p is a matrix valued function. The definition is extended to operators on sections of bundles over a compact manifold using a partition of unity. If p is a polynomial in the ξ variable, the operator P is just a partial differential operator. In general we suppose that p has a decomposition

$$p(x, \xi) = \sigma_P(x, -i\xi) + r(x, \xi),$$

where the 'leading symbol' σ_P is homogeneous of degree k in ξ (and k may be positive or negative) and the remainder r is a lower-order term with $r(x, \xi) = o(|\xi|^k)$ as $\xi \to \infty$. The leading symbol is invariantly defined as a homogeneous function on the cotangent bundle minus its zero section. We say that a pseudo-differential operator over a compact manifold is *elliptic* if the leading symbol is everywhere invertible.

The basic facts about the pseudo-differential operators over compact manifolds are:

(i) An elliptic ψDO of order k defines a Fredholm map between Sobolev spaces L^2_{k+l}, L^2_l.

(ii) Any invertible symbol function comes from some ψDO, and any continuous family of symbols can be lifted to a family of operators.

(iii) The composite of ψDO's is pseudo-differential, with symbol $\sigma_{PQ} = \sigma_P . \sigma_Q$. The formal adjoint of a ψDO is pseudo-differential, and if P is self-adjoint and positive, the square root $P^{1/2}$ is again a ψDO.

The utility of these operators in index theory derives mainly from property (ii). Since the index is a deformation invariant of Fredholm operators, the index of an elliptic ψDO depends only on the homotopy class of the symbol.

We shall now prove the excision theorem (7.1.2) using notation as introduced above. Let D be the operator

$$D = L' \oplus (-1)^{k+1} L^* : \Gamma(\xi' \oplus \eta) \longrightarrow \Gamma(\eta' \oplus \xi), \tag{7.1.5}$$

where L has order k. The index of D is

$$\operatorname{ind} D = \operatorname{ind} L' - \operatorname{ind} L.$$

We let P_0 be the pseudo-differential operator

$$P_0 = (1 + DD^*)^{-1/2}D.$$

The kernels of P_0 and P_0^* are equal to those of D and D^*, so P_0 has the same index as D. The symbol of P_0 is

$$(\sigma\sigma^*)^{-1/2}\sigma$$

where σ is the symbol of D. Now over the open subset V of Z we regard $\xi'$and η' as being canonically identified with ξ, η by the maps α, β. The symbol of P_0, over V, can then be written in the form

$$\begin{pmatrix} 0 & \pi \\ -\pi^* & 0 \end{pmatrix}$$

with π homogeneous of degree 0. There is a homotopy of this symbol over V to the identity map, given by the family

$$\begin{pmatrix} t1 & (1-t)\pi \\ -(1-t)\pi^* & t1 \end{pmatrix}, \tag{7.1.6}$$

for $t \in [0, 1]$. It is easy to see that this is indeed a homotopy through invertible symbols of degree 0. By property (ii) this homotopy of symbols can be lifted to a homotopy of ψDOs, P_t say, with the symbol of P_1 being the identity map over an open set $V^* \Subset V$, containing the complement of U in Z.

The key point is now that we can choose P_1 to be an operator 'equal to the identity' outside a compact subset K of U. Indeed given any choice of P_1 we let $\tilde{P}_1$ be the operator defined by

$$\tilde{P}_1(f) = \psi P_1 \psi f + (1 - \psi^2)f,$$

where ψ is a cut-off function equal to 1 outside some $K \subset U$ and supported in V^*. This has the same symbol as P_1, and hence has the same index. The restriction of $\tilde{P}_1 f$ to $Z \setminus K$ is equal to f. This means that any element of $\ker \tilde{P}_1$ is supported in $K \subset U$. Similarly for the elements of $\ker \tilde{P}_1^*$, representing the cokernel of $\tilde{P}_1$.

The proof of the excision property is now in our hands. If we are given two manifolds Z_1, Z_2 as in Proposition (7.1.2) we can choose operators $P_1^{(1)}, P_1^{(2)}$ say, as above, equal to the identity outside U_1, U_2. Moreover we can choose the operators to agree over U_1, U_2 under the isomorphism τ. Then τ induces an isomorphism between the kernels and cokernels, since these are supported on U_1, U_2. In brief the operators $P_1^{(1)}, P_1^{(2)}$ do indeed depend only on U_1, U_2. Then we have:

$$\operatorname{index} L_1' - \operatorname{index} L_1 = \operatorname{index} P_1^{(1)} = \operatorname{index} P_1^{(2)} = \operatorname{index} L_2' - \operatorname{index} L_2,$$

and the excision property is proved.

7.1.2 *Alternative proof*

We have seen that the excision principle has a rather formal proof, once one is able to appeal to the machinery of pseudo-differential operators. We will now outline another proof, involving more analytical arguments but staying in the realm of partial differential operators. The main idea is to replace the homotopy (7.1.6) of the symbols by a deformation of the lower-order terms under which the kernel and cokernel become approximately localized on U. For simplicity we suppose that L and L' are first-order differential operators.

The first step is again to introduce the 'difference' operator $D = L' \oplus (-L^*)$. To simplify our notation we regard the bundle isomorphisms α and β over V as identities, so over V the operator D has the form:

$$D = \begin{pmatrix} 0 & L \\ -L^* & 0 \end{pmatrix}.$$

In particular for sections f, g of $\xi \oplus \eta$ at least one of which is supported on V, D behaves as a skew adjoint operator:

$$\langle Df, g\rangle = -\langle f, Dg\rangle. \tag{7.1.7}$$

We now choose a pair of cut-off functions, ψ, ϕ, taking values in $[0, 1]$, with ϕ equal to 0 outside U and to 1 outside V; ψ equal to 1 outside U and 0 outside V; and in addition with $\psi = 1$ on the support of $\nabla\phi$ (see Fig. 11). Then for real $u > 0$ let D_u be the operator:

$$D_u = D + u\psi\mathbf{1}: \Gamma(\xi \oplus \eta) \to \Gamma(\xi' \oplus \eta') \tag{7.1.8}$$

where $\mathbf{1}$ denotes the identity operator over V defined by α and β. This definition makes sense since ψ is supported in V. We know that the index of D_u is independent of u. On the other hand we shall see that when u is large the kernels of D_u and D_u^* are concentrated over U. The main step in the argument

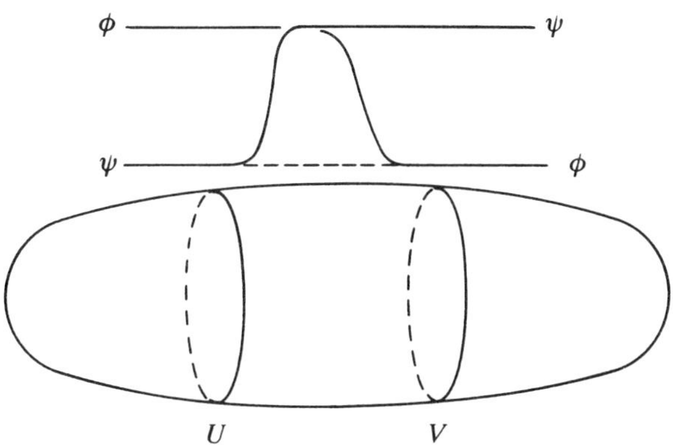

Fig. 11

is contained in the following lemma. To simplify notation we will write E for $\xi \oplus \eta$ and E' for $\xi' \oplus \eta'$, so E and E' are canonically isomorphic over V.

Lemma (7.1.9). *There is a constant c, independent of u, so that for any $\lambda > 0$ and section f of E over Z with*

$$\| D_u f \| \leq \lambda \| f \|$$

we have

$$\| \psi f \|^2 \leq \frac{c + \lambda}{u} \| f \|^2.$$

Proof. Put $g = D_u f$, so $Df = g - u\psi f$. Take the inner product with ψf to get

$$\langle \psi f, Df \rangle = \langle \psi f, g \rangle - u \| \psi f \|^2.$$

Now $\langle \psi f, Df \rangle = -\langle D(\psi f), f \rangle$ by the skew adjoint property (7.1.7) (since ψf is supported in V), but on the other hand

$$D(\psi f) = \psi\, Df + \nabla \psi * f$$

say, where $*$ is the algebraic operation defined by the symbol of D. So $-\langle \psi f, Df \rangle = \langle f, D(\psi f) \rangle = \langle \psi f, Df \rangle + \langle f, \nabla \psi * f \rangle$, and hence $|\langle \psi f, Df \rangle| \leq c \| f \|^2$, for a constant c depending only on ψ and the symbol of D. Then, substituting back into our first formula,

$$\begin{aligned} u \| \psi f \|^2 = \langle g, \psi f \rangle - \langle \psi f, Df \rangle &\leq \| g \| \, \| \psi f \| + c \| f \|^2 \\ &\leq (\lambda + c) \| f \|^2, \end{aligned}$$

as required.

Now suppose we have two such set-ups as above, over manifolds Z_1, Z_2, and an isomorphism τ matching up the data over U_1 and U_2. We define maps

$$\Gamma(E_1) \underset{\tau_2}{\overset{\tau_1}{\rightleftarrows}} \Gamma(E_2)$$

by $\tau_1(f) = \tau(\phi_1 f), \tau_2(f) = \tau^{-1}(\phi_2 f)$, with similar maps, which we also denote by τ_i, on the sections of the bundles E_i'. The idea is that the maps τ_i approximately intertwine the operators over the two manifolds. Precisely, we have the 'commutator formula'

$$(D_{2,u}\tau_1 - \tau_1 D_{1,u}) f = (D_2 \tau_1 - \tau_1 D_1) f = \sigma\{(\nabla \phi_1) * f\},$$

so

$$\| (D_{2,u}\tau_1 - \tau_1 D_{1,u}) f \| = \| \nabla \phi_1 * f \| \leq c \| \psi_1 f \|. \tag{7.1.10}$$

(In this section we will use the convention that c denotes a constant, which may change from line to line, but which is always independent of u.) We have estimates of the same kind as (7.1.9) for each of the four operators— $D_{1,u}, D_{2,u}, D^*_{1,u}, D^*_{2,u}$, and so we get bounds like (7.1.10) for the other commutators between these differential operators and the 'transfer' maps τ_i.

We cannot expect that the τ_i will match up the kernels of the operators exactly, so we compose them with L^2 projection to obtain maps between the kernels. For clarity we state the basic fact we use as a separate lemma.

Lemma (7.1.11). *Suppose B is a linear differential operator between sections of Euclidean vector bundles F_1, F_2 over a compact Riemannian manifold. If $BB^* \geq \mu$, for some $\mu > 0$, and $H \subset \Gamma(F_1)$ is a subspace such that $\|B(h)\| < \mu^{1/2}\|h\|$ for all non-zero h in H, then L^2 projection gives an injective map from H to* $\ker B$.

Proof. The proof is elementary: the projection of a (non-zero) element h to $\ker B$ is given by

$$\pi(h) = h - B^*(BB^*)^{-1}Bh.$$

Now
$$\begin{aligned}\|B^*(BB^*)^{-1}Bh\|^2 &= \langle B^*BB^*)^{-1}Bh, B^*(BB^*)^{-1}Bh\rangle \\ &= \langle (BB^*)^{-1}Bh, Bh\rangle \\ &\leq \mu^{-1}\|Bh\|^2 < \|h\|,\end{aligned}$$
so $\pi(h)$ is non-zero.

We now apply this idea to our problem. To simplify the exposition we suppose first that there is a constant $\mu > 0$ such that $D_{1,u}D_{1,u}^* \geq \mu$ for $i = 1, 2$ and all sufficiently large u. The proof is then very simple; we first observe that $D_{2,u}D_{2,u}^* \geq \frac{1}{2}\mu$ for all large enough u (here we can replace $\frac{1}{2}\mu$ by any number less than μ). Indeed if f is a section, normalized so that $\|f\| = 1$, with

$$\|D_{2,u}D_{2,u}^*f\| < \tfrac{1}{2}\mu,$$

then $\|D_{2,u}^*f\| < (\frac{1}{2}\mu)^{1/2}$ and, by the analogue of (7.1.8) for $D_{2,u}^*$ we have $\|\psi_2 f\| \leq C/u$. Then, by the analogue of (7.1.10), we have

$$\|D_{1,u}^*\tau_2 f\| \leq (C/u) + \|D_{2,u}^*f\| < (C/u) + (\tfrac{1}{2}\mu)^{1/2}.$$

On the other hand, the norm of $\tau_2 f$ is plainly close to 1 for large u since $\|\phi_2 f - f\| \leq \|\psi_2 f\|$. This means that, when u is large,

$$\|D_{1,u}^*\tau_2 f\| < \mu^{1/2}\|f\|,$$

contrary to our hypothesis on the existence of f. So we have $D_{2,u}^*D_{2,u} \geq \frac{1}{2}\mu$.

In particular we have, under our assumption, that the kernels of $D_{i,u}^*$ are both zero. To prove that the indices are the same we construct an isomorphism Π_1 from $\ker D_{1,u}$ to $\ker D_{2,u}$. This is defined by

$$\Pi_1 f = \pi(\tau_1 f), \tag{7.1.12}$$

where π is L^2 projection. We have

$$\|D_{2,u}(\tau_1 f)\| \leq c\|\psi_1 f\| \leq (c/u)\|f\|,$$

so Π_1 is injective for large u, by Lemma (7.1.11), applied to $B = D_{2,u}$. It follows that the dimension of ker $D_{2,u}$ is at least as large as that of ker $D_{1,u}$. By symmetry these dimensions must in fact be equal, and Π_1 is an isomorphism. (One can avoid this dimension argument by defining a map Π_2: ker $D_{2,u} \to$ ker $D_{1,u}$ in the same way as Π_1, and then using the estimates above to check that the composite $\Pi_1 \Pi_2$ is close to the identity.)

We will now remove our assumption on the uniform lower bound for $D_{1,u}D_{1,u}^*$ for large u from the argument. We do this by a 'stabilisation' construction, much as was used to define the index of a family of operators in Section 5.1.3. For a fixed u we can choose a map $S_1: \mathbb{R}^N \to \Gamma(E_1')$ so that $D_{1,u} \oplus S_1$ is surjective. Fix $\mu > 0$ arbitrarily and suppose we choose such a map so that, with respect to a Euclidean metric on $\mathbb{R}^N$:

(i) $$(D_{1,u} \oplus S_1)(D_{1,u} \oplus S_1)^* \geq \mu,$$

i.e.

$$\| D_{1,u}^* f \|^2 + \| S_1^* f \|^2 \geq \mu \| f \|^2$$

and

(ii) $$\| D_{1,u}^* S_1 v \|^2 \leq 2\mu |v|^2.$$

These two properties can be satisfied by, for example, mapping basis vectors of $\mathbb{R}^N$ to a complete set of orthonormal eigenfunctions for $D_{1,u}D_{1,u}^*$ belonging to eigenvalues less than or equal to μ, and taking the metric on $\mathbb{R}^N$ induced by this map from the L^2 metric.

The idea now is that we can carry out the previous argument with the stabilized operator $D_{1,u} \oplus S_1$. We define $S_2: \mathbb{R}^N \to \Gamma(E_2')$ by $S_2(v) = \tau_1(S_1 v)$, so $S_2^* = S_1^* \tau_2$. Then we show, much as before, that $(D_{2,u} \oplus S_2) \times (D_{2,u} \oplus S_2)^* \geq \frac{1}{2}\mu$ for $u \geq u_0$, say. The key point is that u_0 depends only on μ and the constants arising from the cut-off functions; it does not depend on N and the choice of S. We have then to show that for (f, v) in the kernel of $D_{1,u} \oplus S_1$ and for large enough μ we have

$$\| (D_{2,u} + S_2)(\tau_1 f, v) \|^2 < \tfrac{1}{2}\mu \| (f, v) \|^2;$$

then we can appeal again to Lemma (7.1.11) to see that the operator Π_1 defined by

$$\Pi_1(f, v) = (\pi \tau_1 f, v)$$

defines an isomorphism between the kernels of $D_{i,u} \oplus S_i$.

Now if $(D_{1,u} \oplus S_1)(f, v) = 0$ we have

$$\| D_u f \|^2 = \langle f, D_u^* D_u f \rangle = -\langle f, D_u^* S_1 v \rangle \leq \| f \| \, \| D_u^* S_1 v \|$$
$$\leq (2\mu)^{1/2} \| f \| \, |v|,$$

by the property (ii) of S_1. Then Lemma (7.1.8) gives

$$\| \psi_2 f \|^2 \leq u^{-1}(c \| f \|^2 + (2\mu)^{1/2} \| f \| \, |v|) \leq c u^{-1} \| (f, v) \|^2.$$

Now

$$(D_{2,u} \oplus S_2)(f, v) = \tau_1(D_{1,u} \oplus S_1)(f, v) + \tau_1((\nabla \phi_1) * f).$$

The first term on the right is zero so, just as before,

$$\|(D_{2,u} \oplus S_2)(f, v)\|^2 \le c\|\psi_1 f\|^2 \le cu^{-1}\|(f, v)\|^2,$$

and the required property holds for large u.

To obtain the excision formula we use the canonical exact sequences

$$0 \longrightarrow \ker D_{i,u} \longrightarrow \ker(D_{i,u} \oplus S_i) \longrightarrow \mathbb{R}^N \longrightarrow \operatorname{coker} D_{i,u} \longrightarrow 0 \tag{7.1.13}$$

(cf. Section 5.2.1) to deduce that

$$\operatorname{ind} D_{i,u} = \dim \ker(D_{i,u} \oplus S_i) + N,$$

and hence that the two indices are indeed equal.

7.1.3 Excision for families

As we have seen in Section 5.1.3, a family of elliptic operators parametrized by a compact space T has an index which is a virtual bundle over T. There is a version of the excision principle for the indices of families over a manifold $Z = U \cup V$. If two families agree over V then the difference of their virtual index bundles depends only on the data over U. The precise formulation of this generalization, and its proof by either of the approaches discussed above, is a straightforward extension of the discussion of the numerical index of a single operator. Using the excision principle it is easy to verify, on an *ad hoc* basis, the applications of the Atiyah–Singer Index theorem for families, which we have used in Chapter 3 and Chapter 5 ((3.2.16) and (5.2.5)).

We will now illustrate these ideas by giving another description of a line bundle which represents the cohomology classes $\mu(\alpha)$, defined in Chapter 5, over the space of connections on a four-manifold X, where α is a class in $H_2(X)$. In Sections 7.1.4 and 7.1.5 we will use this description to construct line bundles over the compactified moduli spaces.

For simplicity we assume that X is a spin manifold and that α is divisible by two in the homology group, so there is a line bundle L over X with $c_1(L^2)$ the Poincaré dual of α (both of these assumptions can be removed). Our construction uses the four-dimensional Dirac operator, rather than the Dirac operators over two-dimensional surfaces of Section 5.2. We introduce some notation: fix a connection ω on L and for any connection A on a bundle E over X let $A + \omega$ be the induced connection on $E \otimes L$, and $A - \omega$ be the induced connection on $E \otimes L^{-1}$. Let $\Lambda(A + \omega)$ be the determinant line

$$\Lambda(A + \omega) = \Lambda^{\max}(\ker D_{A+\omega})^* \otimes \Lambda^{\max}(\ker D^*_{A+\omega}), \tag{7.1.14}$$

associated with the coupled Dirac operator on $E \otimes L$, and similarly we define $\Lambda(A - \omega)$ to be the determinant line of the Dirac operator on $E \otimes L^{-1}$. For brevity we will often use additive notation in this section, so that, for example, if Λ_1, Λ_2 are two one-dimensional vector spaces we write $\Lambda_1 - \Lambda_2$ for the one-dimensional space $\Lambda_1 \otimes \Lambda_2^*$.

Let $\mathscr{L}_j$ be the line bundle over the space of irreducible $SU(2)$ connections $\mathscr{B}_j^*$, with Chern class j, having fibres

$$\mathscr{L}_{j,A} = \Lambda(A + \omega) - \Lambda(A - \omega). \tag{7.1.15}$$

(Here, just as in Chapter 5, we have to check that this definition does descend to the quotient space, i.e. we have to check the action of isotropy groups Γ_A on the fibres. The argument is essentially the same as in Section 5.2.1.)

Proposition (7.1.16). *The first Chern class of the line bundle $\mathscr{L}_j$ is $\mu(\alpha)$.*

To prove this we choose a surface $\Sigma \subset X$ representing α and a section s of L^2 cutting out Σ. We can regard this as giving a trivialization of L^2 outside a tubular neighbourhood N of Σ. By the general homotopy invariance of the index, the topological type of the determinant line bundle $\mathscr{L}_j$ is independent of the connection ω on L. So we can choose ω to be flat outside N, compatible with the trivialization of L^2. Then for any connection A on a bundle E over X the coupled Dirac operators $D_{A+\omega}$, $D_{A-\omega}$ are isomorphic outside N, intertwined by a bundle map $\sigma = 1 \otimes s$, covering the identity over $X \backslash N$. So we are in just the position envisaged in Section 7.1.1. We can find a family of operators P_t, $(t \in [0, 1])$ such that

$$\ker P_0 = \ker D_{A+\omega} \oplus \ker D_{A-\omega}^*$$

$$\ker P_0^* = \ker D_{A+\omega}^* \oplus \ker D_{A-\omega}$$

and with P_1 equal to the identity outside N. All the choices can be made canonically, so can be carried out in a family, as A varies. Thus we get a line bundle $\det \operatorname{ind} P$ over the product $\mathscr{B}_{j,X}^* \times [0, 1]$. Over $\mathscr{B}_j^* \times \{0\}$ this line bundle is canonically identified with $\mathscr{L}_j$, and over $\mathscr{B}_j^* \times \{1\}$ with the determinant line of a family of operators, $P_{1,A}$ say, equal to the identity outside N. By a standard argument we get an isomorphism, not canonical, between $\mathscr{L}_j$ and the line bundle $\det \operatorname{ind} P_1$. The latter line bundle is however formed from operators $P_{1,A}$, $P_{1,A}^*$ constructed canonically from the connection A and with kernels supported in N. Thus the line $\det \operatorname{ind} P_{1,A}$ depends only on the restriction of A to N. More precisely, the excision principle tells us that, over any family of connections which are irreducible on N, $\mathscr{L}_j$ is isomorphic to a line bundle pulled back from the space $\mathscr{B}_N^*$ of connections over N by the restriction map.

It is now straightforward to check that $\mathscr{L}_j$ has the correct Chern class. One can copy the argument of Section 5.2.1 to see that it suffices to check the degree of $\mathscr{L}_j$ over a standard generator for the homology of $\mathscr{B}_N^*$, and then make a direct calculation in a model case. In fact one can take the model situation when Σ is a complex curve in a complex surface X and connections in the family are compatible with holomorphic structures. Then in this case one can use a canonical isomorphism between $\mathscr{L}_j$ and the determinant line bundle obtained by restriction to Σ; see Section 10.1.3.

7.1.4 Line bundles over the compactified moduli space

Suppose X is an oriented Riemannian four-manifold and the only reducible ASD solution over X is the trivial connection. Then restricting the line bundles associated in Section 7.1.3 with a line bundle L over X, we get line bundles $\mathscr{L}_j$ over the moduli spaces M_j. Now consider the symmetric product $s^l(X)$. This is obtained as a quotient of the l-fold product of X with itself under the action of the permutation group Σ_l. Let $\Pi(L)$ be the line bundle over the product

$$\Pi(L) = \pi_1^*(L) \otimes \ldots \otimes \pi_l^*(L),$$

where $\pi_i : X \times \ldots \times X \to X$ are the projection maps. There is an obvious lift of the action of Σ_l to $\Pi(L)$, and the isotropy groups of points in the product act trivially on the fibres, so we get a quotient line bundle over $s^l(X)$. We denote this line bundle by $s^l(L)$.

Theorem (7.1.17). *There is a line bundle $\bar{\mathscr{L}}$ over the compactified moduli space $\bar{M}_k$ such that the restriction of $\bar{\mathscr{L}}$ to the stratum $\bar{M}_k \cap (M_j \times s^{k-j}(X))$ is isomorphic to $\mathscr{L}_j \otimes s^{k-j}(L)^2$.*

We begin by describing some of the main ideas of the proof of (7.1.17) informally, before moving on to the detailed constructions in Section 7.1.5. Thus we consider an ASD connection A with Chern class j and a point $[A']$ of M_k close to $([A], x_1, \ldots, x_l)$, where $k = j + l$. Let us see first what the ordinary excision property gives in this situation.

Under a suitable bundle isomorphism the connection A' is close to A outside small balls $B_r(x_i)$ about the x_i. So we can deform the connection A' slightly to be equal to A outside these balls. We can then apply the excision construction to obtain isomorphisms

$$\begin{aligned} s_+ : \Lambda(A' + \omega) &\longrightarrow \Lambda(A + \omega) + \Lambda_+, \\ s_- : \Lambda(A' - \omega) &\longrightarrow \Lambda(A - \omega) + \Lambda_-, \end{aligned} \qquad (7.1.18)$$

where Λ_+, Λ_- are determinant lines formed from the indices of pseudo-differential operators P_+, P_-, equal to the identity outside the $B_r(x_i)$. They can thus be expressed (in our additive notation) as sums:

$$\Lambda_+ = \Lambda_+^{(1)} + \ldots + \Lambda_+^{(p)}, \quad \Lambda_- = \Lambda_-^{(1)} + \ldots + \Lambda_-^{(p)}, \qquad (7.1.19)$$

where now we write the multiset $(x_1, \ldots, x_l)$ as a collection of points $x_1, \ldots, x_p$ with multiplicities $n_1, \ldots, n_p$. The isomorphisms (7.1.18) can be extended to any family of connections $[A']$, i.e. to the intersection of the moduli space M_k with a neighbourhood $\bar{N}$ in $\bar{M}_k$ of the ideal ASD connection $([A], x_1, \ldots, x_l)$. (To do this we have to fix a suitable bundle isomorphism away from the points x_i, as in (5.3.7).) Moreover we can deform the connection on L to be flat near the points x_i. Then we can choose the operators P_+, P_- to be equal, in a fixed local trivialization of L. Thus, with this trivialization of L, we get isomorphisms between the lines $\Lambda_+^{(i)}$ and $\Lambda_-^{(i)}$. Composing with the isomorphisms above we deduce then that *$\mathscr{L}_k$ is trivial over $\bar{N} \cap M_k$*. This is certainly a prerequisite for the extension of $\mathscr{L}_k$ over the compactification, but to obtain this extension one needs to prove more. First we should see how the line bundles $s^l(L)$ enter the picture. This is easily done. If V and U are vector spaces and U is one-dimensional, there is a natural isomorphism:

$$\Lambda^d(V \otimes U) = \Lambda^d(V) \otimes U^{\otimes d}. \tag{7.1.20}$$

Since the operators P_+, P_- are equal to the identity outside the disjoint balls $B_r(x_i)$ they can be viewed as collections of operators $P_+^{(i)}$, $P_-^{(i)}$ acting over the individual balls. So $\Lambda_+^{(i)}$, $\Lambda_-^{(i)}$ are the determinant lines of $P_+^{(i)}$, $P_-^{(i)}$. Now the isomorphism between the lines $\Lambda_+^{(i)}$ and $\Lambda_-^{(i)}$ used above depends on a choice of local trivialization of L. Given a flat connection this is provided by a framing for the fibre L_{x_i}, so there are natural isomorphisms

$$\ker P_+^{(i)} = (\ker P_-^{(i)}) \otimes L_{x_i}^2, \tag{7.1.21}$$

and similarly for the formal adjoint operators.

Now we recall that the index of the Dirac operator coupled to an $SU(2)$ bundle on S^4 is minus the first Chern class (cf. Chapter 3). It follows easily that the numerical index of the operator $P_-^{(i)}$ is just the relative Chern class n_i, the multiplicity of the point in our multiset. So, applying (7.1.20) to the kernel and cokernel, we obtain

$$\Lambda_+^{(i)} = \Lambda_-^{(i)} + 2n_i L_{x_i}$$

and so finally we get a natural isomorphism:

$$\det \operatorname{ind} P_+ - \det \operatorname{ind} P_- = s^l(L)^2_{(x_1, \ldots, x_l)}.$$

Of course we can make the same construction on the other strata in the compactified space separately. To sum up then, the excision theory gives us

Proposition (7.1.22). *There is a neighbourhood $\bar{N}$ of $([A], x_1, \ldots, x_l)$ in $\bar{M}_k$ and for each m there is a family of operators P_t $(t \in [0, 1])$ parametrized by the stratum $N_{(m)} = \bar{N} \cap (M_m \times s^{k-m}(X))$, such that the determinant line bundle of P_0 is naturally isomorphic to $\mathscr{L}_m \otimes s^{k-m}(L)^2$ and the determinant line bundle of P_1 is naturally isomorphic to the fixed line $\mathscr{L}_{k-l,A} \otimes L_{x_1}^2 \otimes \ldots \otimes L_{x_l}^2$.*

While it is very suggestive, this falls short of a proof of (7.1.17). We know that the determinant lines of the operators P_0 and P_1 in (7.1.22) are isomorphic. If we pick such an isomorphism we get a candidate for a local trivialization of the line bundle $\bar{\mathscr{L}}$ (with fibres over the individual strata as prescribed in (7.1.17)) over the neighbourhood $\bar{N}$. However if we do not restrict the choice of isomorphism between the determinant lines of P_0 and P_1 in some way, there is no reason why these 'trivializations' need be compatible on the overlap between two such neighbourhoods in $\bar{M}_k$. So to prove (7.1.17) we introduce some more geometric input, specifying the local trivializations more tightly in such a way that they match up on the overlaps. This is more-or-less equivalent to specifying a suitable connection in the determinant line bundle of P over $N_{(m)} \times [0, 1]$, which would give an isomorphism between the line bundle over the ends by parallel transport. However, we shall in fact proceed differently, using another excision mechanism.

7.1.5 Applications of the Weitzenböck formula

In our second proof of the basic excision property (7.1.2) we deformed the differential operators by a large lower-order term to 'localize' the kernel. In this section we will carry out a rather similar analysis for the coupled Dirac operator. In place of the artificial deformations used before we shall now exploit the natural localization mechanism, where the parameter is the 'concentration' of an ASD connection—the 'distance to the boundary' in the compactification. The analysis will be based on the Weitzenböck formula for the spinor Laplacian:

$$D_A^* D_A = \nabla_A^* \nabla_A + F_A^+ + S, \tag{7.1.23}$$

(cf. Section 3.1.1).

Although we will ultimately be interested in the twisted connections $A + \omega$, $A - \omega$ we will, for clarity of exposition, begin by considering the ASD connections themselves. Thus we want to compare directly the kernels of the Dirac operators coupled to A and A', where $[A']$ is sufficiently close to a point $([A], x_1, \dots, x_l)$ in the compactified moduli space, as above. To simplify notation we suppose that all the x_i are equal, so we have a single point x of multiplicity l. We shall see that, roughly speaking, the kernels differ by a contribution to $\ker D_A^*$, from sections localized near x.

For r small (to be chosen later, depending only on A) we let $B = B_x(r)$ be the r-ball about x in X and Ω be the annulus of radii $r, r^{1/2}$. We suppose that, as in (5.3.7), we have fixed a bundle isomorphism so that the restrictions of A and A' to $X \backslash B$ can be regarded as connections on the same bundle, and we write $A' = A + a$, where a is small. For any connection A we let $\mu(A)$ denote the first eigenvalue of the Laplacian $D_A^* D_A$.

Lemma (7.1.24). *$\mu(A')$ tends to $\mu(A)$ as $[A']$ converges to $([A], x, \dots, x)$ in the compactified moduli space.*

Proof. This follows from the Weitzenböck formula. Since A' is ASD the contribution from the bundle curvature vanishes, so if s is an eigenfunction of $D_{A'}^* D_{A'}$ belonging to the first eigenvalue μ we have, integrating the Weitzenböck formula,

$$\| \nabla_{A'} s \|_{L^2}^2 \leq (\mu + \sigma) \| s \|_{L^2}^2,$$

where σ is the supremum of the scalar curvature over X. Applying the Sobolev embedding theorem, as in (6.2.19), we get

$$\|s\|_{L^4}^2 \leq (C_1\mu + C_2)\|s\|_{L^2}^2,$$

for constants C_i depending only on X. We choose a cut-off function ψ, equal to 1 away from x and with $d\psi$ supported in Ω; write $\|d\psi\|_{L^4} = \varepsilon(r)$. We can choose ψ so that ε tends to 0 with r (see the discussion in Section 7.2.2). We now apply the operator D_A to ψs, extending the section by zero near x. This is the same kind of 'transport' procedure as used in Section 7.1.2, though we are now using a more economical notation, suppressing the maps used to identify the different bundles over the open set $X \backslash B$. We have then

$$\|D_A(\psi s)\|_{L^2} \leq \|D_A(s)\|_{L^2} + \|d\psi\|_{L^4}\|s\|_{L^4} + \left\{\sup_{X\backslash B}|a|\right\}\|s\|_{L^2}.$$

This gives

$$\|D_A(\psi s)\|_{L^2} \leq C_3(\mu, r, a)\|s\|_{L^2},$$

where $C_3 = \varepsilon(r)(C_1\mu + C_2)^{1/2} + \mu^{1/2} + \sup|a|$. On the other hand the L^2 norm of ψs differs from that of s by at most

$$\|s\|_{L^4}(\operatorname{Vol} B(2r))^{1/4} \leq \text{Const. } (C_1\mu + C_2)r.$$

By choosing r (and so ε) small and demanding that a be small (i.e. that A' is close to $(A, x, \ldots, x)$) we can obtain from these that $\|D_A s\|_{L^2} \leq (\mu^{1/2} + \delta)\|s\|_{L^2}$, for any preassigned δ. It follows that

$$\liminf_{A' \to (A, x, \ldots, x)} \mu(A') \geq \mu(A).$$

The proof of the reverse inequality, an upper bound on the lim sup, is similar but easier, using the same cut-off function to compare the eigenfunctions.

More generally the same argument shows that the entire spectrum of the spinor Laplacian of A' converges to that of A. The point is that the Weitzenböck formula prevents eigenfunctions of $D_{A'}^* D_{A'}$ becoming concentrated near x. Then the weak convergence is effectively as good as strong convergence, as far as the spectrum goes.

This technique yields a way to compare the eigenfunctions as well as the eigenvalues, in particular we can compare the kernels of the Dirac operators. Let us suppose for simplicity that $\ker D_A = 0$, that is $\mu(A) > 0$. Let u be an element of $\ker D_A^*$. We consider the section ψu as in the proof of (7.1.24). Then we can make $\|D_{A'}^*(\psi u)\|_{L^2}$ arbitrarily small, for suitable choices as above. The L^2 projection of ψu to the kernel of $D_{A'}^*$ is given by

$$p(\psi u) = (\psi u) - D_{A'}(D_{A'}^* D_{A'})^{-1} D_A^*(\psi u). \tag{7.1.25}$$

So

$$\begin{aligned}\|(\psi u) - p(\psi u)\|_{L^2} &= \langle D_{A'}^*(\psi u), (D_{A'}^* D_{A'})^{-1} D_{A'}^*(\psi u)\rangle \\ &\leq \mu(A')^{-1}\|D_{A'}^*(\psi u)\|_{L^2}. \end{aligned} \tag{7.1.26}$$

It follows that, if r is chosen small and A' is close to $(A, x, \ldots, x)$, the linear map

$$i\colon \ker D_A^* \longrightarrow \ker D_{A'}^*, \quad i(u) = p(\psi u), \tag{7.1.27}$$

is an injection.

We will now isolate more precisely the remaining part of the kernel of $D_{A'}^*$. Choose a local trivialization near x for the bundle carrying A, so we get local connection matrices for A and A' over Ω. We apply the cut-off construction of Section 4.4.3 to A', using a cut-off function supported in the twice-sized ball $2B$. We obtain a connection over X which is flat outside $2B$. We then use geodesic coordinates to identify a neighbourhood of x in X with a neighbourhood of 0 in $\mathbb{R}^4$ and (for convenience) compose with a rescaling in the Euclidean space to map B to the unit ball. Transporting the cut-off connection in this way we get a connection, A_0 say, over $\mathbb{R}^4$, flat outside the ball of radius 2. This connection A_0 is not ASD but it is approximately so, in the sense that we can make $\|F^+(A_0)\|_{L^2}$ as small as we like by choosing r small and demanding that A' be close to A away from x. Now regarding A_0 as a connection over the four-sphere we look at the kernel of the coupled Dirac operators. Equivalently, as in Chapter 3, we can look at the L^2 kernels of the Dirac operators D_{A_0}, $D_{A_0}^*$ over $\mathbb{R}^4$. Of course, the metric we get on the ball in $\mathbb{R}^4$, induced from the metric on X, is not flat; however it is nearly so and for simplicity we will ignore the distinction between the two metrics (i.e. we work in the case when X is flat near x; modifications to deal with the general case are straightforward). We apply the Weitzenböck formula again in the following simple lemma on approximately ASD connections over $\mathbb{R}^4$. Let C denote the Sobolev constant on $\mathbb{R}^4$, i.e.

$$\|f\|_{L^4} \le C\|\nabla f\|_{L^2}$$

for functions f in the L_1^2 completion of $C_c^\infty(\mathbb{R}^4)$.

Lemma (7.1.28). *Let A_0 be a connection over $\mathbb{R}^4$ with $\|F^+(A_0)\|_{L^2} \le \frac{1}{2}C^{-2}$, flat outside some ball. Then*

(i) *The kernel of D_{A_0}, acting on L^2 sections, is zero.*

(ii) *If $A_1 = A_0 + \alpha$ is another connection with $\|\alpha\|_{L^4} \le (\sqrt{2}C)^{-1}$, then* $\ker D_{A_1}$ *is zero and L^2 projection gives an isomorphism from* $\ker D_{A_0}^*$ *to* $\ker D_{A_1}^*$.

Part (i) follows from the Weitzenböck formula: if $D_{A_0}s = 0$ we have

$$\nabla_{A_0}^*\nabla_{A_0}s + F_{A_0}^+ s = 0.$$

The discussion of the asymptotics of harmonic spinors in Section 3.3 shows that integration by parts is valid and we get

$$\|\nabla_{A_0}s\|_{L^2}^2 = \|(F_{A_0}^+ s, s)\|_{L^2} \le \|F_{A_0}^+\|_{L^2}\|s\|_{L^4}^2.$$

On the other hand,

$$\|s\|_{L^4}^2 \le C^2\|\nabla_{A_0}s\|_{L^2}^2,$$

so if the L^2 norm of $F^+_{A_0}$ is less than C^{-2} the section s must be zero.

Part (ii) is similar. Let P be the right inverse for $D^*_{A_0}$ with $PD^*_{A_0} - 1$ the L^2 projection onto $\ker D^*_{A_0}$. Then it suffices to show that the L^2 operator norm of $P\alpha$ is less than 1. Now $P\alpha(s) = D_{A_0}t$, where $D^*_{A_0}D_{A_0}t = (\alpha s)$. Then

$$\|t\|^2_{L^4} \leq 2C^2\|D_{A_0}t\|^2_{L^2} = 2C^2\langle \alpha s, t\rangle \leq 2C^2\|\alpha\|_{L^4}\|s\|_{L^2}\|t\|_{L^4},$$

so

$$\|t\|_{L^4} \leq 2C^2\|\alpha\|_{L^4}\|s\|_{L^2}.$$

Substituting back into the formula we get

$$\|P\alpha t\|_{L^2} = \|D_{A_0}t\|_{L^2} \leq \sqrt{2}C\|\alpha\|_{L^4}\|s\|_{L^2},$$

so the operator norm of $P\alpha$ is less than 1 if the L^4 norm of α is less than $(\sqrt{2}C)^{-1}$.

We apply part (i) of this lemma to the connection A_0 obtained from A' and deduce that, for small r and A' close to A, the kernel of D_{A_0} is zero. The dimension of the kernel of $D^*_{A_0}$ must then equal the index, which is given by the Chern class of the bundle over S^4, and this is clearly the multiplicity l.

To sum up thus far, we associate with any connection A' sufficiently close to $(A, x, \ldots, x)$ an l-dimensional space

$$V(A') = \ker D^*_{A_0}. \tag{7.1.29}$$

We now relate this to the kernel of D^*_A in the obvious way. We introduce another parameter Z, which will be large, and consider the behaviour of an element s of $\ker D^*_{A_0}$ over $\mathbb{R}^4$ at distance $O(Z)$ from the origin. If χ is a cut-off function on $\mathbb{R}^4$, vanishing over the two-ball, we can regard χs as a section of the trivial bundle, so

$$\chi s = D^{-1}(D(\chi s))$$

where D is the Dirac operator of the flat connection and D^{-1} is an integral operator with a kernel decaying as $1/|x|^3$. We thus get an estimate at large distances Z,

$$|s| \leq \text{const.}\, Z^{-3}\|s\|_{L^2}, \tag{7.1.30}$$

with a constant independent of the connection A_0. We now introduce a cut-off function θ equal to 1 on the Z-ball and supported in the $2Z$-ball in $\mathbb{R}^4$. Then, reversing the identifications we made before, a rescaled version, $(\theta s)'$ say, of (θs) can be extended by zero over all of X. One finds then from (7.1.30) that $\|D_{A'}(\theta s)'\|_{L^2} \leq \eta\|(\theta s)'\|_{L^2}$, for a constant η which can be made as small as we please by choosing Z large (and our original parameters small). The key point throughout is that the choice of parameters depends only on the original connection A.

Under the same assumption that $\ker D_A = 0$, we can then apply (7.1.11) to the sections $(\theta s)'$. The L^2 projection p defines a linear injection j from $V(A')$ to $\ker D^*_{A'}$, mapping s to $p((\theta s)')$. It is straightforward to show that the images of i and j meet only in 0, using the fact that the sections $(\theta s)'$ and ψt are approximately orthogonal. So we have an injection $i \oplus j$ from $\ker D_A^* \oplus V(A')$ to $\ker D_{A'}^*$. Now we know, by the index theorem or more directly by the excision principle, that the dimensions of the two spaces are equal, so we have defined an isomorphism

$$i \oplus j : \ker D_A^* \oplus V(A') \longrightarrow \ker D_{A'}^*. \tag{7.1.31}$$

This isomorphism can be viewed as a decomposition of $\ker D_A^*$ into the 'global' and 'local' pieces. To sum up so far we have shown:

Proposition (7.1.32). *If $[A]$ is a point in M_j with $\ker D_A = 0$ then for sufficiently small neighbourhood $\bar{N}$ of $([A], x, \ldots, x)$ in $\bar{M}_k$ and for any $[A']$ in $\bar{N} \cap M_k$ the construction above gives an isomorphism,*

$$i \oplus j : \ker D_A \oplus V(A') \longrightarrow \ker D_{A'},$$

where $V(A')$ is a vector space of dimension $l = k - j$ which is determined by the restriction of A' to the $2r$-ball about x.

The importance of this construction is that the map $i \oplus j$ is essentially canonical. In addition to various choices of cut-off functions etc., which can easily be fixed, it depends on the bundle isomorphism used to compare A and A' over $X \setminus B$, which can be fixed as in (5.3.7). It is easy to remove the assumption that $\ker D_A$ is zero. In general we fix a number c less than $\mu(A)$ and work with the spaces $H_{A'}$, $\tilde{H}_{A'}$, spanned by the eigenfunctions of the spinor Laplacians belonging to eigenfunctions less than c. The determinant line of A' is naturally isomorphic to

$$\det H_{A'} - \det \tilde{H}_{A'}$$

(since the non-zero eigenspaces are matched up isomorphically by $D_{A'}^*$). Our arguments adapt easily to give isomorphisms

$$i \oplus j : \ker D_A^* \oplus V(A') \longrightarrow H_{A'},$$
$$\tilde{\imath} : \ker D_A \longrightarrow \tilde{H}_{A'}. \tag{7.1.33}$$

Taking determinants, we have constructed an explicit isomorphism

$$\lambda : \det \operatorname{ind} D_A^* + \det U(A') \longrightarrow \det \operatorname{ind} D_{A'}^*. \tag{7.1.34}$$

We now introduce the twisting by the line bundles L, L^{-1}. The whole discussion above goes through without change for the determinant lines $\Lambda(A \pm \omega)$. While ω need not be an ASD connection, all we really need in the arguments above is, for example, a uniform bound on F^+ over X. Following through the same procedure we construct isomorphisms

$$\lambda^{\pm} : \Lambda(A \pm \omega) + \det V(A' \pm \omega) \longrightarrow \Lambda(A' \pm \omega). \tag{7.1.35}$$

These are our substitutes for the isomorphisms (7.1.18) constructed abstractly using pseudo-differential operators. The next step is to compare $V(A' + \omega)$

and $V(A' - \omega)$. Use parallel transport by ω to identify the fibres of L over $2B$ with the line L_x. If we choose for the moment an isomorphism $\gamma: \mathbb{C} \to L_x$ then the connections A_0 and $(A + \omega)_0$ can be regarded as two connections on the same bundle, E_0 say, over $\mathbb{R}^4$. So we can write

$$(A + \omega)_0 = A_0 + \alpha$$

where α is supported in the ball of radius 2 in $\mathbb{R}^4$ and $|\alpha|$ is $O(r)$. In particular α can be made as small as we please in L^4. So we can apply the second part of (7.1.28) to show that L^2 projection over $\mathbb{R}^4$ gives an isomorphism c_γ from $V(A')$ to $V(A' + \omega)$. This transforms in an obvious way under a change in γ, i.e. $c_{(a\gamma)} = ac_\gamma$. Thus we have defined a canonical isomorphism, independent of γ, from $U(A') \otimes L_x$ to $U(A' + \omega)$. Similarly with L^{-1} and $-\omega$ in place of L and ω.

We combine these isomorphisms with λ^+, λ^- and use (7.1.20) to get finally the desired isomorphism:

$$\rho: \Lambda(A) + 2l\, L_x \longrightarrow \Lambda(A'). \tag{7.1.36}$$

Now as A' varies we get a trivialization of the line bundle $\mathscr{L}_k$ over $\bar{N} \cap M_k$. Similarly for the other strata $N_{(m)} = \bar{N} \cap \{M_m \times s^{k-m}(X)\}$ we proceed as follows: a point of $N_{(m)}$ consists of a pair $([A'], (y_1, \ldots, y_{k-m}))$, where A' is close to A away from x and $y_1, \ldots, y_{k-m}$ are points in X close to x. The construction above then applies to give an isomorphism of the fibre $\mathscr{L}_{m, A'}$ of $\mathscr{L}_m$ at $[A']$ with $\mathscr{L}_{j,A} \otimes L_x^{2(m-j)}$. We use parallel transport in L along radial geodesics to identify the fibres L_{y_i} with L_x, and thus get an isomorphism

$$\rho_m: \mathscr{L}_{j,A} + 2(k-j)L_x \longrightarrow \mathscr{L}_{j,A'} + 2\sum L_{y_i}.$$

Suppose we now define $\bar{\mathscr{L}}$, as a set, by taking the pieces prescribed in (7.1.17) over the individual strata. Then we have defined by the construction above a 'local trivialization' (of sets)

$$\rho_{\bar{N}}: \bar{N} \times \bar{\mathscr{L}}_{(A, x_1, \ldots, x_l)} \longrightarrow \bar{\mathscr{L}}|_{\bar{N}}.$$

By itself this is no more than we could do with the construction of Section 7.1.4. The key final point is contained in the assertion:

Proposition (7.1.37). *There is a unique topology on $\bar{\mathscr{L}}$ with respect to which the maps $\rho_{\bar{N}}$ are homeomorphisms.*

An equivalent and more explicit version of the statement is the assertion that the transition functions $\rho_{\bar{N}_1}^{-1}\rho_{\bar{N}_2}$ are continuous on the intersection of any pair of neighbourhoods $\bar{N}_1$, $\bar{N}_2$ in $\bar{M}_k$.

Written out in full the proof of (7.1.37) is rather long, although not at all difficult. We will be content to describe the salient points, which depend upon two properties of our construction.

The first property is the localization of the kernel. Let A_α be a sequence of connections over $\mathbb{R}^4$ with $\|F^+(A_\alpha)\|_{L^2} \le \frac{1}{2}C$ say, where C is the constant of

(7.1.28). Suppose A_α converges weakly to a limit $(A_\infty, z_1, \ldots, z_l)$, where A_∞ has Chern class zero. Then the index of the Dirac operator of A_∞ is zero, and hence this operator has zero kernel and cokernel. It follows from the argument of (7.1.24) that, as $\alpha \to \infty$, the kernel of $D^*_{A_\alpha}$ becomes localized in small balls about the z_j.

In our application we consider connections over $\mathbb{R}^4$ obtained from connections over X by cutting out over small balls. We can make different choices of cut-off functions and local trivializations, and these give different connections over $\mathbb{R}^4$. The localization principle shows that near the lower strata in $\bar{N}$ the kernels of the Dirac operators over $\mathbb{R}^4$ are essentially independent of the connection outside very small interior balls, and hence that asymptotically in the moduli space the constructions are independent of the choices made. This tells us, for example, that the transition function is continuous on an overlap of the form $\bar{N}_1 \cap \bar{N}_2$ where $\bar{N}_i$ are centred on points of M_k in the same stratum.

The second property concerns the composition of projection maps. Suppose again that A_α are approximately ASD solutions over $\mathbb{R}^4$, but now suppose that they converge to a limit $(A_\infty, z_1, \ldots, z_{l-m})$ where A_∞ has non-zero Chern class. Then, working on $\mathbb{R}^4$, we can split the kernel of $D^*_{A_\alpha}$ into a piece isomorphic to the kernel of $D^*_{A_\infty}$ and pieces localized around the z_v, using the same projection construction as in (7.1.31) above. Now suppose the A_α are obtained from a sequence of connections over X by cutting out a ball $B(r)$ about x. We have two ways of splitting up the kernel of the Dirac operators over X, either by applying the map $(i \oplus j)^{-1}$ and then splitting up the contribution from $B(r)$ as above, or in one step, by regarding the connections as being close to a point of the form $(A, y_1, \ldots, y_{l-m})$, and applying the construction on X to very small balls about the y_v. These two decompositions are not the same but they agree asymptotically in the moduli space. The point is that the projection maps onto the harmonic spinors over $\mathbb{R}^4$ and X are approximately equal on localized elements. This property gives the continuity of the transition functions on $\bar{N}_1 \cap \bar{N}_2$ where $\bar{N}_1$ and $\bar{N}_2$ are centred on points of different strata.

7.1.6 Orientations of moduli spaces

As a final application of the excision principle for linear operators we discuss the orientation of the Yang–Mills moduli spaces. We have seen in Chapter 5 that (at least over simply connected four-manifolds) the moduli spaces are orientable, and that an orientation is induced by a trivialization of a real determinant line bundle $\det \operatorname{ind} \delta$ over the space $\mathscr{B}$ of all gauge equivalence classes of connections. To fix the orientation, in the $SU(2)$ case, we consider a connection A which is flat outside a union of k disjoint balls in X, and has relative Chern class 1 over each ball. Thus we are effectively considering a point 'near to infinity' in the moduli space, although the ASD equations themselves are in fact quite immaterial for this discussion.

We apply the excision principle to compare the determinant line $\det\operatorname{ind}\delta_A$ of A with that of the trivial connection, θ_X say, on the trivial $SU(2)$ bundle over X. We get an isomorphism,

$$\det\operatorname{ind}\delta_A = \det\operatorname{ind}\delta_\theta \otimes \left\{ \bigotimes_{i=1}^{k} \lambda_i \right\}, \tag{7.1.38}$$

where λ_i is the determinant line of an operator equal to the identity outside a small neighbourhood of x_i. Thus λ_i is really independent of the manifold X; by transporting the connections to S^4 and using the same formula there we can identify λ_i with $\det\operatorname{ind}\delta_I \otimes \{\det\operatorname{ind}\delta_{\theta_{S^4}}\}^*$, where I is the standard instanton on S^4. This can in turn be viewed as the determinant of the tangent space to the 'framed' instanton moduli space (with a trivialization at infinity) and we recall that this moduli space is $\mathbb{R}^4 \times \{SO(3) \times \mathbb{R}^+\}$. This has a canonical orientation, so we get a corresponding orientation of λ_i. Thus we deduce that there is a natural isomorphism between the orientation classes of $\det\operatorname{ind}\delta_\theta$ and $\det\operatorname{ind}\delta_A$. More invariantly, we should think of $\det\operatorname{ind}\delta_A$ as being identified with

$$\det\operatorname{ind}\delta_\theta \otimes \left\{ \bigotimes_{i=1}^{k} \det(TX_{x_i} \oplus \mathbb{R} \oplus \Lambda^+_{x_i}) \right\},$$

but the orientation of X gives canonical orientations of all but the first term. Now the kernel and cokernel of δ_θ are formed from the tensor product of the three-dimensional Lie algebra $\mathfrak{su}(2)$ and the homology groups $H^1(X;\mathbb{R})$, $H^0(X;\mathbb{R}) \oplus H^+(X)$. Fixing a standard orientation on the Lie algebra and for H^0 we obtain:

Proposition (7.1.39). *For a compact oriented four-manifold X, an orientation Ω of the space $H^1(X) \oplus H^+(X)$ induces orientations $o_k(\Omega)$ for the determinant line* $\det\operatorname{ind}\delta$ *over the orbit space of $SU(2)$ connections $\mathcal{B}_{k,X}$. If $-\Omega$ is the opposite orientation then $o_k(-\Omega) = -o_k(\Omega)$.*

We use here the fact, mentioned but not proved in Chapter 5, that the determinant lines are orientable even when X is not simply connected.

A point to note here is that the statement as given depends strictly on a metric on X, used to define the positive subspace $H^+(X)$. However the notion of an orientation of $H^+(X)$ (or $H^1(X) \oplus H^+(X)$) is independent of the particular choice of this subspace. This follows from the fact that the subset of the Grasmannian consisting of maximal positive subspaces for the intersection form is contractible, so we can uniquely 'propagate' an orientation from one such subspace to any other.

Proposition (7.1.39) leads to an evaluation of an index of a real elliptic family. Let $f: X \to X$ be an orientation-preserving diffeomorphism. We form the associated fibration $\mathscr{X}_f \to S^1$ with fibre X, and choose a metric on $\mathscr{X}_f$. For any k we can construct an $SU(2)$-bundle $\mathscr{P}$ over $\mathscr{X}_f$ with $c_2 = k$ on the fibres. Choosing a connection on $\mathscr{P}$, and so on the restrictions to the fibres, we get a family of δ operators parametrized by the circle. The determinant line bundle of this family is a real line bundle over S^1, determined by its Stiefel–Whitney class $w_1 \in \{0, 1\}$. Thus we can attach a sign $\alpha_f = \pm 1$ to the diffeomorphism, and (7.1.39) plainly implies:

Corollary (7.1.40). *The sign α_f is $\alpha_f = \beta_f \gamma_f^+$ where β_f is the determinant of the induced action f^* on $H^1(X)$ and γ_f^+ is determined by the induced action of f on $H^2(X)$, via the action on the orientation of maximal positive subspaces for the intersection form.*

The product $\gamma_f^+ \gamma_f^-$ is known as the 'spinor norm' of the automorphism f^* of $(H^2(X), Q)$.

In Chapter 9 we will need the corresponding theory for $SO(3)$ bundles. This is slightly more complicated because we cannot push the discussion down to the trivial connection. The argument above shows that an orientation of the determinant line associated to a bundle V with $\kappa(V) = k$ and $w_2(V) = \alpha$ can be used to fix orientations of the corresponding lines for *all* bundles with $w_2 = \alpha$ (since these differ by the 'addition of small instantons'). Over a simply connected four-manifold we can choose an integral lift $\hat{\alpha}$ of α. Let L be the line bundle with $c_1(L) = \hat{\alpha}$ and put $V = L \oplus \underline{\mathbb{R}}$. For a reducible connection on V the determinant line is the product of a piece from the L factor, which has a canonical orientation from the complex structure on L, and a piece from the $\underline{\mathbb{R}}$ factor which is oriented by an orientation Ω of the cohomology of X, as in the $SU(2)$ case. We obtain then orientations $o_k(\Omega, \hat{\alpha})$ for all bundles with $w_2 = \alpha$. The tricky point is to determine the dependence of this on the integral lift $\hat{\alpha}$. The answer is this: if $\hat{\alpha}_1$, $\hat{\alpha}_2$ are two integral lifts we can write $\hat{\alpha}_1 - \hat{\alpha}_2 = 2\beta$ for some integral class β; then the orientations compare according to the parity of $\beta^2 \in H^4(X, \mathbb{Z}) = \mathbb{Z}$. This is quite easy to prove, and makes a good exercise, in the case when X is a Kähler manifold; for we can then compare each orientation with the canonical one induced by the complex structure on X. (For general X one reduces to this case by an excision argument.) Note in particular that if X is spin, the orientations of the $SO(3)$ moduli spaces depend only on Ω, as in the $SU(2)$ case.

7.2 Gluing anti-self-dual connections

Let X_1, X_2 be compact oriented Riemannian four-manifolds and A_1, A_2 be ASD connections on bundles E_1, E_2 over X_1, X_2 respectively, with the same structure group G. In this section we discuss the parametrization of ASD connections on a 'connected sum' bundle over the connected sum $X = X_1 \# X_2$, close to the A_i on each factor. We begin, in Section 7.2.1, by fixing some definitions and notation. In Section 7.2.2 we move quickly to one of the main steps, the construction of a family of solutions over X. To complete the picture one needs to show that the family constructed gives a complete model of an explicitly defined open set in the moduli space. For technical reasons this is rather more complicated; the proof is spread over Sections 7.2.4 to 7.2.6. While this section is rather long, and the analysis may be found tedious, the basic ideas we use are quite simple, and have the same general flavour as the arguments of Section 7.1: we obtain information about the ASD solutions over X by patching together data over the individual summands, and the key to the theory is an appreciation of the size of the error terms which these cut-offs introduce. As far as the rest of the book goes one of the main applications of these results is the description of neighbourhoods of points at infinity in the compactified ASD moduli spaces, an aspect which will be discussed in Chapter 8.

7.2.1 Preliminaries

We begin with the definition of the connected sum. Choose points x_i in X_i and suppose for simplicity that the metrics on X_i are flat in neighbourhoods of the x_i. (This assumption can easily be removed but is in fact permissible for the applications of the results here that we will make in Chapters 8 and 9.) Using these flat metrics we identify neighbourhoods of the points x_i in X_i with neighbourhoods of zero in the tangent spaces $(TX_i)_{x_i}$. Now let

$$\sigma\colon (TX_1)_{x_1} \longrightarrow (TX_2)_{x_2} \tag{7.2.1}$$

be an orientation-reversing linear isometry. For any real number $\lambda > 0$ we define

$$f_\lambda\colon (TX_1)_{x_1}\backslash 0 \longrightarrow (TX_2)_{x_2}\backslash 0$$

to be the 'inversion' map:

$$f_\lambda(\xi) = \frac{\lambda}{|\xi|^2}\,\sigma(\xi). \tag{7.2.2}$$

We introduce another parameter $N = \exp T > 1$, to be fixed later in the proof but such that $\lambda^{1/2} N \ll 1$. Let $\Omega_i \subset X_i$ be the annulus centred on x_i with inner radius $N^{-1}\lambda^{1/2}$ and outer radius $N\lambda^{1/2}$. The map f_λ induces a diffeomorphism from Ω_1 to Ω_2. We let $X_i' \subset X_i$ be the open set obtained by removing the $N^{-1}\lambda^{1/2}$ ball about x_i (the ball enclosed by Ω_i). Then, in the familiar way, we define the connected sum $X = X(\lambda)$ to be

$$X = X_1' \cup_{f_\lambda} X_2' \tag{7.2.3}$$

where the annuli Ω_i are identified by f_λ; see Fig. 12. Now f_λ is a *conformal* map and this means that X has a natural conformal structure, depending in general upon the parameters σ, λ (but independent of N). We shall often use a cover of X by slightly smaller open sets. We let X_i'' be the complement of the $\frac{1}{2}\lambda^{1/2}$ ball about x_i, so $X = X_1'' \cup X_2''$.

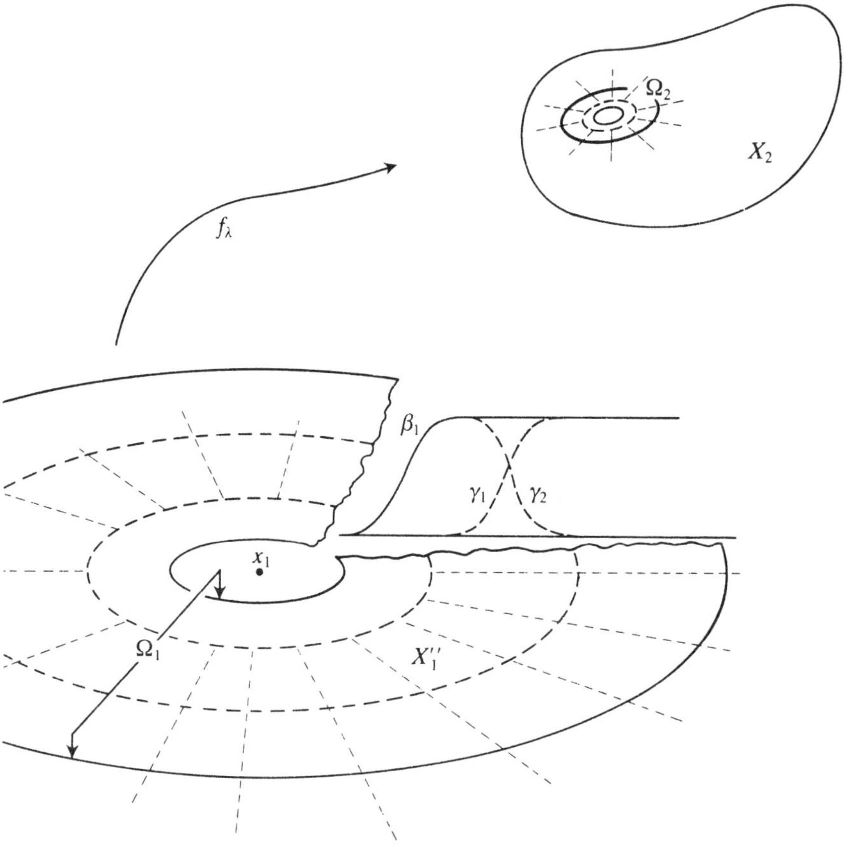

Fig. 12

There is another model for the connected sum which is often useful. This depends on the conformal equivalence

$$e\colon \mathbb{R} \times S^3 \longrightarrow \mathbb{R}^4\backslash\{0\} \tag{7.2.4}$$

given in 'polar coordinates' by $e(t, \omega) = e^t\omega$. Under this map the annulus Ω with radii $N\lambda^{1/2}$, $N^{-1}\lambda^{1/2}$ goes over to the tube

$$e^{-1}(\Omega) = (\tfrac{1}{2}\log\lambda - T, \tfrac{1}{2}\log\lambda + T) \times S^3. \tag{7.2.5}$$

Thus we can think of the connected sum as being formed by deleting the points x_i from X_i, regarding punctured neighbourhoods as half-cylinders and identifying the cylinders by a reflection.

We now turn to the bundles E_i over X_i and connections A_i. Our first move is to replace these by connections A_i' which are flat in neighbourhoods of the x_i. To do this we use the cutting-off construction of Section 4.4.3. We introduce another parameter $b \geq 4N\lambda^{1/2}$ and perform the cutting off over the annulus with radii $\frac{1}{2}b$ and b. We obtain connections A_i' which are flat over the annuli Ω_i and equal to A_i outside the b-balls. As we explained in Section 4.4.2, the construction depends on a choice of local trivialization for the bundles E_i, but it is easy to see that we can choose these trivializations and cut-off functions so that

$$|A_i - A_i'| \leq \text{const.}\, b, \quad |F(A_i')| \leq \text{const.} \tag{7.2.6}$$

and hence so that

$$\|F^+(A_i')\|_{L^2}, \quad \|A_i - A_i'\|_{L^4} \leq \text{const.}\, b^2, \tag{7.2.7}$$

for constants depending only on A_i. (The bounds (7.2.7) follow because $F^+(A_i')$ and $A_i - A_i'$ are supported in an annulus with volume $O(b^4)$, and A_i is ASD.)

Choose a G-isomorphism of the fibres:

$$\rho\colon (E_1)_{x_1} \longrightarrow (E_2)_{x_2}. \tag{7.2.8}$$

Using the flat structures A_i' we can spread this isomorphism out to give a bundle isomorphism g_ρ between the E_i over the annuli Ω_i, covering f_σ. We define a bundle $E(\rho)$ over X using this identification map. Moreover g_ρ respects the flat connections A_i' so we get an induced connection, $A'(\rho)$ say, on $E(\rho)$. The connections $A'(\rho)$, for different ρ, are not in general gauge equivalent (although the bundles $E(\rho)$ are obviously isomorphic). Let

$$\Gamma = \Gamma_{A_1} \times \Gamma_{A_2},$$

where Γ_{A_i} is the isotropy group of A_i over X_i. Define the space of 'gluing parameters' to be:

$$\mathrm{Gl} = \mathrm{Hom}_G((E_1)_{x_1}, (E_2)_{x_2}).$$

The group Γ acts on Gl in an obvious way, and we have:

Proposition (7.2.9). *The connections $A'(\rho_1)$, $A'(\rho_2)$ are gauge equivalent if and only if the parameters ρ_1, ρ_2 are in the same orbits of the action of Γ on* Gl.

So, for example, if A_1, A_2 are irreducible $SU(2)$ connections we get a family of connections $A'(\rho)$ parametrized by a copy of $SO(3)$. The proof of (7.2.9) is left as a simple exercise for the reader. To simplify our notation we will denote $A'(\rho)$ and $E(\rho)$ by A' and E when the gluing parameter ρ is determined by the context.

7.2.2 Constructing solutions

We will now construct a family of ASD connections on X, close to the connections $A'(\rho)$, once the parameter λ defining the conformal structure of the connected sum is small. While the ASD equations are non-linear, we will see that the root of the problem is the solution of the corresponding linearized equation with estimates on the solutions which are independent of λ. We begin with a simple fact about cut-off functions.

Lemma (7.2.10). *There is a constant K and for any N, λ a smooth function $\beta = \beta_{N,\lambda}$ on $\mathbb{R}^4$ with $\beta(x) = 1$ for $|x| \geq N\lambda^{1/2}$, $\beta(x) = 0$ for $|x| \leq N^{-1}\lambda^{1/2}$, and*

$$\|\nabla\beta\|_{L^4} \leq K(\log N)^{-3/4}.$$

To verify this one can just write down a formula for a suitable function. The picture is much clearer in the cylinder model. The key point is that the L^4 norm on 1-forms is conformally invariant (in four dimensions). So we can transform the problem to the cylinder, where we seek a function $\tilde{\beta}(t,\theta)$, equal to zero for $t \leq -\frac{1}{2}\log\lambda - T$ and to one for $t \geq -\frac{1}{2}\log\lambda + T$, where $T = \log N$. We take $\tilde{\beta}$ to be a function of t whose derivative is approximately $(2T)^{-1}$ over the cylinder of volume $8\pi^2 T$, then the L^4 norm of $\nabla\tilde{\beta}$ is approximately $K_0 = (2^{-1/4}\pi^{1/2})T^{-3/4}$, and any constant $K > K_0$ will do.

We will now move on to the core of the argument. Let us suppose for the moment that the cohomology groups $H^2_{A_i}$ are both zero. Thus there are right inverses

$$P_i : \Omega^+_{X_i}(\mathfrak{g}_{E_i}) \longrightarrow \Omega^1_{X_i}(\mathfrak{g}_{E_i}), \tag{7.2.11}$$

to the operators $d^+_{A_i}$. For example we could fix P_i by the condition that $d^*_A P_i \xi = 0$ for all ξ and $P_i\xi$ is orthogonal to the harmonic space, but the particular choice is not important. All we need now is the fact that P_i is a bounded operator over X_i between the Sobolev spaces L^2 and L^2_1. Combined with the Sobolev embedding theorem we get

$$\|P_i\xi\|_{L^4} \leq C_i\|\xi\|_{L^2}, \tag{7.2.12}$$

for some constants C_i. Notice that both the norms appearing in (7.2.12) are conformally invariant. We now fix cut-off functions β_i, γ_i on X_i, where β_i is obtained from the function $\beta_{N,\lambda}$ of (7.2.10) using the local Euclidean co-ordinates near x_i extended by 1 over the rest of X_i. The cut-off function γ_i is less critical; it should be equal to one at points of distance more than $2\lambda^{1/2}$ from x_i, say, and should be supported on the set X_i''. In particular, the support of the derivative $\nabla\gamma_i$ is contained in the region where $\beta_i = 1$, (see Fig. 12). (In fact for many purposes we could take γ_i to be the characteristic function of the complement of the $\lambda^{1/2}$-ball, but it is more convenient to stay with smooth functions.) Moreover we will suppose that γ_i depends on λ only up to a scale, i.e. has the form $f(d(-, x_i)/\lambda^{1/2})$. Then the L^4 norm of $d\gamma_i$ is independent of λ.

Now let Q_i be the operator defined by:

$$Q_i\xi = \beta_i P_i \gamma_i(\xi). \tag{7.2.13}$$

Lemma (7.2.14). *There are constants* $\varepsilon_i = \varepsilon_i(N, b)$, *with* $\varepsilon_i(N, b) \to 0$ *as* $N \to \infty$ *and* $b \to 0$, *such that for any* λ *with* $4N\lambda^{1/2} \leq b$ *and all* ξ *in* $\Omega^+_{X_i}(\mathfrak{g}_{E_i})$,

$$\| \gamma_i \xi - \mathrm{d}^+_{A_i'} Q_i \xi \|_{L^2} \leq \varepsilon_i(N, b) \| \xi \|_{L^2}.$$

Proof. We have, writing $A_i' = A_i + a_i$,

$$\begin{aligned} \mathrm{d}^+_{A_i'}(Q_i \xi) &= (\mathrm{d}^+_{A_i} + [a_i, -\,])(\beta_i P_i(\gamma_i \xi)) \\ &= \beta_i(\mathrm{d}^+_{A_i} P_i(\gamma_i \xi)) + (\nabla \beta_i) P_i(\gamma_i \xi) + [\beta_i a_i, P_i(\gamma_i \xi)]. \end{aligned}$$

The operator P_i is right-inverse to $\mathrm{d}^+_{A_i}$ and $\beta_i \gamma_i = \gamma_i$, since $\beta_i = 1$ on the support of γ_i. So the first term equals $\gamma_i \xi$. We have to estimate the L^2 norm of the other two terms. We have:

$$\begin{aligned} \| (\nabla \beta_i) P_i(\gamma_i \xi) \|_{L^2} &\leq \| \nabla \beta_i \|_{L^4} \| P_i(\gamma_i \xi) \|_{L^4}, \\ &\leq K(\log N)^{-3/4} . C_i \| \xi \|_{L^2}, \end{aligned}$$

and similarly the second term is bounded by

$$C_i \| a_i \|_{L^4} \| \xi \|_{L^4} \leq \text{const.}\, b^2 \| \xi \|_{L^4},$$

by (7.2.6). The result now follows, with $\varepsilon_i = \text{const.}\,(b + (\log N)^{-3/4})$.

We now transport these operators to the connected sum, and the bundle $E = E(\rho)$, for some fixed ρ. We regard X_i' as an open set in X; then for any ξ, $Q_i(\xi)$ is supported in X_i' and so can be regarded as an element of $\Omega^1_X(\mathfrak{g}_E)$. Similarly, we can regard β_i and γ_i as functions on X in an obvious way, extending by zero outside X_i'. Then we can interpret Q_i as an operator

$$Q_i \colon \Omega^+_X(\mathfrak{g}_E) \longrightarrow \Omega^1_X(\mathfrak{g}_E).$$

We may choose the functions γ_i so that

$$\gamma_1 + \gamma_2 = 1 \tag{7.2.15}$$

on X. We put

$$Q = Q_1 + Q_2 \colon \Omega^+_X(\mathfrak{g}_E) \longrightarrow \Omega^1_X(\mathfrak{g}_E). \tag{7.2.16}$$

Now since the L^4 norm on 1-forms and the L^2 norm on 2-forms are conformally invariant we can transfer the result above to X. We put

$$\mathrm{d}^+_{A'} Q = 1 + R,$$

so (7.2.14) and (7.2.15) give

$$\| R(\xi) \|_{L^2} \leq (\varepsilon_1(N, b) + \varepsilon_2(N, b)) \| \xi \|_{L^2}. \tag{7.2.17}$$

Proposition (7.2.18). *If* $H^2_{A_1}$ *and* $H^2_{A_2}$ *are both zero there are constants* C, N_0, b_0 *such that for* $N \geq N_0$, $b \leq b_0$ *and any* λ *with* $4N\lambda^{1/2} \leq b$ *there is a right inverse* P *to the operator* $\mathrm{d}^+_{A'}$ *over* X *with*

$$\| P \xi \|_{L^4} \leq C \| \xi \|_{L^2}.$$

To prove this we choose N_0 and b_0 so that $\varepsilon_i(N, b) \leq \frac{1}{3}$, say. Then the operator norm of R is at most $\frac{2}{3}$, so $1 + R$ is invertible (by the well-known series expansion), and the norm of the inverse is at most 3. Then we put

$$P = Q(1 + R)^{-1}. \tag{7.2.19}$$

Clearly the operator norm of Q_i, from L^2 to L^4, is at most C_i, so we can take $C = 3(C_1 + C_2)$.

Proposition (7.2.18) gives the desired uniform solution to the linearization of the ASD equation over X, with respect to the parameter λ defining the conformal structure, and we now move on to the non-linear problem. We assume the parameters are chosen to satisfy the conditions of (7.2.18). For fixed ρ we seek a solution $A' + a$ to the ASD equations, that is

$$d_{A'}^{+}a + (a \wedge a)^{+} = - F^{+}(A'). \tag{7.2.20}$$

We seek a solution in the form

$$a = P(\xi),$$

so the equation, for ξ in $\Omega_X^{+}(\mathfrak{g}_E)$, becomes

$$\xi + (P\xi \wedge P\xi)^{+} = - F^{+}(A'), \tag{7.2.21}$$

since P is a right inverse for $d_{A'}^{+}$. We write

$$q(\xi) = (P\xi \wedge P\xi)^{+},$$

so q is a quadratic function and, by Cauchy–Schwarz,

$$\| q(\xi_1) - q(\xi_2) \|_{L^2} \leq \sqrt{2} C^2 \| \xi_1 - \xi_2 \|_{L^2} \{ \| \xi_1 \|_{L^2} + \| \xi_2 \|_{L^2} \}. \tag{7.2.22}$$

Now we apply the following simple lemma.

Lemma (7.2.23). *Let $S: B \to B$ be a smooth map on a Banach space with $S(0) = 0$ and $\| S\xi_1 - S\xi_2 \| \leq k\{ \| \xi_1 \| + \| \xi_2 \| \} (\| \xi_1 - \xi_2 \|)$, for some $k > 0$ and all ξ_1, ξ_2 in B_1. Then for each η in B with $\| \eta \| < 1/(10k)$ there is a unique ξ with $\| \xi \| \leq 1/(5k)$ such that*

$$\xi + S(\xi) = \eta.$$

Note that the conditions on ξ imply that $\| \xi \| = \| \eta \| + O(\| \eta \|^2)$, in fact we have $\| \xi \| - \| \eta \| \leq (50/9)k \| \eta \|^2$. Of course the constants here are not optimal, nor particularly important, the key point is that they depend only on k.

The proof of (7.2.23) is a simple application of the contraction mapping principle. We write the equation as $\xi = T\xi$ where $T\xi = \eta - S(\xi)$, and find the solution as the limit

$$\xi = \lim_{n \to \infty} T^n(0).$$

This is the usual proof of the inverse mapping theorem in Banach spaces, which states that the equation can be solved for small enough η. The lemma extends the usual statement of the inverse mapping theorem to give bounds only depending on k.

We can now apply the lemma to our equation (7.2.21), with $S=q$, $\eta = -F^+(A')$ and $k = \sqrt{2}C^2$. We deduce that if $F^+_{A'}$ is small enough in L^2, relative to constants depending only on A_i, there is a unique small solution ξ to the equation. On the other hand we know by (7.2.7) that this condition can be achieved by making b small. So we obtain the following result.

Theorem (7.2.24). *If A_1 and A_2 are ASD connections over X_1, X_2 with $H^2_{A_1} = H^2_{A_2} = 0$, then for all small enough b and λ (with $b > 4N\lambda^{1/2}$ for some $N = N(A_1, A_2)$), and all gluing parameters ρ, there is an L^2_1 ASD connection $A(\rho) + a_\rho$ with $\|a_\rho\|_{L^4} \leq \text{const. } b^2$. Moreover a_ρ is the unique such solution which can be written in the form $P\xi_\rho$. If ρ_1, ρ_2 are in the same orbit under the Γ action on* Gl, *the corresponding ASD connections are gauge equivalent.*

The last statement here follows directly from (7.2.9).

We will now go on to the case when the $H^2_{A_i}$ do not vanish. We use a technique which should by now be familiar, both from the local models in Chapter 4 and, in a linear setting, from the 'stabilization' procedure used in Section 7.1 above.

In the general case, choose once and for all lifts

$$\sigma_i\colon H^2_{A_i} \longrightarrow \Omega^+_{X_i}(\mathfrak{g}_{E_i}),$$

so that the operator

$$d^+_{A_i} \oplus \sigma_i$$

is surjective. We can do this in such a way that the forms $\sigma_i(v)$ are supported in the complement of a small ball about x_i, and choose b always so small that $B_i(b)$ does not meet these supports. So in the notation above we have an obvious map:

$$\sigma = \sigma_1 + \sigma_2\colon H = H^2_{A_1} \oplus H^2_{A_2} \longrightarrow \Omega^+_X(\mathfrak{g}_E). \tag{7.2.25}$$

Now on X_i we have an operator P_i and a finite rank map

$$\pi_i\colon \Omega^+_{X_i}(\mathfrak{g}_{E_i}) \longrightarrow H^2_{A_i}$$

such that

$$\xi = d^+_{A_i} P_i \xi + \sigma_i \pi_i(\xi).$$

Over X we write, just as before,

$$Q\xi = \beta_1 P_1(\gamma_1 \xi) + \beta_2 P_2(\gamma_2 \xi).$$

When we compare $d^+_A Q\xi$ with ξ we obtain, in addition to the previous error terms, a new term:

$$\sigma_1 \pi_1(\gamma_1 \xi) + \sigma_2 \pi_2(\gamma_2 \xi).$$

If we define $\pi'\colon \Omega^+_X(\mathfrak{g}_E) \to H$ by $\pi'(\xi) = \pi_1(\gamma_1 \xi) + \pi_2(\gamma_2 \xi)$, we have

$$\|d^+_A Q\xi + \sigma\pi'(\xi) - \xi\|_{L^2} \leq \tfrac{2}{3}\|\xi\|_{L^2}$$

for suitable choices of parameters. It follows that we can invert $d_A^+ \oplus \sigma$, with a right inverse $P \oplus \pi$ say. Correspondingly we split up our non-linear equation into finite and infinite dimensional parts. We consider the equation for a pair (ξ, h) with h in H:

$$F^+(A' + P\xi) + \sigma(h) = 0, \tag{7.2.26}$$

that is,

$$\xi - \sigma(\pi\xi + h) + q(\xi) = -F^+(A'). \tag{7.2.27}$$

This reduces to the previous equation if we take $h = -\pi\xi$, and just as before there is a small solution (ξ, h) to (7.2.27). The solution is ASD if and only if $\pi(\xi)$ is zero in the finite dimensional space H.

So far we have considered ρ as fixed. We may however vary ρ in the space of gluing parameters Gl. Then $\pi(\xi_\rho)$ varies smoothly in the fixed space H. So we have a map

$$\Psi\colon \mathrm{Gl} \longrightarrow H^2_{A_1} \oplus H^2_{A_2}$$

and the zeros of Ψ represent the ASD solutions in our family, just as in Section 4.2. Moreover, since all our constructions are natural, Ψ is a Γ-equivariant map.

Finally we can combine this construction with that of the deformations of the A_i themselves. Let $T_i \subset H^1_{A_i}$ be neighbourhoods parametrizing Γ_{A_i}—equivariant families of solutions to the infinite dimensional parts of the ASD equations as in Chapter 4. We may then form a parameter space

$$T = T_1 \times T_2 \times \mathrm{Gl}$$

and connections $A[t]$ over X, generalizing the $A(\rho)$, parametrized by T. The group Γ acts in the obvious way.

Proposition (7.2.28). *For small enough neighbourhoods T_i and parameters λ, b with $b \geq 4N\lambda^{1/2}$, and $N \geq N_0$, there is*

(i) *a family of L^2_1 connections $A[t] + a_t$ over X, acted on by Γ and with* $\|a_t\|_{L^4} \leq \text{const. } b^2$;

(ii) *a Γ-equivariant map $\Psi\colon T \to H^2_{A_1} \times H^2_{A_2}$ such that $A[t] + a_t$ is ASD if and only if $\Psi(t) = 0$.*

7.2.3 L^p Theory

In the next three sections we will go beyond the existence results of Section 7.2.2 and consider the moduli problem which the construction naturally suggests. We will show that the description of (7.2.24) and (7.2.28) gives a model for an open subset in the moduli space M_E over X; moreover, and this is the most important point, we will give an explicit characterization of the connections in this open set. This explicit characterization will be used in Section 7.3 where we will see that, for many purposes, these models give a complete description of the solutions over X in terms of solutions over the individual manifolds X_i. For simplicity we shall carry out the proofs under the assumption that the deformation complexes of the two connections A_i are acyclic, $H^*_{A_i} = 0$. We shall state the results for the general situation at the end of Section 7.2—the proofs are not substantially different.

The arguments in this section follow a straightforward pattern. Starting from the simple construction of Section 7.2.2, however, there are a number of technical difficulties. For example, we should clarify the nature of the solutions which have been found. The construction gives L^2_1 solutions of the ASD equations; and as we have mentioned in Chapter 4 the general moduli theory for connections of this class—the construction of a space of gauge orbits—is beset with difficulties. As we have noted in Section 4.4.4, we do know from our sharp regularity theorem that any such solution is in fact equivalent to a smooth connection, so our work above does yield genuine solutions of the ASD equations. However it is conceivable at this stage that the actual connections constructed are not smooth, and it is not very easy to see even that the topology on the family of solutions constructed which is inherited from the model agrees with the intrinsic topology defined in Chapter 4. The same kind of problem appears most seriously in the intrinsic characterisation of the solutions constructed, (more precisely, in proving the analogue of (7.2.41) below, with $p = 2$.)

This technical difficulty is the price to be paid for the simplicity of the construction above, in which we worked always with conformally invariant function spaces. While the L^4 norm on 1-forms is conformally invariant, the functions with derivatives in L^4 are not continuous, and this is the source of the difficulty. Indeed, the failure of the L^4 norm of $d\beta$ in controlling the variation of β is just what we have exploited in our construction. While one can get around many of the problems by various special devices, staying within the conformally invariant framework, the most straightforward and uniform approach is to give up the conformal invariance and work with slightly stronger norms. Previously in this book we have always done this by using the L^2_l norms, for which the basic elliptic theory is comparatively elementary. However these norms are not well adapted to the problem at hand, so in this section we will instead break conformal invariance by modifying the exponents in our norms slightly: that is, we work with L^p_1 connections for some fixed $p > 2$. We begin by recalling the two basic analytical facts about these L^p spaces that we shall need. We fix p with $2 < p < 4$ and let q be defined by

$$1 + 4/q = 4/p. \tag{7.2.29}$$

Thus q lies in the range $(4, \infty)$. The first fact we need is the Sobolev embedding theorem: for functions on any compact domain the L^q_1 norm dominates the C^0 norm. More precisely, suppose $H^0_{A_i} = 0$ and η is a section of the bundle $\mathfrak{g}_{E_i}$ over the subset $X'_i \subset X_i$ (the complement of a small ball about x_i) then we have

$$\sup|\eta| \leq \text{const.}\, \|d_{A_i}\eta\|_{L^q}, \tag{7.2.30}$$

where the constant can be taken to be *independent* of the radius of the ball removed. (More generally, one gets such a uniform bound over domains D each of whose points is the vertex of a cone in D of a fixed solid angle.)

The second fact we need concerns the invertibility of the operators $d^+_{A_i}$ over the compact manifolds X_i. Recall that if $H^2_{A_i} = 0$ we have right inverses P_i, mapping L^2 to L^2_1. To be definite we can take

$$P_i = (d^+_{A_i})^* (\Delta_{A_i})^{-1},$$

where $\Delta = (d^+)(d^+)^*$. The Calderon–Zygmund theory of singular integral operators asserts that the P_i give bounded maps from L^p to L^p_1. In fact all we really need is a rather simpler result obtained by composing with the Sobolev embedding

$$L^p_1 \longrightarrow L^q$$

over X_i. (This is the reason for the choice of q.) Thus we need to know that there are constants $C_i = C_i(p)$ so that

$$\| P_i \xi \|_{L^q} \le C_i \| \xi \|_{L^p}. \tag{7.2.31}$$

The other important point to observe about our function spaces is that Hölder's inequality gives, for 1-forms a, b,

$$\| (a \wedge b)^+ \|_{L^p} \le \sqrt{2} \| a \|_{L^q} \| b \|_{L^4}, \tag{7.2.32}$$

since

$$\frac{1}{p} = \frac{1}{q} + \frac{1}{4}.$$

The explanation for this arithmetical coincidence between the Sobolev and Hölder inequalities is that the L^p norm on 2-forms and the L^q norm on 1-forms have the same conformal weight, so they are related in Hölder's inequality by the conformally invariant L^4 norm on 1-forms. We should remember also that the L^q norm is stronger than the L^4 norm, over a space of finite volume, so (7.2.32) gives

$$\| (a \wedge b)^+ \|_{L^p} \le \text{const.} \| a \|_{L^q} \| b \|_{L^q}.$$

We now follow through the same strategy as in Section 7.2.2, splicing together the estimates over the individual manifolds to obtain estimates over X. The new feature is that we have to choose a metric on X to define the L^p and L^q norms, since these are not conformally invariant. Then one has to consider the effect of the conformal diffeomorphism f_λ, gluing together the manifolds, on these norms. A moment's thought shows however that this effect works in our favour, in the argument of Section 7.2.2. The point is that the L^p norm on 2-forms and L^q norm on 1-forms have negative weight with respect to scale changes—scaling the metric by a factor $c > 1$ scales the norms by $c^{(4/p)-2} = c^{(4/q)-1} < 1$. We fix a metric g on X in the obvious way, a weighted average of the metrics g_1, g_2 on X_1 and X_2, compared by the diffeomorphism f_λ. If $g = m_i g_i$ on X'_i we can arrange that $m_i \ge 1$ (i.e. points are further apart in the X metric), while $m_i \leq 2$, say, on X_i'', containing the support of γ_i.

Then we have, for a 1-form α and 2-form θ,

$$\|\alpha\|_{L^q(X)} \leq \|\alpha\|_{L^q(X_1)}, \quad \|\theta\|_{L^p(X)} \leq \|\theta\|_{L^p(X_1)}; \tag{7.2.33}$$

while in the other direction, for forms supported on supp γ_i, we have:

$$\|\alpha\|_{L^q(X)} \geq 2^{(4/q)-1}\|\alpha\|_{L^q(X_1)}, \quad \|\theta\|_{L^p(X)} \geq 2^{(4/q)-1}\|\theta\|_{L^p(X_1)}. \tag{7.2.34}$$

Now in the argument of Section 7.2.2 we have

$$\begin{aligned}\|d_{A'}^+ Q_i\xi - \gamma_i\xi\|_{L^p(X)} &\leq \|d_{A_i}^+ Q_i\xi - \gamma_i\xi\|_{L^p(X_i)} \quad \text{(by (7.2.33))}\\ &\leq \varepsilon_i\|\xi_i\|_{L^p(X_i)},\end{aligned}$$

where ξ_i is the restriction of ξ to X_i', and $\varepsilon_i = \varepsilon_i(N, b, p)$ tends to zero as $N \to \infty$ and $b \to 0$, using (7.2.31) and (7.2.33). Then we apply (7.2.34) to ξ_i and sum to obtain

$$\|d_{A'}^+ Q\xi - \xi\|_{L^p(X)} \leq 2^{1-(4/q)}(\varepsilon_1 + \varepsilon_2)\|\xi\|_{L^p(X)}.$$

So we obtain an extension of (7.2.18) to L^p norms:

Proposition (7.2.35). *If $H_{A_1}^2 = 0$ and $H_{A_2}^2 = 0$, then for $N \geq N_0$, $b \leq b_0$ and all λ with $4N\lambda^{1/2} \leq b$, the right inverse P to $d_{A'}^+$ over X satisfies*

$$\|P\xi\|_{L^q} \leq C\|\xi\|_{L^p},$$

for a constant C independent of λ, where the metric on X is chosen as above.

We can then carry out the construction using L^p in place of L^2. We have, by (7.2.7),

$$\|F^+(A')\|_{L^p} \leq \text{const. } b^{(4/p)}, \tag{7.2.36}$$

so the same argument applies. We deduce that the solutions $A'(\rho) + a_\rho$ given in (7.2.24) lie in L_1^p, and

$$\|a_\rho\|_{L^q} \leq \text{const. } b^{(4/p)}. \tag{7.2.37}$$

This shift to the L^p framework resolves the first of our difficulties. The techniques of Chapter 4 can be applied to L_1^p connections acted on by L_2^p gauge transformations, and as in (4.2.16), lead to equivalent models for the moduli space. (In fact one can bootstrap from (7.2.21) to show that our solutions are smooth). It is then easy to see that, in the case when the $H_{A_i}^2$ are zero, the construction gives a smooth map

$$\begin{aligned} I\colon \text{Gl} &\longrightarrow M_E\\ I(\rho) &= [A'(\rho) + a_\rho],\end{aligned} \tag{7.2.38}$$

where Gl is the space of gluing parameters $\text{Hom}((E_1)_{x_1}, (E_2)_{x_2})$ and M_E is the moduli space of ASD connections on E, with the conformal structure on X

defined by any given, sufficiently small, λ. From now on we fix the parameter N, large enough to satisfy the conditions of (7.2.35), and we may as well put

$$b = 4N\,\lambda^{1/2}\,,$$

so we only have one parameter λ in the discussion.

7.2.4 *The gauge fixing problem*

We will now characterize the ASD solutions found by our gluing construction. Let d_q be the metric on the space $\mathscr{B}_E$ of connections modulo gauge equivalence given by

$$\mathrm{d}_q([A],[B]) = \inf_{u \in \mathscr{G}} \|A - u(B)\|_{L^q}\,. \tag{7.2.39}$$

Define

$$J\colon \mathrm{Gl} \longrightarrow \mathscr{B}_E,$$

by $J(\rho) = [A'(\rho)]$. For $v > 0$ let $U(v) \subset \mathscr{B}_E$ be the open set

$$U(v) = \{[A] \mid \mathrm{d}_q([A], J(\mathrm{Gl})) < v,\ \|F^+(A)\|_{L^p} < v^{3/2}\}\,. \tag{7.2.40}$$

The solutions we have constructed lie in $U(v)$, if λ is small compared with v. We will now prove a converse result.

Theorem (7.2.41). *If $H^*_{A_1} = 0$ and $H^*_{A_2} = 0$, then for small enough v and $\lambda < \lambda_0(v)$ any point in $U(v)$ can be represented by a connection A of the form $A'(\rho) + P_\rho\xi$, where $\|\xi\|_{L^p} \leq$ const. v.*

This gives

Corollary (7.2.42). *If $H^*_{A_1} = 0$ and $H^*_{A_2} = 0$, then for small enough v and $\lambda < \lambda_0(v)$ the intersection $U(v) \cap M_E$ is equal to the image of $I\colon \mathrm{Gl} \to \mathscr{B}_E$.*

The corollary follows since we have seen in Section 7.2.2 that under the given hypotheses there is a unique small solution ξ to the equation $F^+(A'(\rho) + P_\rho\xi) = 0$.

The definition of the open set $U(v)$ depends on the choice of connections $A'(\rho)$, which involved arbitrary cut-off functions. It is easy to see however that this dependence is not essential. If A is a connection over X write A''_i for the restriction of A to the open set X''_i. Then define another open set $U^*(v) \subset \mathscr{B}_E$ to be the equivalence classes $[A]$ of connections over X such that $\mathrm{d}_q([A''_i], [A_i]) < v$, $\|F^+(A''_i)\|_{L^p} < v^{3/2}$. Here we can use the metrics from X_i to define the norms, although these compare uniformly with those on X over X''_i. Then the open sets $U^*(v)$ are equivalent to the $U(v)$ in the sense that we have:

Lemma (7.2.43). *For any v we can find v_1, v_2 such that $U(v_1) \subset U^*(v)$ and $U^*(v_2) \subset U(v)$.*

We leave this as an exercise. It amounts to the fact that two equivalent connection matrices over the annulus $X''_1 \cap X''_2$ which are small in L^q differ by a gauge transformation with small variation; compare Section 4.4.2. The additional point to note is that if we rescale this small annulus to a standard size, in the manner of Section 4.4.3, the L^q norm on 1-forms decreases.

In the work which follows it will be useful to identify the bundles E_ρ for different ρ, in order to compare the connections A_ρ. Of course, eventually we shall be working in the space of connections modulo gauge transformations where the choice of identification between the E_ρ is irrelevant. But for the analysis it is clearest if we work with connections on a single bundle, using a convenient identification. Let $\rho_0 \in \mathrm{Gl}$ be a given gluing parameter. Points ρ in a small neighbourhood L of ρ_0 in Gl can be written in the form $\rho = \rho_0 \exp(v)$, where $v \in \mathfrak{g}_{E_1, x_1}$. We regard the fibres of $\mathfrak{g}_{E_1}$ and $\mathfrak{g}_{E_2}$ as being identified by ρ_0, so we can think of v as a local section of both $\mathfrak{g}_{E_1}$ and $\mathfrak{g}_{E_2}$, covariant constant with respect to the connections A'_1, A'_2. Consider now the gauge transformations h_i of E_i over X'_i:

$$h_1 = \begin{cases} \exp(\gamma_2 v) & \text{on } \Omega_1 \\ 1 & \text{on } X'_1 \setminus \Omega_1 \end{cases}$$

$$h_2 = \begin{cases} \exp(-\gamma_1 v) & \text{on } \Omega_2 \\ 1 & \text{on } X'_2 \setminus \Omega_2. \end{cases} \tag{7.2.44}$$

Here γ_1, γ_2 are cut-off functions as before, which we regard as being simultaneously defined on X_1, X_2 and X, extending by constants 0, 1 in the obvious way. Note that h_i has a natural extension to a gauge transformation of E_i over all of X_i—equal to $\exp(\pm v)$ on the ball enclosed by the annulus.

Now, under our identification of the bundles and base manifolds over $\Omega = \Omega_1 = \Omega_2$ we have

$$h_1 h_2^{-1} = \exp([\gamma_1 + \gamma_2] v) = \exp(v).$$

So, relative to the flat connections A'_i, h_1 and h_2 differ by a constant bundle automorphism over Ω. So their action on the connection is the same:

$$h_1(A'_{\rho_0})|_\Omega = h_2(A'_{\rho_0})|_\Omega.$$

Thus, while the automorphisms h_i do *not* patch together to give a global automorphism of E_{ρ_0}, their actions on the connection A'_{ρ_0} do. We can define a connection $A'(\rho_0, v)$ on E_{ρ_0} by

$$A'(\rho_0, v) = \begin{cases} h_1(A'_{\rho_0}) & \text{on } X'_1 \\ h_2(A'_{\rho_0}) & \text{on } X'_2. \end{cases}$$

Then we have

Lemma (7.2.46). *If* $\rho = \rho_0 \exp(v)$, *the connections* $A'(\rho_0, v)$ *and* $A'(\rho)$ *are gauge equivalent.*

The proof is a simple exercise in the definitions.

We can now regard our connected sum bundle E as being fixed, with a space of connections $\mathscr{A}$, and construct in this way a cover $\{L_\alpha\}$ of Gl and maps $J_\alpha: L_\alpha \to \mathscr{A}$ such that $J_\alpha(\rho)$ and $J_\beta(\rho)$ are gauge equivalent when ρ is in $L_\alpha \cap L_\beta$. Otherwise stated, we have found explicit local lifts of the canonical map $J: \mathrm{Gl} \to \mathscr{B}_E$.

We now proceed to the proof of Theorem (7.2.41). The problem is one of gauge fixing, similar to those discussed in Chapter 2, where the Coulomb gauge subspace is replaced by the image of the operators P_ρ, together with the freedom to vary the gluing parameter ρ. Our proof will follow the same pattern as that of Uhlenbeck's theorem in Chapter 2, using the method of continuity.

Let B lie in $U(\nu)$, so there is a connection $A' \in J(\mathrm{Gl})$ with

$$\| A' - B \|_{L^q(X)} < \nu .$$

We write $B = A' + b$ and consider the path, for $t \in [0, 1]$,

$$B_t = A' + tb;$$

then $\| A_0 - B_t \|_{L^q} < \nu$ and it is easy to see that, for λ small compared with ν, B_t lies in $U(\nu)$ for all t. We let $S \subset [0, 1]$ be the set of times t for which there is a gauge transformation u_t and an A'_ρ in the image of J such that

$$u_t(B_t) = A'_\rho + P_\rho(\xi),$$

with $\| \xi \|_{L^p} < \delta$, where δ will be chosen below, but we require that $\delta < \frac{1}{2}C^{-2}$ where C is the constant of (7.2.35).

The proof now divides into the usual two parts, showing that S is both closed and open.

To show that S is closed we derive *a priori* bounds. Suppose that t is in S; we may as well take $u_t = 1$. Then the representation $B_t = A'_\rho + P_\rho \xi$ gives:

$$F^+(B_t) = F^+(A') + \mathrm{d}^+_{A'}(P_\rho \xi) + (P_\rho \xi \wedge P_\rho \xi)^+ ,$$

and so

$$\xi = F^+(B_t) - F^+(A') - (P_\rho \xi \wedge P_\rho \xi)^+ .$$

This gives

$$\begin{aligned} \| \xi \|_{L^p} &\leq \| F^+(B_t) \|_{L^p} + \| F^+(A') \|_{L^p} + \| P_\rho \xi \|^2_{L^q}, \\ &\leq \nu^{3/2} + \text{const.}\ \lambda^{2/p} + C^2 \| \xi \|^2_{L^p} . \end{aligned}$$

Then we can rearrange to get a bound on $\| \xi \|_{L^p}$, in the familiar way. One sees then that for small δ, for $\nu \ll \delta$ and $\lambda^{2/p} \ll \nu$, the constraint $\| \xi_t \| < \delta$ implies that $\| \xi_t \| \leq \frac{1}{2}\delta$, so this open condition is also closed.

It is now routine to prove that S is closed. Suppose we have t_i in S, with $t_i \to t$. The parameter space Gl is compact, so we may suppose that the gluing parameters ρ_i used at times t_i converge. For simplicity let us assume these are actually equal to a fixed ρ_0. Thus we have connections $A_i = A'_{\rho_0} + P_{\rho_0} \xi_i$ on

$E = E_{\rho_0}$ and gauge transformations u_i with $u_i(B_{t_i}) = A_i$. By the uniform bound on the ξ_i above we may suppose, taking a subsequence, that the ξ_i converge to a limit ξ, weakly in L^p, with $\| \xi \|_{L^p} < \delta$. Then the connections A_i converge weakly in L_1^p and, just as in Section 2.3.7, we can suppose the gauge transformations u_i converge weakly in L_2^p. The gauge relation is preserved in the limit so

$$u(B_t) = A'_{\rho_0} + P_{\rho_0}\xi ,$$

and we see that t lies in S. (If we want to work with smooth connections, as in Section 2.3, we can bootstrap, using the equation to show that ξ is actually smooth when B is.)

7.2.5 Application of the index formula

We will now proceed to the more interesting part of the proof of (7.2.41), showing that S is open. Of course, we use the implicit function theorem to reduce to a linear problem—proving that a certain linear operator is invertible. To do this we will use the index formula for the virtual dimension of the ASD moduli spaces. It follows immediately from the excision principle that

$$s(E) = s(E_1) + s(E_2) + \dim G . \qquad (7.2.47)$$

This can be interpreted loosely as saying that the number of parameters describing an ASD connection on E is equal to the sum of those describing ASD connections on the E_i plus the number of gluing parameters which is, of course, in line with Corollary (7.2.42). This application of the index formula may seem slightly unsatisfying. It would be tidier to prove the invertibility of the operator over the connected sum directly, by splicing together inverses over the two manifolds X_i, just as we have done for the right inverses of the d^+ operators in Section 7.2.2. This would then give yet another proof of the excision formula for the index in this situation. However, it turns out that this kind of direct approach is not at all as easy as that in Sections 7.2.2 and 7.2.3. The problem is that, if one takes the obvious route, the rescaling behaviour of the L^p norms works in an *unfavourable* direction, in contrast to Section 7.2.3. One can make a direct proof along these lines, by analysing in greater detail the behaviour over the neck, but it is much simpler to invoke the index formula (7.2.47), as we shall do below.

It is in this section that we will make real use of our representation of the connections in the form $A'(\rho_0, v)$. We also need the derivative of this family. For v in $(\mathfrak{g}_{E_1})_{x_1}$, and a given ρ_0, put:

$$j(v) = \frac{\partial}{\partial s} \{A'(\rho_0, sv)\} |_{s=0} \in \Omega^1_X(\mathfrak{g}_E) . \qquad (7.2.48)$$

Thus on X_1 $j(v) = -\,d_{A'}(\gamma_1 v)$, while on X_2 $j(v) = d_{A'}(\gamma_2 v)$, extended in each case by zero outside the annulus Ω. Here we are writing A' for $A'(\rho_0)$, and we regard the fibres of the $\mathfrak{g}_{E_i}$ over the x_i as being canonically identified by ρ_0. To preserve symmetry between the factors we denote this common space by V, so $V = (\mathfrak{g}_{E_1})_{x_1} = (\mathfrak{g}_{E_2})_{x_2}$.

Lemma (7.2.49). *If $H^0_{A_1} = 0$ and $H^0_{A_2} = 0$ there is a constant D, independent of λ, such that for any χ in $\Omega^0_X(\mathfrak{g}_E)$ and v in V we have:*

$$\|\chi\|_{C^0} + |v| \leq D \| \mathrm{d}_{A'}\chi + j(v)\|_{L^q(X)} .$$

This follows from the Sobolev inequalities (7.2.30) over the individual manifolds. Let χ_1 be the section $\chi + (1 - \gamma_1)v$ of $\mathfrak{g}_{E_1}$ over $\mathrm{supp}(\gamma_1) \subset X''_1$, and χ_2 be the section $\chi - (1 - \gamma_2)v$ over $\mathrm{supp}(\gamma_2) \subset X''_2$. Thus

$$\mathrm{d}_{A'}\chi + j(v) = \mathrm{d}_{A'_i}\chi_i$$

over X'_i and $\chi_1 - \chi_2 = v$ over the intersection. So $\|u\|_{C^0}$ and $|v|$ are each bounded above by $\|\chi_1\|_{C^0} + \|\chi_2\|_{C^0}$. On the other hand we have by (7.2.30) that

$$\|\chi_i\|_{C^0} \leq D_i \| \mathrm{d}_{A_i}\chi_i\|_{L^q(X_i)}$$

for some constants D_i. Comparing A_i with A'_i we get

$$\|\chi_i\|_{C^0} \leq D_i \| \mathrm{d}_{A'_i}\chi_i\|_{L^q(X_i)} + \|A_i - A'_i\|_{L^q(X_i)} \|\chi_i\|_{C^0} .$$

We choose λ small enough that $\|A_i - A'_i\|_{L^q(X_i)}$ is less than $\frac{1}{2}$ say; then we can rearrange this to get a uniform bound in terms only of $\mathrm{d}_{A'_i}\chi_i$. For 1-forms supported on $\mathrm{supp}(\gamma_i) \subset X'_i$ the $L^q(X)$ and $L^q(X_i)$ norms are uniformly equivalent and this gives the result.

Supposing that $H^0_{A_i} = H^2_{A_i} = 0$ we now define a map:

$$T\colon \Omega^0_X(\mathfrak{g}_E) \oplus V \oplus \Omega^+_X(\mathfrak{g}_E) \longrightarrow \Omega^1_X(\mathfrak{g}_E) \tag{7.2.50}$$

by

$$T(\chi, v, \xi) = \mathrm{d}_{A'}\chi + j(v) + P\xi ,$$

where $P = P_{\rho_0}$.

We let B_1 be the completion of the domain of T in the norm:

$$\|(\chi, v, \xi)\|_{B_1} = \|\mathrm{d}_{A'}\chi + j(v)\|_{L^q(X)} + \|\xi\|_{L^p(X)} . \tag{7.2.51}$$

This is a norm by (7.2.49), and it dominates the uniform norm of χ. Let B_2 be the completion of the image space $\Omega^1_X(\mathfrak{g}_E)$ in the norm:

$$\|\alpha\|_{B_2} = \|\alpha\|_{L^q(X)} + \|\mathrm{d}^+\alpha\|_{L^q(X)} .$$

Then T is a bounded map from B_1 to B_2 but what we really need is a reverse inequality, uniform in the parameters:

Lemma (7.2.52). *There is a constant K independent of λ, b, N such that*

$$\|(\chi, v, \xi)\|_{B_1} \leq K \|T(\chi, v, \xi)\|_{B_2}$$

for all (χ, v, ξ) in B_1.

Proof. Let $d_{A'}\chi + j(v) + P(\xi) = \alpha$, so that

$$d^+_{A'}\alpha = [F^+(A'), \chi] + \xi .$$

Thus $\|\xi\|_{L^p} \le \|\alpha\|_{B_2} + \|[F^+_{A'}, u]\|_{L^p}$.

Now, recalling that $\|F^+(A')\|_{L^p}$ is $O(b^{4/p}) = O(\lambda^{2/p})$ we have

$$\begin{aligned}\|\xi\|_{L^p} &\le \|\alpha\|_{B_2} + \text{const. } \lambda^{2/p}\|\chi\|_{C^0}\\ &\le \|\alpha\|_{B_2} + \text{const. } \lambda^{2/p}\|d_{A'}\chi + j(v)\|_{L^q},\end{aligned}$$

using (7.2.30). Substituting back into the definition of α, we obtain

$$\begin{aligned}\|\xi\|_{L^p} &\le \|\alpha\|_{B_2} + \text{const. } \lambda^{2/p}\|\alpha - P\xi\|_{L^q}\\ &\le \|\alpha\|_{B_2} + \text{const. } \lambda^{2/p}\{\|\alpha\|_{B_2} + \|\xi\|_{L^p}\},\end{aligned}$$

where in the last step we have used (7.2.35). Thus when λ is small we can rearrange to get a bound $\|\xi\|_{L^p} \le K_1\|\alpha\|_{B_2}$, say. This gives us then

$$\|d_A\chi + j(v)\|_{L^q} \le \|\alpha - P\xi\|_{L^q} \le (1 + K_1)\|\alpha\|_{B_2},$$

which yields the desired bound (with $K = 1 + K_1$).

As an immediate corollary to (7.2.52) we see that the image of T is closed in B_2 and that the kernel of T is zero. We use the index formula to see that T is actually an isomorphism.

Proposition (7.2.53). *If $H^0_{A_i} = H^1_{A_i} = H^2_{A_i} = 0$ the operator T is a surjection from B_1 to B_2, hence a topological isomorphism with operator norm $\|T^{-1}\| \le K$.*

The operator P is a pseudo-differential operator, and it is easy to see that its symbol is homotopic to that of $(d^+_{A'})^*(1 + \Delta_{A'})^{-1}$. It follows that $d_{A'} \oplus P$ is Fredholm and its index equals that of $d_{A'} + \{d^+_{A'}\}^*$—the adjoint of $d^*_{A'} + d^+_{A'} = \delta_{A'}$. Since the indices of $\delta_{A_1}, \delta_{A_2}$ are both zero, by hypothesis, and $\dim V = \dim G$, we have

$$\begin{aligned}\operatorname{index}(T) &= \dim G + \operatorname{index}\delta^*_{A'} = \dim G - \operatorname{index}\delta_{A'}\\ &= \dim G - \{\operatorname{index}\delta_{A_1} + \operatorname{index}\delta_{A_2} + \dim G\} = 0,\end{aligned}$$

where in the second line we use (7.2.47). The result now follows immediately from (7.2.52).

We can now complete the proof of (7.2.41) by the continuity method, showing that the set S for which a solution to the equation exists is open. To simplify notation we may as well prove that if 1 is in S, so $B = A' + P_{\rho_0}\xi$ for some $A' = A'(\rho_0)$ and ξ with $\|\xi\|_{L^p} < \delta$, then there is a small neighbourhood $(1 - \varepsilon, 1]$ in S. In fact we can show that any connection close to B is gauge equivalent to some $A'(\rho) + P_\rho(\xi + \eta)$, with $\rho = \exp(v)\rho_0$ close to ρ_0. So, following the notation used above, we now write $P_{[v]}$ for the operator P_ρ

formed using the connection $A'(\rho_0, v)$ on the fixed bundle $E = E_{\rho_0}$. Define a map

$$M\colon \Omega^0_X(\mathfrak{g}_E) \times V \times \Omega^+_X(\mathfrak{g}_E) \longrightarrow \Omega^1_X(\mathfrak{g}_E)$$

by

$$M(\chi, v, \eta) = (\exp(\chi)(A'(\rho_0, v) + P_{[v]}(\xi + \eta))) - B\,.$$

We need to show that M maps onto a neighbourhood of zero, which follows from the implicit function theorem if we know that the derivative (DM) of M at (0, 0, 0) is surjective. This derivative can be written

$$DM(\chi, v, \eta) = \mathrm{d}_B(\chi) + j(v) + \Pi(v, \xi) + P\eta\,, \tag{7.2.54}$$

where $\Pi(v, \xi)$ is the derivative of $P_{[v]}\chi$ with respect to v, and we write P for $P_{\rho_0} = P_{[0]}$. So we have, writing $B = A'(\rho_0) + \alpha$,

$$DM(\chi, v, \eta) = T(\chi, v, \eta) + [\alpha, \chi] + \Pi(v, \xi)\,.$$

It follows then from Proposition (7.2.53) that DM is invertible provided that the B_1-to-B_2 operator norm of the map τ, with $\tau(\chi, v, \eta) = [\alpha, \chi] + \Pi(v, \xi)$, is less than K^{-1}. So we have to show that the operator norm of τ tends to zero with the parameters δ, ν, λ.

The term $[\alpha, \chi]$ is easily dealt with; we leave it as an exercise. The only hint of difficulty comes in the other term $\Pi(v, \xi)$, since this involves the operators $P_{[v]}$ which are of a global nature. Recall that, in an obvious notation,

$$P_{[v]} = Q_{[v]}(\mathrm{d}^+_{[v]}\, Q_{[v]})^{-1}\,. \tag{7.2.55}$$

We can write

$$Q_{[v]} = Q_{[v],1} + Q_{[v],2}$$

with

$$Q_{[v],i} = h_i Q_i h_i^{-1}$$

and

$$Q_i\xi = \beta_i P_i(\gamma_i\xi).$$

Here h_i are the gauge transformations over the open sets X_i' given by (7.2.44). We now differentiate with respect to v at $v = 0$ to get

$$(\partial Q_{[v]})(\xi) = [(1 - \gamma_1)v, Q_1\xi] + [(1 - \gamma_2)v, Q_2\xi]. \tag{7.2.56}$$

Hence

$$\|\partial Q_{[v]}(\xi)\|_{L^q} \leq \text{const.}\ |v|\ \|\xi\|_{L^p}.$$

Similarly the v-derivative of $\mathrm{d}^+_{[v]} Q_{[v]}(\xi)$ is bounded by const. $|v|\ \|\xi\|_{L^p}$. Now differentiate (7.2.55) with respect to v to get

$$\Pi(v, \xi) = \{(\partial Q_{[v]}) - P(\partial(\mathrm{d}^+_{[v]}\, Q_{[v]}))\}\,(\mathrm{d}^+_{A'} Q)^{-1}.$$

The bounds above on the v-derivatives of $Q_{[v]}$ and $\mathrm{d}^+_{[v]}\, Q_{[v]}$ give $\|\Pi(\xi, v)\|_{L^q} \leq$ const. $|v|\ \|\xi\|_{L^p}$. Similarly, differentiating the identity

$$\mathrm{d}^+_{[v]} P_{[v]} = 1$$

we get

$$\mathrm{d}^+_{A'}(\Pi(v,\xi)) = -\,\partial(\mathrm{d}^+_{[v]})(P\xi) = -\,[j(v), P\xi],$$

so

$$\|\mathrm{d}^+_{A'}(\Pi(v,\xi))\|_{L^p} \le \text{const. } \|j(v)\|_{L^4}\|\xi\|_{L^p} \le \text{const. } |v|\ \|\xi\|_{L^p}$$

(since we may suppose the L^4 norm of $\nabla\gamma_i$ is independent of λ, and hence $\|j(v)\|_{L^4} \le \text{const. } |v|$).

In sum then we obtain

$$\|\Pi(v,\xi)\|_{B_2} \le \text{const. } |v|\ \|\xi\|_{L^p} \le \text{const. } \|(\chi, v, \eta)\|_{B_1},$$

by (7.2.53), and so if $\|\xi\|_{L^p}$ is small (i.e. if δ is small) the operator DM is invertible, and this completes the proof of (7.2.41).

7.2.6 *Distinguishing the solutions*

There is one question left open so far in our model of the moduli spaces on connected sums. We need to show that points in the model represent different gauge equivalence classes of connections unless they lie in the same orbit of the symmetry group $\Gamma = \Gamma_{A_1} \times \Gamma_{A_2}$. Following our usual pattern we will give the proof in the case when this symmetry group is trivial; again the general case presents no interesting extra difficulties.

Proposition (7.2.57). *If $\Gamma_{A_1} = \Gamma_{A_2} = 1$ and $H^2_{A_1} = H^2_{A_2} = 0$ then for small enough λ the map $I: \mathrm{Gl} \to M_E \subset \mathscr{B}_E$ is injective.*

This result should, of course, be seen as an extension of the elementary Proposition (7.2.9). For the proof we consider two points in Gl, which we suppose for simplicity lie in a common coordinate patch L_α. Thus we write the corresponding ASD connections as $A' + a$, $A'_{[v]} + a_{[v]}$, suppressing the fixed gluing parameter ρ_0. Suppose u is a gauge transformation intertwining these two connections over X. We can restrict u to the overlapping punctured manifolds $\mathrm{supp}(\gamma_i)$, on which the connections A', $A'_{[v]}$ are each isomorphic to A'_i, to get automorphisms u_i of E_i. Then if $u'_i = h_i^{-1}u_i$, where h_i is defined by (7.2.44), we have

$$(A'_i + a) = u'_i(A'_i + a_{[v]}) \qquad (7.2.58)$$

over the punctured manifold. We can now appeal to a simple non-linear variant of (7.2.30), whose proof we leave as an exercise, to say that, since Γ_{A_i} is trivial, we can write u'_i as $\exp\chi_i$ where

$$\|\chi_i\|_{C^0} \le \text{const. } \|a - u_i'^{-1}\, a_{[v]}\, u'_i\|_{L^q(X''_i)}, \qquad (7.2.59)$$

with a constant which is independent of λ. Now on the overlap of the X''_i the matching condition for the u_i gives that the u'_i differ by the constant gauge transformation $\exp(v)$. Thus we have:

$$|v| \le \text{const.}\|a - u_i'^{-1}\, a_{[v]}\, u'_i\|_{L^q(X)}.$$

On the other hand, h differs from the identity by $O(|v|)$, so u_i' differs from the identity by $O(|v| + \|\chi_i\|)$. Hence

$$\|a_{[v]} - u_i'^{-1} a_{[v]} u_i'\|_{L^q} \leq \text{const. } \|a(v)\|_{L^q}(|v| + \|\chi_1\| + \|\chi_2\|).$$

Then, substituting back into (7.2.58) we have

$$\|w_1\| + \|w_2\| \leq \text{const. } \|a_{[v]}\|_{L^q}(|v| + \|w_1\| + \|w_2\|).$$

We know that the L^q norm of $a_{[v]}$ is small for small λ and then we can rearrange to get

$$\|w_1\| + \|w_2\| \leq \text{const. } |v|,$$

from which we deduce the inequality

$$|v| \leq \text{const. } \|a - a_{[v]}\|_{L^q(X)}. \tag{7.2.60}$$

The proof is completed by deriving a bound in the other direction. Recall that $a_{[v]} = P_v(\xi_v)$ say, is determined implicitly by the equation:

$$-F^+(A'_{[v]}) = \xi_v + (P_v\xi_v \wedge P_v\xi_v)^+. \tag{7.2.61}$$

We now differentiate this expression with respect to v to estimate the v dependence of ξ_v. The term $F^+(A'_{[v]})$ is independent of v, since it is supported in the region where $h_i = 1$. So the v-derivative $\partial\xi_v$ of ξ_v is $-((\partial P_v)\xi \wedge P\xi + P(\partial\xi_v) \wedge P\xi + P\xi \wedge (\partial P_v)\xi + P\xi \wedge (\partial\xi_v))^+$, and, using the bounds on the derivative of P_v from the previous section, this gives

$$\|\partial\xi_v\|_{L^p} \leq \text{const. } \|P\xi\|_{L^4}(\|\xi\|_{L^p} + \|\partial\xi_v\|_{L^p}).$$

When λ, and hence $P\xi$, is small this gives a bound on the derivative of ξ_v,

$$\|\partial\xi_v\|_{L^p} = O(\lambda^{1+2/p}).$$

We deduce then that the L^q norm of the derivative of $a_{[v]} = P_v\xi_v$ is also $O(\lambda^{1+2/p})$, and integrating along a path in the v-variable that

$$\|a - a_{[v]}\|_{L^q} \leq \text{const. } \lambda^{1+2/p}\,|v|.$$

Combined with the previous estimate (7.2.60) this tells us that, when λ is small, v must be zero, under the hypothesis that the two connections are gauge equivalent.

7.2.7 Conclusions

We can now sum up the results of this section, including the straightforward generalizations to the case when the cohomology groups of the A_i do not vanish.

Theorem (7.2.62). *Let A_1, A_2 be ASD connections over manifolds X_1, X_2. For sufficiently small values of the parameter λ defining the conformal structure on the connected sum $X = X_1 \# X_2$ there is a model for an open set in the moduli* *space of ASD connections on the connected sum bundle over X of the following form.*

(i) *There is a neighbourhood T of* $\mathrm{Gl} \times \{0\}$ *in* $\mathrm{Gl} \times H^1_{A_1} \times H^1_{A_2}$ *and a smooth map* Ψ *from T to* $H^2_{A_1} \times H^2_{A_2}$, *equivariant with respect to the natural action of* $\Gamma = \Gamma_{A_1} \times \Gamma_{A_2}$.

(ii) *There is a map* $I: T/\Gamma \to \mathscr{B}_E$ *which gives a homeomorphism from* $\Psi^{-1}(0)/\Gamma$ *to an open set N in the moduli space* M_E.

(iii) *For any given* ν *the set T can be chosen so that, for all* $\lambda < \lambda_0(\nu)$, *the image N is the set of points* $[A]$ *in* M_E *such that the restriction,* A_i'', *of A to the common open set* $X_i'' \subset X, X_i$ *satisfies* $\mathrm{d}_q([A_i''], [A_i]) < \nu$.

(iv) *The construction gives a model for N as a real analytic space; the sheaf of rings on N inherited from the ASD moduli space is naturally isomorphic to that inherited from* Ψ.

We can extend the construction by allowing the connections over the summands X_i to vary. Let C_1, C_2 be precompact open sets in the moduli spaces M_{E_1}, M_{E_2} and suppose for simplicity that their closures do not contain any reducible connections or singularities. Construct a fibre bundle

$$T(C_1 \times C_2) \longrightarrow C_1 \times C_2,$$

whose points consist of isomorphism classes of triples (A_1, A_2, ρ) where ρ is an identification of the fibres of E_i over the base points. For subsets K_1, K_2 of C_1, C_2 and $\eta > 0$ we let $N(K_1, K_2, \eta)$ be the set of equivalence classes in the moduli space over X which have L^q distance less than η from the constituents A_i in K_i over X_i''. Then we have:

Theorem (7.2.63). *For small enough values of the parameter* λ *there is a homeomorphism from* $T(C_1 \times C_2)$ *to an open set N in the moduli space of ASD connections on the connected sum bundle. For any compact sets* $K_i \subset C_i$ *and* $\eta > 0$ *we have:*

$$N(K_1, K_2, \eta) \subset N' \subset N(C_1, C_2, \eta),$$

once λ *is sufficiently small.*

There are, of course, variants of this result which take into account singularities and reductions. For example, suppose E_2 is the trivial bundle and the moduli space of ASD connections on E_2 is a point, representing the product connection. Then the gluing parameter can be 'cancelled' by the automorphism group of A_2. Suppose in addition that $G_1 \subset M(E_1)$ consists of regular points. The global version of (7.2.62) then takes the following form. Let Ξ be the vector bundle over G_1 associated to the base point fibration by the adjoint representation. For small values of λ there is a section Ψ_λ of $\Xi \otimes \mathscr{H}^+(X_2)$ and a homeomorphism from the zero set of Ψ_λ to a neighbourhood in the moduli space of the connected sum, whose image consists of all ASD connections which are L^q close to the flat connection over X_2'' and to a connection from G_1 over X_1''.

The reader should have no difficulty in supplying the proofs of (7.2.63) and its variants; it is just a matter of carrying through the previous constructions with the A_i now as variables. There is one observation we should make, specially relevant to the models of the ends of moduli spaces considered in Chapter 8. Suppose $[A_2]$ is a regular point of the moduli space M_{E_2} and f_α ($\alpha = 1, \dots, n$) are functions on a neighbourhood of $[A_2]$ in the ambient space $\mathscr{B}_{E_2}$ which restrict to give local coordinates on the moduli space, vanishing at $[A_2]$. Thus n is the dimension of M_{E_2}. Suppose the f_α depend only on the restriction of connections to a compact set in the punctured manifold $X_2 \setminus \{x_2\}$. Then for small λ they define also functions on $\mathscr{B}(E_1 \# E_2)$. We can run our construction to describe solutions A of the ASD equations on X which satisfy the additional constraint $f_\alpha(A) = 0$, $\alpha = 1, \dots, n$. The analytical discussion is essentially unchanged. The effect of the constraint is to replace the moduli space M_{E_2} by a point. One can then use this idea globally by choosing a projection map f from a neighbourhood of M_{E_2} in $\mathscr{B}_{E_2}$ to the moduli space M_{E_2}, depending only on the restriction to a compact set in the punctured manifold. Such a map can be constructed by patching together local coordinates using cut-off functions in $\mathscr{B}_{E_2}$. This discussion will be taken up again in Section 9.3.

7.2.8 Multiple connected sums

There is another generalization of the situation considered above in this chapter, in which we consider multiple connected sums. The input can be described by a collection of summands X_i, each containing some marked points, and a graph with vertices corresponding to the X_i. For each edge in the graph we identify small annuli in the corresponding four-manifolds, with a real parameter measuring the neck size. Then we get a family of conformal structures on a multiple connected sum X, with parameters $(\lambda_1, \dots, \lambda_N)$ say. The techniques used above extend without change to analyse ASD connections on X for small enough λ_j.

In this section we want to mention one technical point which will be rather important in Chapter 8. Suppose, for simplicity, we consider a connected sum $X = X_1 \# Y_1 \# Y_2$, by identifying small regions in X_1 with corresponding regions in the Y_i, using parameters λ_i. Suppose A_1 is an ASD connection over X_1, and B_i are ASD connections over the Y_i and that all the cohomology groups vanish except for $H = H^2_{A_1}$. Then for fixed λ_i we have a model of the form:

$$\psi_{12}\colon \mathrm{Gl}_1 \times \mathrm{Gl}_2 \longrightarrow H.$$

Now the construction varies smoothly with the parameters λ_i and we can extend ψ_{12} to a map, which we denote by the same symbol, from

$\mathrm{Gl}_1 \times (0, \delta) \times \mathrm{Gl}_2 \times (0, \delta)$ to H, thus including the dependence on the λ_i. On the other hand, ignoring the manifold Y_2, we construct a model for ASD connections over $X_1 \# Y_1$, depending on a single gluing parameter. So we have a map:

$$\psi_1 : \mathrm{Gl}_1 \times (0, \delta) \longrightarrow H.$$

Proposition (7.2.64). *In this situation*:

$$\lim_{\lambda_2 \to 0} \psi_{12}(\rho_1, \lambda_1, \rho_2, \lambda_2) = \psi_1(\rho_1, \lambda_1).$$

This continuity property of the construction is intuitively immediately plausible. It generalizes the fact, which we obtain immediately from our construction, that

$$|\psi_{12}| \leq \text{const.}\ (\lambda_1 + \lambda_2). \tag{7.2.65}$$

The proof of the present proposition is just a matter of working through the constructions we have made. For simplicity we will sketch the proof for the corresponding linearized problem, leaving the non-linear terms as an exercise for the reader. We will modify our notation slightly from that used above. Recall that the space H is represented by a fixed set of forms over X_1 supported away from the regions where the connected sums are made. Thus we have a decomposition of bundle-valued self-dual two-forms over X_1:

$$\omega = d^+_{A_1} P_1 \omega + h(\omega),$$

where $h(\omega) \in H$. For the connected sum $X_1 \# Y_1$ we construct a right parametrix $Q^{(1)}$ with

$$d^+_{A'} Q^{(1)} \phi = \phi + \varepsilon(\phi) + h(\gamma \phi),$$

where ε has small operator norm. We then let $P^{(1)} = Q^{(1)}(1 + \varepsilon^{(1)})^{-1}$, so:

$$d^+_{A'} P^{(1)} \phi = \phi + h(\gamma(1 + \varepsilon^{(1)})^{-1} \phi) = \phi + h^{(1)}(\phi),$$

say. We then solve the equation

$$F^+(A' + P^{(1)} \phi) + \psi_1 = 0,$$

for ψ_1 in H. The linearized version of this is to solve

$$d^+_{A'} P^{(1)} \phi + F^+(A') + \psi_1 = 0,$$

and this has the solution

$$\psi_1 = -\, h^{(1)} F^+(A').$$

On the other hand for the double connected sum $X_1 \# Y_1 \# Y_2$ we construct a different connection A'_{12} say, and carry out the same construction, with operators $h^{(12)}$, $\varepsilon^{(12)}$. The self-dual curvature of A'_{12} is supported in two disjoint sets,

$$F^+(A'_{12}) = F^+_1 + F^+_2,$$

say, and we can identify F_1^+ with the term $F^+(A')$ appearing in the previous construction. On the double connected sum we solve the equation

$$F^+(A'_{12} + P^{(12)}\phi) + \psi_{12} = 0;$$

the linearization is to solve

$$F_1^+ + F_2^+ + d_{A'_{12}}^+ P^{(12)}\phi + \psi_{12} = 0,$$

which has solution

$$\begin{aligned}\psi_{12} &= -h^{(12)}(F_1^+ + F_2^+).\\ &= -h^{(12)}(F_1^+) - h^{(12)}(F_2^+).\end{aligned}$$

From (7.2.7) we have:

$$|h^{(12)}(F_2^+)| \leq \text{const.}\, \| F_2^+ \|_{L^2} \leq \text{const.}\, \lambda_2.$$

So we can prove the linearized version of (7.2.64) if we can show that

$$|h^{(12)}(F_1^+) - h^{(1)}(F_1^+)|$$

tends to 0 with λ_2.

Now we can write:

$$h^{(12)} - h^{(1)} = h(\gamma^{(12)}(1 + \varepsilon^{(12)})^{-1} - \gamma^{(1)}(1 + \varepsilon^{(1)})^{-1}),$$

where $\gamma^{(12)}$ is a cut off function vanishing in small balls about both connected sum points and $\gamma^{(1)}$ vanishes in a small ball about just one of these points. So $|\gamma^{(12)} - \gamma^{(1)}|$ is everywhere less than one and is supported on a ball of radius $O(\lambda_2^{1/2})$. Thus the operator norm of $\gamma^{(12)} - \gamma^{(1)}$, mapping L^p to L^2, tends to zero with ε and it suffices to show that

$$\| \gamma^{(12)}((1 + \varepsilon^{(12)})^{-1} - (1 + \varepsilon^{(1)})^{-1}) F_2^+ \|_{L^p} \longrightarrow 0$$

with λ_2. In turn, by the continuity of the inverse map on operators, it suffices to prove that the L^{2+2s} to L^{2+s} operator norm of

$$\gamma^{(12)}(\varepsilon^{(12)} - \varepsilon^{(1)})$$

tends to 0, for some $s > 0$. But this operator has the shape

$$\{\nabla\beta + (A'_1 - A'_{12})\} P_{A_1} \gamma^{(12)} + \{\nabla\beta + A'_{12} - B_2\} P_{A_2} \gamma_2,$$

say, where β and the difference of the connections are determined by the scale λ_2. We estimate this just as in Section 2.2.2, except that now we are allowed to lose a little on the exponent, so for example

$$\begin{aligned}\| \nabla\beta P_1 \phi \|_{L^{2+s}} &\leq \| \nabla\beta \|_{L^{4-u}} \| P_1 \phi \|_{L^{4+v}},\\ &\leq \text{const.}\, \| \nabla\beta \|_{L^{4-u}} \| \phi \|_{L^{2+2s}},\end{aligned}$$

where

$$1 - \frac{4}{2 + 2s} = \frac{-4}{4 + v} \quad \text{(Sobolev embedding theorem)}$$

and

$$\frac{1}{4+u} + \frac{1}{4+v} = \frac{1}{2+s} \quad \text{(Hölder inequality).}$$

Then $u > 0$ and the L^{4-u} norm of $\nabla\beta$ tends to zero with λ_2.

This completes our sketch proof of (7.2.64). Of course, corresponding results hold for connected sums with more summands and in cases when other cohomology groups are present. The general principle is that the model we obtain over the parameter space

$$\{(\lambda_1, \ldots, \lambda_N) | 0 < \lambda_j < \delta\}$$

has a natural extension over the sets where some of the λ_j are zero, and the connected sum degenerates.

7.3 Convergence

7.3.1 The main result

Some more work is required to realize the full scope of the constructions of Section 7.2. Let us approach the matter from the other end and suppose that we have a sequence $\lambda_\alpha \to 0$ and connections $A^{(\alpha)}$ on a fixed bundle E over the connected sum $X = X_1 \# X_2$ which are ASD with respect to the conformal structure defined by λ_α. We want to give simple criteria under which the $A^{(\alpha)}$ are, for large α, contained in one of our models of the previous section. For connections A_i over X_i let us say that the $A^{(\alpha)}$ are L^q-convergent to (A_1, A_2) if the L^q distance between $[A^{(\alpha)}]$ and $[A_i]$ over the subset $X_i''(\lambda_\alpha)$ tends to 0 as α tends to infinity. Here we write $X_i''(\lambda_\alpha)$ for the common subset denoted by X_i'' in Section 7.2. Notice that this notion of convergence depends on the given sequence λ_α. Theorem (7.2.62) asserts that $A^{(\alpha)}$ is contained in a model for large α if the sequence is L^q-convergent. We will now introduce two other notions of convergence. If $(y_1, \ldots, y_l)$ is a multiset in $(X_1 \cup X_2) \backslash \{x_1, x_2\}$ we say that the sequence $A^{(\alpha)}$ is *weakly convergent* to $(A_1, A_2, y_1, \ldots, y_l)$ if the gauge equivalence classes $[A^{(\alpha)}]$ converge to $[A_1], [A_2]$ over compact subsets of $(X_1 \cup X_2) \backslash \{x_1, x_2, y_1, \ldots, y_l\}$, and if the curvature densities $|F(A^{(\alpha)})|^2$ of the $A^{(\alpha)}$ converge to those of the A_i plus $8\pi^2 \Sigma \delta_{y_v}$ over compact subsets of $(X_1 \cup X_2) \backslash \{x_1, x_2\}$. The proof of Uhlenbeck's removal of singularities theorem from Chapter 4 adapts without difficulty to give:

Theorem (7.3.1). *Any sequence $A^{(\alpha)}$ of connections on a bundle E over X, ASD with respect to the conformal structures defined by a sequence $\lambda_\alpha \to 0$, has a weakly convergent subsequence. If the weak limit is $(A_1, A_2, y_1, \ldots, y_l)$ where A_i are connections on bundles E_i we have*

$$\kappa(E_1) + \kappa(E_2) + l \leq \kappa(E).$$

Next, we say that the sequence $A^{(\alpha)}$ is *strongly convergent* to (A_1, A_2) if it is weakly convergent to (A_1, A_2) (i.e. with no exceptional points y_j) and if

$$\kappa(E_1) + \kappa(E_2) = \kappa(E).$$

This condition asserts that no curvature is 'lost' over the neck as the size of the neck shrinks to zero. The result we shall prove in this section is:

Theorem (7.3.2). *A sequence of connections $A^{(\alpha)}$ on X, ASD with respect to parameters $\lambda_\alpha \to 0$, is L^q-convergent if and only if it is strongly convergent.*

In one direction this is rather trivial. If $A^{(\alpha)}$ is L^q-convergent, the restrictions converge in L^q over compact subsets of the punctured manifolds $X_i \backslash \{x_i\}$ and the ASD equation gives C^∞ convergence in a suitable gauge. The force of the result is the converse—C^∞ convergence over compact sets, together with the condition that the curvature is not lost over the neck, implies L^q convergence on the *increasing* series of domains $X_i''(\lambda_\alpha)$, not contained in any compact subset of the punctured manifolds.

The nub of the proof of (7.3.2) is to obtain control of the connections $A^{(\alpha)}$ over the neck region—the complement of compact subsets in the $X_i \backslash \{x_i\}$. The essential result is contained in the next proposition. After proving this proposition we shall return to complete the proof of (7.3.2) in Section 7.3.4.

Recall from Section 7.2.1 that we can model the neck conformally on a tube. For $T > 0$ let us write Z_T for the manifold

$$Z_T = (-T, T) \times S^3,$$

with its standard Riemannian metric.

Proposition (7.3.3). *There are constants η, $C > 0$, independent of T, such that if A is an ASD connection over Z_T with*

$$\|F\|_{L^2}^2 = \int_{Z_T} |F(A)|^2 \, d\mu \leq \eta^2,$$

then

$$|F(A)| \leq C e^{2(|t| - T)} \|F(A)\|_{L^2}$$

at a point (t, θ) of Z_T, for all t with $|t| \leq T - 1$.

We can replace $T - 1$ here by $T - k$, for any fixed constant k, if we adjust C accordingly.

7.3.2 The linearized problem

For purposes of exposition we will first give the proof of (7.3.3) in the abelian, linear case when A is a $U(1)$ connection. We will then come back to the general case which, with the method we use, is not substantially harder, but lacks the cleanness of the linear proof. In either case the following lemma is a basic step in the proof:

Lemma (7.3.4). *A 1-form a over the standard round three-sphere satisfies the inequality:*

$$\int_{S^3} \mathrm{d}a \wedge a \leq \tfrac{1}{2} \int_{S^3} |\mathrm{d}a|^2.$$

The existence of an inequality of this kind with some constant in the place of the factor $\frac{1}{2}$ is a straightforward application of the spectral theory of elliptic operators. If f is a function and a is replaced by $a + \mathrm{d}f$, each of the integrals is unchanged, so we may assume that a satisfies $\mathrm{d}^*a = 0$. Then

$$\int \mathrm{d}a \wedge a \leq \|a\|_{L^2} \|\mathrm{d}a\|_{L^2}$$

and $\int |\mathrm{d}a|^2 = \langle a, \Delta a\rangle$. So $\|\mathrm{d}a\|_{L^2} \geq \lambda \|a\|_{L^2}$, where λ^2 is the first eigenvalue of the Laplacian acting on co-closed 1-forms, and it follows that $\int \mathrm{d}a \wedge a \leq \lambda^{-1} \|\mathrm{d}a\|^2_{L^2}$.

To complete the proof we have to evaluate the eigenvalue λ. Let $H \subset \ker d^* \subset \Omega^1_{S^3}$ be the eigenspace belonging to λ^2. Then the operator $*\mathrm{d}$ takes H to itself and $(*\,\mathrm{d})^2 = \lambda^2$ on H. So there is a decomposition of H into subspaces on which $*\,\mathrm{d} = \pm\lambda$. By symmetry these subspaces have the same dimension. Let a be an element of H with $*\,\mathrm{d}a = \lambda a$, so a is extremal for the inequality, i.e.

$$\int \mathrm{d}a \wedge a = \lambda^{-1} \|\mathrm{d}a\|^2.$$

Now consider the 1-form A on the cylinder $S^3 \times \mathbb{R}$ given by

$$A = e^{-\lambda t} a.$$

We have $\mathrm{d}A = e^{-\lambda t}(\mathrm{d}a + \lambda a \wedge \mathrm{d}t) = \lambda e^{-\lambda t}(*_3 a + a \wedge \mathrm{d}t)$, where we have written $*_3$ for the Hodge $*$ operator on the three-sphere.

Now it is an elementary fact that on $S^3 \times \mathbb{R}$ the anti-self-dual forms are precisely those of the shape $*_3\alpha + \alpha \wedge \mathrm{d}t$. So $\mathrm{d}A$ is an ASD form, i.e. $\mathrm{d}^+ A = 0$. We now use our conformal equivalence between the cylinder and punctured Euclidean space to get a 1-form $A^\#$ over $\mathbb{R}^4 \setminus \{0\}$ satisfying $\mathrm{d}^+ A^\# = 0$. The condition that $\mathrm{d}^* a = 0$ on S^3 gives (in this case) that $\mathrm{d}^* A^\# = 0$ on $\mathbb{R}^4 \setminus \{0\}$, and the exponential form of A on the cylinder translates into the fact that the Euclidean coefficients of $A^\#$ are homogeneous functions of degree $\lambda - 1$.

We claim now that the coefficients of $A^\#$ are in fact linear functions on $\mathbb{R}^4$, so $\lambda - 1 = 1$ and $\lambda = 2$ as required. For if β is a standard cut-off function vanishing on a small ball and $\beta_r(x) = \beta(r^{-1}x)$, the smooth 1-form $\beta_r A^\#$ satisfies

$$|(\mathrm{d}^* + \mathrm{d}^+)(\beta_r A^\#)| = O(r^{\lambda - 2}),$$

and vanishes outside a ball of radius $O(r)$. It follows that, since $\lambda > 0$, $(\mathrm{d}^* + \mathrm{d}^+)(\beta_r A^\#)$ tends to 0 in L^1 as r tends to 0. Thus, as a distribution, $A^\#$

satisfies the equation $(\mathrm{d}^* + \mathrm{d}^+)A^\# = 0$. By the general regularity theory for the elliptic operator $\mathrm{d}^* + \mathrm{d}^+$ we obtain that $A^\#$ is smooth. But $A^\#$ cannot be a constant, since then $\mathrm{d}a$ would be zero, so the order of homogeneity must be at least one, and λ must be at least two. Conversely if we take, for example, the 1-form $x_1\,\mathrm{d}x_2 - x_2\,\mathrm{d}x_1 + x_4\,\mathrm{d}x_3 - x_3\,\mathrm{d}x_4$ and run the argument backwards we see that λ can be at most two.

We now turn to the proof of (7.3.3) in the abelian case. Given a $U(1)$-connection A over Z_T we write, for $0 \le t \le T$,

$$v(t) = \int_{S^3 \times (-t,\, t)} |F|^2\, \mathrm{d}^3\theta\, \mathrm{d}s. \tag{7.3.5}$$

The basic idea of the proof is to use a differential inequality for the function v. We have

$$\frac{\mathrm{d}v}{\mathrm{d}t} = \left(\int_{S^3 \times \{-t\}} + \int_{S^3 \times \{t\}} \right) |F|^2\, \mathrm{d}^3\theta. \tag{7.3.6}$$

The notation here means that we take the pointwise square norm of the curvature tensor in four dimensions, and then integrate this over the boundary three-spheres. Now, at a fixed point we can write

$$F(A) = F(A|_{S^3 \times \{t\}}) + \Phi\, \mathrm{d}t,$$

say, where Φ is a 1-form on the three-sphere. As in the proof of Lemma (7.3.4) the ASD condition takes the form $\Phi = *_3 F(A|_{S^3 \times \{t\}})$, where $*_3$ is the duality operator in three dimensions. In particular the pointwise norms of these two orthogonal components of the four-dimensional curvature tensor are equal and so, if we write A_t for the restriction of A to $S^3 \times \{t\}$, we have

$$\frac{\mathrm{d}v}{\mathrm{d}t} = 2\left(\int_{S^3 \times \{t\}} |F(A_t)|^2\, \mathrm{d}^3\theta + \int_{S^3 \times \{-t\}} |F(A_{-t})|^2\, \mathrm{d}^3\theta \right). \tag{7.3.7}$$

On the other hand we can, using Stokes' theorem, also write v as an integral over the boundary. In this abelian case we can think of A as being an ordinary 1-form. We have then

$$|F(A)|^2 = -F \wedge F = -\mathrm{d}A \wedge \mathrm{d}A = \mathrm{d}(\mathrm{d}A \wedge A).$$

So

$$v(t) = \int_{S^3 \times \{t\}} \mathrm{d}A_t \wedge A_t - \int_{S^3 \times \{-t\}} \mathrm{d}A_{-t} \wedge A_{-t}, \tag{7.3.8}$$

while (7.3.7) takes the form:

$$\frac{\mathrm{d}v}{\mathrm{d}t} = 2\left(\int_{S^3 \times \{t\}} |\mathrm{d}A_t|^2\, \mathrm{d}^3\omega + \int_{S^3 \times \{-t\}} |\mathrm{d}A_{-t}|^2\, \mathrm{d}^3\omega \right). \tag{7.3.9}$$

We now apply Lemma (7.3.4) to compare these boundary integrals and deduce that, in this abelian case, v satisfies the differential inequality:

$$\frac{\mathrm{d}v}{\mathrm{d}t} \geq 4v \tag{7.3.10}$$

or equivalently $\mathrm{d}\{\log v\}/\mathrm{d}t \geq 1$. W can integrate the inequality to get

$$v(t) \leq v(T)e^{4(t-T)}.$$

To complete the proof we put, for $|t| \leq T-1$,

$$E(t) = \int_{S^3 \times (t-1, t+1)} |F|^2 \, \mathrm{d}\mu.$$

Then $E(t)$ is trivially bounded by $v(|t|+1)$. Hence $E(t) \leq e^2 v(T) e^{4(tT)}$. The domain of the integral defining $E(t)$ is a translate of the model 'band' $B = S^3 \times (-1, 1)$. For any harmonic 2-form f over B we have an elliptic estimate:

$$\sup_{S^3 \times \{0\}} |f|^2 \leq \text{const.} \int_B |f|^2. \tag{7.3.11}$$

Applying this to $F = \mathrm{d}A$ on the translated bands we get

$$|F| \leq \text{const.}\, E(t)^{1/2} \leq \text{const.}\, e^{2(|t|-T)} v(T)^{1/2},$$

and the proof of (7.3.3) in the abelian case is complete.

7.3.3 The non-linear case

We now extend the argument to deal with the non-abelian case. With $v(t)$ defined by the integral of the curvature, as before, we still have

$$\frac{\mathrm{d}v}{\mathrm{d}t} = 2\left(\int_{S^3 \times \{t\}} |F(A_t)|^2 + \int_{S^3 \times \{-t\}} |F(A_{-t})|^2 \right),$$

and using the ASD condition we can still express v as a boundary integral of the Chern–Simons form. In terms of a connection matrix A we have:

$$v(T) = \left(\int_{S^3 \times \{t\}} - \int_{S^3 \times \{-t\}} \right) (\mathrm{d}A \wedge A + \tfrac{3}{2} A \wedge A \wedge A).$$

We now use the freedom to choose ε small. For an ASD connection over the standard band B whose curvature is sufficiently small in L^2, we can apply (4.4.10) to choose a correspondingly small connection matrix over an interior domain. So we can suppose that, for each fixed t with $|t| \leq T-1$, the connection A_t can be represented by a C^∞-small connection matrix over $S^3 \times \{t\}$.

We now apply (4.4.11). Let a be a connection over the three-sphere; if the curvature of a is small enough in L^2 we can choose a connection matrix a^τ with

$$\| a^\tau \|_{L^2_1(S^3)} \leq \text{const.} \| F(a) \|_{L^2(S^3)}.$$

The leading (i.e. quadratic) terms in the expressions for $\int |F(a)|^2$ and for the Chern–Simons invariant $T_{S^3}(a) = \int da \wedge a + \frac{2}{3} a \wedge a \wedge a$ are just the same as those considered in the linear case above. Estimating the higher (cubic and quartic) terms, and using the inequality (7.3.4) for the leading term, we get, for any connection a over S^3 with curvature small in L^2,

$$\begin{aligned} T_{S^3}(a) &\leq \tfrac{1}{2} \| F(a) \|^2_{L^2} + \text{const.} \| a^\tau \|^3_{L^2_1}, \\ &\leq \tfrac{1}{2} \| F(a) \|^2_{L^2} + \text{const.} \| F(a) \|^3_{L^2}. \end{aligned}$$

We apply this inequality to the A_t (assuming again that ε is chosen small). Combining with the formulae above we get

$$\frac{dv}{dt} \geq 4v - \text{const.} \left(\frac{dv}{dt} \right)^{3/2}. \tag{7.3.12}$$

We also know that v and its derivative are small. We have an elementary inequality: if $y + Cy^{3/2} \geq 4x$ for some C, and x and y are small, then $y \geq 4x - C'x^{3/2}$ for another constant C'. Hence (7.3.12) gives

$$\frac{dv}{dt} \geq 4v - \text{const.}\, v^{3/2}. \tag{7.3.13}$$

We complete the proof by running twice through the same argument as before. First, with some fixed small δ we choose ε so small that the inequality above gives $dv/dt \geq (4 - \delta)v$, which we can integrate to get an exponential bound

$$v(t) \leq e^{4(t-T)} v(T).$$

Feeding this back into the differential inequality, we get

$$\frac{dv}{dt} \geq 4v - Ce^{(4-\delta)(t-T)} v,$$

say. It follows that

$$\begin{aligned} \log v(t) - \log v(T) &\geq 4 \int_t^T \frac{d\tau}{1 + Ce^{(4-\delta)(\tau - T)}} \\ &\geq 4 \int_t^T (1 - Ce^{(4-\delta)(\tau-T)})\, d\tau \\ &\geq 4(T - t) - \frac{4C}{4 - \delta}. \end{aligned}$$

Taking exponentials, we have the bound

$$\nu(t) \leq K\nu(T)e^{4(t-T)},$$

with $K = \exp(4C/(4-\delta))$. Finally we translate this into a bound on $|F|$ using the elliptic estimates over bands, just as before.

7.3.4 *Completion of proof*

We now go back to prove (7.3.2), using (7.3.3). Consider first an ASD connection A over the cylinder Z_T which satisfies the hypothesis of (7.3.3), and write ν for $\nu(T)$. We construct a connection matrix A^τ for A as follows. First use (4.4.10) (or (4.4.11)), together with the curvature estimate, to construct a connection matrix A_0^τ over $S^3 \times \{0\}$ with

$$\sup |A_0^\tau| \leq \text{const.}\, e^{-2T}\nu^{1/2}.$$

Now use parallel transport along the $\mathbb{R}$ factor in Z_T (the lines 'θ = constant') to get a connection matrix A^τ over the cylinder. We have

$$\frac{\partial A^\tau}{\partial t} = F_{t\theta},$$

so

$$\left|\frac{\partial A^\tau}{\partial t}\right| \leq \text{const.}\, e^{2(|t|-T)}\nu^{1/2},$$

hence

$$|A_{(t,\theta)}| \leq \text{const.}\, e^{2(|t|-T)}\nu^{1/2} + \text{const.}\, e^{-2T}\nu^{1/2} \leq \text{const.}\, e^{2(|t|-T)}\nu^{1/2}.$$

We now transform this result into Euclidean coordinates. Let $e: \mathbb{R} \times S^3 \to \mathbb{R}^4 \backslash \{0\}$ be the standard conformal map and put $\lambda = e^{T/2}$. The composite of e with the map which translates the cylinder by $T + \log \sigma$ gives a conformal map from Z_T to the annulus, $\Omega(\sigma, \lambda)$ say, with radii σ and $\lambda^{-1}\sigma$. Fix $k < 1$ and let $\Omega^{(1)}(\sigma, \lambda) \subset \Omega(\sigma, \lambda)$ be the outer annulus

$$\{x \mid k\sigma\lambda^{1/2} < |x| < \sigma\}.$$

For a connection over $\Omega(\sigma, \lambda)$ with $\|F\|_{L^2} \leq \eta$, the connection matrices constructed in the manner above have

$$|A_x| \leq \text{const.}\, \|F\|_{L^2}|x|$$

over $\Omega^{(1)}(\sigma, \lambda)$, where the constant is independent of λ and σ.

To complete the proof of (7.3.2) we take such a region $\Omega(\sigma, \lambda)$ as a model for the neck in the connected sum. We choose k so that the open set $X_i'' = X_i''(\lambda)$ of X_i is contained in $\{y \in X_i \mid d(x_i, y) \geq k\lambda^{1/2}\}$. So if $X_i^* \subset X_i$ is the complement of the ball of radius $\frac{1}{2}\sigma$, say, about x_i, we have a decomposition into three overlapping open sets

$$X = X_1^* \cup \Omega(\sigma, \lambda) \cup X_2^*,$$

in which, with the obvious notation,

$$X_i'' \subset X_i^* \cup \Omega^{(i)}(\sigma, \lambda).$$

Let $A^{(\alpha)}$ be a strongly convergent sequence, with respect to $\lambda_\alpha \to 0$, as considered in (7.3.2), with limit A_i over X_i. We choose σ so that

$$\int_{B(x_i, \sigma)} |F(A_i)|^2 < \eta.$$

Since no curvature is lost in the limit, we have for large α

$$\int_{\Omega(\sigma, \lambda)} |F(A^{(\alpha)})|^2 \leq \eta,$$

and we can apply (7.3.3). This gives us uniformly small connection matrices for the $A^{(\alpha)}$ over the outer annulus $\Omega^{(i)}(\sigma, \lambda) \subset X_i$. By hypothesis, the connections converge to A_i over the fixed precompact set X_i^* (i.e. independent of λ). We then patch together the trivializations over the fixed overlap $X_i^* \cap \Omega^{(i)}(\sigma, \lambda)$, in the manner of Section 4.4.2, to get a representative $A_i + a_i^{(\alpha)}$ for $A^{(\alpha)}$ over X_i'' with

$$\| a_i^{(\alpha)} \|_{L^q(X_i'')} \longrightarrow 0$$

as α tends to infinity. This proves (7.3.2) (since the L^∞ norm dominates the L^q norm).

Notes

Sections 7.1.1 and 7.1.2

The excision principle was formulated by Atiyah and Singer (1968) and we sketch their proof in Section 7.1.1. There is a detailed exposition in Booss and Bleeker (1985). The proof in Section 7.1.2 was motivated in part by Witten's deformation of the de Rham complex (Witten, 1982). For other excision theorems see Gromov and Lawson (1983) and Teleman (1983).

Section 7.1.5

The basic idea of the analysis here goes back to Taubes (1982), who was working with the Laplacian related to the deformation complex of an ASD connection.

Section 7.1.6

These standard orientations of the determinant line bundles were defined in Donaldson (1987*b*), to which the reader should refer for more details.

Section 7.2

The idea that, under rather general hypotheses, one can glue together solutions of the ASD equations is due to Taubes (1982, 1984). Taubes considered the problem of constructing 'concentrated' ASD solutions, which is contained as a special case in the connected sum framework, as we see in Section 8.2. While we follow Taubes's strategy in general outline, the details are rather simpler in our approach. (Taubes's original set-up is more appropriate for analysing, for example, the metrics on the moduli spaces; see Groisser and Parker (1989).)

The main point not covered explicitly in Taubes's papers is the moduli problem, describing all the concentrated solutions. This was solved in a special case with an *ad hoc* method by Donaldson (1983*b*) and a similar approach was taken by Freed and Uhlenbeck (1984). A general 'gluing' theorem was given by Donaldson (1986), with a rather long proof. That proof does have the aesthetic advantage that it avoids appealing to the index theorem, and so gives another approach to the excision principle, as we described in Section 7.2.5. Donaldson and Sullivan (1990) used this idea to prove the index formula in the setting of 'quasiconformal' base manifolds.

Two versions of these gluing theorems in the framework of holomorphic bundles have appeared in the literature. In Donaldson and Friedman (1989) self-dual connections over the connected sum of 'self-dual manifolds' are studied via holomorphic bundles on the twistor space. The analogue of Taubes's original gluing construction for bundles over algebraic surfaces is developed by Gieseker (1988).

We mention two important aspects of the theory which we have not covered in this chapter. One is to obtain information about the obstruction map Ψ which defines the local model for the moduli space. In Taubes (1984) and Donaldson (1986) it is shown that the leading term in Ψ, as the parameter λ tends to 0, is given by pairing the curvature of one connection A_i at the point x_i with the harmonic forms representing $H^2_{A_j}$ at x_j, using the identification σ of the tangent spaces.

Second, we mention another analytical framework for the theory provided by the use of non-compact four-manifolds with tubular ends. We mentioned this model for the connected sum briefly in Section 7.2.1. Once one has developed the basic elliptic theory on manifolds with tubular ends, see Taubes (1986) and Floer (1989), the analysis of the gluing problem can be simplified at some points. A similar idea is used by Freed and Uhlenbeck (1984). One advantage of this framework is that it extends easily to 'generalized connected sums' in which the three-sphere is replaced by some other three-manifold; see Mrowka (1989). Another advantage is that it extends easily to handle the more complicated problem of gluing self-dual manifolds; see Floer (1990).

Section 7.3

The proof of the decay estimate goes back to Uhlenbeck's (1982*a*) original proof of the removal of singularities theorem, which dealt with general Yang–Mills connections. The simpler version here for ASD connections is much the same as that given by Donaldson (1983*a*, Appendix). Another argument (yielding a weaker decay rate) is given by Freed and Uhlenbeck (1984, Appendix D).

8

NON-EXISTENCE RESULTS

In this chapter we return at last to address some of the problems in four-manifold topology which were raised in Chapter 1. We will prove Theorems (1.3.1) and (1.3.2) on the non-existence of manifolds with certain intersection forms. Throughout the chapter, the strategy of proof of these non-existence results will be the same. Supposing that a manifold X of the type in question exists, one chooses judiciously a bundle E over X and studies the moduli space of ASD connections $M = M_E$. From the known topological features of the moduli space and its embedding in the space $\mathscr{B}_E$ one derives a contradiction. More precisely, one considers a case where the moduli space M is either non-compact or contains some reducible connections. Then one truncates the moduli space to obtain a compact manifold-with-boundary $M' \subset M \cap \mathscr{B}_E^*$, of dimension s. Then for any cohomology class $\theta \in H^{s-1}(\mathscr{B}_E^*)$ we can assert that

$$\langle \theta, \partial M' \rangle = 0.$$

In Section 8.1 we prove Theorem (1.3.1) using a simple approach of Fintushel and Stern (1984), in which attention is focused on the known behaviour of the moduli space around the abelian reductions. This approach does not, however, seem to extend to the indefinite forms considered in (1.3.2). For these we use a more sophisticated proof which takes as its starting point an alternative (and earlier) proof in the definite case, which we give in Section 8.3.1. The extension to indefinite forms can be regarded as a partial stabilization of this proof with respect to the operation of connected sum with $S^2 \times S^2$. Here attention is focused on the non-compact nature of the moduli spaces and a description of the structure in the neighbourhood of points at infinity in the compactification, which we obtain in Section 8.2 as a corollary of our work in Chapter 7.

8.1 Definite forms

8.1.1 The quadratic form $E_8 \oplus E_8$

Rohlin's theorem (1.2.6) tells us that the signature of a spin four-manifold is divisible by 16. Thus the first non-standard quadratic form which is a candidate for the intersection form of a smooth compact four-manifold is $E_8 \oplus E_8$ or equivalently, switching orientations, the negative definite form $-E_8 \oplus -E_8$. We shall show that no such manifold exists, using an argument given by Fintushel and Stern.

Theorem (8.1.1). *There does not exist a smooth, oriented, simply-connected, closed four-manifold with intersection form* $-E_8 \oplus -E_8$.

Suppose X were such a manifold and let e be any class in $H^2(X; \mathbb{Z})$ with $e^2 = -2$. Let L be a complex line bundle over X with $c_1(L) = e$, and let F be the $SO(3)$ bundle $L \oplus \underline{\mathbb{R}}$. Fix a generic Riemannian metric on X, and consider the moduli space M_F. Since $\kappa(F) = -\frac{1}{4}p_1(F) = -\frac{1}{4}e^2 = \frac{1}{2}$, our dimension formula (4.2.22) gives

$$\dim M_F = 8 \cdot \tfrac{1}{2} - 3(1 + 0) = 1. \tag{8.1.2}$$

So the irreducible connections in the moduli space form a smooth one-dimensional manifold. We know from Section 4.3 that the moduli space M_F can be compactified by adjoining ideal ASD connections involving the moduli spaces $M_{F^{(r)}}$ for $SO(3)$ bundles $F^{(r)}$ with

$$w_2(F^{(r)}) = w_2(F) = e(\text{mod } 2), \quad \kappa(F^{(r)}) = \kappa(F) - r,$$

where r is positive. However in this case all the moduli spaces $M_{F^{(r)}}$, for $r > 0$, are *empty*. We can see this in two ways: either from the dimension formula, which tells us that the dimension of these moduli spaces is negative, or more simply from the Chern–Weil formula (2.1.41), since κ is non-negative for bundles which admit ASD connections. So our general compactness theorem simply asserts, in this case, that M_F is compact.

Now the intersection form of X is negative-definite, so we can regard it as defining a Euclidean norm $\|\alpha\|^2 = -(\alpha \cdot \alpha)$ on $H^2(X; \mathbb{R})$. According to (4.2.15) there is exactly one equivalence class of reducible connections in M_F for each pair $\{f, -f\}$ where $f \in H^2(X; \mathbb{Z})$ satisfies

$$f = e(\text{mod } 2), \quad \|f\| = \|e\|.$$

If f is such a cohomology class, the first condition says that the mid-point

$$m = \tfrac{1}{2}(e + f)$$

lies in the integer lattice $H^2(X; \mathbb{Z}) \subset H^2(X; \mathbb{R})$. But we clearly have $\|m\|^2 \leq \|e\|^2 = 2$, with equality if and only if $m = f = e$. So if f does not equal e, the norm of m must be 0 or 1. But the $E_8 \oplus E_8$ lattice is even and hence contains no vector of length one. Therefore we must have $m = 0$ and $f = -e$. Thus we conclude that M_F has exactly one point representing a reducible connection, that corresponding to L and $\{e, -e\}$.

We now have the desired contradiction in our hands: according to (4.3.20) a neighbourhood of the reduction in M_F is modelled on a cone over a complex projective space—in this case a cone over $\mathbb{CP}^0$, i.e. a closed half-line. Thus M_F is a compact one-manifold with boundary, having exactly one boundary point, and this is impossible.

8.1.2 *Other definite forms*

The proof of (8.1.1) used only two properties of the form $E_8 \oplus E_8$, first that it is definite and second that the integer lattice contains a vector of squared-length two which cannot be written as the sum of shorter vectors. There are therefore many other forms to which the same argument applies; for example we see immediately that none of the forms

$$m(1) \oplus n(E_8), \quad n > 0$$

can occur as intersection forms of smooth, simply connected, closed four-manifolds.

Many definite forms, however, do not represent the value two. The first example is the Leech lattice, of rank 24, in which the shortest non-zero vector has squared-length four. A simpler case is the tensor product $E_8 \otimes \ldots \otimes E_8$ of k copies of E_8; this unimodular form takes on no values between zero and 2^k. Forms such as these require some extension of the argument. We will now go on to give a proof of the general result, Theorem (1.3.1).

Consider a general negative definite form Q. The vectors e in the lattice with $Q(e) = -1$ span a sublattice on which q may be diagonalized. So we can write

$$Q = Q' \oplus \mathrm{diag}(-1, \ldots, -1).$$

Here Q' is the 'non-standard' part of the form. Theorem (1.3.1) asserts that if Q arises from a smooth simply-connected four-manifold X then the non-standard part Q' is zero.

Suppose the contrary and choose a non-zero class $e \in H^2(X; \mathbb{Z})$ from the non-standard part of the lattice, with $-Q'(e)$ minimal; say $Q'(e) = -2-d$ with $d \geq 0$. Following the scheme of proof of (8.1.1) we let L be a line bundle with $c_1(L) = e$ and let $F = \underline{\mathbb{R}} \oplus L$. Choose a generic metric on X and consider the moduli space M_F, which has dimension $2d + 1$ and whose only singularities correspond to reductions. The same argument as before shows that the only reduction of F is that corresponding to L. So M_F has exactly one singular point. A neighbourhood of this point is modelled on $\mathbb{C}^{d+1}/S^1$—a cone over $\mathbb{CP}^d$. In the situation considered in (8.1.1), with $d = 0$, the low dimensionality ensured that the moduli space was compact. This will continue to hold so long as $2d + 1 \leq 7$, i.e. $d \leq 3$. For then the lower moduli spaces have negative formal dimension so they contain, generically, no irreducible connections. On the other hand the minimality of $Q'(e)$ ensures that these lower bundles do not admit any reductions either; so the lower moduli spaces are empty and M_F is compact. In the general case, when d is larger than three, we must expect however that the compactification of M_F will involve the lower bundles. Note though that in any case we do not encounter reductions in the lower moduli spaces.

In the compact case, $d \leq 3$, we can finish the proof right away. Let M' be the compact manifold-with-boundary obtained from M_F by removing an open conical neighbourhood of the singular point. Thus M' is an orientable manifold which lies in the space $\mathscr{B}_F^*$ of irreducible connections, and $\partial M' = \mathbb{CP}^d$. So any cohomology class θ over the ambient space must have zero pairing with the boundary. It is here that the analysis from Chapter 5 of the cohomology of the space $\mathscr{B}_F^*$ bears fruit, for it is easy for us to find classes θ which have non-zero pairing with the boundary, and hence derive the desired contradiction. For example, fix a class Σ in $H_2(X; \mathbb{Z})$ with $e(\Sigma)$ non-zero, and consider the cohomology class $\theta = \mu(\Sigma)^d$, with $\mu(\Sigma)$ as defined in Section 5.1. According to (5.1.21), the restriction of the class $\mu(\Sigma)$ to the copy of $\mathbb{CP}^\infty$ linking the reduction is $-\frac{1}{2}e(\Sigma)$ times the standard generator h. The subspace $\mathbb{CP}^d$ is embedded in the link in the standard way so, since h^d is the fundamental class on $\mathbb{CP}^d$, we have

$$\langle \mu(\Sigma)^d, [\partial M'] \rangle = \pm (1/2^d) e(\Sigma)^d,$$

which is non-zero by construction. Here the sign ambiguity depends on our choice of orientation for the moduli space, but is not relevant to the argument. (Note that if d is even we can derive a contradiction using only the 'intrinsic' structure of the moduli space—rather than its 'extrinsic' embedding in the space $\mathscr{B}_F^*$; for in that case the projective space does not bound any manifold. We do not even need orientations.) This completes the proof in the case when $d \leq 3$.

For the proof in the general case we need to add one more ingredient to the theory, an idea which will also be central to the discussion in Chapter 9. We define a subspace of M_F by removing a cone as before. This need not now be compact, so we will denote it by M^0. The same argument goes through if we know that there is a class θ, non-zero on the boundary, which can be represented by a cochain on M^0 whose *support* is compact. In abstract terms we can use the natural pairing ('integration') between a fundamental class of $(M^0, \partial M^0)$ and the compactly supported relative cochains. More explicitly, we truncate M^0 by removing the complement of a large compact subset containing the support of the cochain, to obtain a compact manifold-with-boundary $M' \subset M^0$. While we introduce new boundary components in this way, the pairing of θ with the new components is zero by construction, so we have the same contradiction

$$0 = \langle \theta, \partial M' \rangle = \langle \theta, \mathbb{CP}^d \rangle.$$

The proof of (8.1.1) is thus completed by the following proposition, which is in turn proved in Section 8.1.3.

Proposition (8.1.3). *There is a representative for the cohomology class $\mu(\Sigma)^d$ over M^0 which has compact support.*

8.1.3 Restriction and compactification

We will digress slightly to consider what can be said in general about the support of the cohomology classes $\mu(\Sigma)$ defined in Chapter 5. It is convenient to work with the geometrical representation by codimension-two submanifolds considered in Section 5.2.2, (although these are not strictly cochains). Thus we consider a surface $\Sigma \subset X$, a tubular neighbourhood $\nu(\Sigma)$ and a codimension-two submanifold V_Σ transverse to all the moduli spaces, as in (5.2.12) and (5.2.13). Here we adopt the same abuse of notation as in Section 5.2, regarding the V_Σ as being simultaneously subsets of the different moduli spaces, and this is justified by the fact that they are pulled back by the restriction map from closed subsets of $\mathscr{B}_{\nu(\Sigma)} \cup \{\Theta\}$, in which the distinction between the bundles disappears.

Now suppose that $[A_\alpha] \in M_E$ is a sequence of ASD connections over X which converges weakly to a limit $(A_\infty; x_1, \ldots, x_l)$. Suppose that the $[A_\alpha]$ lie in $V_\Sigma \cap M_E$ for all α. Also suppose that A_∞ is either irreducible or a product connection. (In fact we just need to assume that if A_∞ is reducible the reduction has degree zero over Σ.) Then we have the following elementary alternative:

Lemma (8.1.4). *In this situation, either $[A_\infty]$ lies in V_Σ or one of the points x_r lies in the closure of the tubular neighbourhood $\nu(\Sigma)$.*

The proof of this important lemma is rather trivial. If none of the points x_r lie in the closure of the tubular neighbourhood, the restriction of the connections $[A_\alpha]$ to this neighbourhood converge in C^∞ to $[A_\infty]$. Thus $[A_\infty]$ lies in the closure of V_Σ in $\mathscr{B}_{\nu(\Sigma)}$. Since the trivial connection was chosen not to lie in this closure (see the discussion following (5.2.12)), the limit must be irreducible and hence must lie in V_Σ, since this a closed subset of $\mathscr{B}^*_{\nu(\Sigma)}$.

We will illustrate the utility of this lemma by straightaway deducing (8.1.3). Choose d representatives $\Sigma_1, \ldots, \Sigma_d$ for the same homology class $\Sigma \in H_2(X)$, and small tubular neighbourhoods $\nu(\Sigma_i)$. These may be chosen so that the triple intersections

$$\overline{\nu(\Sigma_i)} \cap \overline{\nu(\Sigma_j)} \cap \overline{\nu(\Sigma_k)}$$

are empty (for i, j, k all distinct). We choose representatives V_{Σ_i} in general position with respect to all the moduli spaces, as in (5.2.12). Now we claim that the intersection

$$V^0 = M^0 \cap V_{\Sigma_1} \cap \ldots \cap V_{\Sigma_d}$$

is compact. To see this, suppose that $[A_\alpha]$ is a sequence in the intersection; taking a subsequence we can suppose it converges weakly to a point $([A_\infty], x_1, \ldots, x_l)$ in the compactified moduli space. Now, as we have noted above, in this situation there are no reductions in any of the lower bundles; so

A_∞ is irreducible and we can invoke the alternative of (8.1.4) for each Σ_i. There are d surfaces in total, and each of the points x_j can lie in at most two of the $\overline{\nu(\Sigma_i)}$. Since there are l points x_j, the connection $[A_\infty]$ must be contained in at least $d - 2l$ of the V_{Σ_i}. But the dimension of the moduli space $M_{F^{(l)}}$ containing $[A_\infty]$ is given by

$$\dim M_{F^{(l)}} = \dim M_F - 8l = 2d + 1 - 8l, \tag{8.1.5}$$

and if l is bigger than zero this is less than $2(d - 2l)$. Since all the multiple intersections with all the lower moduli spaces were chosen to be transverse and the V_{Σ_i} have codimension two, we deduce that l must be zero. Then $[A_\infty]$ represents a limit point of the sequence in $V^0 = M^0 \cap V_{\Sigma_1} \cap \ldots \cap V_{\Sigma_d}$, so this intersection is indeed compact as asserted.

Finally, to prove (8.1.3) and so (8.1.1) one can choose cochains representing the classes $\mu(\Sigma)$ by slightly smoothing out the submanifolds V_{Σ_i}, preserving the compactness of the support. Alternatively, and more directly, we consider the intersection V^0: it is a compact oriented one-manifold whose oriented boundary is (counting algebraically, i.e. with signs) $\langle \mu(\Sigma)^d, \partial M^0 \rangle$ points. Since this is non-zero we obtain our contradiction.

8.2 Structure of the compactified moduli spaces

8.2.1 Scaling

We will now apply the results of Chapter 7 to obtain descriptions of neighbourhoods of points at infinity in the compactified moduli spaces. This uses a simple rescaling construction. Let x be a point in the Riemannian four-manifold X and suppose, for simplicity, that the metric is flat in a neighbourhood of x. Fix a local coordinate system identifying this neighbourhood with a neighbourhood of 0 in $\mathbb{R}^4$. For $\lambda > 0$ let $d_\lambda: \mathbb{R}^4 \to \mathbb{R}^4$ be the dilation map $d_\lambda(y) = \lambda^{-1} y$ and let $m: \mathbb{R}^4 \to S^4 = \mathbb{R}^4 \cup \{\infty\}$ be the standard stereographic map. The composite $m \circ d_\lambda$ maps the r-ball about zero to the complement in S^4 of a ball of radius $O(\lambda r^{-1})$ about the point ∞. Choose conformal coordinates z_i about ∞ related to the coordinates in the finite part by the inversion map, $z_i = (1/|y|^2) y_i$. Then $m \circ d_\lambda$, regarded as a map from a small annulus Ω_1 about x in $\mathbb{R}^4$ to a similar annulus about ∞ in S^4, is the same as the identification map $f_{\sigma, \lambda}$ we used in Section 7.2.1 to construct the connected sum of X and S^4, with σ the natural orientation reversing isometry between the tangent spaces to S^4 at 0 and ∞. On the other hand $m \circ d_\lambda$ extends to the ball enclosed by Ω_1 and this gives a conformal identification of X and $X \# S^4$.

Now suppose that A is an ASD connection on a bundle E over X and I is a non-trivial ASD connection on a bundle V over S^4. We can apply the construction of Section 7.2 to study ASD connections $A(\rho)$ on $X \# S^4$, with λ

small, close to A and I on the two factors. Using the conformal equivalence above, such a solution will give an ASD connection over the original Riemannian manifold X, close to A away from x. On the other hand, viewed on X, the curvature of $A(\rho)$ will be very large near x, in fact $O(\lambda^{-2})$, and it is precisely this kind of connection which we have encountered in our compactification of the moduli spaces. Of course, one can consider connections concentrated near a number of points by allowing multiple connected sums, as in Section 7.2.7. While we have assumed for simplicity that the metric on X is flat near x, this is by no means essential—in general one uses geodesic coordinate systems and readily verifies that the additional terms in the equations cause no new difficulties.

A few remarks are needed to adapt the results of Chapter 7 to this situation, because the parameter λ, and also the point x at which we make the connected sum, are now coupled to the parameters in the moduli spaces over S^4, via the action of the dilations and translations on these moduli spaces. For brevity we shall just consider the case when the structure group is $SU(2)$ or $SO(3)$, and V is a bundle with $\kappa(V) = 1$, so I is a standard one-instanton as described in Section 3.4 and the translations and dilations act transitively on the moduli space. To make our construction we adopt the approach sketched in Chapter 7, using local coordinates on the moduli space over S^4. We need to choose suitable measures of the 'centre' and 'scale' of a connection close to the standard one-instanton (with centre zero and scale one). There are many ways of doing this; convenient measures to take are the centre and radius of the smallest ball in $\mathbb{R}^4$ which contains half of the total energy (integral of $|F|^2$). There will be a unique such minimal ball for connections close to the basic instanton. Or one can mollify this definition by replacing the integrals over balls by integrals weighted by compactly supported functions. The important point is that the definition depends only on the restriction of a connection to a compact subset in $\mathbb{R}^4$. This means that the definition can be taken over to 'concentrated connections' over X. For fixed x and λ we can run our construction to describe connections on X with curvature concentrated near x, with centre x and scale λ. Then we allow x and λ to vary, regarded as smooth parameters in the construction, to describe open sets in the moduli space.

Our discussion of convergence in Section 7.3 can be applied after a similar modification. Suppose that $[A_\alpha]$ is a sequence of ASD connections over X which converge weakly to the ideal point $([A], x)$. Let λ_α be the radius of the smallest ball in X containing A_α-energy $4\pi^2$, and let x_α be the centre of such a minimal ball. Clearly the sequence x_α converges to x. Let χ_α be the map $m \circ d_{\lambda_\alpha}$ formed as above using the centre x_α; it maps a small ball centred on x_α to the complement of a small ball in a copy of the four-sphere. The connections $(\chi^{-1})^*(A_\alpha)$ are ASD connections over an increasing series of domains which exhaust $\mathbb{R}^4$ and it follows from the removability of singularities theorem, together with the normalization chosen, that they converge to the basic instanton I over compact subsets. This implies that the point x_α is unique, for large α, so we can assign a centre and scale to all sufficiently concentrated connections. Suppose then that the centre $x_\alpha = x$ for all α. Then the rescaling map χ_α is the same as that considered in the construction, i.e. we can absorb the rescaling into the neck parameter in the connected sum. We deduce from (7.3.2) that our model describes an entire neighbourhood of the point (A, x) in the compactified moduli space. Similar remarks apply in the case when we have a number of points of concentration in X; for such connections we have a collection of local centres and scales.

Our third and final remark concerns the relation between the gluing parameters and the action of the space rotations on the connections. As we have mentioned in Section 4.1, the standard one-instanton can be obtained as the standard connection on the spin bundle S^- of S^4. The associated $SO(3)$ connection is naturally defined on the bundle $V = \Lambda^-_{S^4}$. To make the connected sum construction one needs to specify an identification ρ of the fibre of $\mathfrak{g}_E$ at x with the fibre of $\Lambda^-_{S^4}$ at ∞. However, the natural orientation-reversing isometry between the tangent spaces to S^4 at the antipodal points 0, ∞ means that we can identify the latter with the fibre of Λ^+_X at x. Our gluing data is the copy of $SO(3)$:

$$\mathrm{Gl}_x = \mathrm{Hom}_{SO(3)}(\Lambda^+_{X,x}, \mathfrak{g}_{E,x}). \tag{8.2.1}$$

8.2.2 *Summary of results*

We give a general result and then illustrate it by a number of examples. Let $x_1, \ldots, x_l$ be distinct points in X, contained in disjoint coordinate neighbourhoods $U_1, \ldots, U_l$. Let A be an ASD connection on an $SU(2)$ or $SO(3)$ bundle E over X and let Ω be the product

$$\Omega = H^1_A \times \prod_{r=1}^{l} (U_r \times \mathbb{R}^+ \times \mathrm{Gl}_{x_r}). \tag{8.2.2}$$

We let $\Omega_\varepsilon \subset \Omega$ be the subset where all the $\mathbb{R}^+$ coordinates lie between 0 and ε, and where the norm of the H^1_A component is less than ε. We write M for the moduli space of ASD connections with $\kappa = \kappa(E) + l$, and $w_2 = w_2(E)$ in the $SO(3)$ case. Then we have:

Theorem (8.2.3). *For small ε there is a smooth map*

$$\Psi : \Omega_\varepsilon \longrightarrow H^2_A,$$

a neighbourhood $\bar{N}$ of $([A], x_1, \ldots, x_l)$ in the compactification $\bar{M}$ and an isomorphism (of ringed spaces) from the quotient $\Psi^{-1}(0)/\Gamma_A$ to $N = \bar{N} \cap M$. Under this isomorphism the projections from Ω_ε onto the U_r and the $\mathbb{R}^+$ factors go over to the r local centre and scale maps.

This follows from (7.2.62), (7.3.2) and the remarks above. We will need a slight extension which is a more-or-less direct consequence of the discussion in Section 7.2.8. Let $\bar{\Omega}$ be the conical completion of Ω: the product of H^1_A with the U_r and cones over the Gl_{x_r}. Let $\bar{\Omega}_\varepsilon$ be the closure of Ω_ε in $\bar{\Omega}$.

Proposition (8.2.4). *The map* Ψ *of* (8.2.3) *extends continuously to a map* $\bar{\Psi}: \bar{\Omega}_\varepsilon \to H^2_A$ *and the isomorphism of* (8.2.3) *extends to a homeomorphism from* $\bar{\Psi}^{-1}(0)/\Gamma_A$ *to a neighbourhood of* $([A], x_1, \ldots, x_l)$ *in* $\bar{M}$.

This gives us a fairly good understanding of the structure of the compactified moduli space. To complete the picture one has to study the diagonals in the symmetric product, and this can be done by an inductive procedure, first compactifying the moduli spaces over S^4. However, the results above will cover all our needs.

We consider some special cases of this general description. First suppose that H^2_A is zero and Γ_A acts trivially. Then with $l = 1$ we see that the compactified moduli space $\bar{M} = \bar{M}_\kappa$ is modelled near the second stratum on a bundle over $M_{\kappa-1} \times X$ whose fibre is a cone over $SO(3)$. The link of the stratum in the moduli space is $SO(3)$. An example of this is provided by Example (iv) of Section 4.1. In that case $M_{\kappa-1}$ is a point, $\bar{M}$ is the symmetric product $s^2(\mathbb{CP}^2)$ and the diagonal has link $SO(3) = S^3/\pm 1$.

The situation we shall be principally concerned with is when A is the product connection θ, so $\Gamma_A = SO(3)$ and $H^2_A = \mathscr{H}^+ \otimes \mathbb{R}^3$, where $\mathscr{H}^+$ is the space of self-dual harmonic forms on X, of dimension $b^+(X)$. We suppose that X is simply connected, so H^1_A is zero. We can represent the gluing factors Gl_{x_r}, locally, by copies of $SO(3)$; then Γ_A acts by left multiplication on these copies of $SO(3)$ and by the standard action on $\mathbb{R}^3$ in $\mathscr{H}^+ \otimes \mathbb{R}^3$. If $l = 1$ we can cancel the gluing factor by the symmetry and represent a neighbourhood in the compactified moduli space by the zeros of a map

$$\bar{\Psi}: U \times [0, \varepsilon) \longrightarrow \mathscr{H}^+ \otimes \mathbb{R}^3,$$

for an open set U in X. More globally, a neighbourhood of $\{\theta\} \times X$ in the compactified moduli space is modelled on the zeros of a section of the vector bundle $\pi_1^*(\mathscr{H}^+ \otimes \Lambda^+_X)$ over $X \times [0, \varepsilon)$. In particular if X has negative definite intersection form we see that $\bar{M}$ is a manifold-with-boundary, containing a collar $X \times [0, \varepsilon)$. This is well illustrated by Examples (i) and (ii) of Section 4.1, and in Example (ii) we can see explicitly how, after rescaling, the concentrated connections on $\overline{\mathbb{CP}}^2$ approach the standard instanton.

More complicated illustrations of the theory are provided by Examples (iii) and (v) of Section 4.1. In these cases $b^+ = 1$ and the compactification involves pairs of points. The link of an ideal connection in the compactified moduli space is a circle, and we can see now that this is obtained as a subset of the full set of gluing data $\mathrm{Gl}_{x_1} \times \mathrm{Gl}_{x_2}/SO(3)$, cut down by the 'obstruction' presented by the self-dual harmonic form.

8.3 Even forms with $b^+ = 0, 1$ or 2

8.3.1 Concentrated connections and definite forms

The remainder of this chapter is taken up with a proof of Theorem (1.3.2), in which we make heavy use of the description of the ends of the moduli spaces from the previous section. We begin by giving another proof of (1.3.1). Consider the moduli space $M_1 = M_1(X)$ of $SU(2)$ connections with $c_2 = 1$, where X is a simply-connected four-manifold with $b^+ = 0$. It is five-dimensional and for a generic metric it will be a smooth manifold except for the points corresponding to reducible connections. There is one reducible connection for each pair $\{e, -e\}$ where $e \in H^2(X; \mathbb{Z})$ has $e^2 = -1$. For each such e a neighbourhood of the corresponding point $[A_e]$ of M_1 is a cone over $\mathbb{CP}^2$. Our picture of the salient features of the moduli space is completed by the description of the boundary in Section 8.2. Let

$$\tau: X \times (0, \varepsilon) \longrightarrow M_1$$

be the map which assigns to (x, λ) the unique equivalence class of ASD connections with centre x and scale λ. The complement of the image of τ is compact. For each e with square -1 let U_e be an open conical neighbourhood of $[A_e]$ in M_1 and let P_e be its boundary. So P_e is a copy of $\mathbb{CP}^2$. Now fix a λ in $(0, \varepsilon)$ and define

$$M' = M_1 \backslash (\bigcup U_e) \backslash \tau(X \times (0, \lambda)).$$

Thus M' is a compact manifold-with-boundary, and the boundary $\partial M'$ is the disjoint union of the P_es and a copy of X—precisely, the manifold $\tau_\lambda(X)$ where τ_λ is the map $x \mapsto \tau(x, \lambda)$ (see Fig. 13).

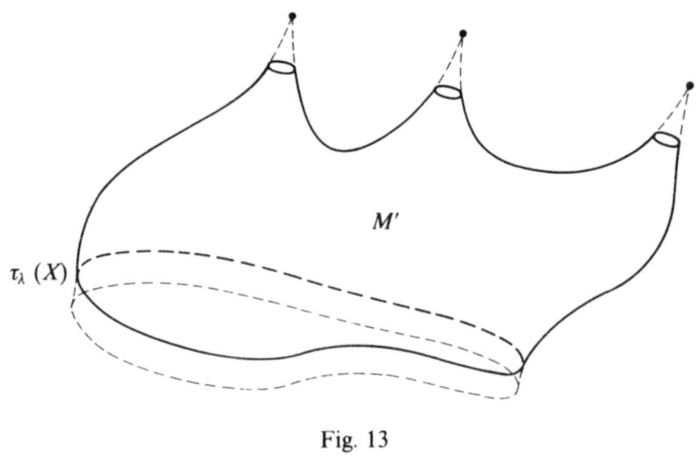

Fig. 13

We now think of this manifold-with-boundary M' as being contained in the ambient space $\mathscr{B}_X^*$. Clearly the fundamental class of the boundary $[\partial M']$ is zero in $H_4(\mathscr{B}_X^*; \mathbb{Z}/2)$. That is

$$[\tau_\lambda(X)] = \sum_{\{e, -e\}, e^2 = -1} [P_e]. \tag{8.3.1}$$

(In fact, since M' is orientable by Section 5.4, this equality also holds with integer coefficients, provided the correct signs are attached to the different terms $[P_e]$.) Now let Σ_1, Σ_2 be two classes in $H^2(X; \mathbb{Z})$. Since the map τ_λ satisfies the hypotheses of the Poincaré duality result, Proposition (5.3.3), we have

$$Q(\Sigma_1, \Sigma_2) = \langle \mathrm{PD}(\Sigma_1) \smile \mathrm{PD}(\Sigma_2), [X] \rangle = \langle \mu(\Sigma_1) \smile \mu(\Sigma_2), [\tau_\lambda(X)] \rangle.$$

Using first the equality (8.3.1) and then the calculation (5.1.21) of the restriction of $\mu(\Sigma)$ to one of the projective spaces, we obtain (modulo 2)

$$Q(\Sigma_1, \Sigma_2) = \sum_{\{e, -e\}, e^2 = -1} \langle \mu(\Sigma_1 \smile \mu(\Sigma_2), [P_e] \rangle = \sum \langle e, \Sigma_1 \rangle \langle e, \Sigma_2 \rangle. \tag{8.3.2}$$

It follows that the classes with square -1 span $H^2(X; \mathbb{Z}/2)$, for otherwise there would be a non-zero class Σ_1 in $H_2(X; \mathbb{Z}/2)$ with $Q(\Sigma_1, \Sigma_2) = 0 (\mathrm{mod}\ 2)$ for all Σ_2, and this is impossible since the pairing Q is perfect (over $\mathbb{Z}/2$, or any field). It is clear then that the classes with square -1 actually span the integral homology—and hence the intersection form is standard; indeed, the form is diagonal in a basis formed from elements of square -1.

8.3.2 Proof of Theorem (1.3.2)

We shall now begin the proof of Theorem (1.3.2) on the intersection forms of simply connected spin four-manifolds with $b^+ = 1$ or 2. In this section we take the argument to the point where the proof is reduced to some topological calculations, which will then be taken up in Section 8.3.4. The motivation behind the proof is this; certainly (1.3.2) implies (1.3.1) in the even case—we just take the connected sum of our definite manifold with copies of $S^2 \times S^2$. So one could expect that a proof of (1.3.2) should reproduce a proof for definite forms in the case when the manifold is such a connected sum. Our proof will indeed have this feature: for connected sums it reduces to the proof given in Section 8.3.1.

Let X be a simply connected, oriented, spin four-manifold. For the moment we impose no restriction on $b^+(X)$; we put $k = b^+(X) + 1$ and consider the moduli space M_k, whose formal dimension is:

$$\dim M_k = 8k - 3(b^+ + 1) = 5k.$$

We may suppose that M_k contains no reducible connections; for $b^+ \geq 1$ this can be achieved by choosing a suitably generic metric (4.3.19), while for

$b^+ = 0$ reductions of the bundle are ruled out on topological grounds, since the form is even and there can be no elements of square -1. Thus a metric can be chosen so that M_k is a smooth $5k$-manifold. Let $\Sigma_1, \ldots, \Sigma_{2k}$ be smooth surfaces in general position in X, and let $\nu(\Sigma_i)$ be a tubular neighbourhood of Σ_i. By making these neighbourhoods sufficiently small we can arrange that any triple intersection of the $\nu(\Sigma_i)$ in X is empty. Now for $i = 1, \ldots, 2k$ let V_i be a codimension-two submanifold of the moduli space, defined by restriction to $\nu(\Sigma_i)$, satisfying the conditions of (5.2.12) and (5.2.13). Put

$$V = M_k \cap V_1 \cap \ldots \cap V_{2k}.$$

By our transversality assumption this intersection is transverse; V is thus a smooth manifold of dimension k. Further, we have arranged that for all subsets $I \subset \{1, \ldots, 2k\}$, the intersections

$$M_j \cap \bigcap_{i \in I} V_i, \quad (j \le k)$$

are transverse also (5.2.12). Let $\bar{V}$ be the closure of V in the compact space $\bar{M}_k$. The next two lemmas concern the possible intersections of $\bar{V}$ with the lower strata in $\bar{M}_k$; the proofs are simple counting arguments (similar to the proof of (8.1.4)).

Lemma (8.3.3). *If $b^+ \le 2$, so $k \le 3$, the closure $\bar{V}$ does not meet the intermediate strata $M_j \times s^{k-j}(X)$, $j = 1, \ldots, k-1$.*

The gist of the counting argument is that $\bar{V}$ has dimension k while the lower strata mentioned in the lemma have codimension at least four (multiples of four in fact); so the intersection will generically be empty if $k \le 3$. For the details, suppose that $[A_\alpha]$ is a sequence in V converging to an ideal ASD connection,

$$([A_\infty]; x_1, \ldots, x_{k-j}) \in M_j \times s^{k-j}(X),$$

with $0 < j < k$. Since no point x_r can lie in more than two of the tubular neighbourhoods $\nu(\Sigma_i)$, there are at least $2j$ surfaces, say $\Sigma_1, \ldots, \Sigma_{2j}$ whose tubular neighbourhoods contain none of the points of concentration x_r. On each of these tubular neighbourhoods the connections $[A_\alpha]$ converge in the C^∞ topology to $[A_\infty]$; that is

$$[A_\alpha|_{\nu(\Sigma_i)}] \longrightarrow [A_\infty|_{\nu(\Sigma_i)}] \text{ in } \mathscr{B}_{\nu(\Sigma_i)}, \quad 1 \le i \le 2j.$$

Since V_i is closed in $\mathscr{B}_{\nu(\Sigma_i)}$, we therefore have:

$$[A_\infty] \in M_j \cap V_1 \cap \ldots \cap V_{2j}.$$

But this intersection was chosen to be transverse, so we must have

$$\dim M_j - 4j = 8j - 3(1 + b^+) = 4j - 3k \ge 0.$$

However, since $j \le k - 1$ this inequality implies that $k \ge 4$, contrary to our hypothesis.

Lemma (8.3.4). *If the point* $([\Theta]; x_1, \ldots, x_k) \in M_0 \times s^k(X)$ *lies in* $\bar{V}$ *then each tubular neighbourhood* $\nu(\Sigma_i)$ $(i = 1, \ldots, 2k)$ *must contain one of the points* x_r.

Suppose on the contrary that $\nu(\Sigma_1)$, say, contains none of the points of concentration. Then there is sequence $[A_\alpha] \in V$ which converges to the trivial connection $[\Theta]$ over compact subsets of $X \backslash \{x_1, \ldots, x_k\}$ and in particular over $\nu(\Sigma_i)$. Thus

$$[A_\alpha|_{\nu(\Sigma_1)}] \longrightarrow [\Theta] \in \mathscr{B}_{\nu(\Sigma_1)}.$$

But this implies that the closed set V_1 contains the trivial connection, contrary to the condition (5.2.13).

The difference between the statements of the two lemmas above stems from the fact that the moduli space M_0 has the 'wrong' dimension; while the dimensions of the intermediate strata decrease in steps of four, the lowest stratum $M_0 \times s^k(X)$ might have dimension $4k$, rather than k which is its 'formal dimension'. Since $\bar{V}$ has dimension k one should expect that $\bar{V}$ may meet the lowest stratum inside the $5k$-dimensional space $\bar{M}_k$. From now on we shall suppose that $k \leq 3$, so that the conclusion of (8.3.3) holds. Let $(x_1, \ldots, x_k)$ be a multiset satisfying the conditions of Lemma (8.3.4). Since there are $2k$ surfaces Σ_i and only k points x_r, each of the x_r must lie in precisely two of the neighbourhoods $\nu(\Sigma_i)$; in particular each of the x_r must lie close to an intersection point of some pair of surfaces. With this in mind let us define:

$$\Delta = \{(x_1, \ldots, x_k) \in s^k(X) | \text{ each } \Sigma_i \text{ contains one } x_r \ (i = 1, \ldots, 2k)\}.$$

Notice that if $\mathbf{x} = (x_1, \ldots, x_k)$ is in Δ then all the points x_r in X are distinct. Each x_r is an intersection point of two surfaces, say $x_r \in \Sigma_i \cap \Sigma_{i'}$. Let U_r be an open neighbourhood of x_r which contains the closure of the component of x_r in $\nu(\Sigma_i) \cap \nu(\Sigma_{i'})$. We can suppose the U_r are disjoint. Then we have a model for a portion of the end of the moduli space $N_{\mathbf{x}}$ as in (8.2.3). Let $\lambda \colon N_{\mathbf{x}} \to \mathbb{R}^+$ be the 'total radius', $\lambda = (\Sigma \lambda_r^2)^{1/2}$, where λ_r are the k local scales. Taken together, the content of the two lemmas is that if v_α is a sequence in V without a cluster point in V, then the sequence ultimately lies in one of the sets $V \cap N_{\mathbf{x}}$, for $\mathbf{x} \in \Delta$, and $\lambda(v_\alpha) \to 0$. Now choose some $\lambda_0 \in (0, \varepsilon)$ and define

$$V' = V \backslash \{v | v \in N_{\mathbf{x}} \text{ for some } \mathbf{x} \in \Delta \text{ and } \lambda(v) < \lambda_0\}.$$

Thus V' is obtained from V by removing the 'ends'. By Sard's theorem we can choose λ_0 so that V' is a manifold-with-boundary; the boundary is the set

$$\partial V' = \bigcup_{\mathbf{x} \in \Delta} L_{\mathbf{x}} \tag{8.3.5}$$

where

$$L_{\mathbf{x}} = \{v \in V \cap N_{\mathbf{x}} | \lambda(v) = \lambda_0\}. \tag{8.3.6}$$

Most important, V' is *compact* (see Fig. 14).

In Section 8.3.4 we shall prove:

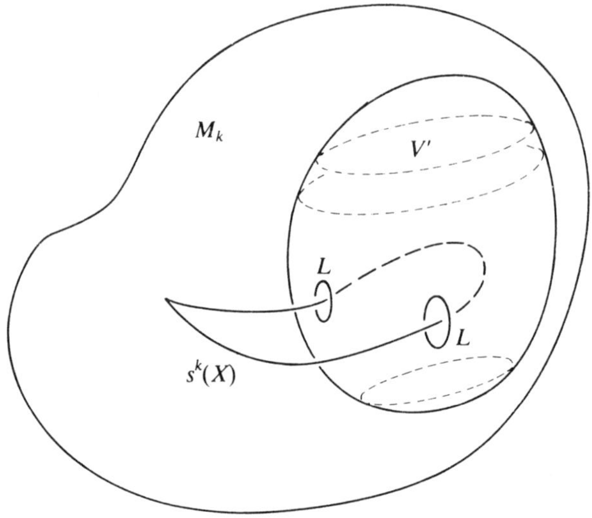

Fig. 14

Proposition (8.3.7). *If $k \leq 3$ there is a class u_{k-1} in $H^{k-1}(\mathscr{B}_k^*, \mathbb{Z}/2)$ such that (for suitably small λ_0 and suitable choices of V_i)*

$$\langle u_{k-1}, [L_{\mathbf{x}}] \rangle = 1$$

for all $\mathbf{x} \in \Delta$.

It is here that the spin condition on the four-manifold X is used. By contrast, the condition $k \leq 3$ is not essential in (8.3.7); it was used in the previous step, Lemma (8.3.3).

We can now make the vital step in the argument, the step in which the existence of the moduli space is used. It is merely the observation that the boundary $\partial V'$ is null-homologous in the ambient space $\mathscr{B}_k^*$ so, applying the cohomology class u_{k-1}, we have:

Corollary (8.3.8). *The cardinality of Δ is even.*

It is now an elementary matter to deduce Theorem (1.3.2). We will first cast the argument in geometric language. Suppose that the conclusion of Theorem (1.3.2) is not satisfied, so the intersection form of X can be written

$$Q = (k-1)\begin{pmatrix} 0 & 1 \\ 1 & 0 \end{pmatrix} \oplus Q',$$

where Q' is a non-trivial negative definite form. (Recall that $k = b^+ + 1$.) Choose surfaces $\Sigma_1, \ldots, \Sigma_{2k-2}$ representing the natural basis for $k-1$

copies of the form $\begin{pmatrix} 0 & 1 \\ 1 & 0 \end{pmatrix}$, i.e.

$$Q(\Sigma_{2i-1}, \Sigma_{2i}) = 1$$

for $i = 1, \ldots, k-1$ and $Q(\Sigma_i, \Sigma_j) = 0$ otherwise. Choose the remaining two surfaces representing classes in the Q' summand. To make the algebra completely transparent we can arrange also that all geometric intersection numbers coincide with the algebraic intersection numbers (by adding handles to the surfaces). With this done, the cardinality of Δ is $|Q'(\Sigma_{2k-1}, \Sigma_{2k})|$. Indeed, if $\mathbf{x} = (x_1, \ldots, x_k)$ is in Δ, we must have one point, x_1 say, equal to the unique intersection point of Σ_1 and Σ_2, one point, x_2 say, of $\Sigma_3 \cap \Sigma_4$ etc. The only choice is in placing the last point x_k on one of the $|Q(\Sigma_{2k-1}, \Sigma_{2k})|$ intersection points of $\Sigma_{2k-1}, \Sigma_{2k}$. We deduce then, from Corollary (8.3.8), that for all homology classes $\Sigma_{2k-1}, \Sigma_{2k}$ in the last factor, the pairing $Q'(\Sigma_{2k-1}, \Sigma_{2k})$ is even. But this contradicts the fact that Q' is a unimodular form, and hence no such four-manifold X can exist (see Fig. 15).

There is, not surprisingly, a more algebraic version of the argument above, which depends neither on the classification of integral quadratic forms nor on the geometry of surfaces in a four-manifold. For each $\mathbf{x} \in \Delta$ let us write

$$\varepsilon(\mathbf{x}) = \prod_{r=1}^{k} \varepsilon(x_r)$$

where $\varepsilon(x_r) = \pm 1$ is the sign of the intersection point of the relevant surfaces. We can then define a multilinear form $Q^{(k)}$ on $H_2(X)$ by

$$Q^{(k)}(\Sigma_1, \ldots, \Sigma_{2k}) = \sum_{\mathbf{x} \in \Delta} \varepsilon(\mathbf{x}).$$

This can be expressed in terms of the intersection form Q as follows:

$$Q^{(k)}(\Sigma_1, \ldots, \Sigma_{2k}) = \frac{1}{2^k k!} \sum_{\sigma \in S_{2k}} Q(\Sigma_{\sigma(1)}, \Sigma_{\sigma(2)}) \times \ldots \times Q(\Sigma_{\sigma(2k-1)}, \Sigma_{\sigma(2k)}).$$

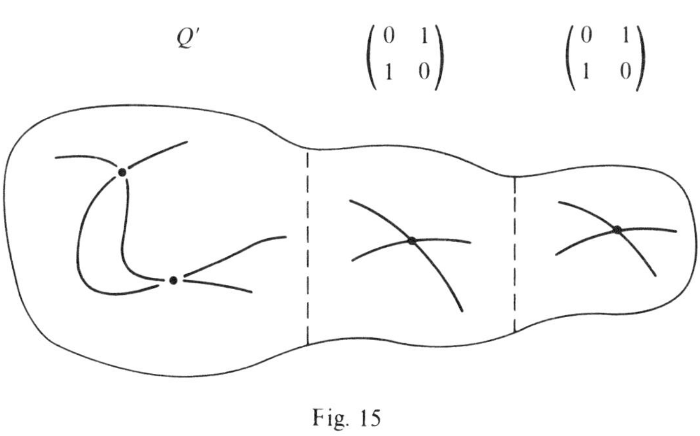

Fig. 15

Thus, for example,

$$Q^{(2)}(\Sigma_1, \dots, \Sigma_4) = (\Sigma_1 . \Sigma_2)(\Sigma_3 . \Sigma_4) + (\Sigma_1 . \Sigma_3)(\Sigma_2 . \Sigma_4) + (\Sigma_1 . \Sigma_4)(\Sigma_2 . \Sigma_3).$$

Corollary (8.3.8) asserts that $Q^{(k)}$ is identically zero mod 2. The last step in the proof of (1.3.2) can then be completed by the following simple algebraic lemma:

Lemma (8.3.9). *If Q is a unimodular even form on $\mathbb{Z}^{2r}$ and $Q^{(s)}$ is zero* mod 2, *then $s > r$.*

This is a mod 2 version of a familiar fact from exterior algebra. If ω is a non-degenerate skew form on $\mathbb{R}^{2n}$ and if the exterior power ω^m is zero, then $m > n$. The proof of the statement of the lemma is much the same as in the exterior case. One establishes the existence of a basis $\alpha_1, \dots, \alpha_{2r}$ for $\mathbb{Z}^{2r}/2\mathbb{Z}^{2r}$ as a vector space over $\mathbb{Z}/2$ such that the mod 2 reduction of Q is represented by the matrix

$$\begin{pmatrix} 0 & 1 \\ 1 & 0 \end{pmatrix} \oplus \cdots \oplus \begin{pmatrix} 0 & 1 \\ 1 & 0 \end{pmatrix} \quad (r \text{ copies});$$

then for any $s \leq r$ we have $Q^{(s)}(\alpha_1, \dots, \alpha_{2s}) = 1 \pmod 2$.

8.3.3 Comments

We collect here some remarks about the proof of the theorem. First, in the case $k = 1$ (that is, when $b^+ = 0$ and Q is negative definite) the argument reduces to that of our previous treatment in Section 8.3.1. There are no reductions since we assume the intersection form to be even, the manifold $V = M_1 \cap V_1 \cap V_2$ is one-dimensional, and we can take the cohomology class $u_0 \in H^0(\mathscr{B}; \mathbb{Z}/2)$ to be the canonical generator. The content of Proposition (8.3.7) in this case is just that V has an odd number of ends associated to each intersection point in $\Delta = \Sigma_1 \cap \Sigma_2$. (We shall soon see that we can arrange for V to have exactly one end for each such point.) The conclusion is, as before, that $Q(\Sigma_1, \Sigma_2) = 0$ for all Σ_1, Σ_2 and this forces $H^2(X)$ to be zero. For larger values of b^+ the structure of the argument is the same as in this basic case but it is complicated by two new features. First, the 'link' of the lowest stratum $s^k(X) \subset \bar{M}_k$ is no longer a point so one must verify that it still carries homological information. Second, questions of compactness become more delicate, since one has to contend with the intermediate strata in the compactification. This second feature leads us to use the explicit representatives V_i for the cohomology classes $\mu(\Sigma_i)$. Our argument certainly fits into the general pattern of this chapter: we truncate the moduli space M_k to obtain a compact manifold-with-boundary M'_k; then

$$Q^{(k)}(\Sigma_1, \dots, \Sigma_{2k}) = \langle u_{k-1} \smile \mu(\Sigma_1) \smile \dots \smile \mu(\Sigma_{2k}), [\partial M'_k] \rangle = 0,$$

since $\partial M'_k$ is a boundary in $\mathscr{B}^*$. From this point of view our use of the representatives V_i is a device for calculating the above pairing with $\partial M'_k$. It is this second feature which shows a change when $b^+(X) \geqslant 3$. Lemma (8.3.3) fails and V will, in general, have other 'ends' associated with the intersection of $\bar{V}$ with the other strata $s^j(X) \times M_{k-j}$. For example, when $b^+ = 3$ and $k = 4$ the dimension of V is 4 and we can expect $\bar{V}$ to have isolated points of intersection with the 16-dimensional stratum $X \times M_3$, inside the 20-dimensional space $\bar{M}_4$.

Finally we consider briefly the case of connected sums mentioned at the beginning of Section 8.3. Suppose, for example, that $X = Y \# (S^2 \times S^2)$ where Y has a negative-definite form. We choose metrics on the connected sum with a small neck, determined by a parameter μ, and take surfaces Σ_1, Σ_2 to be the standard two-sphere generators of $H_2(S^2 \times S^2)$. We then consider another pair of surfaces Σ_3, Σ_4 in the Y factor. If we were carrying out our argument over the definite manifold Y, we would consider a one-dimensional space V_Y say, with ends associated with the intersection points of Σ_3, Σ_4; whereas in the argument over X we use a two-dimensional space V_X say. Using the techniques of Chapter 7, in a way which we will discuss at length in Chapter 9, one can analyse the behaviour of the subset V_X as the parameter μ tends to 0. This will make a good exercise for the reader. For small enough μ one shows that V_X is a circle bundle over V_Y, the ends of V_Y being replaced by the links $L_{\mathbf{x}}$ considered in (8.3.7). The 'stabilization' of the proof amounts to showing that the fibre is non-trivial in homology, so we can make our count of the ends equally well with V_X as with V_Y.

8.3.4 *The homology class of the link*

The proof of Proposition (8.3.7) has two parts: we must identify the $(k-1)$-dimensional homology class carried by the link $L_{\mathbf{x}}$ in the gauge orbit space and we must find a class $u_{k-1} \in H^{k-1}(\mathscr{B}_X^*; \mathbb{Z}/2)$ which 'detects' $[L_{\mathbf{x}}]$. We begin with the first of these tasks. Fix once and for all a point $\mathbf{x} = (x_1, \dots, x_k) \in \Delta$, and let Ω, Ω_ε and Ψ be as in the statement of Theorem (8.2.3). We now write N for $N_{\mathbf{x}}$ and L for $L_{\mathbf{x}}$. Since Ω_ε parametrizes a family of connections it makes sense to talk of the intersections $\Omega_\varepsilon \cap V_i$ etc.

Lemma (8.3.10). *Let x_r be the point of* $\mathbf{x}$ *lying on a surface Σ_i,* let *U_r be its neighbourhood in X and let*

$$p_r : N \longrightarrow U_r$$

be the corresponding local centre map. Then the submanifold V_i can be chosen so that

$$\Omega_\varepsilon \cap V_i = p_r^{-1}(U_r \cap \Sigma_i),$$

if ε is sufficiently small.

This is an adaptation of the local 'Poincaré duality' result (5.3.6) (see also (5.3.7) and (5.3.8)). Provided ε is small enough, the condition that the local centre $p_r(\omega)$ lies in Σ_i depends only on the restriction of A_ω to

$\nu(\Sigma_i)$. So one can certainly find a submanifold $W \subset \mathscr{B}^*_{\nu(\Sigma_i)}$ such that $\Omega_\varepsilon \cap W = p_r^{-1}(U_r \cap \Sigma_i)$. The content of (5.3.6) is that, in the region of $\mathscr{B}^*_{\nu(\Sigma_i)}$ consisting of concentrated connections, this W can be realized as the zero set of a regular section s of $\mathscr{L}_{\Sigma_i}$ which coincides with the standard trivialization of this line bundle near the trivial connection Θ. We can extend this section to all of $\mathscr{B}^*_{\nu(\Sigma_i)}$ using a partition of unity to obtain a V_i with the required properties. The only remaining point is to ensure that the intersections $N \cap p_r^{-1}(\Sigma_i)$ etc. are transverse. This can be achieved simply by perturbing the surface Σ_i slightly within its tubular neighbourhood.

Fix submanifolds V_i which satisfy the conditions of Lemma (8.3.10). Then $N \cap V$ consists of the connections whose local centres are the points $x_1, \dots, x_k$. Appealing to Theorem (8.2.3) we obtain a description of $N \cap V$ as a quotient $\Psi_0^{-1}(0)/SO(3)$, where

$$\Psi_0 \colon \prod_{r=1}^{k} (\mathbb{R}^+ \times SO(3)) \longrightarrow \mathscr{H}_+ \otimes \mathbb{R}^3$$

is the restriction of Ψ. Let us rearrange this description slightly. If we write

$$Z = \left(\prod_{r=1}^{k} (\mathbb{R}^+ \times SO(3)) \right) \Big/ SO(3),$$

then the equivariant map Ψ_0 induces a section (which we again call Ψ_0) of a vector bundle H over Z with fibre $\mathscr{H}_0 \otimes \mathbb{R}^3$, associated with the free $SO(3)$ action. Let $\bar{Z}$ be the natural completion of Z obtained by adjoining the points with zero scales (i.e. extending $\mathbb{R}^+$ to $[0, \infty)$). Thus, thinking of $SO(3)$ as being identified with $\mathbb{RP}^3$, we have:

$$\bar{Z} = \left(\prod_{r=1}^{k} (\mathbb{R}^4/\pm 1) \right) \Big/ SO(3).$$

Now define $Z_{\lambda_0} = Z \cap \{\lambda = \lambda_0\}$ and similarly $\bar{Z}_{\lambda_0} = \bar{Z} \cap \{\lambda = \lambda_0\}$. So $\bar{Z}_{\lambda_0}$ is a quotient of the λ_0-sphere in $\mathbb{R}^{4k}$. The bundle H extends naturally to $\bar{Z}_{\lambda_0}$, and according to Theorem (8.2.3) the section Ψ_0 also has a continuous extension $\bar{\Psi}_0$. We obtain then the following description of the link L.

Proposition (8.3.11). *The link $L \subset Z_{\lambda_0}$ is the zero set of a continuous section $\bar{\Psi}_0$ of a bundle H over $\bar{Z}_{\lambda_0}$.*

Note that Lemma (8.3.3) tells us that the zero set of $\bar{\Psi}_0$ is contained in the open stratum Z_{λ_0}. The link L has a compactly supported dual class:

$$\mathrm{PD}[L] \in H_c^{3k-3}(Z_{\lambda_0}; \mathbb{Z}/2),$$

and the inclusion of Z_{λ_0} in $\bar{Z}_{\lambda_0}$ induces a map from $H_c^{3k-3}(Z_{\lambda_0})$ to $H^{3k-3}(\bar{Z}_{\lambda_0})$.

Lemma (8.3.12). *If $k \leq 3$ the inclusion induces an isomorphism from $H_c^{3k-3}(Z_{\lambda_0})$ to $H^{3k-3}(\bar{Z}_{\lambda_0})$.*

The manifold Z_{λ_0} has dimension $4k - 4$ and its compactification $\bar{Z}_{\lambda_0}$ is a singular complex whose singular set $S = \bar{Z}_{\lambda_0} \backslash Z_{\lambda_0}$ has codimension four. (It may be helpful to observe that $\bar{Z}_{\lambda_0}$ can be identified with the quotient $\mathbb{HP}^{k-1}/(\pm 1)^k$, where $\mathbb{HP}^{k-1}$ is the quaternionic projective space—the quotient of S^{4k-1} by $SU(2)$—and $(\pm 1)^k$ acts by changes of sign on the k homogeneous coordinates, i.e. $[q_1, \ldots, q_k] \mapsto [\pm q_1, \ldots, \pm q_k]$. The singular set S is the image of the union of the coordinate hyperplanes in $\mathbb{HP}^{k-1}$.) In general in such a situation we have:

$$H^p_c(Z_{\lambda_0}) = H^p(\bar{Z}_{\lambda_0}, S).$$

On the other hand the exact sequence of the pair $(\bar{Z}_{\lambda_0}, S)$ tells us that the inclusion gives an isomorphism between $H^p(\bar{Z}_{\lambda_0}, S)$ and $H^p(\bar{Z}_{\lambda_0})$ if $p \geq \dim S + 2$. In our case $p = 3k - 3$ and $\dim S = 4k - 8$, so a sufficient condition for the stated isomorphism is that $k \leq 3$.

Corollary (8.3.13). *Under the isomorphism of Lemma* (8.3.12) *the compact Poincaré dual of* $[L]$ *corresponds to the Euler class of the bundle* $H \to \bar{Z}_{\lambda_0}$.

The force of this corollary is that the homology class of L is determined entirely by the bundle H. Geometrically, it asserts that if s_1, s_2 are sections of H with regular zero sets contained in Z_{λ_0}, then these zero sets are homologous in Z_{λ_0}. So we can use any convenient section to identify the homology class. To construct a suitable section of H we choose $h_1, \ldots, h_k$ in $\mathscr{H}_+$ in such a way that:

(i) $\sum_{r=1}^k h_r = 0$;

(ii) any subset of $k - 1$ members is linearly independent.

(Recall that $\dim \mathscr{H}_+ = k - 1$.) Let (e_1, e_2, e_3) be the standard basis for $\mathbb{R}^3$ and define

$$\Phi: \prod_1^k ([0, \infty) \times SO(3)) \longrightarrow \mathbb{R}^3 \otimes \mathscr{H}_+$$

by

$$\{(\lambda_r, \rho_r)\} \longmapsto \sum_{r=1}^k \lambda_r^2 \rho_r(e_1) \otimes h_r.$$

This is equivariant for the $SO(3)$ actions, so defines a section $\bar{\Phi}$ of $H \to \bar{Z}_{\lambda_0}$. Condition (ii) implies that this section does not vanish on the singular set S (where one of the λ_r is 0). Let $\Gamma \subset Z_{\lambda_0}$ be the zero set of this section. It is the quotient by $SO(3)$ of the set defined by the equations:

$$\sum \lambda_r^2 = \lambda_0^2, \quad \sum \lambda_r^2 \rho_r(e_1) \otimes h_r = 0.$$

The only solutions of these equations occur when

$$\lambda_1 = \ldots = \lambda_k = \frac{1}{\sqrt{k}} \lambda_0, \quad \text{and} \quad \rho_1(e_1) = \ldots = \rho_k(e_1).$$

If we make the obvious identification between $SO(3)^k/SO(3)$ and $SO(3)^{k-1}$ (fixing the last factor), then the second condition becomes $\rho_r(e_1) = e_1$ $(1 \leq r \leq k-1)$. That is, the product of $k-1$ copies of the one-parameter subgroup $\gamma \subset SO(3)$ defined by the condition $\rho(e_1) = e_1$. This circle γ represents the generator of $H_1(SO(3); \mathbb{Z}/2) = \mathbb{Z}/2$. Thus we have, for $k \leq 3$:

Proposition (8.3.14). *Under the natural homotopy equivalence between Z_{λ_0} and $SO(3)^{k-1}$, the homology class of L is the cross-product $\gamma_1 \times \ldots \times \gamma_{k-1}$, where γ_r is the generator of $H_1(SO(3); \mathbb{Z}/2)$ in the rth factor.*

8.3.5 Cohomology classes and the spin condition

Up to this point we have made no use of the spin condition on the four-manifold X. Note that this must enter in the proof, since there certainly are non-spin four-manifolds realizing any values of b^+, b^-. The spin structure comes into the picture now, in the definition of cohomology classes u_{k-1} which detect $[L] = [L_x]$. Recall that, for a spin manifold X, we defined in (5.1.17) cohomology classes $\tilde{u}_i \in H^i(\tilde{\mathscr{B}}_X; \mathbb{Z}/2)$ by:

$$\tilde{u}_i = w_i(\text{ind}(D, \tilde{\mathbb{E}})_{\mathbb{R}}),$$

where $(D_A)_{\mathbb{R}}:(E \otimes S^+)_{\mathbb{R}} \to (E \otimes S^{-1})_{\mathbb{R}}$ is the real part of the Dirac operator coupled to the bundle E. It is not always possible to define corresponding classes over the quotient space $\mathscr{B}_X^*$, because in general there is no $SU(2)$ bundle $\mathbb{E} \to \mathscr{B}_X^* \times X$ analogous to the bundle $\tilde{\mathbb{E}} \to \tilde{\mathscr{B}} \times X$. Recall that $\mathscr{B}^*$ parametrizes only a family of connections $\mathbb{A}$ on an $SO(3)$ bundle $\mathbb{P} \to \mathscr{B}^* \times X$. If $w_2(\mathbb{P})$ is not zero, there is no lift to $SU(2)$. However when $c_2(E)$ is odd and X is spin we shall show that $w_2(\mathbb{P})$ is zero and an $SU(2)$ bundle $\mathbb{E}$ can be constructed, by the following route. Let $\tilde{\Lambda} \to \tilde{\mathscr{B}}_X$ be the real determinant line bundle of the Dirac operator coupled to $\tilde{\mathbb{E}}$:

$$\tilde{\Lambda} = \det(\text{ind}(D, \tilde{\mathbb{E}})_{\mathbb{R}}).$$

Proposition (8.3.15). *If $c_2(E)$ is odd, the $SU(2)$ bundle $\tilde{\mathbb{E}} \otimes_{\mathbb{R}} \tilde{\Lambda} \to \tilde{\mathscr{B}}_X \times X$ descends to an $SU(2)$ bundle $\mathbb{E}$ over $\mathscr{B}_X^* \times X$.*

There is a natural action of $SU(2)$ on $\tilde{\mathbb{E}} \otimes \tilde{\Lambda}$ and the problem, as usual, is that the element $-1 \in SU(2)$ acts trivially on the base $\tilde{\mathscr{B}} \times X$ but not, perhaps, on the fibres. We know, by (5.2.3) that -1 acts as $(-1)^p$ on the fibre of $\tilde{\Lambda}$, where p is the numerical index of $D: E \otimes S^+ \to E \otimes S^-$. According to (5.1.16), this index is

$$p = c_2(E) + 2\hat{A}(X)[X].$$

So -1 acts non-trivially on the fibre precisely when $c_2(E)$ is odd. In this case the non-trivial action on $\tilde{\Lambda}$ cancels that on $\tilde{\mathbb{E}}$ to give a trivial action on $\tilde{\mathbb{E}} \otimes \tilde{\Lambda}$, and we can descend to the quotient $\mathscr{B}^* \times X$.

This result allows us to make the following definition:

Definition (8.3.16). *If X is spin and $c_2(E)$ is odd, let $\mathbb{A}$ be the family of $SU(2)$ connections carried by the bundle $\mathbb{E} \to \mathscr{B}^* \times X$, and put*

$$u_i = w_i(\operatorname{ind}(D, \mathbb{E})_{\mathbb{R}}) \in H^i(\mathscr{B}^*_X; \mathbb{Z}/2).$$

When $c_2(E)$ is even such classes cannot be defined. We may, however, turn this defect into a virtue and still obtain useful cohomology classes. If $c_2(E)$ is even, $-1 \in SU(2)$ acts trivially on the fibres of $\tilde{\Lambda} \to \tilde{\mathscr{B}}_X$, and we can form a quotient line bundle $\Lambda = \tilde{\Lambda}/SU(2)$ over $\mathscr{B}^*_X$. We may therefore make the following definition in this case:

Definition (8.3.17). *If X is spin and $c_2(E)$ is even, define*

$$u_1 = w_1(\Lambda) \in H^1(\mathscr{B}^*; \mathbb{Z}/2).$$

Remark. It follows from this definition that, when c_2 is even, $\tilde{u}_1 = \beta^*(u_1)$, where $\beta: \tilde{\mathscr{B}} \to \mathscr{B}^*$ is the base point fibration. The corresponding assertion is *not* true for the classes u_i defined in (8.3.17), when c_2 is odd.

We return, armed with these cohomology classes, to the situation we have been considering in Section 8.3.2, with connections concentrated at distinct points $x_1, \dots, x_k$ in the four-manifold X. For $r = 1, \dots, k$ let A_r be an $SU(2)$ connection, with curvature supported in a small neighbourhood U_r of x_r, on a bundle with Chern class 1 which is trivialized over the rest of X. Take a base point outside the U_r; we get a family of framed connections, parametrized by $SO(3)^k$, by gluing the connections A_r over $X \setminus (U_1 \cup \dots \cup U_k)$. We obtain a map:

$$\tilde{s}: SO(3)^k \longrightarrow \tilde{\mathscr{B}}_X.$$

Proposition (8.3.18). *The index of the Dirac operator on the family of framed connections parametrized by $\tilde{s}$ is*

$$\operatorname{ind}(D, \tilde{\mathbb{E}})_{\mathbb{R}} = \eta_1 + \dots + \eta_k + 2m.1 \in KO(SO(3))^k,$$

where η_r is the non-trivial real line bundle over the rth copy of $SO(3)$ and m is the numerical index of the Dirac operator $D: S^+ \to S^-$.

This is a formal consequence of the excision property for indices. Before discussing the proof we digress to explain how the result can be confirmed, in the case when X is S^4, by the ADHM construction developed in Chapter 3. Recall from Section 3.3.2 (and (7.1.28)) that if A is an ASD connection over $S^4 = \mathbb{R}^4 \cup \{\infty\}$ then $\ker D_A = 0$, so the index is represented by the kernel of D^*_A, the space $\mathscr{H}_A$ of Section 3.3. From the point of view of the ADHM construction, the framed moduli space $\tilde{M}_k$ of $SU(2)$ connections is obtained by dividing a space, W say, of matrices satisfying the ADHM conditions (3.4.6) and (3.4.7) by the action of $O(k)$. The larger space W can be viewed as the moduli space of equivalence classes of triples (A, ϕ, f) where A is a connection, f is a framing over the base point ∞, and ϕ is an orthonormal, real, basis of $\mathscr{H}_A$. The $O(k)$ action changes the basis ϕ. We have then:

Proposition (8.3.19). *The index of the Dirac family on the framed instanton moduli space $\tilde{M}_k = W/O(k)$ is represented by the real vector bundle associated with the free $O(k)$ action by the fundamental representation of $O(k)$.*

Now a family of framed connections such as $\tilde{s}$ approximates a family of framed ASD connections over S^4 with fixed centres $x_r \in \mathbb{R}^4$ and fixed small scales λ_r. Recall from Section 4.4.3 what the ADHM data looks like for such concentrated connections: with respect to a suitable basis for $\mathscr{H}$, the four endomorphisms $T_i: \mathscr{H} \to \mathscr{H}$ and the map $P: \mathscr{H} \to S^+ \otimes E_\infty$ are within $O(\lambda^2)$ of the simple model:

$$T_i = \operatorname{diag}(x_i^1, \ldots, x_i^k), \quad P = (u_1, \ldots, u_k), \tag{8.3.20}$$

where x_i^r are the coordinates of $x_r \in \mathbb{R}^4$ and $u_r \in S^+ \otimes E_\infty$ is an element of length λ_r. It follows from (8.3.19) that, for the purposes of describing the index, we may as well suppose that the ADHM data has precisely the form above. Then the eigenspaces of the T_i decompose $\mathscr{H}$ into one-dimensional factors $\mathscr{H} = L_1 \oplus \ldots \oplus L_k$, say. Now, as we saw in Section 3.4, the only elements of $O(k)$ which preserve the canonical form (8.3.20) are those of the shape:

$$\operatorname{diag}(\pm 1, \ldots, \pm 1).$$

Such an automorphism fixes the T_i but changes the sign of the terms u_r, and this gives our description of the concentrated, framed, connections parametrized by copies of $SO(3) = \mathbb{RP}^3 = S^3/\pm 1$. Since the rth copy of ± 1 acts non-trivially on the line L_r we get a non-trivial Hopf line bundle on the quotient space, and we see that over the quotient of the matrices in canonical form, a small perturbation of the concentrated connections, there is a canonical isomorphism between the index bundle and a sum $\eta_1 \oplus \ldots \oplus \eta_r$ of real Hopf line bundles.

Having confirmed (8.3.19) for $X = S^4$ the general case can be deduced using the excision principle of Section 7.1, applied to the index of the family. In fact there is no need at all to invoke the ADHM description, or the ASD condition, although these may serve to illuminate the result. A direct proof can be given as follows. Fix any connection I over S^4, on a bundle with Chern class 1, and consider the family of equivalence classes of framed connections $F = \beta^{-1}([I]) \subset \tilde{\mathscr{B}}_{S^4} \cong SO(3)$. A moment's thought shows that the Dirac index over this family is exactly the Hopf line bundle. (It is obtained from the trivial line bundle over $SU(2)$ by dividing by ± 1, and -1 acts non-trivially on the spinors.) Now consider the family of framed equivalence classes over the disjoint union

$$S^4 \amalg \ldots \amalg S^4 \amalg X$$

formed from the product of k copies of F and the trivial connection over X. (We have one base point in each component.) The index is clearly $\eta_1 \oplus \ldots \oplus \eta_r \oplus 2m.1$. Then an application of the excision principle, which we leave as an exercise for the reader, shows that the index of this family equals that of the corresponding family of glued connections over the connected sum

$$S^4 \# \ldots \# S^4 \# X = X.$$

We now pass down to the family of equivalence classes of connections, parametrized by $SO(3)^k/SO(3)$. As before, we identify this quotient with the transversal $SO(3)^{k-1}$ by fixing the last factor. Thus we have a map:

$$s: SO(3)^{k-1} \longrightarrow \mathscr{B}^*_{k,X}.$$

On the transversal, η_k is trivial, so

$$\operatorname{ind}(D, \tilde{\mathbb{E}})_{\mathbb{R}} = \eta_1 + \ldots + \eta_{k-1} + (2m+1)$$

and $\tilde{\Lambda} = \eta_1 \ldots \eta_{k-1}$. When k is odd the bundle $\mathbb{E} \to \mathscr{B}^* \times X$ is obtained by pushing down $\tilde{\mathbb{E}} \otimes \tilde{\Lambda}$. We therefore have

$$\operatorname{ind}(D, \mathbb{E})_{\mathbb{R}} = (\eta_1 + \ldots + \eta_{k-1} + 2m + 1)\eta_1 \ldots \eta_{k-1}$$

on the family s. Similarly, when k is even we have $\Lambda = \eta_1 \ldots \eta_{k-1}$.

Proof of Proposition (8.3.7). The case $k = 1$ is trivial, since the link L is homologous to a point, which is detected by the class $u_0 = 1$. Let $t_r = w_1(\eta_r)$, so that $\langle t_r, \gamma_r \rangle = 1$, for $r = 1, \ldots, k-1$. In the case $k = 2$ we let u_1 be as in Definition (8.3.16). The link L is homologous to γ_1 and we have:

$$\langle u_1, \gamma_1 \rangle = \langle w_1(\Lambda), \gamma_1 \rangle = \langle t_1, \gamma_1 \rangle = 1.$$

Finally, in the case when $k = 3$, let u_2 be as in (8.3.17). We have

$$\begin{aligned} u_2 &= w_2((\eta_1 + \eta_2 + 2m + 1)\eta_1\eta_2) \\ &= w_2(\eta_2 + \eta_1 + (2m+1)\eta_1\eta_2) \\ &= w_1(\eta_2)w_1(\eta_1) + w_1(\eta_2 + \eta_1)w_1(\eta_1\eta_2) \\ &= t_2 t_1 + (t_1 + t_2)^2 \\ &= t_1^2 + t_1 t_2 + t_2^2. \end{aligned}$$

The link L is homologous to $\gamma_1 \times \gamma_2$, so

$$\langle u_2, L \rangle = \langle u_2, \gamma_1 \times \gamma_2 \rangle = \langle t_1 t_2, \gamma_1 \times \gamma_2 \rangle = 1,$$

as required. (One can calculate that, for any value of k, the class $\gamma_1 \times \ldots \times \gamma_{k-1} \in H_{k-1}(\mathscr{B}^*_k)$ is never 0, so long as X is spin. It is detected by u_{k-1} if k is odd and by u_1^{k-1} if k is even.)

Notes

Section 8.1

The simple argument in Section 8.1.1 is taken from Fintushel and Stern (1984); see also Freed and Uhlenbeck (1984). The generalization to other definite forms in Sections 8.1.2 and 8.1.3 is much the same as in Fintushel and Stern (1988), although we use rather different cohomology classes.

Section 8.2

For more details of the definition of the 'centre' and 'scale' of a concentrated connection see Donaldson (1983*b*) and Freed and Uhlenbeck (1984).

Section 8.3.1

The argument given here for definite forms is the modification given by Donaldson (1986) of the original argument, using cobordism theory in Donaldson (1983*b*) and Freed and Uhlenbeck (1984).

Sections 8.3.2–8.3.5

The argument here is essentially the same as that given by Donaldson (1986). The simplification here is that we do not use the 'explicit' description of the ends in terms of harmonic forms, mentioned in the notes to Chapter 7, but only the more general properties of the model for the ends.

9

INVARIANTS OF SMOOTH FOUR-MANIFOLDS

We turn now to the question which formed the second main thread of the topological discussion in Chapter 1, namely the question of distinguishing smooth four-manifolds having the same classical invariants. Our strategy is to define new invariants using the ASD moduli spaces. The ASD equations are not defined until a Riemannian metric g (or rather, a conformal class) is chosen; the space of solutions—the moduli spaces we have been studying—reflect accordingly many properties of the metric. In order to define differential-topological invariants, we must extract some piece of information from the moduli space which is insensitive to a change in the metric and therefore depends only on the underlying manifold.

In Section 9.1 we treat a particularly simple case, showing how a zero-dimensional moduli space may be used to define an integer-valued invariant. Our main purpose here is to provide a guide for the more general constructions in the following section, but these simple invariants do have applications of their own: in Section 9.1.3 we calculate our integer invariant for a $K3$ surface and we show that the result implies the failure of the h-cobordism theorem for cobordisms between four-manifolds. The same calculation, combined with the results of Section 9.3, establishes the indecomposability of the $K3$ surface (Corollary (9.1.7)) by a route which is independent of the results of Chapter 8.

The main work of this chapter is presented in Sections 9.2 and 9.3. In Section 9.2 we use the ASD moduli spaces to construct a range of invariants, which are defined for smooth four-manifolds X with $b^+(X)$ odd. These take the form of homogeneous polynomial functions on the homology $H_2(X)$. A serious application of these polynomial invariants will not be given until Chapter 10, but in Section 9.3 we shall develop one of their key properties, namely that if X can be decomposed as a connected sum $X_1 \# X_2$ with $b^+(X_1)$ and $b^+(X_2)$ both non-zero, then the invariants of X are all identically zero (Theorem (1.3.4)). The proof of this result in its greatest generality involves some technical difficulties which we have chosen not to go into: we content ourselves with giving a complete proof for some simple cases, sufficient for our intended application, and outlining the two important mechanisms which underlie the proof of the general case. Full details can be found in the original references at the end of the chapter.

9.1 A simple invariant

9.1.1 Definition

Let us begin by describing, in broad outline, the strategy we shall use to define invariants. To emphasize the dependence on the metric, we write $M_E(g)$ for the moduli space of connections on a bundle E which are ASD with respect to a metric g. Let us suppose we are in the situation where all solutions are irreducible. Then, as we explained in Chapter 4, we can view the moduli space as the zero-set of a section Ψ_g of a bundle $\mathscr{E}$ of Banach spaces over $\mathscr{B}^*$. Now consider for motivation the analogy of a vector bundle V over a compact, finite-dimensional manifold B and a section s of V. If s vanishes transversely, the zero-set $Z(s)$ is a smooth submanifold of B whose fundamental class represents the Poincaré dual of the Euler class of V. As such, this fundamental class,

$$[Z(s)] \in H_d(B), \quad d = \dim B - \operatorname{rank} V,$$

depends only on the bundle, not on the section s. If V and B are oriented (or more precisely, if we have an orientation of the line bundle $\Lambda^{\max}(TB) \otimes \Lambda^{\max}(V^*)$) then we can use homology with integer coefficients, and in any case we can use mod 2 homology. In the infinite-dimensional case it is reasonable to hope that we may be able to define invariants from the homology class of the moduli space (a finite-dimensional object) in the ambient space $\mathscr{B}^*$. We could think of this as the Euler class of $\mathscr{E}$, in finite-dimensional homology or finite-codimensional cohomology.

This programme for defining invariants can indeed be carried through in some detail. The proofs involve arguments from differential topology which extend easily from the familiar finite-dimensional situation to the setting of Fredolm maps, as well as some more detailed considerations of compactness. In this section we discuss a simple situation, where the formal dimension of the moduli space is zero. In the analogy above we would be considering a bundle V where the fibre and base dimension were equal. Thus we are looking for a class in zero-dimensional homology, i.e. an integer.

Let us suppose then that X is a simply-connected four-manifold with $b^+(X) > 1$, and that $E \to X$ is an $SO(3)$ bundle with $w_2(E)$ non-zero and $8k - 3(1 + b^+) = 0$, where $k = \kappa(E) = -\frac{1}{4}p_1(E)$. Note that this requires that b^+ be *odd*, so in fact we have $b^+ \geq 3$. Then for a generic metric g on X we have the following propositions.

(i) *The moduli space $M_E(g)$ contains only irreducible connections.*
(ii) *The moduli space is a finite set of points, each a transverse zero of F^+.*

Indeed, (i) is true for any manifold with $b^+ > 0$ by the results of Section 4.3, and we know by Corollary (4.3.19) that the moduli space is generically a submanifold cut out transversely in $\mathscr{B}^*$; its dimension is then given by the index formula (4.2.22), which in this case yields zero. On the other hand, for any of the lower bundles $E^{(r)}$ (with $\kappa(E^{(r)}) = \kappa(E) - r$) the dimension formula gives a negative number, so these moduli spaces are generically empty. Thus the compactness theorem (4.4.3) asserts that M is itself compact, and therefore finite.

(iii) *We can attach a sign* ± 1 *to each point of the moduli space.*

These depend on a choice of orientation of the determinant line bundle $\Lambda = \det \operatorname{ind}(\delta, \mathfrak{g}_{\mathbb{E}})$ (see Section 5.4.1). As we have seen in Section 7.1.6, this is in turn fixed by a choice of orientation of $H^+(X)$ and an equivalence class of integral lifts of $w_2(E)$. Having made such a choice we determine a sign $\varepsilon(A)$ at a point $[A]$ of M by comparing the canonical trivialization of the fibre of Λ, which exists by dint of the vanishing of the kernel and cokernel, with the given trivialization over all of $\mathscr{B}^*$.

We now define our invariant in this situation to be the integer

$$q = \sum_{[A] \in M} \varepsilon(A),$$

the 'number of points in the moduli space, counted with signs'. *A priori*, q depends on the metric g, so we will temporarily write $q(g)$. It also depends on $w_2(E)$ and, for its overall sign, on the choice of orientation of $H^+(X)$ and the lift of w_2, but we will not build this into our notation at this stage.

9.1.2 Independence of metric

We now come to the crucial point: the proof that $q(g)$ is an invariant of the underlying four-manifold X.

Proposition (9.1.1). *For any two generic metrics* g_0, g_1 *on* X, *we have* $q(g_0) = q(g_1)$.

To understand this we return to our picture of the universal moduli space, $\mathfrak{M} \subset \mathscr{B} \times \mathscr{C}$, where $\mathscr{C}$ is the space of conformal classes (see (4.3.3)). Let $\pi\colon \mathfrak{M} \to \mathscr{C}$ be the projection map. We want to compare the fibres of π over two general points g_0, g_1 in $\mathscr{C}$ and we do this by considering a suitable path γ in $\mathscr{C}$ from g_0 to g_1. (The space $\mathscr{C}$ is certainly connected, since any two metrics can be joined by a linear path.)

Let us first dispose of the question of reducible connections—the singular points of $\mathscr{B}$. We have seen in Section 4.3 that the metrics for which a given topological reduction of E is realized by an *ASD* connection form a submanifold in $\mathscr{C}$ of codimension $b^+(X)$. Since this is greater than one by hypothesis, we can perturb any path γ slightly to avoid all these submanifolds; for the perturbed path, we do not encounter reductions in any $M_E(\gamma(t))$ (Corollary (4.3.19)).

We postpone for the moment the discussion of the compactness of the solution space in the family and return to our finite-dimensional analogy: how does one prove directly that the algebraic sum of the zeros of a general section defines an invariant of a bundle V over a compact manifold B (with base and fibre dimensions equal)? Extending the analogy with our discussion above we can embed any two general sections in a family s_c parametrized by another finite-dimensional, connected manifold C, so we have a parametrized zero-set Z, which we can suppose to be a submanifold of $B \times C$ of the proper dimension, dim C. For a parameter value $c \in C$, the algebraic sum of the zeros of the corresponding section s_c is the number of points in the fibre $\pi^{-1}(c)$ of the projection map $\pi: Z \to C$, counted with signs. This is just the standard differential topological definition of the *degree* of the (proper) map π, so what we are looking for is the proof that this is a good definition, yielding a number which does not depend on c.

The proof of the invariance of degree in the finite-dimensional situation is based on the idea of transversality which we discussed in Section 4.3. Two general values c_0, c_1 can be joined by a path $\gamma: [0, 1] \to C$ which is transverse to $\pi: Z \to C$. Then the subset

$$W = \{(z, t) | \pi(z) = \gamma(t)\} \subset Z \times [0, 1]$$

is a smooth one-dimensional submanifold with boundary. The boundary components lying over 0 are the points of $\pi^{-1}(c_0)$ and those over 1 are $\pi^{-1}(c_1)$. Moreover, if we have orientations throughout, the oriented boundary of Y is the difference of the algebraic sums over each fibre. So the fact that these sums are equal follows from the fact that the total oriented boundary of a compact one-manifold is zero.

We have seen in Section 4.3 that the key ingredient of this argument carries over to the case of a Fredholm map. That is, the path γ: $[0, 1] \to \mathscr{C}$ which joins the generic metrics g_0, g_1 can be perturbed so as to be transverse to the Fredholm map π: $\mathfrak{M}^* \to \mathscr{C}$ (we are using $\mathfrak{M}^*$ again to denote the universal moduli space of *irreducible* solutions). So for a generic path γ, the parametrized moduli space

$$\mathscr{M}_E = \{([A], t) | [A] \in M_E(\gamma(t))\} \subset \mathscr{B}_E^* \times [0, 1]$$

is a one-manifold with oriented boundary components representing the two moduli sets $M_E(g_0)$, $M_E(g_1)$. The equality of $q(g_0)$ and $q(g_1)$ follows if we know that this one-dimensional parametrized moduli space is *compact*. To prove this we need a rather trivial extension of the compactness theorem (4.4.3) to the case of a family of metrics.

Proposition 9.1.2. *Suppose that we have a sequence g_n of metrics converging (in C^r) to a limit g_∞. Suppose that A_n is a sequence of g_n-ASD connections on E. Then there is a g_∞-ASD connection A_∞ on E or one of the lower bundles $E^{(r)}$, and points x_j in X, such that a subsequence of the A_n converges to $([A_\infty]; \{x_1, \ldots, x_r\})$ in the sense defined in* (4.4.2).

This follows from an obvious extension of the discussion in Section 4.4. Now if γ is a path in $\mathscr{C}$ such that the moduli spaces $M_{E^{(r)}}(\gamma(t))$ are empty for all $t \in [0, 1]$ and $r > 0$, this extended compactness theorem tells us that $\pi^{-1}(\text{Range}(\gamma))$ is a compact subset of $\mathscr{B}$.

To arrange that the lower moduli spaces are empty in the family we use another, auxiliary transversality argument. The lower moduli spaces $M_{E^{(r)}}$ have formal dimension $-8r$ which is negative, but more to the point is less than -1. Our transversality results from Section 4.3 therefore tell us that in generic one-parameter families $\gamma(t)$ the moduli spaces are empty; indeed , this is true for generic seven-parameter families. This completes the proof of Proposition (9.1.1).

9.1.3 Calculation for a K3 surface

Let X be a $K3$ surface and let $\alpha \in H^2(X; \mathbb{Z}/2)$ be a class with $\alpha^2 = 2 \pmod 4$. Consider the $SO(3)$ bundle F with $w_2(F) = \alpha$ and $p_1(F) = -6$. Since the $K3$ surface has $b^+ = 3$, the formal dimension of the moduli space is

$$\dim M_F = -2p_1(F) - 3.(1 + 3) = 0.$$

Thus we are in the position considered above, and we have a numerical invariant q, counting with signs the points in the moduli space. To be more precise however, recall that there is an overall choice of sign involved in the definition, and this can be fixed by a choice of orientation Ω of a positive subspace in $H^2(X)$. (The $K3$ surface has even intersection form, so there is no dependence on a lift of α—see Section 7.1.6.) So we can write our invariant as $q(\alpha, \Omega)$.

Proposition (9.1.3). *There is a class α and orientation Ω such that $q(\alpha, \Omega) = 1$.*

This will be proved shortly, using the general theory developed in Chapter 6. The proposition has an immediate corollary:

Corollary (9.1.4). *There is no diffeomorphism of X which acts trivially on $H^2(X; \mathbb{Z}/2)$ but which reverses the orientation of the positive part of $H^2(X; \mathbb{R})$.*

Proof. This is a consequence of the naturality of the invariant: if $f\colon X \to X$ is any diffeomorphism (necessarily orientation-preserving) then we have

$$q(f^*(\alpha), f^*(\Omega)) = q(\alpha, \Omega). \tag{9.1.5}$$

In particular, if $f^*(\alpha) = \alpha$ then $q(\alpha, \Omega) = \pm q(\alpha, \Omega)$, the sign depending on whether the orientation $f^*(\Omega)$ is equal or opposite to Ω. The existence of such a diffeomorphism which reversed the orientation Ω would therefore imply that $q(\alpha, \Omega) = 0$, contrary to (9.1.3).

Corollary (9.1.6). *There is a simply-connected five-dimensional h-cobordism which is not a product.*

Proof. An h-cobordism W between simply-connected four-manifolds X_1, X_2 induces an isomorphism $f_W: H^2(X_1) \to H^2(X_2)$, preserving the intersection forms. A sharper version of the h-cobordism classification of four-manifolds discussed in Chapter 1 states that any form-preserving isomorphism between the cohomology of simply connected four-manifolds can be realized in this way, as f_W for some W. In particular, we can take X_1 and X_2 both to be the $K3$ surface X and we get an h-cobordism W realizing, say, the map -1 on $H^2(X)$. That is, we have boundary inclusions $i_1, i_2: X \to W$ with $i_1^* = -i_2^*$. But this cannot be a product cobordism, for then there would be a diffeomorphism of X realizing -1, and this would contradict Corollary 9.1.4.

This result shows that our new invariants are far from being trivial: the failure of the h-cobordism theorem for four-manifolds is in contrast to the situation revealed by Smale's theorem in all higher dimensions, and has not so far been established by any other means than the study of the ASD moduli spaces. As another application, we have:

Corollary (9.1.7). *There does not exist a connected sum decomposition* $X = Y \# (S^2 \times S^2)$.

Proof. This can be deduced from the much stronger result (1.3.2), proved in Chapter 8, which shows that no four-manifold Y of the homotopy-type implied by such a decomposition can exist. Using the invariant q however, we have a much more elementary argument. On $(S^2 \times S^2)$ there is a diffeomorphism h inducing the map -1 on homology, for example the product of the antipodal maps of the two factors. In general, given diffeomorphisms h_1 and h_2 of two manifolds Y_1, Y_2, it is easy to construct a 'connected sum' of the diffeomorphisms—a diffeomorphism of $Y_1 \# Y_2$ which induces the map $h_1^* \oplus h_2^*$ on cohomology: it is only necessary to modify the two diffeomorphisms by isotopy until they are equal to the identity map in the small balls in which the sum construction is made. Thus given such a connected sum decomposition of a $K3$ surface we would obtain a diffeomorphism which acted as -1 on one $\begin{pmatrix} 0 & 1 \\ 1 & 0 \end{pmatrix}$ summand and as $+1$ on the complement. This would reverse the orientation of the positive part, contradicting (9.1.4) again.

We shall now carry out the promised calculation establishing Proposition 9.1.3. We take a concrete model of a $K3$ surface as a double cover of $\mathbb{C}P^2$, branched over a smooth curve of degree six, say $\pi: X \to \mathbb{C}P^2$ (see Section 1.1.7). The pull-back $\pi^*(H)$ of the Hopf line bundle H over $\mathbb{C}P^2$ is an ample line bundle—i.e. there is an embedding $X \to \mathbb{C}P^N$ such that the hyperplane class $[X \cap \mathbb{C}P^{N-1}]$ is a positive multiple of $\pi^*(H)$. We calculate moduli spaces with respect to a compatible Kähler metric g—for example the restriction of the Fubini–Study metric on $\mathbb{C}P^N$.

Let $\mathscr{E} \to X$ be the pull-back, $\pi^*(T\mathbb{C}P^2)$, of the tangent bundle of $\mathbb{C}P^2$, and let $\mathscr{F}$ be the associated holomorphic $SO(3, \mathbb{C})$ bundle, the bundle of trace-free endomorphisms $\mathrm{End}_0(\mathscr{E})$; it is the complexification of an $SO(3)$ bundle F. We calculate the Pontryagin class:

$$-p_1(F) = 4c_2(\mathscr{E}) - c_1(\mathscr{E})^2 = 2(4c_2(\mathbb{C}P^2) - c_1(\mathbb{C}P^2)^2)$$
$$= 2.(4.3 - 9) = 6.$$

So F has the topological type considered above. In this case we have $\alpha = w_2(F) = \pi^*(h)$, where h is a generator for $H^2(\mathbb{C}P^2)$. However, we should point out that the result of (9.1.3) holds for *any* class α (and suitable orientation Ω). This follows from the naturality (9.1.5), for it is known that the diffeomorphisms of X act transitively on the set of mod 2 cohomology classes with square 2 mod 4, so as far as differential topology goes, all choices of F are equivalent. The particular choice we have made is adapted however to our choice of geometric model for X.

Now we have seen in Chapter 6 that the moduli space of g-ASD connections on F can be identified with the moduli space of holomorphic two-plane bundles topologically equivalent to $\mathscr{E}$ which are stable with respect to the ample line bundle $\pi^*(H)$. We shall show that this moduli space consists of exactly one point, corresponding to the bundle $\mathscr{E}$ itself.

Lemma (9.1.8). *The tangent bundle $T\mathbb{C}P^2$ is stable.*

Proof. If not, there is a non-trivial map $T\mathbb{C}P^2 \to \mathscr{L}$ for some line bundle $\mathscr{L}$ over $\mathbb{C}P^2$ with $\deg(\mathscr{L}) \leq \frac{1}{2}\deg T\mathbb{C}P^2 = \frac{3}{2}$. But the only line bundles over $\mathbb{C}P^2$ are the powers of the Hopf bundle H, so we would have $\mathscr{L} = H^k$ for some $k \leq \frac{3}{2}$, and since H itself has sections we can assume $k = 1$. So it suffices to show that $\mathrm{Hom}(T\mathbb{C}P^2, H) = \Omega^1 \otimes H$ has only the trivial holomorphic section, which is a standard fact in projective geometry. (For example, we can prove this using the Euler sequence

$$0 \longrightarrow \Omega^1 \otimes H \longrightarrow \underline{\mathbb{C}}^3 \longrightarrow H \longrightarrow 0.)$$

Lemma (9.1.9). *The bundle $\mathscr{E}$ is stable with respect to the ample line bundle $\pi^*(H)$.*

Proof. If the contrary holds there is a rank-one subsheaf $\mathscr{L} \subset \mathscr{E}$ with torsion-free quotient, such that

$$\deg(\mathscr{L}) \geq \deg(\mathscr{E}/\mathscr{L}).$$

Since $\mathscr{E}$ is lifted up from $\mathbb{C}P^2$, the covering involution $\sigma\colon X \to X$ has a tautological lift to $\mathscr{E}$. Let $\mathscr{L}' = \sigma(\mathscr{L})$ and consider the composite

$$\mathscr{L}' \longrightarrow \mathscr{E} \longrightarrow \mathscr{E}/\mathscr{L}.$$

This vanishes over the branch locus of π, because $\mathscr{L}$ and $\mathscr{L}'$ coincide there. Also $\deg(\mathscr{L}') = \deg(\mathscr{L}) \geq \deg(\mathscr{E}/\mathscr{L})$ and it follows that the composite map is everywhere zero, because a torsion-free rank-one sheaf of negative degree has no sections. Thus $\mathscr{L}' \subset \mathscr{L}$, and similarly $\mathscr{L} \subset \mathscr{L}'$, so $\mathscr{L}$ is invariant under σ. We deduce that $\mathscr{L}$ is the pull-back of some subsheaf $\mathscr{L}_0 \subset T\mathbb{C}P^2$ over $\mathbb{C}P^2$, and we have

$$\deg(\mathscr{L}_0) = \tfrac{1}{2}\deg(\mathscr{L}) \leq \tfrac{1}{2}.\tfrac{1}{2}\deg(\mathscr{E}) = \tfrac{1}{2}\deg(T\mathbb{C}P^2),$$

contradicting the stability of $T\mathbb{C}P^2$ established in the previous lemma.

We now consider the deformation theory of this bundle $\mathscr{E}$. We know that the formal dimension of the moduli space it lives in is zero, but as we have explained in Chapter 4, this by itself is no guarantee that the moduli space will consist of isolated points. We need to examine the cohomology groups $H^i(\mathrm{End}_0(\mathscr{E}))$. Now H^0 is zero because $\mathscr{E}$ is stable, and we know from our index formula (or in the algebro-geometric context the Riemann–Roch formula), that

$$\dim H^1 - \dim H^2 = 0.$$

On the other hand Serre duality gives

$$H^2(\mathrm{End}_0\,\mathscr{E}) = H^0(\mathrm{End}_0\,\mathscr{E} \otimes K_X)^*,$$

since $\mathrm{End}_0\,\mathscr{E}$ is self-dual. But the canonical bundle K_X is trivial by the main defining property of a $K3$ surface. So H^2 is the dual space of H^0, and using our index formula we see that all three cohomology groups must vanish.

It follows from this discussion that $[\mathscr{E}]$ is indeed an isolated point in the moduli space, and moreover that the corresponding ASD connection is a transverse zero of the equations. Thus with a suitable orientation Ω we get a contribution one to the invariant q from this solution. The proof of (9.1.3) is completed by showing that there are no other points in the moduli space, a result of Mukai (1984):

Lemma (9.1.10) (Mukai). *Any $\pi^*(H)$-stable holomorphic bundle $\mathscr{E}'$ over X which is topologically equivalent to $\mathscr{E}$ is also holomorphically equivalent.*

Proof. Consider the bundle $\mathrm{End}\,\mathscr{E} = \mathscr{E} \otimes \mathscr{E}^* = \mathrm{End}_0\,\mathscr{E} \oplus \underline{\mathbb{C}}$. The holomorphic Euler characteristic $\chi(E) = \sum(-1)^i \dim H^i(\mathscr{E})$ is

$$\chi(\mathrm{End}\,\mathscr{E}) = \chi(\mathrm{End}_0\,\mathscr{E}) + \chi(\underline{\mathbb{C}}) = 0 + 2 = 2.$$

If $\mathscr{E}'$ is another bundle of the same topological type, then

$$\chi(\mathscr{E}' \otimes \mathscr{E}^*) = \chi(\mathscr{E} \otimes \mathscr{E}^*) = 2,$$

so at least one of $H^0(\mathscr{E}' \otimes \mathscr{E}^*)$, $H^2(\mathscr{E}' \otimes \mathscr{E}^*)$ is non-zero. But, using Serre duality as before, the latter space is dual to $H^0(\mathscr{E} \otimes (\mathscr{E}')^*)$. Thus we conclude that there is either a holomorphic map from $\mathscr{E}$ to $\mathscr{E}'$, or from $\mathscr{E}'$ to $\mathscr{E}$. But any non-zero holomorphic map between stable bundles of the same topological type is an isomorphism (6.2.8), so the two bundles must be the same.

9.2 Polynomial invariants

9.2.1 $SU(2)$ bundles, the stable range

We now return to the general programme outlined at the beginning of the chapter. Recall that we wish to define invariants of a four-manifold X by regarding the moduli space $M \subset \mathscr{B}^*_{E,X}$ as carrying a distinguished homology class, independent of the choice of metric used to define M. The chief obstacle to this programme—an obstacle which is absent in the case that M is zero-dimensional, the situation considered above—is the possible non-compactness of the moduli space. This prevents us from defining a fundamental class $[M]$ as such, but we are nevertheless able to form a well-defined pairing $\langle \beta, [M] \rangle$ for certain cohomology classes $\beta \in H^*(\mathscr{B}^*)$: the idea is that one can find distinguished cochain representatives for the classes β (based on the constructions in Section 5.2), whose restrictions to M are compactly supported; so the pairing $\langle \beta, [M] \rangle$ results from the evaluation $H^*_c(M) \to \mathbb{Z}$ of top-dimensional classes with compact support.

Here we shall carry through this programme in the case that $E \to X$ is an $SU(2)$ bundle and β is a product of classes of the form $\mu(\Sigma)$ (Definition (5.1.11)). We write $k = c_2(E)$ and M_k for the ASD moduli space. Since the classes $\mu(\Sigma)$ have dimension two, it is important that the formal dimension of M_k should be even: we write

$$\dim M_k = 2d,$$

where

$$d = d(k) = 4k - \tfrac{3}{2}(b^+(X) + 1).$$

In order to avoid encountering reducible solutions we will require $b^+(X) > 1$ (cf (4.3.20)). From the formula above it is clear that $b^+(X)$ must be odd if $\dim M_k$ is to be even; as in Section 9.1 therefore, we suppose $b^+(X)$ is odd and ≥ 3. As always, we shall suppose that X is simply-connected.

Let $[\Sigma_1], \ldots, [\Sigma_d]$ be classes in $H_2(X; \mathbb{Z})$, and let $\mu(\Sigma_i) \in H^2(\mathscr{B}^*; \mathbb{Z})$ be the corresponding cohomology classes. The cup-product $\mu(\Sigma_1) \smile \ldots \smile \mu(\Sigma_d)$ has degree $2d$, so we can try to evaluate it on M_k, 'defining' an integer

$$q = \langle \mu(\Sigma_1) \smile \ldots \smile \mu(\Sigma_d), [M_k] \rangle. \tag{9.2.1}$$

If such a definition can be made, we should expect the following properties of q. First, since the orientation of M_k depends on a choice of orientation Ω for $H^+(X)$, we should write q as $q_{k,\Omega}(\Sigma_1, \ldots, \Sigma_d)$. Then we will have:

Conditions (9.2.2).

(i) $q_{k,\Omega}(\Sigma_1, \ldots, \Sigma_d)$ *depends on* Σ_i *only through its homology class* $[\Sigma_i]$;

(ii) $q_{k,\Omega}(\Sigma_1, \ldots, \Sigma_d)$ *is multilinear and symmetric in* $[\Sigma_1], \ldots, [\Sigma_d]$;

(iii) $q_{k,\Omega} = -q_{k,-\Omega}$;

(iv) q_k *is natural, in that if* $f\colon X \to Y$ *is an orientation-preserving diffeomorphism then*

$$q_{k,f^*\Omega}(f(\Sigma_1), \ldots, f(\Sigma_d)) = q_{k,\Omega}(\Sigma_1, \ldots, \Sigma_d).$$

Condition (iv) contains the assertion that q_k is independent of the choice of metric and is an invariant of the oriented diffeomorphism type of X.

The remainder of this section is devoted to showing that such an invariant q_k can indeed be defined, based on the idea expressed by (9.2.1), once k is sufficiently large—in the 'stable range'. The precise condition on k is that

$$d(k) \geq 2k + 1, \tag{9.2.3}$$

or equivalently, $k \geq \frac{1}{4}(3b^+(X) + 5)$. The origin of this constraint is easily understood: it says that the formal dimension, $2d$, of the moduli space exceeds the dimension of the lowest stratum $M_0 \times s^k(X)$ in the compactification $\bar{M}_k$ by at least two: this means that $\bar{M}_k$ is generically a manifold except at a set of codimension two or more, which is the usual condition for a singular complex to possess a fundamental class.

Choose a Riemannian metric g on X for which the ASD moduli spaces have the usual generic properties:

Condition (9.2.4).

(i) *for* $0 \leq j < k$, *the moduli space* M_{k-j} *is a smooth manifold of the correct dimension,* $2d - 8j$, *cut out transversely by the ASD equations, i.e.* $H_A^2 = 0$ *for all* $[A]$;

(ii) *for* $0 \leq j < k$, *there are no reducible connections in* M_{k-j}.

As in Section 8.3.2, choose embedded surfaces Σ_i in general position which represent the homology classes in X, and let $\nu(\Sigma_i)$ be tubular neighbourhoods with the property that the triple intersections are empty:

$$\nu(\Sigma_i) \cap \nu(\Sigma_j) \cap \nu(\Sigma_k) = \varnothing, \quad (i, j, k \text{ distinct}). \tag{9.2.5}$$

For each i choose a section s_i of the line bundle $\mathscr{L}_{\Sigma_i}$ over $\mathscr{B}^*_{\nu(\Sigma_i)}$ and let V_{Σ_i} be its zero-set. We may assume that s_i satisfies the condition (5.2.13), so that the closure of V_{Σ_i} in $\mathscr{B}_\nu(\Sigma_i)$ does not contain the trivial connection:

$$[\Theta] \notin \mathrm{Cl}_{\mathscr{B}}(V_{\Sigma_i}). \tag{9.2.6}$$

Using the transversality argument (5.2.12), we can arrange that:

Condition (9.2.7). *For any* $I \subset \{1, \ldots, d\}$ *and any* j *with* $0 \leq j < k$, *the intersection* $M_{k-j} \cap \left(\bigcap_{i \in I} V_{\Sigma_i} \right)$ *is transverse.*

In particular then, the set

$$M_k \cap V_{\Sigma_1} \cap \ldots \cap V_{\Sigma_d} \tag{9.2.8}$$

consists of isolated points. The proof of the next result is essentially a repetition of the argument used in (8.3.3) and (8.3.4).

Lemma (9.2.9). *The intersection* (9.2.8) *is compact, and hence finite.*

Proof. Let $[A_n]$ be a sequence in (9.2.8). By the compactness theorem, there is a subsequence $[A_m]$ which converges weakly to an ideal ASD connection $([A_\infty]; \{x_1, \ldots, x_l\}) \in M_{k-l} \times s^l(X)$. We have to show that $l = 0$, so that $[A_m]$ actually converges on M_k. The important feature of the subsets V_Σ is the following *alternative*:

Alternative (9.2.10). *For each i, either*

(i) $[A_\infty]$ *is non-trivial and* $[A_\infty] \in V_{\Sigma_i}$, *or*

(ii) $\nu(\Sigma_i)$ *contains one of the points* x_j.

For suppose (ii) does not hold, so $\nu(\Sigma_i)$ contains none of the points x_j. Then $[A_m]$ is converging to $[A_\infty]$ on $\nu(\Sigma_i)$:

$$[A_m|_{\nu(\Sigma_i)}] \longrightarrow [A_\infty|_{\nu(\Sigma_i)}] \quad \text{in } \mathscr{B}_{\nu(\Sigma_i)}.$$

It follows from (9.2.6) that $[A_\infty]$ is non-trivial, and therefore even irreducible by assumption (9.2.4(ii)). So alternative (i) holds, because V_{Σ_i} is closed in $\mathscr{B}^*_{\nu(\Sigma_i)}$.

To show that $l = 0$ let us rule out the other possibilities. Suppose first that $0 < l < k$. Since each x_j lies in at most two of the $\nu(\Sigma_i)$ (by condition (9.2.5)), alternative (i) holds for at least $d - 2l$ surfaces, say $\Sigma_1, \ldots, \Sigma_{d-2l}$. Then

$$[A_\infty] \in M_{k-l} \cap \bigcap_{i=1}^{d-2l} V_{\Sigma_i}.$$

But this is impossible by the transversality condition (9.2.7), for the dimension of this intersection is negative:

$$\dim M_{k-l} - 2(d - 2l) = -4l.$$

The only other possibility is that $l = k$ and A_∞ is the trivial connection Θ. In this case alternative (ii) must hold for all i. But this is impossible again: we have just seen that (ii) must fail for at least $d - 2l$ surfaces, and this number is positive by the inequality (9.2.3), which says that the number of surfaces is more than twice the number of points.

We shall make frequent use of dimension-counting arguments of this sort. The main point can be summarized as follows. If we put $V = M_k \cap \bigcap_{i \in I} V_{\Sigma_i}$ and take the closure $\bar{V}$ in $\bar{M}_k$, then $\bar{V}$ meets only those lower strata which it might be expected to meet on grounds of dimension, based on the dimension of V and the codimension of the strata.

Each point of the intersection (9.2.8) carries a sign ± 1, because both M_k and the normal bundles to the V's are oriented. We define $q_{k,\Omega}$ by counting the points according to their sign, for which we introduce the following notation:

Definition (9.2.11). $q_{k,\Omega}(\Sigma_1, \ldots, \Sigma_d) = \#(M_k \cap V_{\Sigma_1} \cap \ldots \cap V_{\Sigma_d})$.

If M_k is compact, this integer is the pairing (9.2.1). In any case, we have the following result.

Theorem (9.2.12). *Let X be simply-connected, with $b^+(X)$ odd and not less than three. Then if k is in the stable range* (9.2.3), *the integer defined by* (9.2.11) *is independent of the choices made and has the properties* (i)–(iv) *of* (9.2.2).

For the proof, we start with properties (i)–(ii) of (9.2.2), which can conveniently be treated together; the argument also shows that the value of q is independent of the choice of sections s_i. Let $\Sigma_1, \Sigma_1', \Sigma_1''$ be embedded surfaces whose homology classes satisfy $[\Sigma_1] = [\Sigma_1'] + [\Sigma_1'']$. Choose tubular neighbourhoods for each of these and sections of the corresponding determinant line bundles, leading to zero-sets $V_{\Sigma_1}, V_{\Sigma_1'}, V_{\Sigma_1''}$. We suppose that the necessary transversality conditions (9.2.5) and (9.2.7) are satisfied with all three choices ($V_{\Sigma_2}, \ldots, V_{\Sigma_d}$ are fixed), and we put

$$q = \#(M_k \cap V_{\Sigma_1} \cap V_{\Sigma_2} \cap \ldots \cap V_{\Sigma_d})$$
$$q' = \#(M_k \cap V_{\Sigma_1'} \cap V_{\Sigma_2} \cap \ldots \cap V_{\Sigma_d})$$
$$q'' = \#(M_k \cap V_{\Sigma_1''} \cap V_{\Sigma_2} \cap \ldots \cap V_{\Sigma_d}).$$

We shall prove that $q = q' + q''$.

If we are prepared to strengthen slightly the condition (9.2.3) and demand that $d(k) \geq 2k + 2$, the argument is very simple. For in this case the lower strata in $\bar{M}_k$ have codimension ≥ 4, so dimension counting of the sort used in (9.2.9) shows that the two-manifold

$$V_2 = M_k \cap V_{\Sigma_2} \cap \ldots \cap V_{\Sigma_d} \subset \mathcal{B}_X^*$$

is already compact. Thus q, q' and q'' represent honest pairings:

$$q = \langle \mu(\Sigma_1), V_2 \rangle$$
$$q' = \langle \mu(\Sigma_1'), V_2 \rangle$$
$$q'' = \langle \mu(\Sigma_1''), V_2 \rangle,$$

and the required formula follows from the fact that $\mu\colon H_2(X) \to H^2(\mathcal{B}^*)$ is a homomorphism. There remains the borderline case, $d = 2k + 1$. In this case V_2 may be non-compact. If $[A_n] \in V_2$ is a divergent sequence then, after passing to a subsequence, we will have

$$[A_n] \longrightarrow ([\Theta]; x_1, \ldots, x_k)$$

where each x_p lies in one of the intersections $\nu(\Sigma_i) \cap \nu(\Sigma_j)$, for $2 \leq i < j$. (All other possibilities are ruled out on the grounds of dimension.) In particular, because the triple intersections are empty, $[A_n]$ converges to $[\Theta]$ on $\nu(\Sigma_1)$, $\nu(\Sigma_1')$ and $\nu(\Sigma_1'')$. There is therefore an open set $U \subset V_2$, with compact complement, on which the three line bundles $\mathscr{L}_{\Sigma_1}$, $\mathscr{L}_{\Sigma_1'}$, $\mathscr{L}_{\Sigma_1''}$ are canonically trivialized by sections σ, σ', σ'' (see (5.2.8)). The integers q, q' and q'' are the result of pairing the relative Chern classes $c_1(\mathscr{L}_{\Sigma_1}; \sigma)$ etc. with the fundamental class in $H_2(V_2, U)$. The equality $q = q' + q''$ follows as before.

To show, finally, that $q_{k,\Omega}$ is independent of the choice of Riemannian metric, we adapt the argument used in (9.1.1). Let g_0, g_1 be two generic metrics, and put

$$q(g_0) = \#(M_k(g_0) \cap V_{\Sigma_1} \cap \ldots \cap V_{\Sigma_d})$$

$$q(g_1) = \#(M_k(g_1) \cap V_{\Sigma_1} \cap \ldots \cap V_{\Sigma_d}).$$

Join g_0 to g_1 by a path $\gamma(t)$, and consider the parametrized moduli spaces

$$\mathscr{M}_j = \bigcup_t M_j(\gamma(t)) \times \{t\} \subset \mathscr{B}_j^* \times [0, 1].$$

These, we may assume, are manifolds-with-boundary, of the correct dimension. By the transversality argument (5.2.9), we can find perturbations s_i' of the sections s_i so that the zero-sets V_{Σ_i}' have the property that the intersections

$$\mathscr{M}_j \cap \bigcap_{i \in I} V_{\Sigma_i}'$$

are transverse. Replacing s_i by s_i' will not affect either $q(g_0)$ or $q(g_1)$, because we have already seen that these intersection numbers are independent of the choice of sections. Now

$$\mathscr{M}_k \cap V_{\Sigma_1}' \cap \ldots \cap V_{\Sigma_d}'$$

is a one-manifold whose oriented boundary is $q(g_0) - q(g_1)$. It therefore only remains to show that this one-manifold is compact, which can be proved by dimension-counting again, using the compactness theorem (9.1.2).

This concludes the proof of (9.2.12) and with it the definition of the invariants q_k. By polarization, q_k can be uniquely recovered from the corresponding homogeneous polynomial, for which we use the same notation:

$$q_{k,\Omega}: H_2(X) \longrightarrow \mathbb{Z},$$

$$q_{k,\Omega}(\Sigma) = q_{k,\Omega}(\Sigma, \ldots, \Sigma).$$

We can also regard cohomology classes as linear functions on $H_2(X)$, and so write q_k as a homogeneous polynomial expression in the elements of $H^2(X)$.

We shall give one simple example before going on. Take X to be $\mathbb{C}P^2$ with the standard orientation and the Fubini–Study metric, and take $k = 2$. The moduli space M_2 was Example (iii) in Section 4.1: it can be identified with the space of smooth conics C in the dual plane $\mathbb{P}^*$. We need to know how this identification comes about. By the results of Chapter 6, M_2 can be regarded as the moduli-space of stable holomorphic two-plane bundles $\mathscr{E} \to \mathbb{C}P^2$ with

$c_1(\mathscr{E}) = 0$ and $c_2(\mathscr{E}) = 2$. On a generic line $l \subset \mathbb{C}P^2$, such a bundle $\mathscr{E}$ is holomorphically trivial. The lines l for which $\mathscr{E}|_l$ is non-trivial—the jumping lines—are parametrized by a conic $C_{\mathscr{E}}$ in $\mathbb{P}^*$, and the assignment of $C_{\mathscr{E}}$ to $\mathscr{E}$ establishes the correspondence. (This is part of a larger theory; in general, the jumping locus is a curve of degree k in the dual plane.) In any family of lines transverse to $C_{\mathscr{E}}$, the behaviour of $\mathscr{E}$ is of the form given by (5.2.14), i.e. the jumping divisor has multiplicity one. From the discussion at the end of Section 5.2.2, it follows that if $\Sigma \subset \mathbb{C}P^2$ is a projective line, then $\mu(\Sigma)$ can be represented on M_2 by the jumping divisor:

$$\begin{aligned} M_2 \cap V_\Sigma &= \{\mathscr{E} \in M_2 \mid \Sigma \text{ is a jumping line}\} \\ &= \{\text{smooth conics } C \subset \mathbb{P}^* \mid \Sigma \in C\}. \end{aligned}$$

If $\Sigma_1, \ldots, \Sigma_5$ are five lines in general position, then

$$\#(\mathscr{M}_2 \cap V_{\Sigma_1} \cap \ldots \cap V_{\Sigma_5}) = \#\{C \mid \Sigma_1, \ldots, \Sigma_5 \in C\} = 1,$$

since five points in general position determine a unique conic. Another way to put this is to observe that the smooth conics in $\mathbb{P}^*$ are parametrized by the complement of a divisor D in $\mathbb{C}P^5$, and that the conics containing a given point $\Sigma \in \mathbb{P}^*$ form a hyperplane; then the calculation can be rephrased as saying that the five hyperplanes V_{Σ_i} intersect in a unique point, disjoint from D in general.

Although $\mathbb{C}P^2$ does not satisfy our hypothesis $b^+(X) \geq 3$, invariants are still defined. The point is that the hypothesis is used only to avoid the reducible connections, but on a positive-definite manifold such as $\mathbb{C}P^2$, reducible connections are ruled out on topological grounds once k is positive. Thus the discussion above has led to the calculation of q_2 for the manifold $\mathbb{C}P^2$: for a suitable orientation Ω of $H^2(\mathbb{C}P^2)$ and the standard generator H for the homology, we have $q_{2,\Omega}(H) = 1$.

9.2.2 *SO*(3) *bundles*

Polynomial invariants can also be defined using the moduli spaces associated with $SO(3)$ bundles $E \to X$ with $w_2(E)$ non-zero. The construction is even slightly simpler because of the absence of the product connection Θ from the compactification. We shall go through the few points at which changes need to be made.

The first, small point is that the map μ: $H_2(X) \to H^2(\mathscr{B}^*)$ defined by $\mu(\Sigma) = -\frac{1}{4}p_1(\mathbb{P})/[\Sigma]$ does not carry integer classes to integer classes in general: if w_2 is non-zero on Σ, then it is $2\mu(\Sigma)$ which is integral. This only deserves comment as our construction of the representatives V_Σ depended on $\mu(\Sigma)$ being represented by a line bundle. Of course we only have to represent the classes $2\mu(\Sigma)$ by line bundles, count intersections of zero-sets as before, and then divide by the appropriate power of two, to obtain an invariant in $\mathbb{Q}$.

As we have mentioned in Section 7.1.6, the orientation of the moduli space $M_{k,\alpha}$, for a bundle E with $\frac{1}{4}p_1 = -k$ and $w_2 = \alpha$, depends on an orientation Ω for $H^+(X)$ and a choice of equivalence class of integral lifts, $\hat{\alpha} \in H^2(X; \mathbb{Z})$, for α. We shall therefore write the corresponding invariant as $q_{k,\Omega,\hat{\alpha}}$ (sometimes omitting the subscripts if the context allows). In line with the discussion above, if $\Sigma_1, \ldots, \Sigma_d$ represent integral classes, then

$$q_{k,\Omega,\hat{\alpha}}(\Sigma_1, \ldots, \Sigma_d) \in (1/2^d)\mathbb{Z}.$$

Finally, and more importantly, note that the inequality (9.2.3), defining the 'stable range', was used in the definition of q_k only to avoid encountering the trivial connection under weak limits (see the proof of (9.2.1)). When w_2 is non-zero, we can therefore dispense with this condition: each moduli space $M_{k,\alpha}$ defines an invariant, as long as the formal dimension is non-negative.

9.2.3 Extensions and variations of the definition

The use of transversality arguments based on the representatives V_Σ is not the only route to defining the polynomial invariants. Many variations are possible, and here we will point out some alternatives. One motivation for our choice of material in this section is the application we have in mind for Chapter 10, the main points being first that it is inconvenient to have a description of q_k which is so tied to a 'generic' metric, when the metrics for which calculation of moduli spaces is feasible—for example the Kähler metrics on a complex surface—are often very special, and second that it may be hard to work with the restrictions of connections to surfaces in concrete examples.

A conceptually satisfying definition of the polynomial invariants can be obtained from the result of Section 7.1.4 which says that the class $\mu(\alpha)$, for $\alpha \in H_2(X; \mathbb{Z})$, can be extended to the compactified moduli space $\bar{M}_k$ as the first Chern class of a line bundle $\bar{\mathscr{L}}_\alpha$ (Theorem (7.1.17)). Under very mild conditions (see below), the compactified moduli space possesses a fundamental class $[\bar{M}_k]$, and the invariant q can therefore be defined by an 'honest' pairing. One does have to verify that this procedure gives the 'correct' answer, i.e. that the pairing agrees with the definition obtained from the intersection of the representatives V_Σ, but this is quite easily deduced from the discussion below. In practice, calculations are always made with explicit cocycle representatives, and one thing we aim for is some mild conditions on these cocycles which will ensure that we obtain the right pairing, independent of any choices made.

The problem of the restriction maps

The submanifold $M_k \cap V_\Sigma$ was defined using a tubular neighbourhood $\nu(\Sigma)$ and the restriction map $r\colon M_k \to \mathscr{B}^*_{\nu(\Sigma)}$. Recall that restriction to Σ itself was an unsatisfactory procedure because of the possibility that an irreducible ASD connection on X might be reducible when restricted to the surface. It hardly needs pointing out that, should it happen that the restriction maps are 'good' (i.e. $A|_\Sigma$ is irreducible whenever the ASD connection A is), then there is no need to use the neighbourhood. One can instead simply take a subvariety $V_\Sigma \subset \mathscr{B}^*_\Sigma$ and define $M_k \cap V_\Sigma$ using this good restriction map, imposing the usual transversality conditions. The invariant q_k can be calculated using such submanifolds, and the answer will be no different.

The proofs of (4.3.21) and (4.3.25) show that the curvature of an irreducible ASD connection has rank two or more at all points of a dense, open set in X. It follows quite easily that, given a fixed metric g, the set of embeddings for which the restriction map fails to be good is of the first category in the space of all C^r embeddings of Σ in X. This provides an alternative route for defining the submanifolds V_Σ, dispensing altogether with the need to use the tubular neighbourhoods.

Relaxing the transversality requirements

The invariants $q_k(\Sigma_1, \ldots, \Sigma_d)$ can be calculated using a metric g on X which satisfies rather weaker hypotheses than the strict transversality requirements laid down in (9.2.4). We begin with the $SO(3)$ case, for simplicity. Fix a non-zero value for w_2, say $\beta \in H^2(X; \mathbb{Z}/2)$, and let M_k denote the moduli space of ASD connections in a bundle E with $w_2(E) = \beta$ and $-\frac{1}{4}p_1(E) = k$. As usual, let $2d$ be the virtual dimension of M_k. We shall suppose:

Conditions (9.2.13).

(i) *the set $S' \subset M_k$ where H^2_A is non-zero has codimension at least two; that is,* $\dim S' \le 2d - 2$;

(ii) *for all $l \ge 1$, the stratum $M_{k-l} \times s^l(X)$ has codimension at least two in $\bar{M}_k$; that is,* $\dim M_{k-l} \le 2d - 4l - 2$;

(iii) *there are no reducible connections in M_k or M_{k-l} for $l \ge 1$.*

Conditions (i) and (ii) imply that the compactification $\bar{M}_k$ is a manifold of dimension $2d$ outside a singular set $S = S' \cup$ (lower strata) of codimension two.

Since we are not requiring that the moduli spaces are smooth, we should say a few words about the notion of dimension. In practice, all our moduli spaces will be analytic spaces, so the notion of dimension is unambiguous. Generally however, a convenient definition for the discussion below is that of *covering dimension*: a space S is said to have covering dimension $\le n$ if every open cover $\mathfrak{U} = \{U_\alpha\}$ has a refinement $\mathfrak{U}' = \{U'_\beta\}$ for which all the $(n+2)$-fold intersections are empty:

$$U'_B \overset{\text{def}}{=} \bigcap_{\beta \in B} U'_\beta = \varnothing, \quad \text{if } |B| \ge n+2. \tag{9.2.14}$$

Using the Čech definition of cohomology, one immediately sees that $H^m(S) = 0$ if $m > \dim S$. So if S has codimension two in $\bar{M}_k$, one deduces from the exact sequence of the pair that

$$H^{2d}(\bar{M}_k) = H^{2d}(\bar{M}_k, S) = H_c^{2d}(\bar{M}_k \backslash S),$$

from which it follows that $\bar{M}_k$ carries a fundamental class.

In place of the dual submanifolds which we used before, it is convenient to introduce the idea of the *support* of a cochain. In the Čech construction, a p-cochain ν, defined with respect to an open cover $\mathfrak{U}$, is a collection of locally constant functions $\nu_B: U_B \to \mathbb{Z}$, for $|B| = p + 1$, and one defines the support by

$$\operatorname{supp}(\nu) = \operatorname{Cl}\Bigl(\bigcup_B \operatorname{supp}(\nu_B)\Bigr).$$

Evidently, we have

$$\operatorname{supp}(f^*(\alpha)) = f^{-1}(\operatorname{supp}(\alpha))$$

$$\operatorname{supp}(\alpha \smile \beta) \subset \operatorname{supp}(\alpha) \cap \operatorname{supp}(\beta).$$

Furthermore, support decreases on refinement: that is, if $\mathfrak{U}'$ is a refinement of $\mathfrak{U}$ and $r: C^p(\mathfrak{U}) \to C^p(\mathfrak{U}')$ is the corresponding chain map, then $\operatorname{supp}(r(\nu)) \subset \operatorname{supp}(\nu)$.

If $\Sigma \subset X$ is an embedded surface, let $\bar{M}_\Sigma \subset \bar{M}_k$ be the open set

$$\bar{M}_\Sigma = \{([A]; \{x_1, \dots, x_l\}) \mid \text{no } x_j \text{ lies in } \Sigma\}.$$

Let $\tilde{M}_\Sigma$ be the union

$$\tilde{M}_\Sigma = M_k \cup M_{k-1} \cup \dots$$

topologized according to the following rule: a sequence $[A_n]$ in M_{k-l} is said to converge to $[A_\infty] \in M_{k-l-j}$ if $[A_n]$ converges to $([A_\infty]; \{x_1, \dots, x_j\})$ in the usual weak sense *and none of the x_j lie in Σ*. Thus there is a forgetful map $\bar{M}_\Sigma \to \tilde{M}_\Sigma$ (which forgets the points), and $\tilde{M}_\Sigma$ has the quotient topology. The key idea below is to use cochains on $\bar{M}_\Sigma$ which are pulled back from $\tilde{M}_\Sigma$; the dimension counting argument then goes through.

Since the restriction map $\tilde{M}_\Sigma \to \mathscr{B}_\Sigma^*$ is continuous, $\tilde{M}_\Sigma$ parametrizes a family of connections on Σ and there is universal bundle $\mathbb{E} \to \tilde{M}_\Sigma \times \Sigma$; it can be defined as the pull-back of the universal bundle over $\mathscr{B}_\Sigma^* \times \Sigma$. (We continue to assume, as above, that the restriction of any ASD connection to Σ is irreducible.) Thus there is a well-defined two-dimensional class

$$\tilde{\mu} = -\tfrac{1}{4} p_1(\mathbb{E})/[\Sigma] \in H^2(\tilde{M}_\Sigma; \mathbb{Q}). \tag{9.2.15}$$

Now let $\Sigma_1, \dots, \Sigma_d$ be surfaces in general position in X. For each i, let $\tilde{\mu}_i$ be a Čech cocycle on $\tilde{M}_{\Sigma_i}$ representing the class (9.2.15), and let $\bar{\mu}_i$ be its pull-back to $\bar{M}_{\Sigma_i}$, defined using the pulled-back open cover. Suppose that we have the following condition, analogous to the transversality requirement of (5.2.12):

$$\Bigl(\bigcap_{i \in I} \operatorname{supp}(\tilde{\mu}_i)\Bigr) \cap M_{k-l} = \varnothing \quad \text{if } |I| \geq d - 2l \text{ and } l \geq 1.$$

(Each $\tilde{M}_{\Sigma_i}$ contains a copy of M_{k-l}, and it is in this sense that the intersection is to be understood.) The classes $\bar{\mu}_i$ then satisfy a similar *support condition*:

$$\left(\bigcap_{i\in I} \mathrm{supp}(\bar{\mu}_i)\right) \cap (M_{k-l} \times s^l(X)) = \varnothing \quad \text{if } |I| \geq d - 2l \text{ and } l \geq 1. \tag{9.2.16}$$

Proposition (9.2.17). *If the support condition* (9.2.16) *holds, then* $\mathrm{supp}(\bar{\mu}_1 \smile \ldots \smile \bar{\mu}_d)$ *is compact and contained in* M_k.

Proof. This is the usual counting argument. Let $[A_n] \in M_k$ be a sequence in $\mathrm{supp}(\bar{\mu}_1 \smile \ldots \smile \bar{\mu}_d)$ which converges to $([A_\infty]; \{x_1, \ldots, x_l\}) \in M_{k-l} \times s^l(X)$. Suppose $l \geq 1$. Let

$$I = \{i \mid \Sigma_i \text{ contains no } x_j\}.$$

Then $|I| \geq d - 2l$. On the other hand

$$([A_\infty]; \{x_1, \ldots, x_l\}) \in \left(\bigcap_{i\in I} \mathrm{supp}(\bar{\mu}_i)\right) \cap (M_{k-l} \times s^l(X)),$$

contradicting the support condition.

It follows that there is a well-defined pairing

$$\langle \bar{\mu} \smile \ldots \smile \bar{\mu}_d, [\bar{M}_k] \rangle \in \mathbb{Q}. \tag{9.2.18}$$

We shall show that this number is independent of the choices made and coincides with the previous definition of q_k. First let us show that, with a suitable modification, the support condition can always be satisfied. We have to keep a careful track on the open covers used. Let $\tilde{\mathfrak{U}}_i = \{\tilde{U}_{i,\beta}\}$ be the open cover of $\tilde{M}_{\Sigma_i}$ with which the cocycle $\tilde{\mu}_i$ is defined. Because of (9.2.10(ii)), we may assume that the cover satisfies the *dimension condition*

$$\mathrm{Cl}\left(\bigcap_{\beta\in B} \tilde{U}_\beta\right) \cap M_{k-l} = \varnothing, \quad \text{if } |B| \geq 2d - 4l \text{ and } l \geq 1. \tag{9.2.19}$$

For every subset $I \subset \{1, \ldots, d\}$, let $\bar{M}_{(I)}$ be the intersection $\bigcap_{i\in I} \bar{M}_{\Sigma_i}$, and let $\tilde{M}_{(I)}$ be the quotient space in which the points are forgotten. All the spaces $\tilde{M}_{(I)}$ have the same underlying set, and the identity map $\tilde{M}_{(J)} \to \tilde{M}_{(I)}$ is continuous if $I \subset J$, so an open cover of $\tilde{M}_{(I)}$ can be interpreted also as a cover of $\tilde{M}_{(J)}$. For each I, choose an open cover $\tilde{\mathfrak{U}}_{(I)}$ of $\tilde{M}_{(I)}$ such that:

(i) for $I = \{i\}$ the open cover is the original $\tilde{\mathfrak{U}}_i$;

(ii) if $I \subset J$ then $\tilde{\mathfrak{U}}_{(J)}$ is a refinement of $\tilde{\mathfrak{U}}_{(I)}$;

(iii) all the open covers $\tilde{\mathfrak{U}}_{(I)}$ satisfy the dimension condition (9.2.19).

If $i \in I$, the cocycle $\tilde{\mu}_i$ gives rise to a cocycle on $\tilde{M}_{(I)}$, defined using the refined cover $\tilde{\mathfrak{U}}_{(I)}$. Put

$$\tilde{\mu}_{(I)} = \prod_{i\in I} \tilde{\mu}_i \quad \text{on } \tilde{M}_{(I)},$$

and let $\bar{\mu}_{(I)}$ be the pull-back of this cocycle to $\bar{M}_{(I)}$. The dimension condition (9.2.19) means that, without option,

$$\operatorname{supp}(\bar{\mu}_{(I)}) \cap (M_{k-l} \times s^l(X)) = \varnothing \quad \text{if } |I| \geq d - 2l \text{ and } l \geq 1. \tag{9.2.20}$$

This is weaker than the support condition (9.2.16), but it is all that is needed for the proposition above: condition (9.2.20) implies that $\bar{\mu}_{(1\dots d)}$ has compact support.

The number defined by (9.2.14) is certainly independent of the choice of refinements $\tilde{\mathfrak{U}}_{(I)}$, since any two choices will have further refinements in common. More importantly, the pairing is independent of the chosen representatives $\tilde{\mu}_i$. For suppose that $\tilde{\mu}_1$ and $\tilde{\mu}'_1$ are two different choices for the first cocycle. We may assume that

$$\tilde{\mu}_1 - \tilde{\mu}'_1 = \delta\theta,$$

and that $\tilde{\mu}_1, \tilde{\mu}'_1$ and θ are all defined using the same cover $\tilde{\mathfrak{U}}_1$ of $\tilde{M}_{\Sigma_1}$. Let Θ be the cochain

$$\Theta = \theta \smile \tilde{\mu}_2 \smile \dots \smile \tilde{\mu}_d$$

on $\bigcap_i \tilde{M}_{\Sigma_i}$ and $\bar{\Theta}$ its pull-back to $\bigcap_i \bar{M}_{\Sigma_i}$. As in the proof of the proposition, the dimension condition implies that $\bar{\Theta}$ is compactly supported in M_k; and since

$$\delta\bar{\Theta} = (\bar{\mu}_1 \smile \dots \smile \bar{\mu}_d) - (\bar{\mu}'_1 \smile \dots \smile \bar{\mu}_d),$$

it follows that

$$\langle \bar{\mu}_1 \smile \dots \smile \bar{\mu}_d, [\bar{M}_k] \rangle = \langle \bar{\mu}'_1 \smile \dots \smile \bar{\mu}_d, [\bar{M}_k] \rangle.$$

The pairing is also independent of the choice of metric g. If g_0 and g_1 are two metrics satisfying (9.2.13 (i)–(iii)), they may be joined by a path g_t such that

(i) for $l \geq 1$, the parametrized moduli space $\mathscr{M}_{k-l} = \bigcup_t (M_{k-l}(g_t) \times \{t\})$ has dimension $\leq 2d - 4l - 1$:

(ii) $\mathscr{M}_k$ is a manifold cut out transversely, except at a singular set of codimension at least two;

(iii) there are no reducible connections in $\mathscr{M}_k$ or $\mathscr{M}_{k-l}$ for $l \geq 1$.

These conditions ensure that the usual compactification $\bar{\mathscr{M}}_k$ gives a homology between the fundamental classes of the two ends. The whole construction can now be repeated for these parametrized moduli spaces, starting with the definition of the spaces $\tilde{\mathscr{M}}_{\Sigma_i}$ etc., and finishing with a cocycle $\bar{\mu}_1 \smile \dots \smile \bar{\mu}_d$, which, by virtue of a support condition just like (9.2.16), will be compactly supported in $\mathscr{M}_k$. It follows that

$$\langle \bar{\mu}_1 \smile \dots \smile \bar{\mu}_d, [\bar{M}_k(g_0)] \rangle = \langle \bar{\mu}_1 \smile \dots \smile \bar{\mu}_d, [\bar{M}_k(g_1)] \rangle.$$

Finally, it is now easy to prove that this definition coincides with the old definition of $q_k(\Sigma_1, \ldots, \Sigma_d)$. Choose a metric g satisfying the original non-degeneracy conditions (9.2.4), and take transverse submanifolds $M_k \cap V_{\Sigma_i}$ as before. Then the cocycles $\bar{\mu}_i$ can be chosen so as to be supported in the neighbourhood of $\bar{V}_{\Sigma_i} \subset \bar{M}_{\Sigma_i}$; the transversality conditions for the submanifolds imply the support conditions for the cocycles, provided the neighbourhoods are sufficiently small. For such a choice of $\bar{\mu}_i$, the equality of the two definitions is apparent.

We can also now see that the right pairing is also obtained from the first Chern classes of the line bundles $\bar{\mathscr{L}}_\alpha \to \bar{M}_k$ defined in Chapter 7. Indeed, from the definition of $\bar{\mathscr{L}}_\alpha$ given in Section 7.1.5 it follows quite easily that if $\Sigma \subset X$ represents the class α then $c_1(\bar{\mathscr{L}}_\alpha) = -\frac{1}{4}p_1(\mathbb{E})/[\Sigma]$ on the subspace $\bar{M}_\Sigma \subset \bar{M}_k$. The arguments of the previous paragraphs show that this is all that is important.

A generalization

In Chapter 10 we will be concerned with describing some of the ASD moduli spaces for a very particular four-manifold. It turns out that the spaces which are easiest to describe are not the moduli spaces M_k themselves but slightly larger spaces N_k, which parametrize pairs consisting of an ASD connection together with an extra piece of data (actually a section of an associated holomorphic vector bundle). We wish to abstract this situation here and show that the invariants can be correctly calculated in terms of such N_k. We continue to deal with the case $w_2 \neq 0$, as above.

Suppose that for $l \geq 0$ we are given a space N_{k-l} and a surjective map $\rho_{k-l}: N_{k-l} \to M_{k-l}$. Suppose also that we are given a compact Hausdorff space $\bar{N}_k$ which is a union of subspaces

$$\bar{N}_k = N_k \cup \left(\bigcup_{l \geq 1} N_{k-l} \times s^l(X) \right).$$

Let $\bar{\rho}: \bar{N}_k \to \bar{M}_k$ be the total map formed from the ρ_{k-l} and let the topology of $\bar{N}_k$ be such that $\bar{\rho}$ is continuous. We continue to make the regularity assumptions (9.2.13(i)–(iii)) concerning the moduli spaces, and we make the following assumptions concerning $\bar{N}_k$:

(i) the space N_k is a $2d$-manifold outside a singular set of dimension at most $2d-2$, and $\rho_k: N_k \to M_k$ is generically one-to-one;
(ii) for $l \geq 1$, the dimension of the stratum $N_{k-l} \times s^l(X)$ is at most $2d-2$.

These imply that $\bar{N}_k$ has a fundamental class and that $\bar{\rho}^*: H^{2d}(\bar{M}_k) \to H^{2d}(\bar{N}_k)$ is an isomorphism. If $\Sigma_1, \ldots, \Sigma_d$ are embedded surfaces, one has open subspaces $\bar{N}_{\Sigma_i} = \bar{\rho}^{-1}(\bar{M}_{\Sigma_i}) \subset \bar{N}_k$ and quotient spaces $\tilde{N}_{\Sigma_i}$ just like $\bar{M}_{\Sigma_i}$ and $\tilde{M}_{\Sigma_i}$. Over $\tilde{N}_{\Sigma_i} \times \Sigma_i$ there is a universal bundle $\mathbb{E}$, so a two-dimensional class $\mu(\Sigma_i) \in H^2(\tilde{N}_{\Sigma_i})$ is defined. Let $\tilde{\mu}_i$ be a cocycle representing this class and let $\bar{\mu}_i$ be its pull-back to $\bar{N}_{\Sigma_i}$. Exactly as before, we have:

Proposition (9.2.21). *Suppose that the cocycles satisfy the support condition*

$$\mathrm{Cl}\left(\bigcap_{i \in I} \mathrm{supp}(\bar{\mu}_i)\right) \cap (N_{k-l} \times s^l(X)) = \varnothing \quad \textit{if } |I| \geq d - 2l \textit{ and } l \geq 1.$$

Then $\bar{\mu}_1 \smile \ldots \smile \bar{\mu}_d$ is compactly supported in N_k, and $\langle \bar{\mu}_1 \smile \ldots \smile \bar{\mu}_d, [\bar{N}_k] \rangle$ is equal to $q_k(\Sigma_1, \ldots, \Sigma_d)$.

Proof. The pairing is well-defined and independent of the choice of representative cocycles, by the same argument as before. In particular, one can choose representatives which are pulled back from $\tilde{M}_{\Sigma_i}$, in which case the construction plainly agrees with the previous one.

The $SU(2)$ case

Only a minor change needs to be made in the case of an $SU(2)$ bundle. Generally there will be no universal bundle $\mathfrak{g}_{\mathbb{E}} \to \tilde{M}_\Sigma \times \Sigma$ because of the problem of the trivial connection; thus $\mathfrak{g}_{\mathbb{E}}$ exists only on $(\tilde{M}_\Sigma \backslash M_0) \times \Sigma$. However, since the class $\mu(\Sigma) \in H^2(\mathscr{B}_\Sigma^*)$ has an extension across a neighbourhood of the trivial connection in $\mathscr{B}_\Sigma$ (see Section 5.2.2), there are well-defined two-dimensional classes $\tilde{\mu}_i \in \tilde{M}_{\Sigma_i}$ for which one may take cocycle representatives as before. Given the conditions (9.2.13(i)–(iii)) on the moduli spaces and the support condition (9.2.16) for the cocycles, the whole construction goes through without alteration.

If we have larger spaces N_{k-l} mapping onto the moduli spaces, it may happen that there does exist an $SO(3)$ bundle $\mathfrak{g}_{\mathbb{E}} \to \tilde{N}_{\Sigma_i} \times \Sigma_i$ carrying the correct family of connections, even though such a bundle fails to exist downstairs. This occurs in our application in Chapter 10, and in such a case the class $\tilde{\mu}_i$ can be correctly calculated as $-\frac{1}{4}p_1(\mathfrak{g}_{\mathbb{E}})/[\Sigma_i]$.

The four-dimensional class

Thus far, the only cohomology classes which we have attempted to evaluate on the ASD moduli spaces are those in the polynomial algebra generated by the $\mu(\Sigma_i)$. In Chapter 5 it was shown that the rational cohomology of $\mathscr{B}_X^*$ contains a four-dimensional generator ν, coming from the Pontryagin class of the base-point fibration. In the case of an $SO(3)$ bundle with $w_2 \neq 0$, this class can also be used to define invariants.

The idea is that, just as the two-dimensional classes may be obtained via the restriction maps to the neighbourhood of a surface, so the four-dimensional class may be obtained via the restriction to the neighbourhood of a point. Suppose that the dimension of the $SO(3)$ moduli space M_k is divisible by 4, so $d = 2e$ say. Let $y_1, \ldots, y_e$ be distinct points in X and let $W_1, \ldots, W_e$ be disjoint neighbourhoods, each one a small ball. The orbit space $\mathscr{B}^*_{W_i}$ of irreducible connections has the weak homotopy-type of $BSO(3)$, for it is the quotient of the weakly contractible space $\tilde{\mathscr{B}}^*_{W_i}$ by the $SO(3)$ action. There is therefore a class $v_i \in H^4(\mathscr{B}^*_{W_i})$ corresponding to $-\frac{1}{4}p_1$. We should expect that if $\tilde{v}_i$ is a generic cocycle representing this class and $\bar{v}_i$ is its pull-back to M_k via the restriction map $r_i\colon M_k \to \mathscr{B}^*_{W_i}$, then the support of $\bar{v}_1 \smile \ldots \smile \bar{v}_e$ should be compact, so that there is a pairing

$$\langle \bar{v}_1 \smile \ldots \smile \bar{v}_e, [M_k] \rangle \in \mathbb{Q}. \tag{9.2.22}$$

This idea can be made to work along the lines we have just outlined for the two-dimensional case. Let $\bar{M}_{(i)} \subset \bar{M}_k$ be the set

$$\{([A]; \{x_1, \ldots, x_l\}) \,|\, \text{no } x_j \text{ lies in } W_i\},$$

and let $\tilde{M}_{(i)}$ be the quotient space in which the points are forgotten. If the metric on X is such that conditions (9.2.13(i)–(iii)) hold, then the counting argument goes through: we find representatives $\bar{v}_i$ on $\bar{M}_{(i)}$ satisfying the support condition

$$\operatorname{supp}\left(\prod_{i \in I} \bar{v}_i\right) \cap (M_{k-l} \times s^l(X)) = \varnothing \quad \text{if } |I| \geq e - l \text{ and } l \geq 1,$$

and the counting argument shows that $\bar{v}_1 \smile \ldots \smile \bar{v}_e$ is then compactly supported in M_k. The key point is that if a sequence $[A_n]$ in M_k converges to $([A_\infty]; \{x_1, \ldots, x_l\})$, then $[A_n] \to [A_\infty]$ in all but at most l of the spaces $\bar{M}_{(i)}$.

As before, one proves that the pairing (9.2.22) is independent of the choice of metric, so an invariant is defined.

Cohomology classes of the form $v^n \smile \mu(\Sigma_1) \smile \ldots \smile \mu(\Sigma_m)$, with $2n + m = d$, can similarly be paired with the moduli space. Since such classes span all of the rational cohomology algebra (5.1.15), this shows, in effect, that when $w_2 \neq 0$ the moduli space M_k defines at least a rational homology class in $\mathscr{B}^*_X$.

The problem in the $SU(2)$ case is that the class v does not extend across a neighbourhood of the trivial connection Θ. Indeed, if W is a ball, then $\mathscr{B}^*_W \cup \{\Theta\}$ is contractible, (cf. the step in the proof of Uhlenbeck's theorem in Section 2.3.9). This is in contrast to the situation described in (5.2.8). The classes $v^n \smile \mu(\Sigma_1) \smile \ldots \smile \mu(\Sigma_m)$ can nevertheless be evaluated on the moduli space M_k as long as $2m$ exceeds the dimension of the lowest stratum $M_0 \times s^k(X)$. This inequality is the same one that defined the 'stable range' for the ordinary polynomial invariants; its role is to ensure that the closure of the cocycle representing $\Pi\, \mu(\Sigma_i)$ avoids the lowest stratum.

9.3 Vanishing theorems

9.3.1 Vanishing by automorphisms

In Section 9.1 we saw how the non-vanishing of a moduli-space invariant implied constraints on the possible action of the diffeomorphism group on the homology of a $K3$ surface and, as a corollary, the non-existence of a particular connected sum decomposition (Corollary 9.1.7). This idea has applications also for the polynomial invariants and can be used to show that $q_k = 0$ if the diffeomorphisms of X induce a sufficiently large group of automorphisms of $H^2(X)$. Consider for example the manifold

$$X = m\mathbb{C}P^2 \# n\overline{\mathbb{C}P}^2,$$

(with m odd and ≥ 3). Complex conjugation provides a diffeomorphism of $\mathbb{C}P^2$ inducing the map -1 on $H^2(\mathbb{C}P^2)$. Thus, if $e_1, \ldots, e_m$ and $f_1, \ldots, f_n$ are the natural generators for $H^2(X; \mathbb{Z})$, it is easy to construct a diffeomorphism h of X which changes the sign of any one generator, inducing (say) the automorphism

$$e_1 \mapsto -e_1 \quad e_i \mapsto e_i \quad (i \neq 1),$$
$$f_i \mapsto f_i \quad (\text{for all } i).$$

(See the proof of (9.1.7).) This particular automorphism reverses the orientation of $H^+(X)$, so q_k must change sign under this substitution. (We are considering only the $SU(2)$ case.) It follows that when q_k is expanded as a polynomial in e_i and f_i, each non-zero term contains e_1 with an *odd* exponent. The same applies to all the other e_i. By similar reasoning, applied to a diffeomorphism which changes the sign of one f_i, we see that every term in q_k contains f_i with an *even* exponent (possibly zero). Since m is odd, it follows that q_k is either zero or has odd degree. However, the degree d is odd only if $m \equiv 1 \pmod 4$, independent of k. So we have proved:

Proposition (9.3.1). *If $m \equiv 3 \pmod 4$, the polynomial invariants q_k for the manifold $m\mathbb{C}P^2 \# n\overline{\mathbb{C}P}^2$ are all zero.*

The same conclusion ($q_k = 0$) can be drawn also when $m \equiv 1 \pmod 4$, provided n is positive and $m \geq 5$, but for this one needs to construct some less obvious automorphisms of the manifold. These arguments may seem *ad hoc* as they stand, but they fall into place in the light of a theorem due to Wall (1964*a*), which shows that a connected sum decomposition with one standard piece can be used, quite generally, to obtain a large collection of diffeomorphisms:

Theorem (9.3.2) (Wall). *Suppose that the simply-connected manifold X admits a connected sum decomposition $X = Y \# (S^2 \times S^2)$, and that Y has indefinite intersection form. Then every form-preserving automorphism of the lattice*

$H_2(X; \mathbb{Z})$ is realized as the map induced by some orientation-preserving diffeomorphism of X.

We mention also that if the intersection form of Y is odd, then $Y \# (S^2 \times S^2)$ is diffeomorphic to $Y \# (\mathbb{C}P^2 \# \overline{\mathbb{C}P}^2)$, (just as in two dimensions, $S \# [\text{torus}] \cong S \# [\text{Klein bottle}]$ if S is non-orientable). So the conclusion of (9.3.2) applies also to $m\mathbb{C}P^2 \# n\overline{\mathbb{C}P}^2$ for example, if $m, n \geq 2$. Wall's theorem implies a more general vanishing theorem than (9.3.1):

Proposition (9.3.3). *If $X = Y \# (S^2 \times S^2)$ as in* (9.3.2), *with $b^+(X)$ odd, then the polynomial invariants q_k are all identically zero.*

Proof. If $v \in H_2(X; \mathbb{Z})$ satisfies $\langle v, v\rangle = 1$ or 2, then the reflection in the hyperplane orthogonal to v defines an automorphism r_v of $H_2(X; \mathbb{Z})$:

$$r_v(x) = x - 2\left(\frac{\langle v, x\rangle}{\langle v, v\rangle}\right)v.$$

Here $\langle\,,\rangle$ denotes the intersection form. This automorphism reverses the orientation of H^+, so if we regard q_k as a polynomial function on the real homology,

$$q_k\colon H_2(X; \mathbb{R}) \longrightarrow \mathbb{R},$$

we will have $q_k(x) = -q_k(x) = 0$ whenever $\langle v, x\rangle = 0$. Thus q_k vanishes on the set $K \subset H_2(X; \mathbb{R})$ which is the union of hyperplanes

$$K = \bigcup_{\substack{v \in H_2(X; \mathbb{Z}) \\ \langle v, v\rangle = 1 \text{ or } 2}} \langle v\rangle^{\perp}.$$

The proposition now follows, for K is dense in $H_2(X; \mathbb{R})$: this is a property shared by all unimodular forms whose maximal positive and negative subspaces have dimension two or more. If we are prepared to use the classification of indefinite forms (Section 1.1.3), then it is enough to check this for the forms $(1) \oplus (1) \oplus (-1) \oplus (-1)$ and

$$\begin{pmatrix} 0 & 1 \\ 1 & 0 \end{pmatrix} \oplus \begin{pmatrix} 0 & 1 \\ 1 & 0 \end{pmatrix},$$

where the arithmetic is not hard. To illustrate this, take the bilinear form $\langle\,,\rangle$ given by

$$\begin{pmatrix} 0 & 1 \\ 1 & 0 \end{pmatrix} \oplus \begin{pmatrix} 0 & 1 \\ 1 & 0 \end{pmatrix}.$$

It is quite enough to show that the closure of K contains all integer vectors $a = (a_1, a_2, a_3, a_4)$ with a_2, a_4 coprime. Given such an a, take s and t with $sa_2 + ta_4 = 1$, let N be any large integer, and put

$$a' = (a_1, a_2, a_3, (a_4 - \delta))$$

where

$$\delta = (\langle a, a\rangle + 2)/2(t - Na_2).$$

Clearly $a' \to a$ as $N \to \infty$. On the other hand $a' \in K$, for a' is orthogonal to an integer vector v with $\langle v, v\rangle = 2$, namely the vector

$$v = (s + Na_4, a_2, t - Na_2, a_4).$$

As we said in Chapter 1, this result ties in very well with Wall's theorem (1.2.4). On the one hand that theorem shows that the *stable* invariants of four-manifolds (i.e. stable with respect to connected sums with $S^2 \times S^2$) are all contained in the intersection form. On the other hand the invariants derived from the ASD moduli spaces (which certainly go beyond the classical invariants, as the example of the $K3$ surface already indicates) are shown to be unstable by the proposition above.

9.3.2 *Shrinking the neck*

We now turn to the proof of the more general vanishing theorem promised in Chapter 1 (see Theorem (1.3.4)):

Theorem (9.3.4). *Let X be a simply-connected four-manifold with $b^+(X)$ odd, and suppose that X is a connected sum $X_1 \# X_2$ with $b^+(X_i) > 0$, $i = 1, 2$. Then the $SU(2)$ invariants q_k (for k in the stable range) and the $SO(3)$ invariants $q_{k,\hat{\alpha}}$ (for all k when $\alpha \neq 0$) are identically zero.*

This theorem contains Proposition (9.3.3) as a special case, but its proof requires more than the simple considerations of naturality which we exploited before. Two.mechanisms are involved in the argument, and as these might be obscured in any detailed account, we shall concentrate on simple cases which isolate and illustrate the main ideas.

The first main idea rests on the relation between the dimension of the moduli spaces for X as compared with the dimension of the moduli spaces for the two summands: namely, if $k = k_1 + k_2$ then

$$\dim M_k(X) = \dim M_{k_1}(X_1) + \dim M_{k_2}(X_2) + 3. \qquad (9.3.5)$$

This follows directly from the index formula (4.2.21), for every term in the formula is additive under connected sums except for the extra '3' which comes from the term $3(b^+ + 1)$. In Chapter 7 we saw that the relation above can be interpreted quite simply in terms of a gluing construction: if the geometry of the connected sum is suitable, an open set $\mathscr{U}$ in $M_k(X)$ can be constructed by joining ASD connections on the two summands; the extra three parameters represent the freedom to choose the gluing parameter ρ by which the fibres of the two bundles are identified.

Consider again the simplest type of invariant, an integer q obtained by counting with signs the points of a zero-dimensional moduli space $M_{k,\alpha}$, with

$\alpha = w_2(E)$ non-zero. In addition to the hypotheses of (9.3.4), we make the following simplifying assumption:

Hypothesis (9.3.6). *The Stiefel–Whitney class α is non-zero on both X_1 and X_2.*

We are going to prove that $q = 0$. The strategy exploits our freedom to choose a metric with which to calculate the invariant. As in Chapter 7 we consider forming the connected sum by means of a small neck. So let g^1 and g^2 be metrics on the two summands and let $\lambda > 0$ be given. We shall suppose that the two metrics are flat in the neighbourhood of chosen base-points x_1 and x_2, so that the small geodesic balls are Euclidean. Then a metric g_λ can be defined on the connected sum by the procedure of Section 7.2.1, using the parameter λ which we are going to make small. The details of this construction are actually quite unimportant for our first argument: the only feature of g_λ which we shall appeal to is that it contains disjoint, isometric (or conformal) copies of the two sets $X_1 \backslash B_1$ and $X_2 \backslash B_2$, where B_i is a ball around x_i whose radius goes to zero with λ. We shall require that the metrics g^i have the usual generic properties: there are no non-trivial, reducible ASD connections on X_i, and the dimensions of the moduli spaces $M_{j,\alpha}(X_i)$ agree with their virtual dimensions. (Because of the unique continuation argument (4.3.21), this requirement need not conflict with our assumption that g^i is flat near x_i.) Under these assumptions we shall prove:

Proposition (9.3.7). *There exists λ_0 such that for all $\lambda < \lambda_0$ the moduli space $M_{k,\alpha}(X, g_\lambda)$ is empty.*

The argument will show *a fortiori* that the lower moduli spaces $M_{k-j,\alpha}$ are empty also; this means that the metric g_λ has all the generic properties required to calculate the invariant by counting points. It therefore follows that $q = 0$.

There is little to the proof of (9.3.7) but the compactness theorems and a dimension count. Suppose that, contrary to (9.3.7), we have a sequence λ_n approaching zero and for each n a connection $A^{(n)}$ in the bundle E which is ASD with respect to g_{λ_n}. By Theorem (7.3.1), some subsequence $A^{(n')}$ has a weak limit, which is an ideal ASD connection on the disjoint union $X_1 \cup X_2$, say

$$[A^{(n')}] \xrightarrow{\text{weak}} ([A_1, A_2]; y_1, \ldots, y_l),$$

and we have an inequality

$$\kappa(E) \geq \kappa(E_1) + \kappa(E_2) + l, \tag{9.3.8}$$

where E_i is the bundle which carries A_i. This inequality between the characteristic classes, combined with the basic formula for the index, gives

$$\dim M_E \geq \dim M_{E_1} + \dim M_{E_2} + 8l + 3,$$

(compare the relation (9.3.5)). Since $\dim M_E$ is zero by hypothesis, we have

$$\dim M_{E_1} + \dim M_{E_2} \leq -3.$$

So at least one of the two moduli spaces has negative virtual dimension, and should therefore be empty because of our assumptions concerning the metrics g^i. This contradicts the existence of either A_1 or A_2, so proving our proposition. It is in the last step that we used the hypothesis (9.3.6). Since the second Stiefel–Whitney class is preserved under weak limits, neither E_1 nor E_2 is the trivial bundle. The virtual and actual dimensions of the moduli spaces therefore coincide.

This argument readily extends to the case of polynomial invariants $q_{k,\hat{\alpha}}$ defined using higher-dimensional $SO(3)$ moduli spaces (with w_2 non-zero), as long as the hypothesis (9.3.6) holds. Because the invariant is multi-linear, it is enough to show that

$$q_{k,\hat{\alpha}}(\Sigma_1, \ldots, \Sigma_{d_1}, \Sigma'_1, \ldots, \Sigma'_{d_2}) = 0$$

in the case that the surfaces Σ_i are contained in $X_1 \backslash B_1$ and Σ'_i are contained in $X_2 \backslash B_2$. The argument is to show that the intersection

$$M_E(g_\lambda) \cap V_{\Sigma_1} \cap \ldots \cap V_{\Sigma_{d_1}} \cap V_{\Sigma'_1} \cap \ldots \cap V_{\Sigma'_{d_2}}$$

is empty when λ is sufficiently small. Suppose the contrary. Then as in (9.3.7) there are connections $[A^{(n)}]$ on X which are ASD with respect to the degenerating family of metrics g_{λ_n} and which converge weakly as $n \to \infty$:

$$[A^{(n)}] \xrightarrow{\text{weak}} ([A_1, A_2]; y_1, \ldots, y_l),$$

where A_i are ASD connections in bundles E_i. Write $l = l_1 + l_2$ according to the number of the points y_j which lie in each of the two manifolds. Each y_j lies in at most two of the neighbourhoods $\nu(\Sigma_i)$; so after perhaps relabeling the surfaces we will have

$$[A_1] \in M_{E_1} \cap V_{\Sigma_1} \cap \ldots \cap V_{\Sigma_{e_1}} \quad \text{with} \quad e_1 \geq d_1 - 2l_1,$$
$$[A_2] \in M_{E_2} \cap V_{\Sigma'_1} \cap \ldots \cap V_{\Sigma'_{e_2}} \quad \text{with} \quad e_2 \geq d_2 - 2l_2$$

by the alternative (9.2.10). Counting dimensions, using the inequality (9.3.8), we find

$$\dim(M_{E_1} \cap V_{\Sigma_1} \cap \ldots \cap V_{\Sigma_{e_1}}) + \dim(M_{E_2} \cap V_{\Sigma'_1} \cap \ldots \cap V_{\Sigma'_{e_2}}) \leq -3.$$

So we have a contradiction: one of the intersections has negative virtual dimension and will be empty if the V_Σ are transverse.

Again, the joining construction of Chapter 7 provides an intuitive guide for this argument. If $k = k_1 + k_2$ and we are given precompact open sets $\mathcal{U}_1 \subset M_{k_1}(X_1)$, $\mathcal{U}_2 \subset M_{k_2}(X_2)$, then for sufficiently small λ there is an open set $\mathcal{U} \subset M_k(X)$ (the moduli space for the metric g_λ) and a fibration

$$p \colon \mathcal{U} \longrightarrow \mathcal{U}_1 \times \mathcal{U}_2$$

with fibres $SO(3)$ (the gluing parameter). On such an open set $\mathscr{U}$, the classes V_Σ are pulled back from the base of the fibration since the restriction maps to the surfaces $\Sigma \subset X_i$ factor through p. So a product of these classes in the top dimension must be zero, for it will be zero on the lower-dimensional base.

Were it the case that the moduli space M_k actually fell into finitely many compact components $\mathscr{U}$ corresponding to the different partitions of k, then it would follow quite rigorously from this argument that $q = 0$. As it is, the argument is only a guide: in general the joining construction can only produce large open sets in the moduli space.

9.3.3 *The second mechanism*

When the second Stiefel–Whitney class of the bundle is zero on either of the two summands, there is more to the proof of the vanishing theorem (9.3.4) than the dimension-counting argument used above. Once again, consider a zero-dimensional moduli space M_E for an $SO(3)$ bundle $E \to X$, and suppose now that the class $\alpha = w_2(E)$ is zero on (say) the first summand:

Hypothesis (9.3.9). *The Stiefel–Whitney class α is zero on X_1 and non-zero on X_2.*

Let g_λ be the same degenerating family of metrics as before. It is no longer true that the moduli space $M_E(g_\lambda)$ need be empty when λ is small; the argument breaks down because we may obtain the trivial connection on X_1 in the limit. The proof of (9.3.7) does however rule out all other possibilities:

Lemma (9.3.10). *If $\{\lambda_n\}$ is a sequence approaching zero, and $[A^{(n)}]$ is an ASD connection in E for the metric g_{λ_n}, then there is a subsequence $\{[A^{(n')}]\}$ which converges weakly, and in the limit the connection on X_1 will be the trivial connection Θ_1:*

$$[A^{(n')}] \xrightarrow{\text{weak}} ([\Theta_1, A_2]; y_1, \ldots, y_l).$$

Proof. A weakly convergent subsequence will exist in any case; that the limiting connection on X_1 is trivial follows by dimension counting, as in (9.3.7).

In the situation of the lemma above, let $L = \kappa(E) - \kappa(E_2)$ be the amount of action that is 'lost' in the limit. Thus L is the sum of three terms: the number of ideal points of concentration in the two halves of the connected sum, say $l = l_1 + l_2$, plus the amount we 'lose' in the neck region in the limit. The dimension formula gives

$$\begin{aligned}\dim M_E - \dim M_{E_2} &= 8L - 3(b^+(X) - b^+(X_2)) \\ &= 8L - 3b^+(X_1),\end{aligned}$$

and since M_E is zero-dimensional this becomes

$$\dim M_{E_2} = 3b^+(X_1) - 8L. \tag{9.3.11}$$

The existence of the (irreducible) connection A_2 implies that this dimension is non-negative, so L is bounded in terms of $b^+(X_1)$. The situation is particularly simple if $b^+(X_1)$ is equal to one or two. In this case L must be zero and $\kappa(E) = \kappa(E_2)$. This is the case we shall discuss:

Hypothesis (9.3.12). *$b^+(X_1)$ is equal to one or two.*

Note that we do not include the case $b^+(X_1) = 0$. The hypothesis $b^+(X_i) > 0$ contained in Theorem (9.3.4) is essential to the argument.

The dimension formula now says that M_{E_2} has dimension $3b^+(X_1)$, which is either three or six. In either case the lower moduli spaces are empty, so M_{E_2} is compact. Since L is zero, the sequence $A^{(n')}$ in (9.3.10) is actually converging strongly to (Θ_1, A_2) in the sense of Section 7.3. It follows that for sufficiently large n', the connection $A^{(n')}$ lies in that part of the moduli space $M_E(g_{\lambda_{n'}})$ which is modelled by the gluing construction (see (7.2.63) and (7.3.2)). So, because of the compactness of M_E, we can find a λ sufficiently small that *every* connection in $M_E(g_\lambda)$ is in the domain of the gluing construction, formed from the trivial connection on X_1 and some connection A_2 on X_2.

Let us apply Theorem (7.2.62) to this situation. Since A_1 is trivial, we have $H^1_{A_1} = 0$ and $\Gamma_{A_1} = SO(3)$, while $H^2_{A_1}$ can be identified with $\mathscr{H}^+ \otimes \mathfrak{so}(3)$. Here $\mathscr{H}^+$ is the space of self-dual harmonic forms on X_1. Our smoothness assumption for M_{E_2} means that a neighbourhood $\mathscr{U}_2$ in this moduli space is isomorphic to an open neighbourhood of zero in $H^1_{A_2}$. Thus Theorem (7.2.62) says that an open set in $M_E(g_\lambda)$ is modelled on $\Psi^{-1}(0)/SO(3)$ for some smooth map $\Psi: \mathscr{U}_2^+ \to \mathscr{H}^+ \otimes \mathfrak{so}(3)$, where $\mathscr{U}_2^+$ parametrizes points of $\mathscr{U}_2$ together with gluing data.

We may think of $\mathscr{U}_2^+$ as a principal bundle over $\mathscr{U}_2$, in which case it is naturally the base-point fibration β. Thus if we write $\mathscr{H} \to M_{E_2}$ for the associated vector bundle with fibre $\mathscr{H}^+ \otimes \mathfrak{so}(3)$, then the local model is the zero set of a local section $\bar{\Psi}: \mathscr{U}_2 \to \mathscr{H}$. We need a global version of this statement:

Proposition (9.3.13). *For λ sufficiently small, the moduli space M_E, for the metric g_λ can be identified with the zero set of a smooth section $\bar{\Psi}$ of the bundle $\mathscr{H} \to M_{E_2}$.*

This follows easily enough, using the techniques of Chapter 7. The point to watch is that the local models referred to above do not fit together to give a global section $\bar{\Psi}$, because the construction of each local model, as described, incorporates a gauge-fixing condition which depends on A_2. One way around this is not to use the local models as such, but rather to treat the moduli space 'point by point' as discussed in Section 7.2.7. Thus we can fix functions (f_α) on $\mathscr{B}_{E_2}$ which depend only on the restriction of a connection to

$X_2 \backslash B_\varepsilon(x_2)$ and which give local coordinates on M_{E_2}. We then arrive at the description of $M_E \cap f^{-1}(c)$ as the zero set of a 'section' $\Psi_c: M_{E_2} \cap f^{-1}(c) \to \mathscr{H}$ which depends continuously on c. To make something global out of this, we need to replace the local coordinate functions $f: \mathscr{B}_{E_2} \to \mathbb{R}^n$ by a single map $F: \mathscr{B}' \to M_{E_2}$ which is defined on an open set $\mathscr{B}' \subset \mathscr{B}_{E_2}$ and is equal to the identity on M_{E_2}. Such an F can be constructed, for example, by extending the collection of functions f_α until they define an embedding $\bar{f}: M_{E_2} \to \mathbb{R}^N$ and then defining $F = p \circ \bar{f}$, where p is the projection of a tubular neighbourhood in $\mathbb{R}^N$ onto M_{E_2}.

The fibre of $\mathscr{H}$ has dimension $3b^+(X_2)$, which is the same as the dimension of the base; that is as it should be, for the virtual dimension of M_E is zero. The usual transversality condition—that $H_A^2 = 0$ at all points of the moduli space—is equivalent to the condition that Ψ vanishes transversely, and it can be achieved by perturbing the metric g_λ without essentially altering the description we have obtained. The integer invariant q, which is defined by counting with signs the points of M_E, now appears with another interpretation: it is the Euler number of the bundle $\mathscr{H}$—the number of zeros of a transverse section. Theorem (9.3.4), which asserts the vanishing of q in this case, will therefore follow if we know that this Euler number is zero. But $\mathscr{H}$ is a sum of $b^+(X_1)$ copies of the $\mathbb{R}^3$ bundle associated to β; so in the cohomology ring of the base, its Euler class is equal to the Euler class of the $\mathbb{R}^3$ bundle raised to the power $b^+(X_1)$. Since the Euler class of any oriented $\mathbb{R}^3$-bundle is trivial in rational cohomology it follows that $q = 0$. This completes the proof of (9.3.4) in the case of a zero-dimensional moduli space under the additional hypotheses (9.3.9) and (9.3.12).

It is interesting to note that the topological mechanism which is involved here—the vanishing of an Euler class—is something which is not special to the group $SO(3)$ or $SU(2)$. If we considered an arbitrary compact group G we would be led to the bundle $\mathfrak{g}_\beta$ associated with a principal G-bundle $\beta: P \to M$, and the point is that the Euler class of such a bundle is always zero over $\mathbb{Q}$. This can be proved by the 'splitting principle'. Let $\pi: Z \to M$ be the associated bundle with fibre G/T (where T is a maximal torus). The Leray–Hirsch theorem implies that $H^*(Z; \mathbb{Q})$ is a free module over $H^*(M; \mathbb{Q})$, so in particular the pull-back map is injective. Thus it is enough to show that $\pi^*(\mathfrak{g}_\beta)$ has vanishing Euler class. But the structure group of $\pi^*(\beta)$ reduces to the torus T, so $\pi^*(\mathfrak{g}_\beta)$ has a trivial subbundle, and hence also a non-vanishing section. This fact strongly indicates that, were this theory developed for more general structure groups, a vanishing theorem such as (9.3.4) would continue to hold for the corresponding invariants.

At this point we have an alternative proof of the indecomposability of the $K3$ surface. For suppose a $K3$ surface X had a connected sum decomposition $X_1 \# X_2$ which was non-trivial in homology. Since no non-trivial, even, definite form can occur as an intersection form (by the results of Section 8.1), it must be that $b^+(X_i) > 0$ for $i = 1$ and 2. We know that there is an $\alpha \in H^2(X; \mathbb{Z}/2)$ such that the integer invariant $q_{6,\alpha}$ is non-zero. But if α is non-zero on both the summands then the hypothesis (9.3.5) holds, while if α is zero say on X_1 then the hypotheses (9.3.9) and (9.3.12) hold; and in either case we have shown that $q = 0$.

In fact, as we mentioned earlier, it is known that the diffeomorphism group of a $K3$ surface acts transitively on the classes α with $\alpha^2 \equiv 2$, so $q_{6,\alpha}$ is non-zero for any such class. In particular we could always choose α so that it is non-zero on both X_i, and thereby avoid the use of the Euler class argument, making the proof genuinely more elementary.

Finally we note that if $b^+(X_1)$ is zero then the Euler class mechanism does not come into play. In this case, when λ is small, the moduli space M_E again consists of connections which are close to trivial on X_1, but now there is no obstruction, and M_E is isomorphic to M_{E_2} (in the zero-dimensional case). This leads, for example, to the following result, which we state for the case of the $SU(2)$ polynomial invariants:

Proposition (9.3.14). *Let $b^+(X_1) = 0$ and $b^+(X_2)$ be odd and not less than three. Then the polynomial invariants of X_2 are related to those of the connected sum X by*

$$q_k(X)|_{H_2(X_2)} = q_k(X_2).$$

9.3.4 The general case

The two topological ideas illustrated in Sections 9.3.2 and 9.3.3 provide the heart of the proof of Theorem (9.3.4) in the general case, but there are considerable complications which have to be dealt with when one moves away from the simple situations we have considered. Let us stay with the case of a zero-dimensional moduli space M_E under Hypothesis (9.3.9), but let us try and understand the situation when $b^+(X_1)$ is larger than two. Lemma (9.3.10) still applies: this is the dimension counting argument which tells us that, when λ is small, the moduli space $M_E(g_\lambda)$ consists of connections which are close to the trivial connection on X_1 in the *weak* sense. The dimension count does not rule out the possibility that there are points of concentration y_j in X_1; it only provides a bound through the inequality (9.3.11). For example, if $b^+(X_1)$ is three, then we must consider contributions to the invariant q which come from connections which are flat on X_1 except in the neighbourhood of a single point y_1 and which have charge $k_2 = k - 1$ on X_2.

Recall from Chapter 7 the nature of the local model when A_1 is a connection with $H^2_{A_1}$ non-zero, (such as the trivial connection if $b^+(X_1) > 0$). What was described first was a family of solutions of the extended equations $F^+(A' + a(t)) + \Psi(t) = 0$, with Ψ taking values in an *ad hoc* subspace $W \subset \Omega^+(X_1, \mathfrak{g}_{E_1})$. The true moduli space M_E then appeared as the zero set of

Ψ. It was this description which provided the mechanism for the Euler-number argument above. As it stands, however, it is not suited to the global argument we have in mind. The problem is that the subspace W is not defined in a gauge-invariant manner, so that the family of connections $A' + a(t)$ is not the solution set of any intrinsically defined problem, but depends on A_1 and A_2. This was the awkward point in (9.3.13), but whereas an *ad hoc* construction was satisfactory in this simple case, where we had strong convergence as λ went to zero, such constructions would be unwieldy as the situation became more complicated.

The solution is to define a subspace $W \subset \Omega^+(X_1, \mathfrak{g}_{E_1})$, in an intrinsic, gauge-invariant manner. This W should consist of forms supported in the punctured manifold $Z_1 = X_1 \setminus B_\rho(x_1)$ and should depend only on the restriction of the connection A to Z_1. It should be defined whenever A is sufficiently close to the trivial connection (in the *weak* sense) on Z_1, and it should be well-behaved under weak limits. This entails defining some real-valued function $h_1(A)$ which measures the distance between $A|_{Z_1}$ and the trivial connection and is continuous in the weak topology. There is then a well-defined problem, and a solution set

$$L_E = \{(A, w) | F^+(A) + w = 0 \text{ on } X \text{ and } h_1(A) < \varepsilon\}.$$

Here w lies in the space $W = W_A$ which is understood to depend on $A|_{Z_1}$, and ε is a small parameter. The virtual dimension of L_E is $\dim M_E + \dim W$, and the true moduli space appears as the zero set, $\{w = 0\}$. This is a situation in which, given some compactness properties and transversality, we can hope to show that $q = 0$ by the Euler number argument.

A suitable construction for W is the following, based on a technique of Taubes. One begins by constructing a three-dimensional subspace $\Xi_A \subset \Omega^0(Z_1, \mathfrak{g}_E)$ consisting of sections which are approximately constant on most of Z_1 (where A is approximately flat). For any A with $h_1(A) < \varepsilon_0$ such a Ξ_A can be defined as the span of the eigenfunctions belonging to the three smallest eigenfunctions of the Laplacian Δ_A on Z_1 with suitable boundary conditions. This procedure is compatible with weak limits, so that as A approaches the trivial connection weakly on Z_1 the three smallest eigenvalues approach zero and the eigenfunctions approach the covariant constant sections except at finitely many points. Given P, a finite-dimensional space of self-dual two-forms supported in Z_1, one can then define W_A as $\{\xi . \omega | \xi \in \Xi_A, \omega \in P\}$. If P has dimension $b^+(X_1)$—for example, if P is constructed by cutting off the harmonic forms on X_1 in the neighbourhood of the base-point—then L_E will have the same dimension as $M_k(X_2)$, in close imitation of the previous local model for the case of strong convergence. However, we will not achieve compactness until we cut down the solution set by the constraints coming from all but one of the harmonic forms on X_1. We therefore make P one-dimensional, so that W_A has dimension three, and M_E appears as the zero set of a section of a three-plane bundle W over the three-manifold L_E.

Some routine work is now needed to adapt the transversality and compactness results to the extended equations. For the former, since the theorem of Freed and Uhlenbeck is no longer applicable, one must construct abstract perturbations of the equations. For the latter, the weak compactness results can be carried over, and these establish that L_E is compact except for a reason which is plainly unavoidable: the definition of L includes the constraint $h_1(A) < \varepsilon$, so we must expect that L contains sequences on which h_1 approaches ε from below. Thus if we temporarily write $L(\varepsilon)$ for L, we can expect $L(\varepsilon)$ to have compact closure as a manifold with boundary in $L(2\varepsilon)$, say.

The Euler number argument will not apply to an open manifold, so to complete the proof that $q = 0$ one must show that, if λ is small enough, the components of L which reach the level $h_1 = \varepsilon$ do not contain points of M_E. Now certain components of L_E arise from the following construction. If A_1 is a solution of the extended equations on X_1 and A_2 is an ASD connection on X_2, and if both connections are transverse points of a zero-dimensional solution set, then the gluing construction of Chapter 7 adapts to give us a family of solutions on X parametrized by the gluing parameter $SO(3)$. If $h_1(A_1) = \varepsilon$ then this copy of $SO(3)$ will form a complete component of $L(2\varepsilon)$, lying partly above and partly below the level $h_1 = \varepsilon$; and as $\lambda \to 0$ the variation of h_1 on the $SO(3)$ goes to zero—so, with ε fixed and λ approaching zero, this component eventually lies, say, between the levels $\frac{3}{4}\varepsilon$ and $\frac{5}{4}\varepsilon$. In this way $L(\varepsilon)$ can contain a piece which is non-compact. Such a component, however, cannot contain points of M_E, because we know that $\max(h_1)$ goes to zero on M_E as λ goes to zero (the connections converge weakly to the trivial connection on X_1), so h_1 is eventually less than $\frac{3}{4}\varepsilon$ when λ is small enough. Thus the key to the last step in the proof is to see that these $SO(3)$s are the *only* components of $L(2\varepsilon)$ to reach the level $h_1 = \varepsilon$. This follows from consideration of the weak limit as $\lambda \to 0$, as in the first mechanism of Section 9.3.2. A more detailed summary of this argument, as well as the proof of the compactness, transversality and gluing results for the extended equations, can be found in Donaldson (1990*a*).

So far we have been considering only the case of a zero-dimensional moduli space with w_2 non-zero. The proof of the vanishing theorem (9.3.4) for the more general polynomial invariants proceeds by applying the above analysis to the cut-down moduli space $M_E \cap V_{\Sigma_1} \cap \ldots \cap V_{\Sigma_d}$ in a manner which will by now be familiar; only the counting arguments are slightly altered. In the case that w_2 is everywhere zero, the counting argument from Section 9.3.2 tells us only that, when λ is small, the cut-down moduli space consists of connections A which are close to the trivial connection *either* on X_1 *or* on X_2 in the weak sense. The point here is that the inequality (9.2.3) defining the stable range forbids weak convergence to the trivial connection on *both* pieces simultaneously, so for small λ the invariant q is a sum of two separate terms, and the Euler number argument will show that both are zero.

Notes

Section 9.1.1

The idea that our invariants can be regarded as the Euler class of an infinite dimensional bundle is reinforced by work of Witten (1988), who shows that, at least formally, the invariants can be defined by integrals over the space of all connections. In an analogous finite-dimensional problem Witten's formulae give de Rham representatives for the Euler class. These developments are discussed briefly by Donaldson (1990*c*).

Section 9.1.2

This application is intended as a partner to the proof in Section 8.1.1, showing how very simple arguments with Yang–Mills moduli spaces can give highly non-trivial information. Some applications of the result are given by Donaldson (1990*a*). Lemma (9.1.10) is taken from Mukai (1984) which contains many other interesting results about the moduli spaces for $K3$ surfaces.

Section 9.2

The first definition of the invariants follows Donaldson (1990*a*), but the treatment here of the restriction maps is much cleaner. The alternative definition in Section 9.2.3 is new and seems to be preferable from some points of view, although it is rather cumbersome. Probably the best definition of the invariants remains to be found. Similarly we have not taken the opportunity here to define invariants in the widest possible generality, for example using other structure groups, principally because we do not know of any applications for these generalizations. For the definition of covering dimension used in Section 9.2.3, see Engelking (1977).

Section 9.3

The general vanishing theorem (9.3.4) is proved by Donaldson (1990*a*). The simple cases discussed here suffice for most applications. The result on self-diffeomorphisms is proved by Wall (1964*a*). For the 'splitting principle' for characteristic classes see, for example, Husemoller (1966).

10

THE DIFFERENTIAL TOPOLOGY OF ALGEBRAIC SURFACES

Let us now review our position with regard to the overall aims of this book. In Chapter 1 we set up the twin problems in four-manifold theory: deciding the existence and uniqueness of manifolds with given intersection forms. In Chapter 8 we have shown how arguments with Yang–Mills moduli spaces give strong non-existence results, and in Chapter 9 we have developed invariants of four-manifolds aimed at the complementary question of uniqueness. However we have not yet given any full-blooded example of the application of these invariants, i.e. shown that they can distinguish differentiable four-manifolds with the same intersection form, and this is the purpose of the present chapter. We will describe the main ideas in the proof of the general Theorem (1.3.5), and give a complete proof of the special result (1.3.6).

We have seen in Chapter 9 that for most four-manifolds which admit connected sum decompositions the new invariants are trivial. We can construct connected sums which realize many homotopy types, so finding an example of the kind we want is essentially equivalent to showing that the invariants are not always trivial. In this chapter we shall see that, for rather basic reasons, the invariants are indeed non-trivial for complex algebraic surfaces. This general theory is explained in Section 10.1.

Another problem taken up in this chapter is that of the calculation of the invariants in any kind of generality. While we have defined invariants for a large class of four-manifolds, we have given little indication as to how they may be calculated in practice. The definition involves the solution of the non-linear ASD partial differential equations, and this can certainly not be done in general in any explicit form. However, for the particular class of complex surfaces, we have seen in Chapter 6 that the ASD connections can be identified with stable holomorphic bundles, and in this chapter we continue the same line of ideas by describing techniques which can be applied fairly generally to analyse moduli spaces of stable bundles. The basic construction is reviewed, within a differential-geometric framework in Section 10.2, and in Section 10.3 we illustrate its application in a particular case: moduli spaces of bundles over a 'double plane' (branched cover). These geometric calculations are then applied in Section 10.4 to calculate some of the Yang–Mills invariants for this four-manifold.

10.1 General theory

10.1.1 Statement of results

Let S be a smooth, simply connected, complex algebraic surface. As we have seen in Section 1.1.7, the underlying differentiable four-manifold has a standard orientation and $b^+(S) = 1 + 2p_g(S)$. We will henceforth assume that $p_g(S) > 0$, so $b^+ \geq 3$ and the various different polynomial invariants on $H_2(S)$ of Chapter 9 are defined. Moreover the complex structure fixes a natural orientation on the moduli space and so there are no ambiguities of sign (see Section 6.4.2). Here we want to have in mind primarily $SU(2)$-bundles, but our results apply equally to $SO(3)$-bundles where w_2 is the reduction of the first Chern class of a holomorphic line bundle over S (i.e. an integral (1, 1) class). In particular we can consider bundles with $w_2 = c_1(K_S)$ mod 2. (The point of this condition is that we can then lift up to rank-two vector bundles over S, as explained in Section 6.1.4.)

The surface S can be holomorphically embedded in projective space. Such an embedding defines a 'hyperplane class' h in $H^2(S)$, the restriction of the standard generator for $H^2(\mathbb{CP}^N)$. Geometrically, this is the class realized by the intersecton of S with a general hyperplane or, in other language, a complex curve C in the linear system $|\mathcal{O}(1)|$ over S. The main general theorem we have is then:

Theorem (10.1.1). *For any simply connected complex algebraic surface S with $p_g(S) > 0$ and any hyperplane class h in $H_2(S)$ there is a $k_0 = k_0(S, h)$ such that for $k \geq k_0$ and any Stiefel–Whitney class α which is the reduction of a (1, 1) class $\hat{\alpha}$, we have $q_{k,\hat{\alpha}}(h) > 0$.*

This obviously implies Theorem (1.3.5), and combined with the results of Chapter 9 gives

Corollary (10.1.2). *No simply connected complex algebraic surface S can be written as a smooth connected sum $X_1 \# X_2$ with $b^+(X_1)$ and $b^+(X_2)$ both positive.*

From this we immediately obtain many examples of distinct smooth four-manifolds with the same intersection forms. Indeed we have:

Corollary (10.1.3) {= (1.3.7)}. *For any simply connected complex algebraic surface S with $b^+(S) \geq 5$, there is a smooth four-manifold $X(S)$, homotopy equivalent to S but not diffeomorphic to S, nor to any complex surface.*

For manifolds with odd forms this is immediate; by the classification of forms we can take

$$X(S) = \mathbb{CP}^2 \# \mathbb{CP}^2 \# \mathbb{CP}^2 \# \dots \# \mathbb{CP}^2 \# \overline{\mathbb{CP}}^2 \# \dots \# \overline{\mathbb{CP}}^2 \# \overline{\mathbb{CP}}^2$$

(and we need only assume $b^+ \geq 3$). In the other case, we consider a connected sum of the form $X = lK \# m(S^2 \times S^2)$ or $l\bar{K} \# m(S^2 \times S^2)$, where K is the $K3$ surface and $\bar{K}$ is the same manifold with reversed orientation. By Rohlin's theorem and the classification of even forms we can arrange X to have the same homotopy type as S so long as S satisfies the '11/8 inequality',

$$b^+ + b^- \leq (11/8)|b^+ - b^-|.$$

This inequality can be deduced from standard results about surfaces (an observation of Moishezon). One easily reduces to minimal surfaces of general type, for which $c_1^2 \geq 0$, and $c_1^2 \leq 3c_2$. These give $b^+ - 2b^- \leq 1$ and $5b^+ - b^- \geq -4$ (using the formulae of Section 1.1.7) which, together with the special consideration of small values of b^+, b^-, imply the result.

In any case, S cannot be diffeomorphic to $X(S)$, by Corollary (10.1.3). Of course, as well as these general results we have many explicit examples, for example the hypersurfaces S_d ($d \geq 5$) and the branched covers R_p ($p \geq 4$) of Section 1.1.7.

Notice that, combining (10.1.2) with the non-existence results of Chapter 8, in any smooth connected sum splitting of a simply connected algebraic surface one of the summands must have intersection form $\mathrm{diag}(-1, -1, -1, \ldots, -1)$, i.e. the intersection form of a connected sum of $\overline{\mathbb{CP}}^2$s. Conversely, this situation certainly can occur: we can take S to be the multiple blow up of another complex surface S'; then, as we have seen in Chapter 1:

$$S = S' \# \overline{\mathbb{CP}}^2 \# \overline{\mathbb{CP}}^2 \ldots \# \overline{\mathbb{CP}}^2.$$

Now let us go back to our remarks about Wall's theorem on the stable classification of four-manifolds from Chapter 1. We know that for any simply connected complex surface S and sufficiently large integers m, n, the manifold $S \# m\mathbb{CP}^2 \# n\overline{\mathbb{CP}}^2$ is diffeomorphic to $X(S) \# m\mathbb{CP}^2 \# n\overline{\mathbb{CP}}^2$. But we see now that we cannot take $m = 0$, however large n may be, since the connected sum of S with $n\overline{\mathbb{CP}}^2$ is still an algebraic surface. In the reverse direction it has been shown by Mandelbaum (1980) and Moishezon (1977) that for many surfaces S, for example the hypersurfaces in $\mathbb{CP}^3$, we can take $n = 0$ and $m = 1$; i.e. adding a single $\mathbb{CP}^2$ kills the 'exotic' property of the differentiable structure on S. (Such surfaces are called 'almost completely decomposable' by Mandelbaum and Moishezon, and they conjecture that in fact all surfaces have this property.) Thus we see that there is a radical difference between the differential-topological effect of the addition of $\mathbb{CP}^2$s and $\overline{\mathbb{CP}}^2$s. Of course this is quite in line with our methods, based on the ASD equation which involves a definite choice of orientation. Indeed we see that for almost completely decomposable surfaces S with b^+, b^- both odd the ASD Yang–Mills invariants of $\bar{S}$ are all zero (by (9.3.14)) while those of S are not.

10.1.2 The main idea

Let S be a complex algebraic surface as above and fix a Kähler metric on S, for simplicity a 'Hodge metric', compatible with a projective embedding. Thus the de Rham cohomology class of the metric 2-form ω is Poincaré dual to the hyperplane section class ω. (For example we could take the pull-back of the Fubini–Study metric on $\mathbb{CP}^N$.) We shall now explain why (10.1.1) is true in a favourable (but rare) case when we have a moduli space M of ASD connections, defined relative to this Kähler metric, which is non-empty, compact and regular, i.e. $H_A^2 = 0$ for all points in M. The argument is very simple: the number $q(h)$ is defined by pairing the relevant power of the cohomology class $\mu(h)$ with the fundamental cycle of M:

$$q(h) = \langle \mu(h)^d, [M] \rangle. \tag{10.1.4}$$

Now we know on the one hand, from (5.2.19), that $\mu(h)$ can be represented by the 2-form on M,

$$\Omega(a, b) = \frac{1}{8\pi^2} \int_S \mathrm{Tr}(a \wedge b) \wedge \omega,$$

since the metric form ω is self-dual. Thus, using de Rham cohomology,

$$q(h) = \int_M \Omega^d. \tag{10.1.5}$$

On the other hand we know that M is a complex manifold and that Ω is the two-form of the natural Kähler metric on M (see Section 6.5.3). It follows then that Ω^d is $(d-1)!$ times the Riemannian volume element on M, so

$$q(h) = (d-1)!\,\mathrm{Vol}(M) > 0,$$

as required.

This proof does not use the fact that the metric is a Hodge metric; it applies to any Kähler surface. However in the algebraic case we can carry through a parallel algebro-geometric argument. We know that $-2\pi i\Omega$ is the curvature form of a line bundle $\mathscr{L}$ over M, the determinant line bundle of the restriction of the connections to a (real) surface representing h. Now since $\mathscr{L}$ is a 'positive' line bundle, we know by Kodaira's embedding theorem that the holomorphic sections of some positive power $\mathscr{L}^{\otimes n}$ give a projective embedding $j\colon M \to \mathbb{CP}^r$. So the image $j(M)$ is a complex algebraic subvariety of $\mathbb{CP}^r$, and the restriction of the Hopf line bundle on $\mathbb{CP}^r$ to $j(M)$ is isomorphic to $\mathscr{L}^n$. It follows then that $q(h)$ is $1/n^d$ times the *degree* of $j(M)$: the degree being the number of points in the intersection of $j(M)$ with d general hyperplanes in $\mathbb{CP}^r$. Since the degree of a non-empty variety is always positive we reach the same conclusion. Notice that if we can construct the projective embedding directly, without recourse to Kodaira's theorem, we obtain an independent algebro-geometric proof, and we shall use this approach in Section 10.1.4 to handle the technical difficulties of singularities and non-compact moduli spaces which occur in realistic problems. First we digress to describe projective embeddings of moduli spaces of bundles, beginning with bundles over curves.

10.1.3 Gieseker's projective embedding

Let C be an algebraic curve (compact Riemann surface) of genus g and fix a line bundle $\mathcal{O}(1)$ of degree one over C. Let p be a positive integer, to be fixed below, and consider two moduli spaces W_0, W_1 of rank-two stable holomorphic bundles $\mathcal{E}$ over C with determinant $\mathcal{O}(2p)$ (the tensor power $\mathcal{O}(1)^{2p}$) and $\mathcal{O}(2p+1)$ respectively. The operation of tensoring with $\mathcal{O}(1)$ shows that these are independent of p. Each can be described in terms of projectively flat unitary connections; the space W_0 is the same as the space denoted W_C in Section 6.1.4. Both W_0 and W_1 are complex manifolds of complex dimension $3g-3$, and W_1 is also compact. We want to construct projective embeddings of W_0 and W_1 by sections of the determinant line bundle $\mathcal{L}$. In the case of W_1 we get such an embedding by Kodaira's theorem, since we know that $\mathcal{L}$ is a positive line bundle, but as it stands this does not cover the non-compact space W_0. We will describe here an algebro-geometric construction due to Gieseker which is rather more explicit and has the advantage that it works equally well in the non-compact case. In this construction we shall see another way in which the notion of stability for bundles can be fitted into the general theory described in Section 6.5, involving quotients by linear actions. As in the case of the ADHM construction of instantons, the relevant symmetry group is the automorphism group of the cohomology of a bundle.

Recall that, topologically, any bundle can be induced from a universal bundle U over a Grassman manifold by a suitable map. Gieseker's construction takes as starting point the analogue of this for holomorphic bundles. Suppose a rank-two bundle $\mathcal{E}$ over C is *generated by its global sections*, i.e. for each point x in C the evaluation map $e_x\colon H^0(\mathcal{E}) \to \mathcal{E}_x$ is surjective. If we write H for $H^0(\mathcal{E})$ and $\underline{H}$ for the corresponding trivial bundle over C we get a surjective bundle map

$$e\colon \underline{H} \longrightarrow \mathcal{E}.$$

Thus the bundle $\mathcal{E}$ is completely described by giving a family of two-dimensional quotients of the fixed vector space H, or equivalently by a family of two-dimensional subspaces of the dual space H^*. We get a canonical map to the Grassmannian of two-planes:

$$f\colon C \longrightarrow \mathrm{Gr}_2(H^*), \tag{10.1.6}$$

with $f(x)$ the annihilator of $\ker e_x$. The universal bundle U over the

Grassmannian (the dual of the tautological bundle) is holomorphic, and $\mathscr{E}$ is canonically isomorphic to the pull-back $f^*(U)$.

We next give a rather different description of f. First we apply the standard Plücker embedding i of a Grassmannian. This maps $\mathrm{Gr}_2(H^*)$ to the projective space $\mathbb{P}(\Lambda^2 H^*)$, with

$$i(\mathrm{Span}(\varepsilon_1, \varepsilon_2)) = [\varepsilon_1 \wedge \varepsilon_2].$$

Thus we have a composite map

$$g = if\colon C \longrightarrow \mathbb{P}(\Lambda^2 H^*).$$

Now, maps to a projective space correspond to linear systems of sections of line bundles. The pull back by g of the Hopf line bundle over the projective space is canonically isomorphic to the line bundle $\Lambda^2\mathscr{E}$ over C. The map g must therefore be induced from the universal map,

$$u\colon C \longrightarrow \mathbb{P}(H^0(\Lambda^2\mathscr{E})^*)$$

by a linear mapping:

$$\sigma^T\colon H^0(\Lambda^2\mathscr{E})^* \longrightarrow (\Lambda^2 H)^*.$$

More precisely, any such linear map induces a rational map on the projective spaces and this becomes a well-defined mapping on the one-dimensional curve C (provided σ^T is not identically zero). So we have a commutative diagram:

$$\begin{array}{ccccc} C & \xrightarrow{f} & \mathrm{Gr}_2(H^*) & \xrightarrow{i} & \mathbb{P}(\Lambda^2 H^*) \\ & {\scriptstyle u}\searrow & & \nearrow{\scriptstyle \mathbb{P}(\sigma^T)} & \\ & & \mathbb{P}(H^0(\Lambda^2\mathscr{E}))^* & & \end{array}$$

Lemma (10.1.7). *The map σ^T is the transpose of the tautological map $\sigma\colon \Lambda^2 H = \Lambda^2(H^0(\mathscr{E})) \to H^0(\Lambda^2\mathscr{E})$.*

This is just a matter of checking definitions, and we can safely omit the proof.

It follows then that our bundle is completely determined by the associated map σ. To apply this to moduli problems we consider families of bundles $\mathscr{E}$ with the same determinant $\Lambda^2\mathscr{E}$, and with a fixed dimension, N say, of $H^0(\mathscr{E})$. Then to each bundle we can associate an orbit in the vector space $\mathrm{Hom}(\Lambda^2\mathbb{C}^N, H^0(\Lambda^2\mathscr{E}))$ under the natural action of $GL(N, \mathbb{C})$. For bundles which are generated by their global sections, this orbit determines the bundle up to isomorphism.

Lemma (10.1.8). *A stable bundle $\mathscr{E}$ over C with $\deg(\mathscr{E}) \geq 4g + 2$ is generated by its global sections and has $\dim H^0(\mathscr{E}) = \deg(\mathscr{E}) - 2(g - 1)$.*

Proof. For any point x in C we have an exact sequence:

$$0 \longrightarrow \mathscr{E} \otimes [-x] \longrightarrow \mathscr{E} \xrightarrow[e_x]{} \mathscr{E}_x \longrightarrow 0,$$

where $[-x]$ denotes the line bundle of degree -1 defined by x. The long exact cohomology sequence shows that e_x is surjective if $H^1(\mathscr{E} \otimes [-x]) = 0$. But this space is dual to $H^0(\mathscr{E}^* \otimes K_C \otimes [x])$, i.e. to the maps from $\mathscr{E}$ to the line bundle $K_C \otimes [x]$, which has degree $2(g-1)+1$. By the definition of stability, such maps exist only if $\frac{1}{2}\deg\mathscr{E} < 2(g-1)+1$. Similarly, when $\deg\mathscr{E} \geq 4g-4$ we have $H^1(\mathscr{E}) = 0$ and $\dim H$ is given by the Riemann–Roch formula.

We thus obtain the following proposition, in which we let W denote either of the moduli spaces W_0, W_1, we write $\mathbb{C}^q$ for $H^0(\mathcal{O}(2p))$, $H^0(\mathcal{O}(2p+1))$ and put $N = 2p - 2(g-1)$, $(2p+1) - 2(g-1)$ respectively, where p is any integer bigger than $2g-1$.

Proposition (10.1.9). *There is a natural injection γ of W into the set of orbits of $SL(N, \mathbb{C})$ in the projective space $\mathbb{P}(\mathrm{Hom}(\Lambda^2\mathbb{C}^N, \mathbb{C}^q))$.*

Gieseker's approach thus arrives squarely in the class of problems we discussed in Section 6.5.2. If we restrict the $SL(N, \mathbb{C})$ action to the open set of stable orbits in the projective space, the quotient U^* becomes a complex manifold (or orbifold) and the invariant polynomials, of fixed large degree s say, induce an embedding of U^* into a projective space $\mathbb{CP}^r$. What needs to be shown is that the injection γ maps W to the subset U^*, i.e. that stable bundles $\mathscr{E}$ have stable maps $\sigma_{\mathscr{E}}$. We omit the proof of this and refer to the very readable account by Gieseker (1977). In outline, one applies the Hilbert criterion to show that if $\sigma_{\mathscr{E}}$ is not stable and is destabilized by the one-parameter subgroup associated with weights w_i and a basis s_i for $H^0(\mathscr{E})$, then for all pairs with $w_i + w_j > 0$, s_i and s_j lie in a common line sub-bundle. Then the Riemann–Roch formula can be used to show that the sub-line bundle associated to the largest weight destabilizes $\mathscr{E}$ as a bundle. The conclusion is that there is a projective embedding $j\colon W \to \mathbb{CP}^r$ obtained by composing γ with the embedding of U^*. (Similarly one finds that semi-stable bundles $\mathscr{E}$ have semi-stable maps $\sigma_{\mathscr{E}}$.)

We want now to show that the line bundle associated to this Gieseker embedding j can be identified with the power $\mathscr{L}^s$ of the determinant line bundle over W. We consider the even case for simplicity. Recall that we defined the determinant line bundle over the moduli space of bundles V with $\Lambda^2 V$ trivial by the index of the Dirac operator, which could be identified with the $\bar{\partial}$ operator after twisting by a square root of the canonical bundle. We will now clarify the role of this twisting. Observe first that the moduli space of bundles with $\Lambda^2 V$ trivial, up to bundle isomorphism, is identical with the moduli space of pairs (V, ψ) where ψ is an isomorphism from $\Lambda^2 V$ to $\mathcal{O}$. This is because the scalars act transitively on the maps ψ. So when classifying such bundles we may suppose we have a definite trivialization of Λ^2. Now for any line bundle θ over C we can assign to the pair (V, ψ) the line obtained from the cohomology of $V \otimes \theta$. If V is stable, say, the only automorphisms of (V, ψ) are ± 1, and since the numerical Euler characteristic of V is even this acts trivially on the determinant line. So, as in Chapter 5, we obtain a line bundle, $\mathscr{L}_\theta$ say, over the moduli space of stable holomorphic bundles with a trivialization of Λ^2. At first sight it may seem that we are obtaining different line bundles in this way, since the cohomology groups of $V \otimes \theta$ are not isomorphic for different θ. In reality, however, all these line bundles are isomorphic. To see this we represent line bundles by divisors on C. Suppose for example that $\theta = [P]$ is the line bundle associated with a point P in C. Then we have an exact sequence:

$$0 \longrightarrow V \longrightarrow V \otimes \theta \longrightarrow V \otimes \theta|_P \longrightarrow 0. \qquad (10.1.10)$$

The resulting long exact cohomology sequence, and the usual property of the 'multiplicative Euler characteristic' in exact sequences, gives a natural isomorphism between the determinant line of the cohomology of $V \otimes \theta$ and the tensor product of the determinant line of the cohomology of V and $\Lambda^2(V \otimes \theta)_P$. Fix a trivialization of the fibre of θ over P. Then the trivialization of $\Lambda^2 V$ gives us an isomorphism between the two determinant lines, and hence an isomorphism between $\mathscr{L}_\theta$ and $\mathscr{L}_0$. Repeating this procedure, we see that all the line bundles $\mathscr{L}_\theta$ are isomorphic. So we can use any twist to describe our basic line bundle $\mathscr{L}$ over the moduli space W_0.

With this preliminary observation we return to the line bundle induced by Gieseker's embedding—the pull-back by j of L_U, the quotient of the Hopf line bundle over $\mathbb{P}(\mathrm{Hom}(\Lambda^2\mathbb{C}^N, \mathbb{C}^q))$. We let $\mathscr{E} = V \otimes \mathscr{O}(p)$, and by the remarks above we suppose we have a fixed isomorphism between $\Lambda^2\mathscr{E}$ and $\mathscr{O}(2p)$. Now we introduce the moduli space of triples (V, ψ, f) where ψ is a trivialization of $\Lambda^2 V$ and f is an isomorphism:

$$f\colon H^0(V \otimes \mathscr{O}(p)) \longrightarrow \mathbb{C}^N.$$

Given such a triple we get a point $\sigma_{V,\psi,f}$ in $\mathrm{Hom}(\Lambda^2\mathbb{C}^N, \mathbb{C}^q)$ whose orbit is $j([V])$. Changing f to ζf, where ζ is in $\mathbb{C}^*$, changes $\sigma_{V,\psi,f}$ to $\zeta^{-1}\sigma_{V,\psi,f}$. Here we come to a point which we hurried past in Section 6.5. The centre $\mathbb{Z}/N$ of $SL(N, \mathbb{C})$ acts trivially on the projective space, so the Hopf line bundle need not descend to the quotient. However its N'th power will do, i.e. the line bundle L_U over the quotient lifts to this power of the Hopf bundle. So, by the discussion above, the choice of f gives a basis element in the fibre of the pull-back of L_U, and multiplication of f by ζ multiplies this basis element by ζ^N. On the other hand, the choice of f gives a basis element in the determinant line $\det H^0(V \otimes \mathscr{O}(p))$, and multiplication of f by ζ changes this by ζ^{-N}.

We deduce that there is a natural isomorphism between the line bundles $j^*(L_U)$ and the $\mathscr{L}_{\mathscr{O}(p)}$, the latter being the bundle with fibres $\det H^0(V \otimes \mathscr{O}(p))$ (since the other cohomology group is zero). In sum then we obtain a projective embedding j of W_0 by holomorphic sections of some high power sN of the determinant line bundle $\mathscr{L}$ over W_0.

There is one last fact to mention. The points of W_0 represent stable bundles and the holomorphic sections which give Gieseker's projective embedding can be viewed as holomorphic sections of a line bundle $\mathscr{L}$ over an open subset $\mathscr{A}_s$ of the space of connections ($\bar{\partial}$ operators) over C, invariant under the natural action of the complexified gauge group $\mathscr{G}^c$. However, these sections can all be extended holomorphically over the whole space $\mathscr{A}$. One can see this by examining the definition of the sections carefully or, more directly, by using the fact that (if $g > 0$) the complement of $\mathscr{A}_s$ has complex co-dimension two or more. We know then that for any stable point in $\mathscr{A}$ there is a holomorphic, invariant, section which does not vanish at that point. The last fact we need is that the same is true for the trivial connection or, more generally, for any connection which defines a semi-stable holomorphic bundle over C. The point $\sigma_{V \otimes \mathcal{O}(p)}$ defined by such a connection is semi-stable for the $SL(N, \mathbb{C})$ action, so this follows from the corresponding piece of theory for the finite-dimensional quotient.

10.1.4 Technical facts about moduli spaces

We will now begin our description of the detailed proof of (10.1.1) which will rely on a number of general facts about moduli spaces, which we will marshall in this section. The first point is that the ASD moduli spaces are *non-empty* when k is sufficiently large. This is true for any Riemannian four-manifold and was proved by Taubes (1984) for $SU(2)$-bundles and (as a special application of the results) by Taubes (1989) for $SO(3)$-bundles. In the algebro-geometric context the existence of stable holomorphic $SL(2, \mathbb{C})$ bundles was proved by Gieseker (1988); results on general two-plane bundles can be obtained using the Serre construction discussed in Section 10.2.

The next point concerns the dimension of the moduli space of stable bundles, i.e. the ASD connections relative to a Kähler metric. We know that for *generic* Riemannian metrics the moduli space of irreducible connections is regular, and is therefore smooth and of the proper 'virtual' dimension. But the Kähler metrics are not generic and it certainly may happen that we encounter moduli spaces with singularities of various kinds, or with components which have dimension larger than the virtual dimension. In the latter case the argument we have given above for the positivity of $q(h)$ is certainly not valid. The situation is quite analogous to the familiar intersection theory of subvarieties: if P, Q are complex subvarieties of an ambient compact complex manifold V, with $\dim V = \dim P + \dim Q$, and if the intersection of P and Q consists of isolated points, then the topological intersection number $[P].[Q]$ is positive. In fact $[P].[Q]$ is at least the number of intersection points. But if $P \cap Q$ has dimension one or more it may be that $[P].[Q]$ is negative. For example, we could take $P = Q$ to be an 'exceptional curve' on a surface, with self-intersection -1.

To avoid this difficulty we look at bundles with k large. Then it can be proved that each component of the moduli space has the proper dimension.

Precisely, let us define for the fixed Hodge metric:

$$\Sigma_{k,\alpha} = \{[A] \in M_{k,\alpha} \mid H_A^2 \neq 0\}.$$

Σ is a complex-analytic subspace of the complex space M_k and we have:

Proposition (10.1.11). *There are constants B_1, B_2 such that for all k and all reductions α of* (1, 1) *classes*:

$$\dim \Sigma_{k,\alpha} \leq 3k + B_1 k^{1/2} + B_2.$$

This is proved by Donaldson (1990*a*) for $SU(2)$ bundles, but the argument extends easily to the general case.

Thus the dimension of Σ grows more slowly than the virtual complex dimension $4k - 3(1 + p_g)$ of M_k. On the other hand, if the dimension of a component of M exceeds the virtual dimension, then this component is contained in Σ. It follows that for large k the moduli space does indeed have the proper dimension. Moreover the singular set has large codimension in M_k.

The third fact we want is that the moduli spaces, viewed as moduli spaces of stable bundles on our projective algebraic surface S, are themselves naturally complex varieties, in the sense of abstract algebraic geometry. This is the algebro-geometric analogue of the theory we studied from the transcendental point of view in Section 6.4, that the moduli spaces are Hausdorff complex spaces. It is a generalization of the theorem that holomorphic bundles over S are necessarily algebraic. What is asserted is, first, that the local versal deformations of algebraic vector bundles over S can be realized as algebraic vector bundles over $S \times V_\lambda$, where V_λ is a 'quasi-affine' variety (the difference $V \setminus W$ of affine varieties), and that the gluing maps

$$\begin{array}{ccc} \phi_{\lambda\mu}\colon V_{\lambda\mu} & \longrightarrow & V_{\mu\lambda} \\ \cap & & \cap \\ V_\lambda & & V_\mu \end{array}$$

are represented by rational functions. Second, the statement that M_k is a complex variety asserts that M_k has 'finite type', i.e. that it can be covered by a finite number of quasi-affine patches V_α. This is closely related to our compactness theorem for moduli spaces of ASD connections. For proofs of these assertions see Maruyama (1977) and Gieseker (1977).

Next we want to study the restriction of stable holomorphic bundles on S to curves. The main theorem we need is the following result from Mehta and Ramanathan (1984).

Proposition (10.1.12) (Mehta and Ramanathan). *For any stable bundle $\mathcal{E}$ on S there is an integer p_0 such that, for $p \geq p_0$ and generic curves C in the linear system $|\mathcal{O}(p)|$ on S, the restriction of $\mathcal{E}$ to C is also stable.*

Now the stability condition is itself open, so for each (smooth) curve C in a linear system $|\mathcal{O}(p)|$ we have a Zariski open subset $U_C \subset M_k$ consisting of bundles whose restriction to C is stable. The proposition asserts that the union of the U_C, over all p and C, is the whole of M_k. But the finite type condition asserts that M_k is compact in the Zariski topology, so we can find a finite cover:

$$M = \bigcup_{i=1}^{R} U_i, \quad U_i = U_{C_i}. \tag{10.1.13}$$

Moreover we can suppose that the curves C_i are in the same linear system $|\mathcal{O}(p)|$ and are in general position, with all triple intersections empty. Similarly we can suppose that for all pairs $[\mathcal{E}]$, $[\mathcal{F}]$ in M_k the cohomology group $H^1(\mathrm{Hom}(\mathcal{E}, \mathcal{F})(-p))$ is zero. Finally, replacing the given projective embedding by that defined by $|\mathcal{O}(p)|$, we may as well suppose that $p = 1$ and the curves C_i are hyperplane sections.

10.1.5 Restriction to curves

We can now construct a projective embedding of M_k, using restriction to complex curves in the surface S. We first consider the situation from the point of view of determinant line bundles. Suppose first that S is spin, so there is a square root $K_S^{1/2}$ of the canonical bundle. Let C be a curve in the linear system $|\mathcal{O}(2d)|$. There is an induced square root of K_C and an exact sequence:

$$0 \longrightarrow \mathcal{O}(-d) \otimes K_S^{1/2} \longrightarrow \mathcal{O}(d) \otimes K_S^{1/2} \longrightarrow K_C^{1/2} \longrightarrow 0. \tag{10.1.14}$$

For any bundle $\mathcal{E}$ over S we get, just as for (10.1.10), an induced isomorphism of determinant lines:

$$\{\det H^*(\mathcal{E} \otimes K_S^{1/2} \otimes \mathcal{O}(d))\}\{\det H^*(\mathcal{E} \otimes K_S^{1/2} \otimes \mathcal{O}(-d))\}^{-1}$$
$$= \det H^*(\mathcal{E}|_C \otimes K_C^{1/2}).$$

The term on the right is the fibre of the determinant line bundle defined by restriction to C, whereas the expression on the left is independent of C. We deduce that restriction to different curves gives isomorphic line bundles over the ASD moduli space. This is an algebro-geometric version of Proposition (7.1.16). It is easy to remove the spin condition, and the requirement that the homology class of C be even, by introducing twisting factors just as in Section 10.1.3.

To sum up then, for each i we have a smooth moduli space $W_i = W(C_i)$, a determinant line bundle $\mathcal{L}_i$ over W_i and a projective embedding:

$$J_i \colon W_i \longrightarrow \mathbb{P}_i = \mathbb{P}(G_i^*)$$

associated with a vector space G_i of sections of $\mathcal{L}_i^N$. Now the restriction map

$$r_i \colon U_i \longrightarrow W_i$$
$$r_i(\mathcal{E}) = \mathcal{E}|_{C_i}$$

is regular, so we get composite maps $J_i r_i: U_i \to \mathbb{P}_i$. On M_k we have the fixed line bundle $\mathscr{L}$, and a holomorphic isomorphism from $\mathscr{L}$ to $\mathscr{L}_i$. So we can regard the G_i as spaces of sections of $\mathscr{L}^N$ over U_i, and these extend holomorphically over M_k by the remark at the end of Section 10.1.3. Hence we now have a space $G = \bigoplus G_i$ of sections of $\mathscr{L}^N$ over M_k, and for each point $\mathscr{E}$ in M_k there is a section in G not vanishing at $\mathscr{E}$. But this just means that we have an induced map,

$$J: M_k \longrightarrow \mathbb{P}(G^*) \quad \text{with } J^*(\mathcal{O}(l)) = \mathscr{L}^N.$$

This also holds with $k = 0$, when we have the moduli space with one point representing the trivial bundle, by the remark at the end of Section 10.1.3. The maps $J_i r_i$ are of course the composites of J with the projection maps to the individual factors. They are rational maps on M_k, regular on U_i.

To see that J is an embedding we use the vanishing of the cohomology groups $H^1(\mathrm{Hom}(\mathscr{E}, \mathscr{F})(-1))$. This implies that any two bundles $\mathscr{E}, \mathscr{F}$ which become isomorphic when restricted to some C_i are already isomorphic over S. Then the fact that J is injective follows from the corresponding property of the J_i. Similarly, taking $\mathscr{E} = \mathscr{F}$, we see that J is an immersion. As a final technicality, we can arrange the given set of surfaces to have the desired properties for all the finite number of moduli spaces M_j for $j \leq k$. So we have embeddings, which we will still denote by J, of M_j in the same projective space, and similarly $J_i: M_j \to \mathbb{P}_i$.

10.1.6 The detailed argument

By examining Gieseker's algebraic construction one sees that the embedding J of the algebraic variety is defined by a rational map. It is a general fact that the image of an abstract complex variety under a projective embedding by a rational map is a *quasi-projective* variety, i.e. the difference $V_1 \setminus V_2$ of projective varieties. For any quasi-projective variety $Y \subset \mathbb{CP}^r$ we define the degree $\deg(Y)$ to be the degree of the projective variety $\bar{Y}$; it is the number of intersection points of a generic $\overline{\mathbb{CP}^q}$ $(q = r - \dim \Omega)$ with Y, and if Y is non-empty the degree is always strictly positive. So the proof of our main theorem (10.1.1) is completed by the next lemma:

Lemma (10.1.15).

(i) *Let $J: M_k \to \mathbb{P}(G)$ be the projective embedding, defined by sections of $\mathscr{L}^N$, constructed above. Then the degree of the image $J(M_k)$ is independent of the choice of curves C_i.*

(ii) *For sufficiently large k, $q_k([C_i]) = (1/N^d)\deg(J(M_k))$.*

Proof. To prove (i) it suffices to show that if we extend a given collection of curves $C_1, \ldots, C_L$ by another one C_{L+1} then the degree of the resulting projective embeddings is unchanged. But this follows immediately from the general fact that if a projection $\mathbb{P}^{n+s} \to \mathbb{P}^n$ (a rational map) restricts to a subvariety $Q \subset \mathbb{P}^{n+s}$ to give an embedding $\pi: Q \to \mathbb{P}^n$, then the degrees of Q and its image $\pi(Q)$ are equal. To prove (ii) we trace through the definition of our polynomial invariant q_k using restriction to surfaces, and make an appropriate choice of the curves C_i. Let us suppose inductively that we have chosen curves C_i $(i = 1, \ldots, l)$ in general position, and sections g_i in G_i so that the common zero set

$$Z_l^{(k)} = Z(g_1, \ldots, g_l) = \{[\mathscr{E}] \in M_k \,|\, g_i(\mathscr{E}) = 0, \quad i = 1, \ldots, l\}$$

has the following good properties:

(i) $Z_l^{(k)}$ has the correct dimension $d - l$,

(ii) $Z_l^{(k)} \cap \Sigma$ has the correct dimension $\dim \Sigma - l$.

(iii) On a dense open subset of $(M_k \backslash \Sigma) \cap Z_l^{(k)}$ the zeros of g_i are transverse.

(iv) For all the lower moduli spaces M_i the zero sets

$$Z_l^{(i)} = \{[\mathscr{E}] \in M_i \mid g_j(\mathscr{E}) = 0, \quad j = 1, \ldots, l\}$$

have dimension $\dim M_i - l$.

To pass from l to $l + 1$ we note that, as quasi-projective varieties, all of the $Z_l^{(i)}$ and $Z_l^{(k)} \cap \Sigma$ have finitely many components. We choose a point in each component, i.e. a finite set of bundles $\mathscr{E}_\lambda$. Then we can, by (10.1.12), choose a curve C_{l+1} such that all the $\mathscr{E}_\lambda$ are stable on C_{l+1}. Thus the generic hyperplane section of $\mathbb{P}_{l+1}$ induces a section g_{l+1} which does not vanish at any of the $\mathscr{E}_\lambda$; then the zero sets $Z_{l+1}^{(i)}$ do not contain any component of the $Z_l^{(i)}$ and it follows that they have the proper dimension. So, inductively, we can choose a set of curves $C_1, \ldots, C_d$ and sections $g_1, \ldots, g_d$ such that the common zero set in M_k is a finite set of points $\{E_\mu\}$, none contained in Σ, and with the general position properties (i)–(iv) with respect to the lower moduli spaces.

Now on the one hand we can extend this collection of curves, if necessary, to get a projective embedding of M_k. The g_i represent hyperplane sections of $J(M_k)$ and the number of points in the intersection, counted with appropriate multiplicities, represents the degree of $J(M_k)$ *provided* there are no 'zeros at infinity', i.e. no common zeros of all the g_i in $\overline{J(M_k)} \backslash J(M_k)$. On the other hand, from the point of view of the general set-up in Chapters 5 and 9, each g_i represents a section of the determinant line bundle $\mathscr{L}_i$, and its zero set represents a codimension-two submanifold V_{C_i} in the space of connections. The intersection $M_k \cap V_{C_1} \cap \ldots \cap V_{C_d}$ is by construction a finite set of points $\{A_\mu\}$. To prove the equality of $q_k(\alpha)$ and $\deg(J(M_k))$ we have to check three things:

(*a*) That the multiplicities with which we count the points, regarded either as intersections in the projective space P or in the space of connections, agree. That is, if we perturb the Kähler metric to a nearby generic metric and perturb the V_{C_i} to be transverse, as in the definition of Section 9.2.1, the intersections A_μ split up into the correct number of nearby transverse intersection points.

(*b*) That when we perturb the metric and V_{C_i} as in Section 10.1.1, all the new intersection points are close to the A_μ.

(*c*) That in the projective space $\mathbb{P}$ there are no common zeros of the g_i in $\overline{J(M_k)}\backslash J(M_k)$.

Of these (*a*) is quite straightforward; it is just the assertion that the local multiplicity of the zero set equals the multiplicity of any transverse deformation. Points (*b*) and (*c*) are more interesting and the proof is much the same in each case. We begin with (*c*), and suppose that on the contrary there is a zero in $\overline{J(M_k)}\backslash J(M_k)$. Then we can find a sequence of holomorphic bundles $\mathscr{E}_\alpha$ over S without convergent subsequences but with all $g_i(\mathscr{E}_\alpha)$ tending to zero as α tends to ∞. Regarded as ASD connections A_α we can apply our compactness theorem and, without loss of generality, suppose that A_α tends to a limiting connection A_∞ on the complement of a finite set of points $\{x_\nu\}$. The limiting connection has Chern class $l < k$ say, and there are at most $k - l$ points x_ν. We now apply our familiar argument: for any curve C_i which does not contain any of the points x_ν the restriction of the connections converges in C^∞; it follows that for any such i, $g_i(A_\infty)$ is zero. But we have arranged that all the multiple zeros of the g_i in the lower moduli spaces are of the appropriate dimension (property (iv) above), so if d_l is the complex dimension of M_l we must have:

$$d_l + 2(k - l) \geq d = 4k - 3(1 + p_g). \tag{10.1.16}$$

In the present situation we do not know that all the lower moduli spaces have the correct (virtual) dimension. But we do know that this holds for large enough values $j \geq k_0$, of the Chern class, by (10.1.11). We put

$$D = \max_{j \leq k_0} \dim_C M_j.$$

Then if $2k > D + 3(1 + p_g)$ (say) the inequality (10.1.16) cannot hold, thus verifying property (*c*). Precisely the same argument shows that for a sequence of metrics $g^{(\alpha)} \to g$ and sections $g_i^{(\alpha)} \to g_i$, the common zeros in the perturbed moduli space $M_k(g^{(\alpha)})$ converge to the A_μ, verifying property (*b*).

10.2 Construction of holomorphic bundles

10.2.1 Extensions

In this section we will describe general techniques for constructing holomorphic bundles out of linear data. Consider first a complex manifold Z and exact sequences of holomorphic bundles over Z:

$$0 \longrightarrow \mathscr{E}' \xrightarrow[i]{} \mathscr{E} \xrightarrow[p]{} \mathscr{E}'' \longrightarrow 0. \tag{10.2.1}$$

We say that the bundle $\mathscr{E}$ is given as an extension of $\mathscr{E}''$ by $\mathscr{E}'$ and we say that two such sequences, with fixed end terms, are equivalent if there is a commutative diagram:

$$\begin{array}{ccccccccc} 0 & \longrightarrow & \mathscr{E}' & \xrightarrow{i_1} & \mathscr{E}_1 & \xrightarrow{p_1} & \mathscr{E}'' & \longrightarrow & 0 \\ & & \| & & \downarrow & & \| & & \\ 0 & \longrightarrow & \mathscr{E}' & \xrightarrow[i_2]{} & \mathscr{E}_2 & \xrightarrow[p_2]{} & \mathscr{E}'' & \longrightarrow & 0. \end{array} \tag{10.2.2}$$

With any extension we associate a class in the cohomology group $H^1(\mathrm{Hom}(\mathscr{E}'', \mathscr{E}'))$ as follows. Applying Hom $(\mathscr{E}'', \;-)$ to the sequence (10.2.1) gives:

$$0 \longrightarrow \mathrm{Hom}(\mathscr{E}'', \mathscr{E}') \longrightarrow \mathrm{Hom}(\mathscr{E}'', \mathscr{E}) \longrightarrow \mathrm{Hom}(\mathscr{E}'', \mathscr{E}'') \longrightarrow 0.$$

Now take the induced boundary map on cohomology

$$\partial \colon H^0(\mathrm{Hom}(\mathscr{E}'', \mathscr{E}'')) \longrightarrow H^1(\mathrm{Hom}(\mathscr{E}'', \mathscr{E}')) \tag{10.2.3}$$

and evaluate it on the identity to get the extension class $\partial(1)$ in $H^1(\mathrm{Hom}(\mathscr{E}'', \mathscr{E}'))$.

Proposition (10.2.4). *There is a natural bijection $(\mathscr{E}, i, p) \mapsto \partial(1)$ between the equivalence classes of extensions of $\mathscr{E}''$ by $\mathscr{E}'$ and the cohomology group $H^1(\mathrm{Hom}(\mathscr{E}'', \mathscr{E}'))$.*

To understand this it is convenient to introduce two concrete representations for the extension class. First, using Čech cohomology, we choose a cover $Z = \bigcup U_\alpha$ by open sets over each of which the sequence splits, so we have isomorphisms

$$j_\alpha \colon \mathscr{E}|_{U_\alpha} \longrightarrow \mathscr{E}'|_{U_\alpha} \oplus \mathscr{E}''|_{U_\alpha}$$

compatible, in the obvious sense, with i and p. Then on each overlap $U_\alpha \cap U_\beta$ we can write

$$j_\alpha = j_\beta a_{\alpha\beta}$$

where $a_{\alpha\beta}$ is an automorphism of $\mathscr{E}' \oplus \mathscr{E}''$ over $U_\alpha \cap U_\beta$ of the form

$$a_{\alpha\beta} = 1 + \begin{pmatrix} 0 & \chi_{\alpha\beta} \\ 0 & 0 \end{pmatrix},$$

so $\chi_{\alpha\beta}$ is a holomorphic bundle map from $\mathscr{E}''|_{U_\alpha \cap U_\beta}$ to $\mathscr{E}'|_{U_\alpha \cap U_\beta}$. On a triple overlap $U_\alpha \cap U_\beta \cap U_\gamma$ we have:

$$\begin{aligned} j_\gamma &= j_\beta a_{\gamma\beta} \\ &= j_\alpha a_{\gamma\alpha} = (j_\beta a_{\alpha\beta}) a_{\gamma\alpha}, \end{aligned}$$

so $a_{\gamma\beta} = a_{\alpha\beta} a_{\gamma\alpha}$, which gives the cocycle relation $\chi_{\gamma\beta} = \chi_{\alpha\beta} + \chi_{\gamma\alpha}$. The extension class $\partial(1)$ is represented by the Čech cocycle $(\chi_{\alpha\beta})$ on this cover.

For the second approach we use Dolbeault cohomology and the operator $\bar\partial = \bar\partial_{\mathscr{E}}$ defining the structure. We choose over all of Z a C^∞ splitting of the sequence, for example by choosing a Hermitian metric on $\mathscr{E}$ and taking the orthogonal complement of $\mathscr{E}'$. This splitting can be represented by a map $l\colon \mathscr{E} \to \mathscr{E}'$, such that $p \circ l$ is the identity.

Now, for any section s of $\mathscr{E}''$, $\bar\partial(l(s)) - l(\bar\partial s)$ lies in the image of i and we can define a tensor β in $\Omega_Z^{0,1}(\mathrm{Hom}(\mathscr{E}'', \mathscr{E}'))$ by

$$\bar\partial(l(s)) - l(\bar\partial s) = i(\beta(s)). \tag{10.2.5}$$

This bundle-valued form is annihilated by $\bar\partial$:

$$\begin{aligned} i[\bar\partial(\beta(s))] &= \bar\partial(i(\beta(s))) \\ &= \bar\partial(\bar\partial l - l\bar\partial)s = -\bar\partial l \bar\partial s. \end{aligned}$$

But on the other hand:

$$\begin{aligned} i[\bar\partial(\beta(s))] &= i[(\bar\partial\beta)s] - i[\beta(\bar\partial s)] \\ &= i[(\bar\partial\beta)s] - (\bar\partial l - l\bar\partial)\bar\partial s \\ &= i[(\bar\partial\beta)s] - \bar\partial l \bar\partial s. \end{aligned}$$

So $i[(\bar\partial\beta)s]$ vanishes for all s, and hence $\bar\partial\beta = 0$. This tensor β is then a Dolbeault representative for the extension class. We can express this construction in terms of connections by choosing unitary connections A', A'' on $\mathscr{E}'$, $\mathscr{E}''$ and comparing a connection on $\mathscr{E}' \oplus \mathscr{E}''$ with $A' \oplus A''$. For any tensor β in $\Omega_Z^{0,1}(\mathrm{Hom}(\mathscr{E}'', \mathscr{E}'))$, the unitary connection

$$A = A' \oplus A'' + \begin{pmatrix} 0 & \beta \\ -\beta^* & 0 \end{pmatrix} \tag{10.2.6}$$

has

$$F_A^{0,2} = \begin{pmatrix} F_{A'}^{0,2} & \bar\partial_{A'*A''}\beta \\ 0 & F_{A''}^{0,2} \end{pmatrix}.$$

So A is integrable if and only if A', A'' are, and if $\bar\partial_{A'*A''}\beta = 0$. (Compare the discussion in the proof of Proposition (6.2.25)).

It is a straightforward exercise to show that these Čech and Dolbeault classes do indeed represent $\partial(1)$ and, using either definition, to verify Proposition (10.2.4). This proposition gives a technique for constructing holomorphic bundles out of bundles of lower rank and suitable cohomology classes. In particular we can construct rank-two bundles $\mathscr{E}$ starting from complex line bundles $\mathscr{E}'$, $\mathscr{E}''$. If the base space Z is a complex curve, any rank-two bundle can be constructed in this way. For given any bundle $\mathscr{F}$ we can tensor with a large multiple of a positive line bundle $\mathcal{O}(1)$, as in Section 10.1.3, so that $\mathscr{F} \otimes \mathcal{O}(N)$ has a non-trivial section s. If s vanishes on a positive divisor D, the bundle $\mathscr{E} = \mathscr{F} \otimes \mathcal{O}(N) \otimes [-D]$ has a nowhere vanishing section, which yields an extension

$$0 \longrightarrow \mathcal{O} \longrightarrow \mathscr{E} \longrightarrow \mathscr{L} \longrightarrow 0,$$

for a line bundle $\mathscr{L} = \Lambda^2 \mathscr{E}$. Undoing the twist by the line bundle we get

$$0 \longrightarrow \mathcal{O}(-N) \otimes [D] \longrightarrow \mathscr{F} \longrightarrow (\Lambda^2 \mathscr{F}) \otimes \mathcal{O}(N) \otimes [-D] \longrightarrow 0.$$

So, in principle, complete knowledge of the line bundles over a curve, and of their cohomology groups, gives complete information about the rank-two bundles. The difficulty with this approach, from the point of view of moduli problems, is that there will in general be many different ways of representing the same rank-two bundle as an extension.

10.2.2 *Rank-two bundles over surfaces and configurations of points*

We will now go on to consider a more sophisticated version of the construction in Section 10.2.1. Suppose $\mathscr{E}$ is a rank-two bundle over a compact complex surface S and s is a holomorphic section of $\mathscr{E}$ having isolated zeros $\{x_i\}$ in S. We have then holomorphic bundle maps,

$$\mathcal{O}_S \longrightarrow \mathscr{E} \longrightarrow \Lambda^2 \mathscr{E}, \tag{10.2.7}$$

given by the wedge product with s. Away from the zeros, these express $\mathscr{E}$ as an extension of the line bundle $\Lambda^2\mathscr{E}$ by the trivial line bundle. We assume that all of the zeros are transverse (so the number of zeros equals $c_2(\mathscr{E})$). Then near any zero we can choose local coordinates and a trivialization of $\mathscr{E}$ so that the sequence (10.2.7) is represented by the standard Koszul sequence

$$\mathcal{O}_S \xrightarrow{\begin{pmatrix} z_1 \\ z_2 \end{pmatrix}} \mathcal{O}_S \oplus \mathcal{O}_S \xrightarrow{(-z_2 \;\; z_1)} \mathcal{O}_S, \tag{10.2.8}$$

which we have met in Chapter 3. This gives a resolution of the ideal-sheaf of holomorphic functions vanishing at the origin. Globally our sequence (10.2.7) gives an exact sequence of sheaves,

$$0 \longrightarrow \mathcal{O}_S \longrightarrow \mathscr{E} \longrightarrow \Lambda^2 \mathscr{E} \otimes \mathscr{I} \longrightarrow 0, \tag{10.2.9}$$

where $\mathscr{I}$ is the ideal sheaf of functions vanishing on the set of points $\{x_i\}$ in S. The third term $\Lambda^2\mathscr{E} \otimes \mathscr{I}$ is thus the sheaf of sections of $\Lambda^2\mathscr{E}$ vanishing at all the x_i. We say that (10.2.9) represents $\mathscr{E}$ as an extension of $\Lambda^2\mathscr{E} \otimes \mathscr{I}$ by $\mathcal{O}_S$ and define the equivalence of extensions just as before. We will now give an explicit description of the equivalence classes of such extensions, generalizing (10.2.4). For brevity we denote the line bundle $\Lambda^2\mathscr{E}$ by L.

First, away from the zeros x_i, (10.2.9) represents an extension of bundles, just as we considered in Section 10.2.1. So the restriction to $S\backslash\{x_i\}$ is classified by an element of $H^1(S\backslash\{x_i\}; L^*)$. Taking the Dolbeault approach we can represent this by an element β of $\Omega^{0,1}(L^*)$ defined over the punctured manifold, with $\bar\partial\beta = 0$. Turning now to the zeros, we observe that the section s defines local invariants at each point x_i, as follows. The derivative of s at x_i is an intrinsically defined map,

$$(Ds)_{x_i}\colon (TS)_{x_i} \longrightarrow \mathscr{E}_{x_i}, \tag{10.2.10}$$

which is an isomorphism by hypothesis. So we have an inverse map from the fibre of $\mathscr{E}$ to the tangent space and, taking the induced map on Λ^2, an element:

$$r_i = \Lambda^2(Ds)_{x_i}^{-1} \in (\Lambda^2\mathscr{E})^*_{x_i} \otimes (\Lambda^2 TS)_{x_i} = (L\otimes K_S)_{x_i}^{-1}, \tag{10.2.11}$$

where K_S is the canonical line bundle of S. We will call the r_i the 'residue data' associated with the section. We shall now see that they describe the singularities of the (0, 1)-form β at the zeros x_i.

We recall that the space $\mathscr{D}_S^{0,q}(L^{-1})$ of distributional L^{-1}-valued $(0, q)$-forms is defined to be the topological dual of the vector space $\Omega_S^{2-q,0}(L)$; it contains the space of smooth forms $\Omega_S^{0,q}(L^{-1})$ and the $\bar\partial$ operator extends to the distributions. We say that a lifting $l\colon L \to \mathscr{E}$ over the punctured manifold is admissible if l is $O(r^{-1})$ near the x_i, where r is the distance to a point x_i in a local coordinate system, and its derivative is $O(r^{-2})$. (These conditions are satisfied by any splitting coming from a metric on $\mathscr{E}$.) Then the representative β is $O(r^{-3})$, so β is integrable and defines a distribution in $\mathscr{D}^{0,1}(L^{-1})$.

Proposition (10.2.12).

(i) *For any admissible splitting, the form β in $\mathscr{D}^{0,1}(L^{-1})$ satisfies the equation*

$$\bar\partial\beta = 4\pi^2 \sum_i r_i \delta_{x_i}.$$

(ii) *If β_1, β_2 are the forms corresponding to two admissible splittings, there is a distributional section $\gamma \in \mathscr{D}^{0,0}(L^{-1})$ with $\bar\partial\gamma = \beta_1 - \beta_2$.*

The equation in (i) has the following meaning. For each point x in S there is a delta distribution δ_x in $\mathscr{D}^{0,2}\otimes(K_S)_x$: just the evaluation of a (2, 0) form at x. The coefficients r_i lie in the lines $(L\otimes K_S)_{x_i}^{-1}$ and $r_i\delta_{x_i} \in \mathscr{D}^{0,2}(L^{-1})$ denotes the natural product. In the equation we apply the $\bar\partial$-operator on distributions,

$$\bar\partial\colon \mathscr{D}^{0,1}(L^{-1}) \longrightarrow \mathscr{D}^{0,2}(L^{-1}).$$

Explicitly, the equation asserts that for any smooth θ in $\Omega^{0,2}(L)$ we have

$$\int_S \beta\,\partial\bar\theta = \sum_i \langle r_i, \theta(x_i)\rangle.$$

To prove (10.2.12(i)) we can work locally around a point x_i in a standard coordinate system and a holomorphic trivialization of $\mathscr{E}$ in which the sequence is represented by the Koszul complex. We consider first the splitting defined by the flat metric on $\mathscr{E}$, in this trivialization. The lift of the local generator '1' of L is

$$l(1) = r^{-2}\begin{pmatrix} -\bar{z}_2 \\ \bar{z}_1 \end{pmatrix},$$

so

$$\bar{\partial}\{l(1)\} = r^{-4}(\bar{z}_2 \mathrm{d}\bar{z}_1 - \bar{z}_1 \mathrm{d}\bar{z}_2)\begin{pmatrix} z \\ w \end{pmatrix} = r^{-4}(\bar{z}_2 \mathrm{d}\bar{z}_1 - \bar{z}_1 \mathrm{d}\bar{z}_2) i(1).$$

So, in these trivializations,

$$\beta = r^{-4}(\bar{z}_2 \mathrm{d}\bar{z}_1 - \bar{z}_1 \mathrm{d}\bar{z}_2).$$

But this is $\frac{1}{2}\bar{\partial}^*(r^{-2}\mathrm{d}\bar{z}_1 \mathrm{d}\bar{z}_2)$ so $\bar{\partial}\beta = \Delta(r^{-2}) = 4\pi^2\delta_0$ (β is the same as the form denoted l_ρ in Section 3.3.6, with $\rho = \mathrm{d}\bar{z}_1 \mathrm{d}\bar{z}_2$).

Now let l' be another admissible splitting in these local coordinates. Then, away from the singularity, $l' - l = i\gamma$, where γ is $O(r^{-2})$ and $\beta' - \beta = \bar{\partial}\gamma$ away from the singularity. For any smooth test form θ,

$$\int_{r \geq \varepsilon} \theta \wedge (\beta - \beta') = \int_{r \geq \varepsilon} \theta \wedge \bar{\partial}\gamma = \int_{r \geq \varepsilon} \bar{\partial}\theta \wedge \gamma - \int_{r = \varepsilon} \theta \wedge \gamma.$$

The boundary term tends to zero with ε so

$$\lim_{\varepsilon \to 0} \int_{r \geq \varepsilon} \theta \wedge \beta' = \int \bar{\partial}\theta \wedge \gamma + \int \theta \wedge \beta$$

and the equation $\beta' - \beta = \bar{\partial}\gamma$ holds distributionally. Since $\bar{\partial}^2 = 0$ this shows that (i) holds for any admissible splitting, and also proves (ii).

We now come to the generalization of (10.2.4) to extensions of the form (10.2.9):

Proposition (10.2.13). *Let L be a line bundle over the surface S, $\mathscr{I}$ be the ideal sheaf defined by a set of points x_i in S, and r_i be a non-zero element of $(K_S \otimes L)^{-1}_{x_i}$. There is a one-to-one correspondence between the set of equivalence classes of extensions*

$$0 \longrightarrow \mathcal{O} \longrightarrow \mathscr{E} \longrightarrow L \otimes \mathscr{I} \longrightarrow 0$$

with a vector bundle $\mathscr{E}$ as middle term and residue data (r_i), and the equivalence classes of solutions to the distributional equation $\bar{\partial}\beta = \sum_i r_i \delta_{x_i}$ modulo the equivalence relation $\beta \sim \beta + \bar{\partial}\gamma$.

The proof of this is straightforward checking of the definitions. Suppose we are given a solution β to $\bar{\partial}\beta = \sum r_i \delta_{x_i}$, with non-zero r_i. By the distributional $\bar{\partial}$-Poincaré Lemma we can, changing β by a $\bar{\partial}$-boundary, suppose that β is smooth away from the singular points and that in some local coordinates on a neighbourhood N_i of the singular point x_i and local trivialization of L, β is given by the standard fundamental solution $r^{-4}(\bar{z}_2 \mathrm{d}\bar{z}_1 - \bar{z}_1 \mathrm{d}\bar{z}_2)$ obtained in (10.2.12). Using β in (10.2.4) we construct a bundle $\mathscr{E}_0$ over the punctured manifold as an extension

$$0 \longrightarrow \mathcal{O} \longrightarrow \mathscr{E}_0 \longrightarrow L|_{S\setminus\{x_i\}} \longrightarrow 0.$$

There is a splitting l_0 of this sequence which gives the form β. On each neighbourhood N_i we take the standard Koszul complex:

$$0 \longrightarrow \mathcal{O} \longrightarrow \mathcal{O} \oplus \mathcal{O} \longrightarrow \mathscr{I} \longrightarrow 0,$$

with the orthogonal splitting l_1. We now compare these two over the punctured neighbourhood $N_i\setminus\{x_i\}$, using the local trivialization σ_i of L. The extension forms are equal, so l_0 and l_1 induce a holomorphic isomorphism,

$$\nu_i \colon \mathcal{O} \oplus \mathcal{O} \longrightarrow \mathscr{E}_0|_{N_i\setminus\{x_i\}},$$

fitting into a commutative diagram:

$$\begin{array}{ccccccccc}
0 & \longrightarrow & \mathcal{O} & \longrightarrow & \mathcal{O} \oplus \mathcal{O} & \longrightarrow & \mathscr{I}|_{N_i\setminus\{x_i\}} = \mathcal{O} & \longrightarrow & 0 \\
 & & \| & & \downarrow{\scriptstyle \nu_i} & & /\!\!/ & & \\
0 & \longrightarrow & \mathcal{O} & \longrightarrow & \tilde{\mathscr{E}} & \longrightarrow & \mathcal{O} & \longrightarrow & 0.
\end{array}$$

Then we construct our holomorphic bundle $\mathscr{E}$ over S by gluing together $\mathscr{E}_0$ and the trivial bundle over $N_i\setminus\{x_i\}$, using ν_i. It is straightforward to check that this construction is indeed inverse to the previous one, which obtained a form β from an exact sequence and admissible C^∞ splitting.

We will now give a more explicit description of these extensions in terms of cohomology. We use the fact that the $\bar{\partial}$-cohomology computed using distributions agrees with the ordinary Dolbeault cohomology. For a given set of points in S we define a complex vector space V to be the solutions β of $\bar{\partial}\beta = 4\pi^2 \sum r_i \delta_{x_i}$, for some r_i, modulo $\beta \sim \beta + \bar{\partial}\gamma$. The residues r_i define a linear map

$$\delta \colon V \longrightarrow \bigoplus_i (K_S \otimes L)^{-1}_{x_i}. \tag{10.2.14}$$

(From a more abstract point of view, V is the 'global Ext' group $\mathrm{Ext}^1(L \otimes \mathscr{I}, \mathcal{O})$.) Proposition (10.2.13) can be restated in the form:

Corollary (10.2.15). *There is a one-to-one correspondence between*

(i) *The space of equivalence classes of triples* $(\mathscr{E}, \psi, s)$, *where* $\mathscr{E}$ *is a rank-two bundle over* S, ψ *is a holomorphic isomorphism* $\psi \colon \Lambda^2\mathscr{E} \to \mathcal{O}$, *and* s *is a* *holomorphic section of $\mathscr{E}$ having regular zeros at the points x_i, but otherwise non-vanishing.*

(ii) *The set of elements $\mathscr{V}$ in V such that all coordinates of $\delta(\mathscr{V})$ in $\bigoplus (K_S \otimes L)_{x_i}^{-1}$ are non-zero.*

On the other hand, we have an evaluation map

$$\theta: H^0(K_S \otimes L) \longrightarrow \bigoplus (K_S \otimes L)_{x_i}, \tag{10.2.16}$$

whose transpose e^{T} can be viewed as a map from the direct sum of the $(K_S \otimes L)_{x_i}^{-1}$. We have:

Proposition (10.2.17). *There is an exact sequence:*

$$0 \longrightarrow H^1(L^{-1}) \xrightarrow{\ j\ } V \xrightarrow{\ \delta\ } \bigoplus (K_S \otimes L)_{x_i}^{-1} \xrightarrow{\ e^{\mathrm{T}}\ } H^0(K_S \otimes L)^*.$$

Here, of course, j is the inclusion map defined by setting all the r_i to zero and taking Dolbeault representatives of the classes in $H^1(L^{-1})$. The essence of the proposition is the assertion that for any (r_i), a necessary and sufficient condition for the existence of a solution to the equation $\bar{\partial}\beta = \sum r_i \delta_{x_i}$ is that $\sum r_i \sigma(x_i) = 0$ for every holomorphic section σ of $K_S \otimes L$. This is just the Fredholm alternative for the $\bar{\partial}$ operator on distributions (which is easily deduced from the corresponding alternative for smooth forms—given χ in $\Omega^{0,2}(L^{-1})$ we can solve the equation $\bar{\partial}\beta = \chi$ if and only if $\langle \chi, \sigma \rangle = 0$ for every holomorphic section σ of the dual bundle. This amounts to the Serre duality $H^2(L^{-1}) = H^0(K_S \otimes L)^*$.)

The point here is that the Ext group includes extensions where the middle term is a sheaf rather than a bundle; for example, zero corresponds to the direct sum $\mathcal{O} \oplus (L \otimes \mathscr{I})$.

10.2.3 Moduli problems

The construction of Section 10.2.2 yields a powerful tool for analysing rank-two bundles over algebraic surfaces. Given any bundle $\mathscr{F}$ we can twist by a sufficiently positive line bundle $\mathcal{O}(N)$ so that $\mathscr{F} \otimes \mathcal{O}(N)$ has a non-trivial section s. If s vanishes on a divisor D we can twist by $[-D]$ to get a section of $\mathscr{E} = \mathscr{F} \otimes \mathcal{O}(N) \otimes [-D]$ with only isolated zeros, which can then be analysed by the above procedure. The only gap in the discussion of Section 10.2.2 is the case when the section has isolated but non-transverse zeros, and one has to introduce more complicated ideal sheaves. We will not go into this in any detail for lack of space. We will take the naïve attitude that such multiple or 'infinitely near' zeros should behave as natural degenerations of the transverse case, referring the interested reader to other, more systematic, accounts. In brief, for an ideal sheaf $\mathscr{I}$ such that $\mathcal{O}/\mathscr{I} = \bigoplus R_i$ is supported on a finite set $\{x_i\}$ the generalization of the residue data is a collection of elements in the dual spaces $\mathrm{Hom}(R_i, \mathcal{O})$, which generate these spaces as $\mathcal{O}$-modules. In fact our calculations in Section 10.4 will not depend in any important way on the detailed theory of these multiple points.

We will now set down some generalities about the moduli problem, in preparation for the detailed work in Sections 10.3 and 10.4. For a given line bundle L and positive integer p we let $\hat{N} = \hat{N}_{L,p}$ be the moduli space of equivalence classes of triples $(\mathcal{E}, \psi, s)$, where $\mathcal{E}$ is a rank-two bundle with $c_2(E) = p$, ψ is an isomorphism from $\Lambda^2\mathcal{E}$ to L, and s is a holomorphic section of $\mathcal{E}$ with isolated zeros. We let $N = N_{p,L}$ be the quotient of $\hat{N}$ obtained by imposing the equivalence relation $(\mathcal{E}, \psi, s) \sim (\mathcal{E}, \psi, \zeta s)$ for non-zero ζ in $\mathbb{C}^*$. Thus $\hat{N}$ is a $\mathbb{C}^*$-bundle over N, a subset of the associated 'tautological' line bundle $U \to N$ (the total space of U is defined in the same way as $\hat{N}$, but with $s = 0$ allowed). Notice that $(\mathcal{E}, \psi, -s)$ is already equal to $(\mathcal{E}, \psi, s)$ in $\hat{N}$.

Now let M be the moduli space of stable bundles of the given topological type, relative to some hyperplane class, and let $N^+ \subset N$ be the subset corresponding to triples where the bundle $\mathcal{E}$ is stable. There is a natural forgetful map

$$\rho\colon N^+ \longrightarrow M, \tag{10.2.18}$$

which maps onto a subset $M^+ \subset M$. The fibre of ρ over a stable bundle $\mathcal{E}$ is an open subset of the projective space $\mathbb{P}(H^0(\mathcal{E}))$ (representing the sections with isolated zeros). The construction will be most useful in cases where the image of ρ in M and the subset N^+ in N are dense and ρ is generically one-to-one, i.e. the generic bundle $\mathcal{E}$ has $\dim H^0(\mathcal{E}) = 1$. In this case ρ gives a birational isomorphism from N to M.

The space N can be described rather explicitly, using our construction of Section 10.2.2. Suppose, for simplicity, that $H^1(L^{-1}) = 0$. Let $\hat{R} = \hat{R}_{p,L}$ be the set of configurations of p points x_i in S and residue data (r_i), with each r_i non-zero. We should include here multiple points, using the theory alluded to above. There is an obvious map from $\hat{R}$ to the symmetric product $s^p(S)$, with generic fibre a copy of $\mathbb{C}^p$ minus a union of p coordinate hyperplanes. Let $R = R_{p,L}$ be the quotient of R obtained by identifying (r_i) with (χr_i) for χ in $\mathbb{C}^*$, so $\hat{R}$ is a $\mathbb{C}^*$-bundle over R. The space R maps to the symmetric product, with generic fibre a projective space minus a union of hyperplanes. We have a diagram:

$$\begin{array}{ccc} \hat{N} & \xrightarrow{\hat{i}} & \hat{R} \\ \downarrow & & \downarrow \\ N & \xrightarrow{i} & R. \end{array} \tag{10.2.19}$$

Notice that $i(\mathcal{E}, \psi, \zeta s) = (\zeta^2 r_i)$ if $\hat{i}(\mathcal{E}, \psi, s) = (r_i)$. We denote the complex line bundle over R associated to the $\mathbb{C}^*$ action on R also by U^*, so U is a line bundle over R extending the line bundle we defined before over N, and which restricts to the hyperplane bundle on the space of residues (r_i) with fixed points x_i. The image of N in R is the subvariety defined by the constraints $\sum r_i \sigma(x_i) = 0$. More precisely, for any section σ of $K_S \otimes L$ over S there is an associated section r_σ of the line bundle U^2, represented by $(r_i) \to \sum r_i \sigma(x_i)$, and N is the common zero set of the r_σ, as σ runs over the holomorphic sections of $K_S \otimes L$.

Finally we want to consider the construction in a family. Again we suppose for simplicity that $H^1(L^{-1}) = 0$, so the data is determined by the residues. We consider a parameter space $\hat{T}$ and a family of points and residues depending on T. We wish to construct a universal bundle $\hat{\mathbb{E}}$ over $\hat{T} \times S$. We may set this up in any of the different categories of bundles; smooth, continuous or holomorphic depending on the context. The universal case is to take $\hat{T} = \hat{N}$; any other case is induced from this by a map.

First, while the points of V represent equivalence classes of singular forms β, it is easy to fix a preferred representative in the equivalence class. For example we can take the Hodge-theory representative β with $\bar{\partial}^* \beta = 0$, relative to some fixed metric on L. So in our family we have a continuously varying family of forms β_t. For such a form we let $\mathscr{E}_t$ be the holomorphic bundle over S minus the singular set Z_t of β_t, given by the $\bar{\partial}$ operator

$$\bar{\partial}_t = \bar{\partial} + \begin{pmatrix} 0 & \beta_t \\ 0 & 0 \end{pmatrix}$$

on the underlying smooth bundle $\mathcal{O} \oplus L$. The sheaf of holomorphic sections of $\mathscr{E}_t$ defines a canonical extension to S. Let $\hat{Z} \subset \hat{T} \times S$ be the parametrized singular set: $\hat{Z} = \{(t, x) | \, x \in Z_t\}$. Since the extension of $\mathscr{E}_t$ for each t is canonical, it extends to a family and we get a universal bundle $\hat{\mathbb{E}}$ over $\hat{T} \times S$, with an isomorphism $\mathbf{\Psi}$: $\Lambda^2 \hat{\mathbb{E}} \to \pi_2^*(L)$ and a section $\mathbf{s}$ of $\hat{\mathbb{E}}$ vanishing on $\hat{Z}$.

We now go on to consider the $\mathbb{C}^*$ action. Suppose $\hat{T}$ is preserved by the $\mathbb{C}^*$ action on the residues, so we have a quotient space T (the universal example being the quotient N of $\hat{N}$). Consider the situation over a single $\mathbb{C}^*$ orbit: the automorphism $\begin{pmatrix} 1 & 0 \\ 0 & \lambda \end{pmatrix}$ of $\mathcal{O} \oplus L$ over $S \backslash Z_t$ intertwines $\bar{\partial}_\beta$ and $\bar{\partial}_{\lambda\beta}$, and this gives a lift of the $\mathbb{C}^*$ action to $\hat{\mathbb{E}}$, which preserves the section $\mathbf{s}$ but not the isomorphism $\mathbf{\Psi}$, which is transformed to $\lambda \mathbf{\Psi}$. In the universal case we get a bundle $\mathbb{E}$ over $N \times S$, with a section s vanishing on the quotient Z of $\hat{Z}$, and

$$\Lambda^2 \mathbb{E} = \pi_1^*(L) \otimes \pi_2^*(U). \tag{10.2.20}$$

10.2.4 Digression: monads and spectra

It is instructive to relate the theory developed above to the ideas used in Chapter 3. Recall from Section 3.3.4 that we can introduce the notion of a bundle on $\mathbb{C}^2$ which is trivialized at infinity; for example by taking bundles on the projective plane trivialized over the line at infinity. In Section 3.3.5 we showed that any such bundle could be represented by a 'monad' and, as we have mentioned in Section 3.4.2, for rank-two bundles this monad can be taken to be symmetric. Thus we have matrix data comprising $k \times k$ symmetric matrices τ_1, τ_2 and a $k \times 2$ matrix σ satisfying the condition

$$[\tau_1, \tau_2] = \sigma \varepsilon \sigma^{\mathrm{T}}. \tag{10.2.21}$$

where $\varepsilon = \begin{pmatrix} 0 & 1 \\ -1 & 0 \end{pmatrix}$.

We also have the non-degeneracy condition, that for all (z_1, z_2) in $\mathbb{C}^2$ the linear map

$$\alpha_{z_1, z_2} = \begin{pmatrix} \tau_1 - z_1 \\ \tau_2 - z_2 \\ \varepsilon \sigma^{\mathrm{T}} \end{pmatrix} : \mathbb{C}^k \longrightarrow \mathbb{C}^k \oplus \mathbb{C}^k \oplus \mathbb{C}^2$$

should be injective. We then obtain a bundle $\mathscr{E}$ over $\mathbb{C}^2$,

$$\mathscr{E}_{(z_1, z_2)} = \ker \alpha^{\mathrm{T}}_{z_1, z_2} / \operatorname{im} \alpha_{z_1, z_2},$$

where the transpose is taken with respect to the natural skew form on $\mathbb{C}^k \oplus \mathbb{C}^k \oplus \mathbb{C}^2$. (This description has a holomorphic extension to $\mathbb{CP}^2$, as we saw in the proof of (6.3.13).)

In this section we want to discuss a particularly simple class of solutions to the monad equations, when the matrix σ has rank one. We can then choose bases so that:

$$\sigma = \begin{pmatrix} \sigma_1 & 0 \\ \vdots & \vdots \\ \sigma_k & 0 \end{pmatrix}. \tag{10.2.22}$$

The product $\sigma \varepsilon \sigma^{\mathrm{T}}$ is zero, so our equation (10.2.21) comes down to the requirement that τ_1 and τ_2 *commute*.

Generic commuting, symmetric matrices are conjugate by orthogonal transformations to the obvious examples:

$$\tau_1 = \operatorname{diag}(\lambda_1, \ldots, \lambda_k), \quad \tau_2 = \operatorname{diag}(\mu_1, \ldots, \mu_k); \tag{10.2.23}$$

so let us consider a bundle $\mathscr{E}$ defined by a matrix data of this form, depending on parameters $(\lambda_1, \ldots, \lambda_k)$, $(\mu_1, \ldots, \mu_k)$, $(\sigma_1, \ldots, \sigma_k)$. One can interpret the pairs (λ_i, μ_i) as points p_i in $\mathbb{C}^2$. It is easy to check that the non-degeneracy conditions for these monads come to:

(i) the points p_i are distinct, and

(ii) the σ_i are all non-zero.

We see then that we have precisely the same data, non-zero complex numbers attached to distinct points in the base space, that we considered in our construction in Section 10.2.2. It is thus very plausible that the two constructions correspond, and we will now verify that this is indeed the case.

Let $\mathscr{E}$ be the bundle over $\mathbb{CP}^2$ constructed from the monad data as above. There is an obvious holomorphic section s of $\mathscr{E}$. Under the skew form on $\mathbb{C}^{2k+2}$ the first basis vector, e say, in $\mathbb{C}^2 \subset \mathbb{C}^{2k+2}$, annihilates the image of α_{z_1,z_2} for all (z_1, z_2). So e lies in the kernel of $\alpha^{\mathrm{T}}_{z_1,z_2}$ for all z_1, z_2 and, projecting e to $\mathscr{E}$, we define the section s. This section vanishes at points (z_1, z_2) for which e lies in the image of α_{z_1,z_2}, say $\alpha_{z_1,z_2}(v) = e$, where $v = \begin{pmatrix} v_1 \\ v_2 \\ \vdots \\ v_k \end{pmatrix}$. We require then that $\sum v_i\sigma_i = 1$, and $(\tau_1 - z_1)v = (\tau_2 - z_2)v = 0$. This occurs precisely when (z_1, z_2) is one of the pairs (λ_i, μ_i), say (λ_1, μ_1), and we can take $v_1 = \theta_1^{-1}$, $v_i = 0$ for $i > 1$. So the eigenvalue pairs (λ_i, μ_i) do indeed give the zeros of s. It remains to identify the residue terms. For this we can suppose that one of the zeros, say (λ_1, μ_1), is the origin (0, 0). Recall that the residue term

$$r_i \in \Lambda^2 T\mathbb{C}^2$$

is defined by the determinant of the derivative,

$$(Ds)_0 \colon T\mathbb{C}^2 \longrightarrow \mathscr{E}_0,$$

using the isomorphism $\Lambda^2\mathscr{E}_0 = \mathbb{C}$. This isomorphism in turn is induced by the skew form on $\mathbb{C}^{2k+2}$. We will now display these explicitly in terms of our matrices.

Let e_1', e_2' be the first basis elements of the two copies of $\mathbb{C}^k$ in $\mathbb{C}^{k+2}$, so the skew pairing $\langle e_1', e_2' \rangle = 1$. These are annihilated by α^{T} and their images give a basis for the fibre $\mathscr{E}_0$. For non-zero z_1, z_2, e_1' and e_2' need not lie in the kernel of $\alpha^{\mathrm{T}}_{z_1,z_2}$, but we can modify them by terms $\varepsilon_1(z), \varepsilon_2(z)$, vanishing at the origin, so that $s_1 = e_1' + \varepsilon_1$, $s_2 = e_2' + \varepsilon_2$ define a local trivialization of the bundle $\mathscr{E}$ near 0. Now let w be the column vector:

$$w = \begin{pmatrix} \theta_1^{-1} \\ 0 \\ \vdots \\ 0 \end{pmatrix}$$

Then $e - \alpha_{z_1,z_2}(w) = \theta_1^{-1}(z_1 e_1' + z_2 e_2')$, so that for the sections of the bundle $\mathscr{E}$ we have:

$$s = \theta_1^{-1}(\lambda s_1 + \mu s_2) + Q(\lambda, \mu)$$

where Q is of order two in λ, μ. It follows that, with respect to the basis e'_1, e'_2, the derivative of s at 0 is represented by the θ_1^{-1} times the identity matrix, and hence the residue term $r_1 = \det(Ds)^{-1}$ is θ_1^2. (We use here the fact that the basis e'_1, e'_2 is normalized with respect to the symplectic form.) This calculation verifies that the two constructions do agree and we sum up the conclusions in the next proposition.

Proposition (10.2.24). *If* τ_1, τ_2, σ *are matrices of the forms given in* (10.2.22) *and* (10.2.23) *with* θ_i *not equal to zero and the points* $p_i = (\lambda_i, \mu_i)$ *in* $\mathbb{C}^2$ *all distinct, the* $SL(2, \mathbb{C})$ *bundle* $\mathscr{E}$ *over* $\mathbb{CP}^2$ *defined by the corresponding maps* α_{z_1, z_2} *has a holomorphic section s vanishing precisely at the points* p_i *in* $\mathbb{C}^2 \subset \mathbb{CP}^2$ *and with* $\det(Ds)_{p_i}^{-1} = \theta_i^2$.

Note that the square of the θ_i are the natural parameters, since the action of the diagonal group $\{\mathrm{diag}(\pm 1, \ldots, \pm 1)\}$ shows that changing the sign of any θ_i does not affect the cohomology bundle.

One can see the monad construction in this special case as a form of spectral construction. This generalizes to arbitrary 'regular' ideal sheaves $\mathscr{I}_Z$ in $\mathcal{O}_{\mathbb{C}^2}$, i.e. we allow infinitely near points (which correspond to matrices which are not diagonalizable). In one direction we start with $\mathscr{I}_Z$ and the corresponding quotient ring

$$A = \mathcal{O}_Z = \mathcal{O}_{\mathbb{C}^2}/\mathscr{I}_Z.$$

As a vector space, A is k-dimensional; on the other hand it is generated as a ring by the coordinate functions z_1, z_2 on $\mathbb{C}^2$. We obtain vector space endomorphisms τ_1, τ_2 of A as the action of multiplication by z_1 and z_2. One can verify that these correspond to the monad matrices one obtains via the vector bundles. (The intrinsic symmetric form on A is defined by the 'residue pairing'; see Griffiths and Harris, (1978, Chapter 5).) In the other direction, starting from a pair of commuting $k \times k$ matrices (τ_1, τ_2)—endomorphisms of $\mathbb{C}^k$—we form the joint spectrum Z of the pair in $\mathbb{C}^2$, i.e. the spectrum of the ring A generated by the matrices. This gives the set of points in $\mathbb{C}^2$.

The other observation we should make is that the non-degeneracy condition for the monad is equivalent to the condition that the vector $\chi = (\sigma_1, \ldots, \sigma_k)$ be a generator for $\mathbb{C}^k$ as a module over A. This gives us a neat algebraic description of the spaces $\hat{R}$ and R in the case when the base space is $\mathbb{C}^2$: $\hat{R}$ is the quotient by $O(k, \mathbb{C})$ of the set of triples (τ_1, τ_2, χ) where τ_i commute and χ generates $\mathbb{C}^k$ over $\mathbb{C}[\tau_1, \tau_2]$, and R is the corresponding projective quotient. Note that these points are stable for the $O(k, \mathbb{C})$ action.

10.3 Moduli spaces of bundles over a double plane

In this section we will apply the techniques of Section 10.2 to describe moduli spaces of bundles over a particular algebraic surface S, the double cover of $\mathbb{CP}^2$ branched over a smooth curve B of degree eight. This surface was denoted by R_4 in Section 1.1.7 where we saw that S is simply connected and

$$b^+(S) = 7, \quad b^-(S) = 37. \tag{10.3.1}$$

Moreover the intersection form is odd, so S has the homotopy (and homeomorphism) type of:

$$X(S) = 7\mathbb{CP}^2 \# 37\overline{\mathbb{CP}}^2.$$

The surface S is of general type, and is a prototype for this large class of complex surfaces.

10.3.1 General properties

We begin by marshalling some simple algebro-geometric facts about S. We denote the branched covering map by $\pi: S \to \mathbb{CP}^2$, and identify the branch curve B with its preimage in S. The canonical bundle K_S is isomorphic to the lift $\pi^*(\mathcal{O}(1))$ of the Hopf bundle. It is an ample line bundle over S (i.e. the sections of a positive power define a projective embedding of S). Thus we can choose our Kähler class ω equal to $\pi^*(h)$, where h is the standard generator of $H^2(\mathbb{CP}^2)$. We know that $b^+(S)$ is $1 + 2p_g$ (where $p_g = \dim H^0(K_S)$), so $p_g = 3$. We can see explicitly these three sections of K_S—they are the lifts of the sections of $\mathcal{O}(1)$ over $\mathbb{CP}^2$. Thus our general formulae tell us that π^* induces an isomorphism between $H^0(\mathbb{CP}^2; \mathcal{O}(1))$ and $H^0(S; \pi^*(\mathcal{O}(1)))$. The same is true for the low powers of the canonical bundle; we have:

Lemma (10.3.2). *For $s = 1, 2, 3$, π^* induces an isomorphism between $H^0(\mathbb{CP}^2; \mathcal{O}(s))$ and $H^0(S; \pi^*(\mathcal{O}(s)))$. The higher cohomology groups $H^1(\pi^*(\mathcal{O}(s)))$ vanish and $H^2(\pi^*(\mathcal{O}(s)))$ vanishes for $s = 2, 3$.*

Proof. By Serre duality $H^2(\pi^*(\mathcal{O}(s)))$ is dual to $H^0(\pi^*(\mathcal{O}(-s))) \otimes K_S)$ $= H^0(\pi^*(\mathcal{O}(1-s)))$ and this certainly vanishes for all $s > 1$ since the bundle has negative degree. Now consider the involution σ^* mapping $H^0(\pi^*(\mathcal{O}(s))$ to itself, induced by the covering map $\sigma: S \to S$. This decomposes the space of sections into the direct sum of $+1$ and -1 eigenspaces. The $+1$ eigenspace corresponds to the sections lifted up from $\mathbb{CP}^2$. But a section f of $\pi^*(\mathcal{O}(s))$ in the -1 eigenspace of σ^* must vanish on the fixed point set B of σ. Now the divisor B on S represents the line bundle $\pi^*(\mathcal{O}(4))$. If g is a defining function for B in $\mathbb{CP}^2$, of degree 8, the expression $g^{1/2}$ represents a section of $\pi^*(\mathcal{O}(4))$ vanishing to order one on B in S. So $f/g^{1/2}$ represents a section of $\pi^*(\mathcal{O}(s-4))$ and if $s < 4$ this must be identically zero. This proves that all the sections of $\mathcal{O}(s)$ are lifted up from $\mathbb{CP}^2$ when $s = 1, 2, 3$. (Conversely, when $s = 4$ we see that the -1 eigenspace is one-dimensional, generated by $g^{1/2}$.) Lastly to get the vanishing of the H^1s, we use the Riemann–Roch formula for a line bundle $\mathcal{L}$ over S, which reads:

$$\chi(\mathcal{L}) = \tfrac{1}{2}c_1(\mathcal{L})(c_1(\mathcal{L}) - c_1(K_S)) + 4;$$

so

$$\chi(K_S^s) = s(s-1) + 4.$$

We see that $\chi(K_S) = 4$, $\chi(K_S^2) = 6$ and $\chi(K_S^3) = 10$, so the H^1s vanish since the space of sections of $\mathcal{O}(1)$, $\mathcal{O}(2)$, $\mathcal{O}(3)$ on $\mathbb{CP}^2$ have dimension 3, 6, 10 respectively.

In fact, it is true in general that $H^1(K_S^s)$ is zero for any minimal surface S of general type and $s \neq 0, 1$. For $s = 0, 1$ the dimensions are both equal to half the first Betti number, by the Hodge decomposition.

Another fact that will be useful to us is:

Proposition (10.3.3). *For a generic choice of branch curve B the Picard group of isomorphism classes of line bundles over S is generated by K_S.*

This is a variant of the Noether theorem (for hypersurfaces in $\mathbb{CP}^3$). It can be proved by analysing the periods of the holomorphic forms on S and their dependence on the branch curve, or using the fact that S has a large monodromy group. We will just refer to Griffiths and Harris (1985), and Friedman, Moishezon and Morgan (1987) here. From now on we assume that B is a generic curve, as in (10.3.3).

10.3.2 Conics and configurations of points

In this subsection we consider stable rank-two bundles over S with $c_1(\mathcal{E}) = K_S$, and $c_2(\mathcal{E}) = p$, for $1 \leq p \leq 7$. These will go over to ASD $SO(3)$ connections, with $w_2 = K_S(\text{mod } 2)$ and $\kappa = p - \frac{1}{2}$. To simplify our notation we will just write $M_{p,\alpha}$ for the moduli space of such bundles. The virtual dimension of the moduli space $M_{p,\alpha}$ is

$$\text{virtual dim}_{\mathbb{C}} M_{p,\alpha} = 4p - 14, \tag{10.3.4}$$

so we have invariants defined once $p \geq 4$.

Proposition (10.3.5). *Any stable bundle $\mathcal{E}$ with $c_1(\mathcal{E}) = K_S$ and $c_2(\mathcal{E}) \leq 7$ has a holomorphic section vanishing at isolated points in S.*

Proof. On the one hand if $\mathcal{E}$ is such a bundle the Riemann–Roch formula gives $\chi(\mathcal{E}) = 8 - c_2(E)$; so if $c_2 \leq 7$ either $H^0(\mathcal{E})$ or $H^2(\mathcal{E})$ is non-zero. We now use the stability condition: the dual of $H^2(\mathcal{E})$ is $H^0(\mathcal{E}^* \otimes K_S)$, the bundle maps from $\mathcal{E}$ to K_s, but if $\mathcal{E}$ is stable this must be zero since $\frac{1}{2}\deg(\mathcal{E}) = \frac{1}{2}\deg K_S < \deg K_S$. So we conclude that if $c_2 \leq 7$ our bundle has a non-trivial section. On the other hand this section cannot vanish on a divisor D. For by (10.3.3), $[D] = K_S^d$ for some $d \geq 1$, and $\mathcal{E} \otimes [-D]$ has a non-trivial section; but this again contradicts stability, since $\deg[D] > \frac{1}{2}\deg \mathcal{E}$.

Any stable bundle $\mathcal{E}$ of the kind considered in (10.3.5) can thus be written as an extension:

$$0 \longrightarrow \mathcal{O}_S \longrightarrow \mathcal{E} \longrightarrow K_S \otimes \mathcal{I} \longrightarrow 0, \tag{10.3.6}$$

where $\mathcal{I}$ is an ideal sheaf, with $\mathcal{O}_S/\mathcal{I}$ supported on a finite set. Conversely we have:

Lemma (10.3.7). *Any bundle $\mathscr{E}$ over S constructed from an extension of the form* (10.3.6) *is stable.*

The proof is easy; since the only line bundles over S are the powers of the canonical bundle it suffices to check that $H^0(\mathscr{E} \otimes K_S^{-1})$ is zero, and this follows from the exact cohomology sequence of (10.3.6).

To sum up, in notation like that of Section 10.2.3, but now with $N_{p,\alpha}$ denoting the moduli space of triples $(\mathscr{E}, \psi, s)$ with $c_2(\mathscr{E}) = p \leq 7$, divided by scalars, we see that the subset $N_{p,\alpha}^+$ of $N_{p,\alpha}$ is in this case equal to $N_{p,\alpha}$, and the resulting forgetful map

$$\rho_p \colon N_{p,\alpha} \longrightarrow M_{p,\alpha}$$

is surjective.

We now use our general theory to analyse extensions of the form (10.3.6) in the case when the ideal sheaf $\mathscr{I}$ is defined by p points of multiplicity one. First, $H^1(K_S^{-1})$ is dual to $H^1(K_S^2)$, and hence is zero by (10.3.2). Thus the space V which parametrizes extensions is a subset of the direct sum $\oplus (K_S^{-2})_{x_i}$, the kernel of the map:

$$e^{\mathrm{T}} \colon \oplus (K_S^{-2})_{x_i} \longrightarrow H^0(K_S^2)^*,$$

or equivalently the annihilator of the image of the evaluation map e. Second, the obstruction space $H^0(K_S^2)$ can be identified, by (10.3.2), with the six-dimensional space of sections of $\mathcal{O}(2)$ over $\mathbb{CP}^2$, i.e. with the polynomials of degree two, cutting out conic curves in the plane. The space V is non-zero precisely when e is not surjective, i.e. when there is a family of curves π^{-1} (conic) of dimension at least $6 - p$ through the p points x_i. To construct bundles we restrict attention to the subset of V where the 'weights' r_i in the residue data are non-zero. The condition imposed on the points for the existence of such elements of V can be neatly summed up by saying that

> *any curve of the form π^{-1} (conic) which passes through $p - 1$ of the points x_i should also pass through the remaining point.*

10.3.3 Two remarks

Before getting on with the analysis of the various configurations of points which may occur, for different values of p, we begin with two observations. First, we have seen that the obstructions $H^0(K_S^2)$ are all lifted up from $\mathbb{CP}^2$. This means that given any solution to the constraints, based on p points x_i in S, we can obtain another solution based on $p + 2$ points (in fact a one-parameter family of solutions) by taking the extra points to be a pair $\pi^{-1}(y)$ and opposite weights λ, $-\lambda$ at these points. Similarly, if y is in $\mathbb{CP}^2$, and not in the branch curve B, and if there is just one of the original set of points x_i in the fibre $\pi^{-1}(y)$, we can construct solutions with $p + 1$ points by adding the other point of $\pi^{-1}(y)$ and 'redistributing' the original weight at x_1 over this pair of points.

Our second observation concerns the non-degeneracy condition, that the weights r_i should be non-zero. Suppose we have an element $(r_i)_{i=1}^p$ in the kernel of

$$e^{\mathrm{T}}: \bigoplus_{i=1}^{p} (K_S^{-2})_{x_i} \longrightarrow H^0(K_S^2)^*$$

but some of the r_i are zero, say $r_{q+1}, \ldots, r_p$, for $1 \leq q < p$. Then we can remove the points $x_{q+1}, \ldots, x_p$ from our configuration and obtain a new solution based on a configuration of q points, which does satisfy the non-degeneracy condition, and hence yields a bundle with $c_2 = q$. So a good way to proceed in analysing the moduli spaces is to find all solutions of the closed constraints $e^{\mathrm{T}}(r_i) = 0$, for different configurations of points in S, and then stratify them according to the Chern class of the bundle constructed, i.e. according to the number of vanishing weights. More formally we can construct a completion of our moduli spaces which, as we will see in Section 10.4, is closely related to our compactification of ASD moduli spaces. We sketch how this may be done, reverting to the general case of a bundles with $\Lambda^2 = L$ for clarity. We first introduce a space $\hat{\bar{R}}_p$, whose points consist of a regular ideal $\mathscr{I}$ of multiplicity p on S together with certain extra data. If $\mathscr{I}$ is the ideal of a set of p distinct points then the extra data is a non-zero element $\bigoplus (K_S \otimes L)_{x_i}^{-1}$. In general, for multiple points, we can use the geometric invariant theory quotient of the semi-stable set in the linear algebra description of Section 10.2.4. We let $\bar{R}_p$ be the quotient of $\hat{\bar{R}}$ by the $\mathbb{C}^*$ action, so there is a map from $\bar{R}_p$ to $s^p(S)$ with generic fibre a copy of $\mathbb{CP}^{p-1}$. Now $\bar{R}_p$ is stratified according to the number of zero weights into a union of pieces

$$\bar{R}_p = R_p \cup S \times R_{p-1} \cup s^2(S) \times R_{p-2} \cdots \tag{10.3.8}$$

The line bundle U over R_p extends to $\bar{R}_p$, as do the sections r_σ of U we associated with holomorphic sections σ of $K_S \otimes L$ over S. The common zero set of the r_σ is a compact subvariety $\bar{N}_p$ of $\bar{R}_p$, which inherits a stratification

$$\bar{N}_p = N_p \cup S \times N_{p-1} \cup s^2(S) \times N_{p-2} \cdots . \tag{10.3.9}$$

With these observations in mind we can say that the core of the problem of describing the spaces N_p is to find configurations of points y_i ($i = 1, \ldots, p$) in $\mathbb{CP}^2$ which lie on a $(6 - p)$-dimensional family of conics. Once we have these configurations we can go back to find the solutions $\{r_1, \ldots, r_p\}$, with r_i non-zero, based on configurations $\{x_1, \ldots, x_p\} \subset \pi^{-1}(\{y_1, \ldots, y_p\})$. (And for a full description we would also, of course, have to take account of multiple points.)

10.3.4 Description of moduli spaces

For $p = 1$, there is clearly no solution to the constraints. For $p = 2$ all solutions are obtained from the trivial solution by the mechanism in our first observation, i.e. we take two points x_1, x_2 with $\pi(x_1) = \pi(x_2) = y$ and residues r, $-r$. There is a two-dimensional space of sections of K_S vanishing at the two points, hence the exact sequence (10.3.6) gives that the resulting bundle, $\mathscr{E}_y$ say, has a three-dimensional space of sections. It follows easily that all the $\mathscr{E}_y$ are isomorphic, so $N_{2,\alpha}$ is a copy of $\mathbb{CP}^2$ and $M_{2,\alpha}$ is a point. In fact this bundle is the lift of the tangent bundle of $\mathbb{CP}^2$ to S, tensored by K_S^{-1}. However, in contrast to the case of the $K3$ surface discussed in Chapter 9, the virtual dimension of the moduli space $M_{2,\alpha}$ is now negative, and it does not define any invariants. For $p = 3$ one finds again that the moduli space is empty. For $p = 4$ we do have solutions: if four points in $\mathbb{CP}^2$ lie on a two-dimensional family of conics they must lie on a line, so our zeros x_i in S are constrained to lie on a curve π^{-1} (line). Conversely for any such set of four points we can construct a bundle; indeed, four points on a line L in $\mathbb{CP}^2$ lie on a two-parameter family of reducible conics $L + L'$. We have then a moduli space N_4 of complex dimension $2 + 4 = 6$ (two parameters for the line in the plane and four for a choice of four points on the line). The fibre of the natural map $\rho_4 \colon N_{4,\alpha} \to M_{4,\alpha}$ over a bundle $\mathscr{E}$ is identified with the projective space $\mathbb{P}(H^0(\mathscr{E}))$ and, referring again to the exact sequence of (10.3.6), the dimension of the projective space is the number of independent lines through any defining configuration of points y_i. In this case there is just one line through the four points, so the moduli space $M_{4,\alpha}$ has complex dimension five, and in fact it fibres over the space of lines in $\mathbb{CP}^2$.

Moving on now to the case $p = 5$, one again finds that the points must lie on a line and the kernel of the restriction map is now two-dimensional, so $N_{5,\alpha}$ has dimension $2 + 5 + 1 = 8$, and is a $\mathbb{CP}^1$ bundle over $M_{5,\alpha}$, which itself fibres over the $\mathbb{CP}^2$ of lines. When $p = 6$ we need six points in $\mathbb{CP}^2$ lying on a conic; generically they will lie on a single conic and not on any line. So $\rho_6 \colon N_{6,\alpha} \to M_{6,\alpha}$ is generically one-to-one and $M_{6,\alpha}$ is birationally equivalent to a 2^6-fold cover of the space of configurations of six points on a conic in $\mathbb{CP}^2$. Thus M_6 has dimension $5 + 6 = 11$ (five for the conic and six for the points on it). The cover enters here of course because for a general set of six points y_i in $\mathbb{CP}^2$ we have 2^6 choices of points x_i in S.

The moduli space we shall use for the purposes of calculating invariants is $M_{7,\alpha}$. When $p = 7$, any general set of points can be used, since the evaluation map e from $\mathbb{C}^6$ to $\mathbb{C}^7$ cannot be surjective. The moduli space $N_{7,\alpha}$ is birationally equivalent to the seven-fold symmetric product $s^7(S)$ and, in turn, the map $\rho_7 \colon N_{7,\alpha} \to M_{7,\alpha}$ is again generically one-to-one, and a birational isomorphism. So both $N_{7,\alpha}$ and $M_{7,\alpha}$ have dimension 14.

The conclusions of the above discussion can be summarized by Table 1 in which we list the complex dimensions of $M_{p,\alpha}$ and $N_{p,\alpha}$ alongside the 'virtual' complex dimension $4p - 14$ of $M_{p,\alpha}$.

Table 1 illustrates the general pattern asserted by Proposition (10.1.11); for small values of p the dimension of the moduli space exceeds its virtual dimension, so the general positivity argument given in Section 10.1 does not

Table 1

p	Configurations	dim $N_{p,\alpha}$	dim $M_{p,\alpha}$	Virtual dim $M_{p,\alpha}$
1	None	–	–	−10
2	Pairs in a fibre	2	0	−6
3	None	–	–	−2
4	Four points on line	6	6	2
5	Five points on line	8	7	6
6	Six points on conic	11	11	10
7	Seven general points	14	14	14

apply, but when $p = 7$ the dimensions agree. (It is plausible that the dimensions agree for all $p \geq 7$ but this would require a more extensive analysis.) Actually the moduli space $M_{7,\alpha}$ does not exhibit all the properties of the 'large p' range given by (10.1.11) since, as we shall see in Section 10.3.6, the cohomology spaces $H^2(\mathrm{End}_0\,\mathscr{E})$ are not zero at the generic point, and the solutions of the ASD equations are not regular. However, we shall nevertheless be able in Section 10.4 to calculate some invariants of S using this description of the moduli space.

10.3.5 Bundles with $c_1 = 0$

We now turn to our second example, bundles over S with $c_1 = 0$, which will correspond to $SU(2)$ connections. We begin again by considering the Riemann–Roch formula for such a bundle $\mathscr{F}$:

$$\chi(\mathscr{F}) = 8 - c_2(\mathscr{F}).$$

If $c_2(\mathscr{F}) = k$ is less than eight, we deduce that either $\mathscr{F}$ or $\mathscr{F} \otimes K_S$ has a non-trivial section, and if the bundle is stable the first possibility is ruled out. Moreover, for the same reasons as before, the section of $\mathscr{F} \otimes K_S$ must vanish at isolated points. Writing $\mathscr{E} = \mathscr{F} \otimes K_S$, we have to consider extensions

$$0 \longrightarrow \mathcal{O} \longrightarrow \mathscr{E} \longrightarrow K_S^2 \otimes \mathscr{I} \longrightarrow 0 \tag{10.3.10}$$

where $\mathscr{I}$ is an ideal sheaf associated to a configuration of $p = c_2(\mathscr{E} \otimes K_S) = c_2(\mathscr{E}) + c_1(K_S)^2 = k + 2$ points in S. For the purposes of defining invariants we require $k \geq 7$, so we are restricted in this approach to the single case $k = 7$, but it is still interesting to look at the lower moduli spaces of bundles obtained in this way. The obstructions to constructing extensions (10.3.10) are given by the holomorphic sections of K_S^3 over S. By (10.3.2) these are all lifted up from sections of $\mathcal{O}(3)$ over $\mathbb{CP}^2$, so we have to carry out much the same analysis as before with cubic curves in place of conics; that is, we consider configurations of p points in the plane which lie on a $(10 - p)$-

dimensional family of cubics. (Note that there is a ten-dimensional space of cubic polynomials in three variables.)

This example is more complicated than that considered above in one respect: the bundles constructed from extensions (10.3.10) are not guaranteed to be stable, unlike those of the form (10.3.6). Arguing as in (10.3.7) we see that the bundle is stable if and only if the configuration of points does not lie on a line. In sum, we now have surjective maps, for $p \leq 9$

$$\rho_{p-2}: N^{+}_{p-2} \longrightarrow M_{p-2}, \tag{10.3.11}$$

but N^{+}_{p-2} will in general be a proper subset of N_{p-2}. (The slightly cumbersome indexing of the moduli spaces is used to fit in with our practice in the rest of the book.) We will return to discuss the extension of ρ_{p-2} to N_{p-2} in Section 10.4.

We will now run through the various possibilities for different values of p, much as in Section 10.3.4. When p is 1 or 3 there are no solutions of the constraining equations, so the moduli spaces are empty. When $p = 2$ or 4 we can construct bundles using the 'doubling' construction observed above, but these are not stable since the points lie on a line in $\mathbb{CP}^2$. When $p = 5$ the moduli space N_3 is obtained from configurations of five points y_i on a line $\mathbb{CP}^2$, hence has dimension seven. These do not, however, give stable bundles; N_3^+ is empty. For $p = 6$ the moduli space N_4 has two components (in the algebro-geometric sense). We can take six general points on a line to get a component of dimension nine (one parameter in the choice of residues satisfying the constraints); these bundles are not stable. On the other hand we can apply the doubling construction, taking the six points to be three pairs of preimages. This gives us an eight-dimensional component, in which the stable part N_4^+ is open and dense, and the generic fibre of ρ has dimension three. When $p = 7$ we again get points on a line, and hence no stable bundles.

When $p = 8$ we can satisfy the constraints by a set of eight points on a conic in the plane. We get a component of the moduli space N_6 of dimension $5 + 8 = 13$ (five parameters for the conic and eight for the points on the conic). The generic bundle $\mathscr{E}$ obtained in this way is stable and has two independent sections, so we get a 12-dimensional dense open set in the moduli space M_6 (this contains in its closure the configurations obtained by the doubling construction, taking four pairs of preimages in S).

The moduli space we are really interested in is M_7, when $p = 9$. Then we want nine points in $\mathbb{CP}^2$ which lie on two independent cubic curves. Such configurations are easily found: any eight general points in $\mathbb{CP}^2$ lie on two cubics, C_1, C_2 say; but C_1 and C_2 intersect in nine points—the eight given ones plus one extra. A configuration of nine points found in this way satisfies the constraints, so we have a 16-dimensional moduli space N_7 with local coordinates on an open set given by the choice of eight arbitrary points in $\mathbb{CP}^2$.

This discussion can be summarized by Table 2.

Table 2

k	$p = k + 2$	Configuration	dim N_k	dim M_k	Virtual Dim M_k
-1	1	None	–	–	-16
0	2	Pairs in pre-image	2	Empty	-12
1	3	None	–	–	-8
2	4	Pairs in pre-image	4	Empty	4
3	5	Five points on line	7	Empty	0
4	6	Six points on line,	9		
		Three pairs in preimage	5	4	4
5	7	Seven points on line	11	Empty	8
6	8	Eight points on conic	13	12	12
7	9	Intersection of cubics	16	16	16

10.3.6 Deformations and multiplicities

We will now discuss the local structure of the moduli space $M = M_7$ constructed in the first example, Section 10.3.4. Let $\mathscr{E}$ be a bundle constructed from an extension (10.3.6) based on seven general points in S. We begin by calculating the cohomology group $H^2(\mathrm{End}_0\,\mathscr{E})$ or rather its Serre dual $H^0(\mathrm{End}_0\,\mathscr{E} \otimes K_S)$. Note that $\mathscr{E}^*$ is isomorphic to $\mathscr{E} \otimes K_S^{-1}$, so $\mathrm{End}_0\,\mathscr{E} \otimes K_S$ can be identified with the kernel of the wedge product map:

$$w\colon \mathscr{E} \otimes \mathscr{E} \longrightarrow \Lambda^2\mathscr{E} = K_S. \tag{10.3.12}$$

(Recall that End_0 denotes the trace-free endomorphisms.) Now (10.3.6) induces maps:

$$\begin{array}{ccccccccc} \mathbb{C} = H^0(\mathscr{E}) & \longrightarrow & H^0(\mathscr{E} \otimes \mathscr{E}) & \longrightarrow & H^0(\mathscr{E} \otimes K_S \otimes \mathscr{I}) & \longrightarrow & H^1(\mathscr{E}) & \longrightarrow & \dots \\ & & {\scriptstyle w}\big\downarrow & & \big\uparrow{\scriptstyle s \otimes 1_{K_S}} & & & & \\ & & H^0(K_S) & & H^0(K_S) & & & & \end{array} \tag{10.3.13}$$

It is easy to see that the image of $s \otimes 1_{K_S}$ in $H^0(\mathscr{E} \otimes K_S \otimes \mathscr{I})$ lifts to $H^0(\mathscr{E} \otimes \mathscr{E})$ and maps by w isomorphically back to $H^0(K_S)$. So we conclude that $H^0(\mathrm{End}_0\,\mathscr{E})$ is isomorphic to $\mathbb{C}$, generated by the section $s \otimes s$ of $\mathscr{E} \otimes \mathscr{E}$. Dually, we have an isomorphism:

$$v\colon H^2(\mathrm{End}_0\,\mathscr{E}) \longrightarrow \mathbb{C},$$

induced by the map $A \to As \wedge s$ from $\mathrm{End}_0\,\mathscr{E}$ to K_S, followed by the integration over the fundamental class: $H^2(K_S) \to \mathbb{C}$.

We deduce then that the Zariski tangent space to the moduli space, $H^1(\mathrm{End}\,\mathscr{E})$, must have dimension 15, one more than the actual dimension of the moduli space. That is, one direction in the Zariski tangent space of infinitesimal deformations is obstructed. We can see this more clearly by returning to the exact sequences induced by (10.3.6). First, by considering the long exact cohomology sequence of

$$0 \longrightarrow K_S^{-1} \longrightarrow \mathscr{E} \otimes K_S^{-1} \longrightarrow \mathscr{I}_Z \longrightarrow 0, \qquad (10.3.14)$$

one finds that $H^1(\mathscr{E} \otimes K_S^{-1})$ is zero. Then, taking the tensor product with $\mathscr{E}$, and the long exact sequence, one gets:

$$0 \longrightarrow H^1(\mathscr{E} \otimes \mathscr{E}^*) \longrightarrow H^1(\mathscr{E} \otimes \mathscr{I}_Z) \longrightarrow 0. \qquad (10.3.15)$$

Finally, we have an exact sequence:

$$0 \longrightarrow H^0(\mathscr{E} \otimes \mathcal{O}_Z) \longrightarrow H^1(\mathscr{E} \otimes \mathscr{I}_Z) \longrightarrow H^1(\mathscr{E}) \longrightarrow 0. \qquad (10.3.16)$$

Here, the first term is the direct sum of the fibres of $\mathscr{E}$ over the points $\{x_i\}$, which can be identified, using the derivative of the section s, with the direct sum of the tangent spaces to S at these points. This is just the part of the Zariski tangent space corresponding to the actual deformations of the bundle—moving the points x_i in S. The other term, $H^1(\mathscr{E})$, makes up the one-dimensional space of obstructed infinitesimal deformations (one can verify, using the exact cohomology sequence yet again, that $H^1(\mathscr{E})$ is one-dimensional).

Recall from our account of the general deformation theory in Section 6.4.1 that the moduli space near $\mathscr{E}$ is modelled by the zeros of a map

$$\psi : H^1(\operatorname{End} \mathscr{E}) \longrightarrow H^2(\operatorname{End}_0 \mathscr{E}),$$

and that the quadratic part of ψ is induced by the cup product (Proposition (6.4.3(ii)). On the other hand, the cup product gives, by Serre duality, a dual pairing:

$$H^1(\mathscr{E}) \otimes H^1(\mathscr{E}^* \otimes K_S) \longrightarrow H^2(K_S) = \mathbb{C}. \qquad (10.3.17)$$

But $\mathscr{E}^* \otimes K_S$ is isomorphic to $\mathscr{E}$, so this says that the combination of the cup product on cohomology and wedge product gives a non-degenerate symmetric form:

$$\sigma : H^1(\mathscr{E}) \otimes H^1(\mathscr{E}) \longrightarrow H^2(K_S) = \mathbb{C}. \qquad (10.3.18)$$

Proposition (10.3.19). *There is a commutative diagram:*

$$\begin{array}{ccccc} H^1(\operatorname{End} \mathscr{E}) \otimes H^1(\operatorname{End} \mathscr{E}) & \xrightarrow{\smile} & H^2(\operatorname{End}_0 \mathscr{E}) & \xrightarrow{\nu} & C \\ {\scriptstyle s \otimes s} \uparrow & \nearrow {\scriptstyle 2\sigma} & & & \\ H^1(\mathscr{E}) \otimes H^1(\mathscr{E}) & & & & \end{array}$$

Proof. This follows from the naturality properties of the cup product and the following elementary identity. Let V be a two-dimensional vector space and $w : V \otimes V \to L$ a non-degenerate skew form with values in a one-dimensional space L. Suppose A and B are trace-free endomorphisms of V, then:

$$w(([A, B]s) \otimes s) = 2w(A(s) \otimes B(s))$$

for all s in V.

It follows then that the quadratic part of ψ is non-degenerate on $H^1(\mathscr{E}) \subset H^1(\mathrm{End}\,\mathscr{E})$. This means that, with a suitable choice of local model, we can suppose that all the higher order terms in ψ are zero. So a local model for our moduli space is given by:

$$\{(z_0, z_1, z_2, \ldots, z_{14}) \in \mathbb{C}^{15} \mid z_0^2 = 0\}. \tag{10.3.20}$$

The given Kähler metric on S is not generic in the sense of Chapter 4, since the ASD equations do not cut out the moduli space transversely. In the complex-analytic framework of moduli of holomorphic bundles, we would say that the moduli space is not reduced, its structure sheaf contains the nilpotent element z_0. However, as in Section 6.4.3 Example (i), it is easy to see how to calculate with the given moduli space, taking account of this lack of transversality. For a nearby generic metric g', the perturbed moduli space $M(g')$ can be represented by means of a smooth function $\varepsilon(z_1, \ldots, z_{14})$ as:

$$\{(z_0, z_1, \ldots, z_{14}) \mid z_0^2 = \varepsilon(z_1, \ldots, z_{14})\}. \tag{10.3.21}$$

Thus the moduli space splits into two sheets in a neighbourhood of a generic point $[\mathscr{E}]$. This means that, as in Section 6.4.3, we can calculate invariants with the given moduli space but must multiply the 'apparent' fundamental class by two.

Our second example, bundles with $c_1 = 0$ and $c_2 = 7$, is simpler in this respect. By analysing the long exact cohomology sequences associated with (10.3.10) one finds that the obstruction space $H^2(\mathrm{End}_0\,\mathscr{F})$ is zero in this case, for generic bundles $\mathscr{F}$, so generic points in the moduli space are regular, and we can calculate invariants without introducing any extra factors.

10.4 Calculation of invariants

10.4.1 Statement of results

In this section we will achieve our final goal, the calculation of certain components of the polynomial invariants defined by the moduli spaces which we now denote by $M_{7,\alpha}$ of connections with $w_2 = K_S$, $\kappa = 6\frac{1}{2}$ (considered in Section 10.3.4), and M_7 of connections with $c_1 = 0$, $\kappa = 7$ (considered in Section 10.3.5).

Recall that our invariants are polynomial, or symmetric, functions on the homology groups $H_2(S)$. We begin with a general remark. The set of such polynomial functions can be made into a ring in two different natural ways. On the one hand we can think of functions on $H_2(S)$, multiplied by the rule

$$(q_1 q_2)(\beta) = q_1(\beta) q_2(\beta). \tag{10.4.1}$$

On the other hand we can think of multilinear functions $q(\beta_1, \ldots, \beta_d)$, multiplied by the rule

$$(q_1 q_2)(\beta_1, \ldots, \beta_{d_1+d_2}) = \frac{1}{d_1!\,d_2!} \sum_{\sigma} q_1(\beta_{\sigma(1)}, \ldots, \beta_{\sigma(d_1)}) \times q_2(\beta_{\sigma(d_1+1)}, \ldots, \beta_{\sigma(d_1+d_2)}), \tag{10.4.2}$$

where σ runs over the permutations of $(1 \ldots (d_1 + d_2))$. If we associate the function $q(\beta, \ldots, \beta)$ with a symmetric multilinear function, these rules differ by a combinatorial factor, although the resulting rings are isomorphic.

We find it more convenient to use the second rule here. This is in line with our discussion in Chapter 8. We have a distinguished polynomial Q, the intersection form of the manifold,

$$Q(\beta_1, \beta_2) = \beta_1 . \beta_2,$$

and the second rule determines the powers Q^d of Q. We write $Q^{(d)} = (1/d!)Q^d$, so for example,

$$Q^{(2)}(\beta_1, \ldots, \beta_4) = (\beta_1 . \beta_2)(\beta_3 . \beta_4) + (\beta_1 . \beta_3)(\beta_2 . \beta_4) + (\beta_1 . \beta_4)(\beta_2 . \beta_3). \tag{10.4.3}$$

Geometrically, if $\Sigma_1, \ldots, \Sigma_{2d}$ are surfaces in a four-manifold X in general position, then $Q^{(d)}(\Sigma_1, \ldots, \Sigma_{2d})$ represents the number of ways of placing d points on the intersections of pairs of surfaces, counting with suitable signs, as in Section 8.7.2. In more abstract terms if we associate with a class β in $H^2(X)$ a 'symmetric sum' β' in $H^2(s^d(X))$ (the first Chern class of the line bundle denoted by $s^d(L)$ in Section 7.1.5, if L is the line bundle with $c_1(L) = \beta$), then

$$Q^{(d)}(\beta_1, \ldots, \beta_d) = \langle (\beta'_1 \smile \ldots \smile \beta'_d), [s^d(X)] \rangle. \tag{10.4.4}$$

(The symmetric product carries a fundamental class, even though it is not a manifold.)

For an algebraic surface like S we have another preferred polynomial, the class k_S given by

$$k_S(\beta) = \langle c_1(K_S), \beta \rangle. \tag{10.4.5}$$

In this notation we will calculate the polynomial invariants defined by the two moduli spaces, modulo the ideal generated by k_S. Otherwise stated, we will calculate the restriction of the functions to the hyperplane $k_S^{\perp}$ in $H_2(S)$ consisting of classes annihilated by k_S. The results are as follows:

Theorem (10.4.6).

(i) *The moduli space of ASD connections on an $SO(3)$ bundle over the four-manifold S with $w_2 = K_S$ and $\kappa = 6\frac{1}{2}$ (i.e. $p_1 = -26$) defines a polynomial invariant of the form*

$$2Q^{(7)} + k_S . F_1,$$

where F_1 is a polynomial of degree 13.

(ii) *The moduli space of ASD connections on an $SU(2)$ bundle over S with $\kappa = 7$ defines a polynomial invariant of the form*

$$2Q^{(8)} + k_S . F_2,$$

where F_2 is a polynomial of degree 15.

While these answers may not seem especially exciting, they do show, in conjunction with the vanishing theorems of Chapter 9, that S is not diffeomorphic to a connected sum of $\mathbb{CP}^2$s and $\overline{\mathbb{CP}}^2$s, and hence they do fulfil the main goal of this book. (Note that this case is covered by the very simple vanishing theorem (9.3.1).)

In fact it can be shown on grounds of symmetry that all the polynomial invariants of S are polynomials in Q and k_S. Given this fact the invariants are determined by, respectively, eight and nine numerical coefficients, in the general expressions $\sum a_i Q^i k_S^{14-2i}, \sum b_i Q^i k_S^{16-2i}$. Our work amounts to calculations of the leading terms a_7, b_8.

10.4.2 *Compactification*

To perform these calculations we will make use of our definition in Section 9.2.3 of the invariants, which was tailored precisely for this application. Our first task is to understand the compactification of the moduli spaces. We will consider first the simpler case of bundles with $w_2 = K_S$, with moduli spaces $M_{p,\alpha}$, $N_{p,\alpha}$. We now regard the moduli spaces $M_{p,\alpha}$ as parametrizing ASD $SO(3)$-connections, or better, $U(2)$-connections with a fixed central part and ASD semisimple part. Thus we can regard our moduli spaces $N_{p,\alpha}$ as equivalence classes of data (E, ψ, A, s) where E is a rank-two complex vector bundle, ψ is in isomorphism from $\Lambda^2 E$ to K_S, A is such a connection on E and s is a (non-trivial) section of E which is holomorphic with respect to the holomorphic structure defined by A.

We put a topology on the union,

$$\bigcup_p N_{p,\alpha} \times s^{7-p}(S),$$

as follows. We say that a sequence (E_n, ψ_n, A_n, s_n) converges to a limit $((E, \psi, A, s), x_1, \ldots, x_r)$ if there are bundle maps χ_n from E to E_n on the complement of $\{x_1, \ldots, x_r\}$, compatible with ψ_n and ψ, and scalars ζ_n such that

(i) The connections $\rho^*(A_n)$ converge to A and the sections $\rho^*(s_n)$ converge to s, in C^∞ on compact subsets of the punctured manifold.

(ii) The curvature densities $|F(A_n)|^2$ converge to $|F(A)|^2 + 8\pi^2 \sum \delta_{x_i}$.

We let $\bar{N}_\alpha$ be the closure of $N_{7,\alpha}$ in this disjoint union of strata with respect to the resulting topology. It is clear that $\bar{N}_\alpha$ is compact and that the individual maps ρ_p on the $N_{p,\alpha}$ fit together to define a continuous map

$$\bar{\rho}: \bar{N}_\alpha \longrightarrow \bar{M}_\alpha \tag{10.4.7}$$

where $\bar{M}_\alpha$ is the ordinary compactification of the moduli space $M_{7,\alpha}$. Thus we are in the position envisaged in Section 9.2.3.

In the second case, for bundles with $c_1 = 0$, things are a little more complicated. The moduli spaces N_k contain points representing unstable bundles which do not admit ASD connections. So we define a compactification by another route. We say that a sequence of triples $(\mathscr{E}_n, \psi_n, s_n)$ converges to a limit $((\mathscr{E}, \psi, s), x_1, \ldots, x_r)$ if there are bundle maps $\chi_\alpha: \mathscr{E} \to \mathscr{E}_\alpha$ over the punctured manifold $S\backslash\{x_1, \ldots, x_r\}$, compatible with ψ_n and ψ, and scalars ζ_n such that:

(i) The pull-back of the $\bar{\partial}$ operators of the $\mathscr{E}_n$ by χ_n converge to the $\bar{\partial}$ operator of $\mathscr{E}$.

(ii) The pull-back of $\zeta_n s_n$ converges to s.

(iii) The zero sets of the s_n converge to that of s plus the multi-set $(x_1, \ldots, x_r)$, (including multiplicities, in the obvious way).

Of course, we can use this definition also in the previous case of bundles with $c_1 = K_S$ and one can verify that it is equivalent to the other definition, involving the convergence of the ASD connections.

We now have to define a map $\bar{\rho}: \bar{N}_k \to \bar{M}_k$, where $\bar{N}_k$ is the compactification defined by the above topology. On the open subsets $N_j^+ \times s^{k-j}(S)$ we can use the maps ρ_j; the difficulty comes again with the bundles which are not stable. However, we know that they are semi-stable, destabilized by a section vanishing on a multiset $(z_1, \ldots, z_j)$. We define $\bar{\rho}$ on $N_j \times s^{k-j}(S)$ by mapping $((\mathscr{E}, \psi, s), y_1, \ldots, y_{k-j})$ where $\mathscr{E}$ is such a semi-stable bundle, to $([\theta], z_1, \ldots, z_j, y_1, \ldots, y_{k-j}) \in M_0 \times s^k(S)$. Proposition (6.2.25) shows that the image lies in $\bar{M}_k$ and that the resulting map $\bar{\rho}$ from $\bar{N}_k$ to $\bar{M}_k$ is continuous.

We note in passing that these compactifications of the moduli spaces $N_{p,\alpha}$ and N_p agree with those obtained by the construction we sketched in Section 10.3.4—taking the closures in $\bar{R}_p$. Thus it is in a sense easier to describe the compactified spaces than the actual moduli spaces, since this allows us to ignore the open constraint on the weights r_i.

10.4.3 The universal bundle and slant product

We will complete the proof of (10.4.6) by constructing suitable cocycles over the moduli spaces. Again we begin with the case of bundles with $w_2 = K_S$, which is slightly simpler. Recall from Section 10.2.3 that there is a universal bundle, $\mathbb{E} = \mathbb{E}_{p,\alpha}$ say, over the product $N_{p,\alpha} \times S$, with

$$c_1(\mathbb{E}) = \pi_2^*(c_1(K_S)) + \pi_1^*(c_1(U)), \tag{10.4.8}$$

and a section $\mathbf{s}$. If Σ is a surface in S, the cohomology class $\mu(\Sigma)$ is defined to be the slant product

$$\mu(\Sigma) = -\tfrac{1}{4}p_1(\mathfrak{g}_{\mathbb{E}})/[\Sigma],$$

where $\mathfrak{g}_{\mathbb{E}}$ is the associated $SO(3)$ bundle. Now

$$p_1(\mathfrak{g}_{\mathbb{E}}) = c_1(\mathbb{E})^2 - 4c_2(\mathbb{E}).$$

So we have

$$\mu(\Sigma) = c_2(\mathbb{E})/[\Sigma] - \tfrac{1}{2}\langle c_1(K_S), \Sigma\rangle c_1(U). \tag{10.4.9}$$

It is here we see the reason why we have restricted attention in (10.4.6) to classes $[\Sigma]$ annihilated by $k_S = c_1(K_S)$. For such classes the second term in the slant product drops out and we are left with the simple formula

$$\mu(\Sigma) = c_2(\mathbb{E})/[\Sigma];$$

and $c_2(\mathbb{E})$ is represented by the zero set $Z \subset N \times S$ of the section $\mathbf{s}$. More precisely, this is true on the dense open subset $\pi_1^{-1}(G)$ of $N \times S$ which is a smooth manifold and on which $\mathbf{s}$ vanishes transversely. The intersection of $\pi_1^{-1}(G)$ with Z is a smooth four-dimensional submanifold dual to $c_2(\mathbb{E})$. But in any case, $c_2(\mathbb{E})$ can be represented by a cocycle supported in an arbitrarily small neighbourhood of Z.

Let us now recall what the slant product means for classes represented by dual submanifolds. Suppose we have a pair of manifolds X and Y, that R is a submanifold of the product $X \times Y$ and P is a submanifold of Y. Let M^* denote the cohomology class dual to a manifold M; then if P is in general position we have:

$$(R^* \backslash P^*) = (\pi_1(R \cap \pi_2^{-1}(P)))^*. \tag{10.4.10}$$

To apply this in the product $N \times S$ we write, for any surface Σ in S,

$$W_\Sigma = \{(\mathscr{E}, \psi, s) \in N \mid s \text{ vanishes at some point of } \Sigma\}. \tag{10.4.11}$$

Then we have:

Proposition (10.4.12). *If Σ is a surface in S with $k_S(\Sigma) = 0$, then $\mu(\Sigma)$ can be represented on N by a cocycle supported in an arbitrarily small neighbourhood of W_Σ and, over the dense open subset G of N, W_Σ is a submanifold Poincaré dual to $\mu(\Sigma)$.*

We will now proceed with our calculation. For the sake of exposition we will first proceed rather formally to get a numerical solution, and then backtrack to justify our arguments rigorously.

Suppose that $\Sigma_1, \ldots, \Sigma_{14}$ are surfaces in general position in S, each annihilated by the canonical class k_S. We consider the subsets W_{Σ_i} as representatives for the cohomology classes $\mu(\Sigma_i)$. So the cup product is represented by the common intersection $I = W_{\Sigma_1} \cap \ldots \cap W_{\Sigma_{14}}$ in $N = N_{7,\alpha}$. But we have seen in Section 10.3.4 that the projection map from $N_{7,\alpha}$ to the symmetric product $s^7(S)$ is generically one-to-one. So, for surfaces in general position, the points of I correspond to configurations of seven points in S lying on intersections of pairs of surfaces. Thus, counting with signs, we get

$$\# I = Q^{(7)}(\Sigma_1, \ldots, \Sigma_{14}). \tag{10.4.13}$$

Now we recall from Section 10.3.6 that the apparent fundamental class of the moduli space must be multiplied by two in calculating the invariants. So we arrive at the answer:

$$q(\Sigma_1, \ldots, \Sigma_{14}) = 2Q^{(7)}(\Sigma_1, \ldots, \Sigma_{14}), \tag{10.4.14}$$

as given in (10.4.6).

The justification for this calculation is provided by Section 9.2.3. The stratified space $\bar{N}_\alpha = \bar{N}_{7,\alpha}$ satisfies the conditions laid down before (9.2.21). We define quotient spaces $\tilde{N}_{\Sigma_i}$ by the procedure of Section 9.2.3. Then it is clear that the universal complex two-plane bundles over the strata fit together to define a universal bundle over $N_{\Sigma_i} \times \Sigma_i$, and moreover that the sections of this bundle over the strata yield a continuous section $\mathbf{s}_{(i)}$ of this bundle. We can take a cocycle $\tilde{\mu}_i$ representing the slant product which is supported in a suitably small neighbourhood of the projection of the zero set of $\mathbf{s}_{(i)}$ to $\tilde{N}_{\Sigma_i}$. There is no real loss in replacing the support of the cocycles in the argument by these projections of the zero sets. Then, unwinding the definitions, the support condition of (9.2.21) comes down to the condition that for any point $(\mathscr{E}, \psi, s)$ of $N_{7-j,\alpha}$ the zero set of s should not meet any $14 - j$ of the surfaces Σ_i.

It is now easy to check from our description of the moduli spaces that this condition is indeed satisfied, for surfaces Σ_i in general position. When $j = 1$ we require that no set of six intersection points of the surfaces should lie on a conic. When $j = 2$ we require that no set of five intersection points should lie on a line, and so on. We see then that the invariant can be calculated from cocycles on $N_{7,\alpha}$ supported in small neighbourhoods of the W_{Σ_i}, and the cup product from a class supported in a small neighbourhood of the intersection I. We can also arrange that I lies in the open dense subset G where the W_{Σ_i} are submanifolds dual to the cohomology classes, and it is then clear that the homological calculation is indeed represented by counting, with signs, the points of I. This completes the proof of (10.4.6(i)).

We now turn to the other case, of bundles with $c_1 = 0$. The theoretical part of the discussion goes through much as before. We have a universal bundle over the sets $\tilde{N}_\Sigma \times \Sigma$, although these do not descend to the $\tilde{M}_\Sigma \times \Sigma$ because of the reducible connection. This is just the situation envisaged in Section 9.2.3. We conclude that for surfaces Σ_i in general position with $k_S(\Sigma_i) = 0$, the invariant $q(\Sigma_1, \ldots, \Sigma_{16})$ can be calculated by counting configurations of points $\{x_1, \ldots, x_9\}$ satisfying the conditions:

(i) every surface Σ_i contains one of the points x_ν;

(ii) the nine points $y_\nu = \pi(x_\nu)$ are an intersection of cubics in $\mathbb{CP}^2$.

There are two classes of such configurations; *either*:

(*a*) eight of the points, say $x_1, \ldots, x_8$, lie on intersection points of the surfaces and the ninth point is unconstrained by the surfaces. However the pencil of cubics through $p(x_1), \ldots, p(x_8)$ has a single further intersection point, y say, (which we can suppose does not lie in the branch locus) so condition (ii) shows that there are exactly two possibilities for x_9: the two points of $p^{-1}(y)$.

or:

(*b*) seven of the points, say $x_1, \ldots, x_7$, lie on intersections of the Σ_i leaving two surfaces, say Σ_{15}, Σ_{16}, unaccounted for, and the remaining two points lie on these surfaces, say x_8 on Σ_{15} and x_9 on Σ_{16}.

The contribution at the level of homology from configurations (*a*) is clearly $2Q^8(\Sigma_1, \ldots, \Sigma_{16})$, the factor 2 coming from the two choices of x_9.

We shall now argue that, for surfaces annihilated by k_S, the contribution to the total intersection number from configurations of type (*b*) is zero. Let $z_1, \ldots, z_7$ be generic points in $\mathbb{CP}^2$. For any other point x in $\mathbb{CP}^2$, distinct from the z_i, we can find a pencil of cubics through $(z_1, \ldots, z_7, x)$ meeting in a further point, y say. Thus we can define a map:

$$\phi: \mathbb{CP}^2 \setminus \{z_1, \ldots, z_7\} \longrightarrow \mathbb{CP}^2$$

by $\phi(x) = y$. (In fact this is a rational map on $\mathbb{CP}^2$ and extends to a smooth map of the plane blown up at the points z_i.) Taking the z_i to be intersection points of seven disjoint pairs of the surfaces we can express the contribution from these configurations to the intersection number as a sum of multiples of terms of the form:

$$\pi(\Sigma_i) \,.\, \phi(\pi(\Sigma_j)),$$

i.e. the algebraic intersection number in $\mathbb{CP}^2 \setminus \{z_1, \ldots, z_7\}$. However, if $\langle c_1(K_S), [\Sigma_i] \rangle = 0$, the surface $\pi(\Sigma_i)$ is null-homologous in $\mathbb{CP}^2$, hence also in $\mathbb{CP}^2 \setminus \{z_1, \ldots, z_7\}$, so this intersection number vanishes. Thus we can restrict attention to the configurations (*a*), and we obtain

$$q(\Sigma_1, \ldots, \Sigma_{16}) = 2Q^{(8)}(\Sigma_1, \ldots, \Sigma_{16}),$$

verifying (10.4.6(ii)).

The techniques we have used in this chapter can be applied to calculate some of our invariants for other surfaces. As an exercise the reader may show that the polynomial invariant defined by the moduli space of ASD $SU(2)$

connections over a $K3$ surface with $c_2 = 5$ is $Q^{(7)}$, where Q is the intersection form of the $K3$ surface.

Notes

Section 10.1.1

The observation that any complex surface with even form satisfies the '11/8 inequality' is taken from Friedman and Morgan (1988*b*). For the results of Mandelbaum and Moishezon on almost completely decomposable surfaces see Mandelbaum (1980) and Moishezon (1977).

Section 10.1.2

For Kodaira's theorem on projective embedding see Griffiths and Harris (1978, Chapter 1) or Wells (1980).

Section 10.1.3

Gieseker's construction in the case of curves is given by Gieseker (1982). There is a version which works directly for bundles over surfaces, but the notion of stability is a little different, see Gieseker (1977). Projective embeddings of moduli spaces of bundles over curves had been constructed earlier, by a different approach; see Mumford and Fogarty (1982), but it is harder to identify the hyperplane bundle in this approach.

Section 10.1.4

Taubes (1989) proves a more general theorem on the homotopy groups of moduli spaces for large κ. The existence result used here follows by considering π_0. The proof of Proposition (10.1.11) is given by Donaldson (1990*a*); an alternative approach to some parts of the argument has been outlined by Friedman (1989).

Section 10.2.2

This construction appears in many places in the literature on holomorphic bundles, in various different guises. For a complete treatment of the construction see Griffiths and Harris (1978, Chapter 5), and for generalizations to higher dimensions see Okonek *et al.* (1980) and Hartshorne (1978) (who attribute the construction to Serre (1961)). An equivalent procedure, in the surface case, is to blow up the points and construct a bundle on the blown-up surface by the simpler extension construction of (*a*); see Schwarzenberger (1961).

Section 10.3.1

For general facts about the cohomology of pluricanonical bundles see Barth *et al.* (1984). The Noether–Lefschetz property follows from the fact that these surfaces have 'large monodromy groups' in the sense of Friedman *et al.* (1987)—acting irreducibly on the part of the cohomology orthogonal to the canonical class.

Section 10.3.3

We have been slightly vague about the definition of the compactified space $\bar{R}_p$, since this seems to be a part of the theory which has not been worked out in detail in the literature. One can expect that, rather generally, the problem of describing some of the Yang–Mills invariants of surfaces can be translated into calculations in the cohomology ring of the associated spaces $\bar{R}_p$; for a discussion of this see Donaldson (1990*c*).

Section 10.3.6

The calculation which shows that $M_{7,\alpha}$ has multiplicity two is not special to this one case; essentially the same calculation has been given by Kotschick (1990).

Section 10.4.1

The fact that the invariants are polynomials in k_S and Q follows from consideration of the large group of diffeomorphisms of the surface obtained from monodromy around loops in the parameter space of branch curves; see Friedman *et al.* (1987). It follows that for large classes of surfaces the canonical class is, up to sign, a differential topological invariant. This can be used to give examples of homeomorphic but non-diffeomorphic surfaces; see Salvetti (1989), Eberlein (1990) and Friedman *et al.* (1987).

Recently Friedman and Morgan (1989) have calculated terms in the polynomial invariants for elliptic surfaces and deduced that there are infinitely many surfaces homeomorphic, but not diffeomorphic, to the $K3$ surface.

APPENDIX

In this Appendix we gather together facts from analysis which are used throughout the book. With one exception these are all very standard, and we do not give any proofs. Our purpose is merely to summarize some of the ideas we take for granted in the text.

I Equations in Banach spaces

See, for example, Lang (1969).

Let f be a map from an open set in a Banach space E to a Banach space F. The map is *differentiable* at a point x in its domain if there is a bounded linear map $(Df)_x : E \to F$ such that

$$\lim(f(x+h) - f(x) - (Df)_x h)/\|h\| = 0,$$

as h tends to 0 in E. If f is differentiable at every point we can regard its derivative as a map into the Banach space $\mathrm{Hom}(E, F)$, with the operator norm. This allows us to define, inductively, the notion of a C^r map f, and so of a smooth (i.e. C^∞) map. If E and F are complex Banach spaces we have the obvious notion of a holomorphic map—a smooth map whose derivative is complex linear.

(A1) (Inverse function theorem in Banach spaces). *Suppose $f: U \to F$ is a smooth map and x_0 is a point of U at which the derivative of f is an isomorphism from E to F. Then there is an open neighbourhood U' of x_0 such that the restriction of f to U' has a smooth two-sided inverse*

$$f^{-1}: f(U') \longrightarrow U'.$$

Moreover if f is holomorphic then so is f^{-1}.

This result asserts that for y near to $f(x_0)$ there is a unique solution x near x_0 to the equation $f(x) = y$, and the solution varies smoothly with y.

We recall the proof of the inverse function theorem in outline. Without loss we can assume $E = F$, $x_0 = f(x_0) = 0$ and that the derivative of f at 0 is the identity. Then write $f(x) = x + R(x)$, where $\|R(x)\| = o(\|x\|)$. The equation $f(x) = y$ can be written

$$x = T(x)$$

where $T(x) = y - R(x)$. One shows that, for small x, T is a contraction and applies the contraction mapping principle to obtain a solution, if y is small. Thus the solution is found in the form:

$$x = \lim T^n(0).$$

There are a number of other results which are easy corollaries of the inverse function theorem. For example:

(A2). *If the derivative $(Df)_{x_0}$ is surjective and admits a bounded right inverse $P: F \to E$, then there is a neighbourhood N of x_0 and a smooth map g mapping N homeomorphically to a neighborhood of x_0 such that*

$$f(x) = (Df)_{x_0}(g(x)).$$

In particular for y near $f(x)$ the equation $f(x) = y$ has a solution x near x_0.

(A3) (Implicit function theorem in Banach spaces). *Suppose E is a product of Banach spaces E_1, E_2 and write $D_1 f, D_2 f$ for the partial derivatives of a smooth map f. If the partial derivative $(D_2 f)$ at a point (ξ_1, ξ_2) is an isomorphism from E_2 to F there is a smooth map h from a neighbourhood of ξ_1 in E_1 to a neighbourhood of ξ_2 in E_2 such that*

$$f(\eta, h(\eta)) = f(\xi_1, \xi_2).$$

(Moreover for any (η_1, η_2) near (ξ_1, ξ_2) with $f(\eta_1, \eta_2) = f(\xi_1, \xi_2)$ we have $\eta_2 = h(\eta_1)$.) If f is holomorphic then so is h.

For our applications we often need a variant of (A3), along the lines of (A2): if the partial derivative $D_2 f$ is surjective and admits a bounded right inverse then for all η_1 near ξ_1 there is a solution η_2 to the equation $f(\eta_1, \eta_2) = f(\xi_1, \xi_2)$.

II Sobolev spaces

See, for example, Gilbarg and Trudinger (1983), Wells (1980), Griffiths and Harris (1978), Stein (1970), Aubin (1982) and Warner (1983).

For $k \geq 0$ the space $L^2_k(\mathbb{R}^n)$ is defined to be the completion of the space of smooth, compactly-supported functions on $\mathbb{R}^n$ under the norm:

$$\| f \|_{L^2_k} = \left(\sum_{i=0}^{k} \| \nabla^i f \|^2_{L^2} \right)^{1/2},$$

where ∇^i denotes the tensor of all ith derivatives of f.

There are many natural equivalent norms, for example:

$$(\| f \|^2_{L^2} + \| \nabla^k f \|^2_{L^2})^{1/2}.$$

Another variant of the definition is to require that f be an L^2 function whose distributional derivatives up to order k are in L^2 (Gilbarg and Trudinger, 1983, Chapter 7).

These norms are also defined, in an obvious way, on functions whose domain of definition is an open subset of $\mathbb{R}^n$. A function f on an open subset $\Omega \subset \mathbb{R}^n$ is said to be *locally in* L^2_k if each point of Ω is contained in a neighbourhood over which the L^2_k norm is finite.

Now let X be a compact manifold and V be a vector bundle over X. We can define spaces $L^2_k(X; V)$ of 'L^2_k sections of V' by two approaches:

(i) Choose local coordinates and bundle-trivializations and define a section s to be in $L^2_k(X; V)$ if it is represented by locally L^2_k functions in these trivializations.

(ii) Choose a metric on X, and a fibre metric and compatible connection on V. Then take the completion of the smooth sections of V in the norm,

$$\|s\|_{L^2_k} = \|s\|_{L^2_k(X;\, V)} = \left(\sum_{i=0}^{k} \int_X |\nabla_A^{(i)} s|^2 \, d\mu \right)^{1/2}.$$

These are equivalent definitions. We can define norms which make L^2_k into a Hilbert space either using the integrals in the second approach or using a partition of unity to reduce to coordinate patches, as in the first approach. (In general in this book we just write L^2_k, leaving the bundle V and base manifold X to be understood from the context.)

There are two standard results about these Sobolev spaces, over compact base manifolds.

(A4) (Rellich Lemma). *The inclusion $L^2_{k+1} \to L^2_k$ is compact.*

(A5) (Sobolev embedding theorem). *If* $\dim X = n$ *then there is a natural bounded inclusion map from L^2_k into the Banach space C^r (of r-times continuously differentiable sections) provided*

$$k - \frac{n}{2} > r.$$

Hence a function which lies in L^2_k for all k is smooth.

In most of our applications we can equally well use (A4) or the *Ascoli–Arzela* theorem, which implies that the inclusion of C^{r+1} in C^r is compact.

The embedding theorem (A5) is essentially equivalent to the *Sobolev inequality*:

(A6). *If $k - n/2 > 0$ there is a constant C such that for all smooth sections s of V*

$$\max_{x \in X} |s(x)| \leq C \,.\, \|s\|_{L^2_k}.$$

The maximum value on the left-hand side is the C^0 or L^∞ norm, which we denote equivalently by $\|s\|_{C^0}$ or $\|s\|_{L^\infty}$. The constant C depends of course on the manifold X, and the particular choice of L^2_k norm which we have made.

III Elliptic operators

See for example Wells (1980), Warner (1983) and Booss and Bleeker (1985).

Let $D: \Gamma(V_1) \to \Gamma(V_2)$ be a linear differential operator of order l between sections of bundles over a compact base manifold X. For each k the operator obviously extends to a bounded linear map from L^2_{k+l} to L^2_k. The highest-order term in D defines the 'symbol' σ_D; for each cotangent vector ξ at a point x in X the symbol gives a linear map $\sigma_D(\xi)$ from the fibre of V_1 at x to that of V_2. The operator is *elliptic* if $\sigma_D(\xi)$ is invertible for all non-zero ξ. Examples of elliptic operators are:

(i) The Laplace operator $\Delta = \mathrm{d}^*\mathrm{d}$. We have:

$$\sigma_\Delta(\xi)(v) = -|\xi|^2 v.$$

(ii) The Cauchy–Riemann operator $\bar\partial$ on (say) functions over a Riemann surface. In standard coordinates the symbol is

$$\sigma_{\bar\partial}(\xi_1, \xi_2) = \begin{pmatrix} \xi_1 & -\xi_2 \\ \xi_2 & \xi_1 \end{pmatrix}.$$

(iii) The Dirac operator, specifically the Dirac operator over a four-manifold which we use in Chapter 3.

(iv) Suppose we have a sequence of operators of the same order:

$$\Gamma(V_0) \xrightarrow{D} \Gamma(V_1) \xrightarrow{D} \Gamma(V_2) \ldots.$$

These give a sequence of symbols $\sigma_D(\xi) \in \mathrm{Hom}(V_i, V_{i+1})$. If the symbol sequence is exact then the operator $D + D^*$, mapping the direct sum of the even terms to the direct sum of the odd terms, is elliptic. (Here D^* is the *formal adjoint*, defined relative to some metrics.)

This set-up applies to:

(*a*) The de Rham complex;

(*b*) The Dolbeault complex over a complex manifold;

(*c*) The two-step complex defined by the operators d and d^+ over an oriented Riemannian four-manifold. (Exercise: verify the exactness of the symbol sequence for this complex.)

We can also take any of these operators coupled to a connection on an auxiliary bundle.

The main result about elliptic operators over compact manifolds is the following (which we sometimes refer to, slightly inaccurately, as the 'Fredholm alternative'):

(A7). *Let $D: \Gamma(V_1) \to \Gamma(V_2)$ be an elliptic operator over a compact manifold X. Suppose V_1 and V_2 have metrics. Then the formal adjoint D^* is also elliptic and:*

(i) *the kernels of D and D^* are finite dimensional;*

(ii) *a section s of V_2 is in the image of D if and only if the L^2 inner product $\langle s, \alpha \rangle$ vanishes for all α in* Ker D^*.

The foundation of the proof of (A7) is the fundamental inequality for elliptic operators on Sobolev spaces:

(A8). *If D is an elliptic operator of order l then for each $k \geq 0$ there is a constant C_k so that for all sections of V_1,*

$$\| s \|_{L^2_{k+l}} \leq C_k (\| Ds \|_{L^2_k} + \| s \|_{L^2}).$$

If we work over, say, a bounded domain Ω in $\mathbb{R}^n$ there are similar inequalities, but we lose control at the boundary: for any $\Omega' \Subset \Omega$ we have an inequality of the form:

$$\| s \|_{L^2_{k+l}(\Omega')} \leq C(\| Ds \|_{L^2_k(\Omega)} + \| s \|_{L^2(\Omega')}).$$

The properties of elliptic operators on Sobolev spaces over compact manifolds can be summed up by saying that D defines a *Fredholm map* from L^2_{k+l} to L^2_k.

Remarks.

(i) The requirement of ellipticity naturally splits into two parts: the injectivity and surjectivity of the symbol. Roughly speaking, half of the results of the theory only require the injectivity and half only require the surjectivity. For example, the inequality (A8) holds for operators whose symbol is injective (sometimes called overdetermined elliptic), while the 'solubility criteria' (A7 (ii)) holds if the symbol is surjective.

(ii) The formal adjoint is not the same as the adjoint of D, regarded as a bounded map between the Hilbert spaces L^2_{k+l}, L^2_k.

(iii) In the inequality (A8), if we suppose that s is L^2 orthogonal to the kernel of D then we can omit the term $\| s \|_{L^2}$ on the right-hand side.

IV. Sobolev inequalities and non-linear problems

See Gilbarg and Trudinger (1983), Aubin (1982), Freed and Uhlenbeck (1984) and Palais (1968).

We have encountered in (A8) the first of a wide range of 'Sobolev inequalities'. It is simplest to set the scene for these now by mentioning that for any exponent $p > 1$ we can define spaces L^p_k, just as for the case of L^2_k above, replacing the L^2-norm by the L^p norm. (It is often convenient to fit the C^r norms into the picture by taking $p = \infty$, but many of the results do *not* extend to this case; we always take p finite below.) In a given base dimension n we define the 'scaling weight' of the function space L^p_k to be the number:

$$w(k, p) = k - \frac{n}{p}.$$

This is the weight by which the leading term $\|\nabla^k f\|_{L^p}$ transforms under dilations of $\mathbb{R}^n$. Roughly speaking, over a compact manifold, larger values of w correspond to stronger norms. More precisely:

(A9). *Let V be a bundle over a compact manifold X. If $k > l$ and $w(k, p) \geq w(l, q)$ there is a bounded inclusion map*

$$L^p_k(X; V) \longrightarrow L^q_l(X; V).$$

This can be deduced fairly easily from a basic Sobolev inequality for functions on $\mathbb{R}^n$. For given $p < n$ put $q = np/(n - p)$, so $w(1, p) = w(0, q)$.

(A10). *There is a constant $C(n, p)$ such that for any smooth compactly supported function f on $\mathbb{R}^n$*

$$\|f\|_{L^q} \leq C(n, p)\|\nabla f\|_{L^p}.$$

The most geometric proof reduces this inequality to the *isoperimetric inequality*, using the 'co-area formula' (see Aubin (1982)).

Remark.
If strict inequality $w(k, p) > w(l, q)$ holds in (A9) the embedding is compact.

The embeddings (A9) lead to multiplication properties for Sobolev spaces. For simplicity we consider now dimension $n = 4$. Then we have an embedding $L^2_1 \to L^4$, and this immediately tells us that multiplication gives a bounded bilinear map:

$$L^2_1 \times L^2_1 \longrightarrow L^2.$$

On the other hand we recall from (A5) that for $k \geq 3$ there is a bounded inclusion $L^2_k \to C^0$. Expanding by the Leibnitz rule this tells us that if $k \geq l$ and $k \geq 3$ then multiplication is bounded:

$$L^2_k \times L^2_l \longrightarrow L^2_l \quad (k \geq l, k \geq 3).$$

In the intermediate cases we obtain similar but less tidy results: for example we have

$$\|fg\|_{L^2_1} \leq \text{const.}(\|f\|_{L^2_1}\|g\|_{L^2_2} + \|f\|_{L^2_2}\|g\|_{L^2_1}).$$

These multiplication results allow one to define the action of certain *non-linear* differential operators on Sobolev special L^2_k, for large enough k. For example a map of the form:

$$N(f) = Df + p(f),$$

where D is first order and linear, and p is a polynomial of order d will map L^2_k to L^2_{k-1} provided $k \geq 1$ for $d = 2$, $k \geq 2$ for $d = 3$ and $k \geq 3$ for any d. The same picture holds for general non-linear operators. We refer to Palais (1968) for a comprehensive theory. The main ingredient is the composition property:

(A11). *If $H: \mathbb{R} \to \mathbb{R}$ is a smooth function and f is in $L^2_k(X) = L^2_k(X; \mathbb{R})$, with $k - n/2 > 0$ then the composite Hf is again in L^2_k, and the operation of composition with H defines a smooth map from $L^2_k(X)$ to itself.*

Of course there are corresponding results for sections of bundles.

As an exercise the reader can use this and the multiplication properties to prove that the operator:

$$N(\chi) = \mathrm{d}^*(\mathrm{d}gg^{-1}), \quad \text{where } g = \exp(\chi)$$

for χ an $\mathfrak{su}(2)$ valued function on X, which appears in the gauge fixing problem of Section 2.3, yields a smooth map from L^2_l to L^2_{l-2}, for $l \geq 3$ (if $\dim X = 4$).

If we have extended a non-linear operator N to Sobolev spaces the Fréchet derivative DN is represented by a linear differential operator. If this linearization is elliptic the theory of III above can be applied. If the linear operator is invertible we can apply the inverse and implicit function theorems in Banach (i.e. Sobolev) spaces of I to obtain 'local' results about the non-linear operator. The reader will find many examples of the use of this standard technique in the body of the book.

V. Further L^p theory; integral operators

These results are only used in Section 7.2.

The Sobolev embedding theorems are completed by the generalization of (A5) to L^p spaces. The basic result, in dimension n, is:

(A12). *Over a compact base manifold X of dimension n there is a bounded inclusion map $L^p_1 \to C^0$ if $p > n$ (i.e. $w(1, p) > 0$).*

The proof is not difficult: the crux of the matter is to prove an inequality

$$f(0) \leq \text{const.} \, \| \nabla f \|_{L^p} \tag{A13}$$

for compactly supported functions on the open ball in $\mathbb{R}^n$. This follows by integrating ∇f along rays to get an inequality:

$$|f(0)| \leq \int_{B^n} \frac{1}{|x|^{n-1}} |\nabla f| \, \mathrm{d}\mu_x.$$

Then use Hölder's inequality. (Note that it suffices to integrate over any cone centred at zero and this allows extensions of the results to manifolds with boundary which satisfy a 'uniform cone' condition.) A refinement of the argument gives a Hölder bound on f, with exponent $w(1, p)$; see Gilbarg and Trudinger (1983, p. 162).

Finally we extend our elliptic theory part-way to these L^p spaces. Let D be an elliptic operator over a compact manifold X, and for simplicity suppose D is a first-order operator. Let p and q be related as before, i.e. $q = np/(n-p)$. Then we have:

(A14). *There is a constant C such that*

$$\| f \|_{L^q} \leq C(\| Df \|_{L^p} + \| f \|_{L^p}).$$

As before, if we restrict to sections f which are L^2-orthogonal to the kernel of D we can omit the term $\| f \|_{L^p}$.

To prove this one can reduce by standard arguments to the case of a constant coefficient operator D_0 over $\mathbb{R}^n$. Such an operator has a fundamental solution $L(x)$ which, for scaling reasons, must be homogeneous of order $-(n-1)$. If f has, say, compact support we can write:

$$f(x) = \int_{\mathbb{R}^n} L(x-y)g(y)\,dy, \quad \text{i.e. } f = L * g,$$

where $g = D_0(f)$. The result follows easily from the following theorem on convolution operators (see Stein (1970), pp. 118–122):

(A15). *Let L be any function on* $\mathbb{R}^n \setminus \{0\}$ *such that*

$$|L(x)| \leqslant A./|x|^{n-1}$$

for some constant A. Then there is a $C = C(A)$ *such that*

$$\| L * f \|_{L^q} \leq C \| f \|_{L^p},$$

for all f in $L^p(\mathbb{R}^n)$.

We note that this can be used to give another proof of the Sobolev inequality (A10), using the integral representation as in (A12) above, see Aubin (1982).

If (as is the case in our application in Chapter 7) D_0 has the property that $D_0^* D_0$ is the standard Laplacian $\nabla^* \nabla$ one can give a simple proof of (A14) by working with the integral of

$$(|f|^{b-1} f, D_0^* D_0 f),$$

with $b = (n-2)q/2n$. One integrates by parts in two different ways and rearranges, using the Sobolev and Hölder inequalities.

Finally, although it will not be used in this book, we should mention for completeness that there is a stronger extension of the elliptic theory to L^p spaces, derived from the Calderon–Zygmund theory of singular integral operators; see Stein (1970) and Gilbarg and Trudinger (1983). This leads to inequalities:

$$\| s \|_{L^p_{k+1}} \leq C(\| Ds \|_{L^p_k} + \| s \|_{L^p}). \tag{A16}$$

REFERENCES

Agmon, S. and Nirenberg, L. (1967). Lower bounds and uniqueness theorems for solutions of differential equations in Hilbert spaces. *Communications in Pure and Applied Mathematics*, **20**, 207–29.

Akbulut, S. and McCarthy, J. (1990). *Cassons's invariant for homology 3-spheres.* Princeton University Press.

Aronszajin, N. (1957). A unique continuation theorem for solutions of elliptic partial differential equations or inequalities of the second order. *Journal de Mathématiques Pures et Appliquées* (9) **36**, 235–49.

Atiyah, M.F. (1967). *K-theory*. Benjamin, New York.

Atiyah, M.F. (1970). Global theory of elliptic operators. In *Proceedings of the International Congress on Functional Analysis and Related Topics.* University of Tokyo Press.

Atiyah, M.F. (1979). *The geometry of Yang–Mills fields.* Fermi lectures. Scuola Normale, Pisa.

Atiyah, M.F. (1988). New invariants for 3 and 4 dimensional manifolds. In *The mathematical heritage of Hermann Weyl. Proceedings of Symposia in Pure Mathematics*, **48**, 285–99.

Atiyah, M.F. and Bott, R. (1982). The Yang–Mills equations over Riemann surfaces. *Philosophical Transactions of the Royal Society of London, Series A*, **308**, 523–615.

Atiyah, M.F., Drinfeld, V., Hitchin, N.J., and Manin, Yu.I. (1978*a*). Construction of instantons. *Physics Letters*, **65A**, 185–87.

Atiyah, M.F. and Hitchin, N.J. (1989). *The geometry and dynamics of magnetic monopoles.* Princeton University Press.

Atiyah, M.F., Hitchin, N.J., and Singer, I.M. (1978*b*). Self-duality in four dimensional Riemannian Geometry. *Proceedings of the Royal Society of London, Series A*, **362**, 425–61.

Atiyah, M.F. and Jones, J.D.S. (1978). Topological aspects of Yang–Mills theory. *Communications in Mathematical Physics*, **61**, 97–118.

Atiyah, M.F. and Singer, I.M. (1968). The index of elliptic operators. I. *Annals of Mathematics*, **87**, 484–530.

Atiyah, M.F. and Singer, I.M. (1971). The index of elliptic operators. IV. *Annals of Mathematics*, **93**, 119–38.

Atiyah, M.F. and Singer, I.M. (1984). Dirac operators coupled to vector potentials. *Proceedings of the National Academy of Sciences of the U.S.A.*, **81**, 2597–2600.

Aubin, T. (1982). *Nonlinear analysis on manifolds.* Springer, New York.

Barth, W. (1977). Moduli of bundles on the projective plane. *Inventiones Mathematicae*, **42**, 63–91.

Barth, W., Peters, C. and Van de Ven, A. (1984). *Compact complex surfaces.* Ergebnisse der Mathematik. Springer-Verlag, Berlin.

Birkes, D. (1971) Orbits of linear algebraic groups. *Annals of Mathematics* (2) **93**, 459–75.

Bismut, J.M. and Freed, D.S. (1986). The analysis of elliptic families I, metrics and connections on determinant line bundles. *Communications in Mathematical Physics*, **106**, 159–76.

Bismut, J.M., Gillet, H., and Soulé, C. (1988). Analytic torsion and holomorphic determinant bundles I–III. *Communications in Mathematical Physics*, **115**, 49–126, 301–51.

Booss, B. and Bleeker, D.D. (1985). *Topology and analysis: the Atiyah–Singer index formula and gauge theoretic physics*. Springer, Berlin.

Bourguignon, J-P. (1981). Formules de Weitzenböck en dimension 4. In *Géometrie Riemannienne de dimension 4*. CEDIC, Paris.

Bourguignon, J-P, and Lawson, H.B. (1982). Yang–Mills theory: its physical origins and differential geometric aspects. In *Seminar on Differential Geometry*, Annals of Mathematics Studies, **102** (ed. S-T. Yau). Princeton University Press.

Braam, P.J. and van Baal, P. (1989). Nahm's transformation for instantons. *Communications in Mathematical Physics*, **122**, 267–80.

Bradlow, S.B. (1990). Special metrics and stability for holomorphic bundles with global sections. Preprint.

Buchdahl, N. (1986). Instantons on $\mathbb{CP}^2$. *Journal of Differential Geometry*, **24**, 19–52.

Buchdahl, N. (1988). Hermitian Einstein connections and stable vector bundles over compact complex surfaces. *Mathematische Annalen*, **280**, 625–48.

Chern, S.S. (1979). *Complex manifolds without potential theory*. Springer, Berlin.

Corlette, K. (1988). Flat *G*-bundles with canonical metrics. *Journal of Differential Geometry*, **28**, 361–82.

Corrigan, E. and Goddard, P. (1984). Construction of instanton and monopole solutions and reciprocity. *Annals of Physics*, **154**, 253–79.

Deturck, D.M. (1983). Deforming metrics in the direction of their Ricci tensors. *Journal of Differential Geometry*, **18**, 157–62.

Dold, A. and Whitney, H. (1959). Classification of oriented sphere bundles over a 4-complex. *Annals of Mathematics*, **69**, 667–77.

Donaldson, S.K. (1983*a*). A new proof of a theorem of Narasimhan and Seshadri. *Journal of Differential Geometry*, **18**, 269–78.

Donaldson, S.K. (1983*b*). An application of gauge theory to four dimensional topology. *Journal of Differential Geometry*, **18**, 279–315.

Donaldson, S.K. (1984*a*). Instantons and Geometric Invariant Theory. *Communications in Mathematical Physics*, **93**, 453–60.

Donaldson, S.K. (1984*b*). Nahm's equations and the classification of monopoles. *Communications in Mathematical Physics*, **96**, 387–407.

Donaldson, S.K. (1985*a*). Anti-self-dual Yang–Mills connections on complex algebraic surfaces and stable vector bundles. *Proceedings of the London Mathematical Society*, **3**, 1–26.

Donaldson, S.K. (1985*b*). Vector bundles on the flag manifold and the Ward correspondence. In *Geometry Today, Giornata di Geometria*, (ed. Arbarello *et al.*) *Roma, 1984*. Birkhäuser, Boston.

Donaldson, S.K. (1986). Connections, cohomology and the intersection forms of four manifolds. *Journal of Differential Geometry*, **24**, 275–341.

Donaldson, S.K. (1987*a*). Irrationality and the *h*-cobordism conjecture. *Journal of Differential Geometry*, **26**, 141–68.

Donaldson, S.K. (1987*b*). The orientation of Yang–Mills moduli spaces and 4-manifold topology. *Journal of Differential Geometry*, **26**, 397–428.

Donaldson, S.K. (1987*c*). The geometry of 4-manifolds. *Proceedings of the International Congress of Mathematicians, Berkeley, 1986*. Vol. 1. pp. 43–53. American Mathematical Society, Providence, RI.

Donaldson, S.K. (1987*d*). Infinite determinants, stable bundles and curvature. *Duke Mathematical Journal*, **54**, 231–247.

Donaldson, S.K. (1990*a*). Polynomial invariants for smooth 4-manifolds. *Topology*, **29**, 257–315.

Donaldson, S.K. (1990*b*). Compactification and completion of Yang–Mills moduli spaces. *Proceedings of the International Symposium on Differential Geometry, Peniscola, 1988*, Lecture Notes in Mathematics, **1410** (ed. Carreras *et al.*). Springer-Verlag, Berlin.

Donaldson, S.K. (1990*c*). Instantons in Yang–Mills theory. *Proceedings IMA conference on Geometry and Particle Physics, Oxford, 1988* (ed. F. Tsou). Oxford University Press.

Donaldson, S.K., and Friedman, R. (1989). Connected sums of self-dual manifolds and deformations of singular spaces. *Nonlinearity*, **2**, 197–239.

Donaldson, S.K. and Sullivan, D.P. (1990). Quasiconformal four-manifolds. *Acta Mathematica*, **163**, 181–252.

Donaldson, S.K., Furuta, M., and Kotshick, D. (1990). *Floer homology groups in Yang–Mills theory*. In preparation.

Douglas, R.G. (1980). *C*-algebra extensions and K-homology*. Annals of Mathematics Studies, **95**. Princeton University Press.

Eberling, W. (1990). An example of two homeomorphic, nondiffeomorphic complete intersection surfaces. *Inventiones Mathematicae*, **99**, 651–4.

Eells, J. (1966). A setting for global analysis. *Bulletin of the American Mathematical Society*, **72**, 751–807.

Eells, J. and Sampson, J.H. (1964). Harmonic mappings into Riemannian manifolds. *American Journal of Mathematics*, **86**, 109–160.

Engelking, R. (1977). *General topology*. Polish Scientific Publishers, Warsaw.

Fintushel, R. and Stern, R.J. (1984). $SO(3)$ connections and the topology of 4-manifolds. *Journal of Differential Geometry*, **20**, 523–39.

Fintushel, R. and Stern, R.J. (1985). Pseudo-free orbifolds. *Annals of Mathematics* (2) **122**, 335–364.

Fintushel, R. and Stern, R.J. (1988). Definite 4-manifolds. *Journal of Differential Geometry*, **28**, 133–41.

Floer, A. (1989). An instanton invariant for 3-manifolds. *Communications in Mathematical Physics*, **118**, 215–40.

Floer, A. (1990). Self-dual conformal structures on lCP^2. *Journal of Differential Geometry*, **33**, 551–73.

Folland, G. B. and Kohn, J.J. (1972). *The Neumann problem for the Cauchy–Riemann complex*. Annals of Mathematics Studies, **75**. Princeton University Press.

Forster, O. (1977). Power series methods in deformation theory. *Proceedings of*

Symposia in Pure Mathematics, **30**, pt. 2, 199–217. Amer. Math. Soc. Providence, RI, 1977.

Freed, D.S. (1986). Determinants, torsion, and strings. *Communications in Mathematical Physics*, **107**, 483–513.

Freed, D.S. and Uhlenbeck, K.K. (1984). *Instantons and four-manifolds.* M.S.R.I. Publications, Vol. 1. Springer, New York.

Freedman, M.H. (1982). The topology of four-dimensional manifolds. *Journal of Differential Geometry*, **17**, 357–454.

Freedman, M.H., and Kirby, R. (1978). A geometric proof of Rohlin's theorem. *Proceedings of Symposia in Pure Mathematics* (Amer. Math. Soc., Providence, RI 1978) 32, Vol. 2, 85–98.

Freedman, M.H. and Taylor, L. (1977). Λ-splitting 4-manifolds. *Topology*, **16**, 181–4.

Freedman, M.H. and Quinn, F. (1990). *The topology of 4-manifolds.* Princeton University Press.

Friedman, R. (1989). Rank two vector bundles over regular elliptic surfaces. *Inventiones Mathematicae*, **96**, 283–332.

Friedman, R. and Morgan, J.W. (1988*a*). On the diffeomorphism types of certain algebraic surfaces I, II. *Journal of Differential Geometry*, **27**, 297–369.

Friedman, R. and Morgan, J.W. (1988*b*). Algebraic surfaces and 4-manifolds: some conjectures and speculations. *Bulletin of the American Mathematical Society*, **18**, 1–19.

Friedman, R. and Morgan, J.W. (1989). Complex versus differentiable classification of algebraic surfaces. *Topology and its Applications*, **32**, 135–9.

Friedman, R., Moishezon, B. and Morgan, J.W. (1987). On the C^∞ invariance of the canonical class of certain algebraic surfaces. *Bulletin of the American Mathematical Society*, **17**; 357–454.

Furuta, M. (1987). Perturbation of moduli spaces of self-dual connections. *Journal of the Faculty of Science, University of Tokyo, Section 1A*, **34**, 275–97.

Furuta, M. (1990). *On self-dual connections on some orbifolds.* Preprint.

Gieseker, D. (1977). On moduli spaces of vector bundles on an algebraic surface. *Annals of Mathematics*, **106**, 45–60.

Gieseker, D. (1982). Geometric invariant theory and applications to moduli problems. In: *Lecture Notes in Mathematics*, **996** (ed. F. Gherardelli). Springer-Verlag, Berlin.

Gieseker, D. (1988). A construction of stable bundles on an algebraic surface. *Journal of Differential Geometry*, **27**, 137–54.

Gilbarg, D. and Trudinger, N.S. (1983). *Elliptic partial differential equations of second order.* (2nd edn) Springer, Berlin.

Gilkey, P. (1984). *Invariant theory, the heat equation and the Atiyah–Singer index theorem.* Mathematics Lecture Notes Series, No. 11, Publish or Perish 1984.

Gompf, R. (1983). Three exotic R^4's and other anomalies. *Journal of Differential Geometry*, **18**, 317–28.

Gompf, R. (1985). An infinite set of exotic R^4's. *Journal of Differential Geometry*, **21**, 283–300.

Gompf, R. (1990). Nuclei of elliptic surfaces. *Topology* (In press).

Greene, R. and Wu, H. (1988). Lipschitz convergence of Riemannian manifolds. *Pacific Journal of Mathematics*, **131**, 119–141.

Griffiths, P. and Harris, J. (1978). *Principles of algebraic geometry*. Wiley, New York.

Griffiths, P. and Harris, J. (1985). On the Noether–Lefschetz theorem and some remarks on codimension-2 cycles. *Mathematische Annalen*, **271**, 31–51.

Groisser, D. and Parker, T. (1987). The Riemannian geometry of the Yang–Mills moduli space. *Communications in Mathematical Physics*, **112**, 663–89.

Groisser, D. and Parker, T. (1989). The geometry of the Yang–Mills moduli space for definite manifolds. *Journal of Differential Geometry*, **29**, 499–544.

Gromov, M. and Lawson, H.B. (1983). Positive scalar curvature and the index of the Dirac operator on complete Riemannian manifolds. *Inst. Hautes Etudes Scientifiques Publ. Math.* **58**, 295–408.

Guillemin, V. and Sternberg, S. (1982). Geometric quantization and multiplicities of group representations. *Inventiones Mathematicae*, **67**, 515–38.

Hamilton, R. (1975). *Harmonic maps of manifolds with boundary*. Lecture Notes in Mathematics, **471**. Springer-Verlag, Berlin.

Hamilton, R. (1982). Three-manifolds with positive Ricci curvature. *Journal of Differential Geometry*, **17**, 255–306.

Hartshorne, R. (1978). Stable bundles of rank 2 on $\mathbb{CP}^2$. *Mathematische Annalen*, **238**, 229–80.

Hirsch, M.W. (1976). *Differential topology*. Springer, Berlin.

Hirzebruch, F. and Hopf, H. (1958). Felder von Flächenelementen in 4-dimensionalen Mannigfaltigkeiten. *Mathematische Annalen*, **136**, 156–72.

Hitchin, N.J. (1974). Harmonic spinors. *Advances in Mathematics*, **14**, 1–54.

Hitchin, N.J. (1980). Nonlinear problems in geometry. *Proceedings of the VI International Symposium, Division of Mathematics*. The Taniguchi Foundation, Japan.

Hitchin, N.J. (1987). The self-duality equations on a Riemann surface. *Proceedings of the London Mathematical Society*, **55**, 59–126.

Hitchin, N.J., Karhlede, A., Lindström, U., and Roček, M. (1987). Hyperkähler metrics and supersymmetry. *Communications in Mathematical Physics*, **108**, 535–89.

Hodge, W.V.D. (1989). *The theory and application of harmonic integrals*. Cambridge University Press.

Hörmander, L. (1973). *An introduction to complex analysis in several variables* (2nd. edn). North-Holland, New York.

Husemoller, D. (1966). *Fibre bundles*. Springer, Berlin.

Husemoller, D. and Milnor, J. (1973). *Symmetric bilinear forms*. Springer, Berlin.

Itoh, M. (1983). Geometry of Yang–Mills connections over a Kähler surface. *Proceedings of the Japan Academy*, **59**, 431–433.

Jaffe, A. and Taubes, C.H. (1980). *Vortices and Monopoles*. Progress in Physics 2. Birkhauser, Boston.

Jost, J. (1988). *Nonlinear Methods in Riemannian and Kählerian Geometry*, DMV Seminar, Band 10, Birkhäuser, Basel.

Jost, J. and Karcher, H. (1982). Geometrische methoden zür Gewinnung von A-priori-shranken für harmonische Abbildungen. *Manuscipta Mathematica*, **40**, 27–77.

Kazdan, J.L. (1981). Another proof of the Bianchi identity in Riemannian geometry. *Proceedings of the American Mathematical Society*, **81**, 341–42.

Kempf, G. and Ness, L. (1978). On the lengths of vectors in representation spaces. In: Lecture Notes in Math., **732**, pp. 233–242. Springer-Verlag.

Kervaire, M. and Milnor, J. (1958). Bernoulli numbers, homotopy groups, and a theorem of Rohlin. *Proceedings of the International Congress of Mathematicians, Edinburgh, 1958*. pp. 454–458. Cambridge University Press.

King, A.D. (1989). Instantons and holomorphic bundles on the blown-up plane. D.Phil. Thesis, Oxford.

Kirby, R.C. (1978*a*). A calculus for framed links in S^3. *Inventiones Mathematicae*, **45**, 35–56.

Kirby, R.C. (1978*b*). Problems in low-dimensional manifold theory. *Proceedings of Symposia in Pure Mathematics*, **32** Vol. 2 (1978) 273–322, Amer. Math. Soc. Providence, RI.

Kirby, R. and Siebenmann, L. (1977). *Foundational essays on topological manifolds, smoothings and triangulations*. Annals of Mathematics Studies, **88**. Princeton University Press.

Kirwan, F.C. (1984). *Cohomology of quotients in symplectic and algebraic geometry*. Mathematics Notes, **31**. Princeton University Press.

Kobayashi, S. (1980). First Chern class and holomorphic tensor fields. *Nagoya Mathematical Journal*, **77**, 5–11.

Kobayashi, S. (1987). *Differential geometry of complex vector bundles*. Iwanami Shoten and Princeton University Press.

Kobayashi, S. and Nomizu, K. (1963, 1969). *Foundations of differential geometry*. Vols. I, II. Wiley, New York.

Kodaira, K. (1963). On compact analytic surfaces, III. *Annals of Mathematics*, **78**, 1–4.

Kodaira, K. and Spencer, D.C. (1988). On deformations of complex analytic structure, I, II. *Annals of Mathematics*, **67**, 759–896.

Koszul, J-L. and Malgrange, B. (1958). Sur certaines structures fibrés complexes. *Arch. Math.* **9**, 102–9.

Kotschick, D. (1989). On manifolds homeomorphic to $\mathbb{CP}^2 \# 8\overline{\mathbb{CP}}^2$. *Inventiones Mathematicae*, **95**, 591–600.

Kotschick, D. (1990). On the geometry of certain 4-manifolds. D.Phil. thesis, Oxford.

Kuranishi, M. (1965). New proof for the existence of locally complete families of complex structures. In *Proceedings of Conference on Complex Analysis, Minneapolis*. Springer-Verlag, New York.

Kronheimer, P.B. (1989). The construction of ALE spaces as hyperkähler quotients. *Journal of Differential Geometry*, **29**, 665–89.

Kronheimer, P.B. (1990). A hyper-Kählerian structure on co-adjoint orbits of a semi-simple complex group. *Journal of the London Mathematical Society*, (2), **42**, 193–208.

Lang, S. (1969). *Real Analysis*. Addison-Wesley, New York.

Lawson, H.B. (1985). *The theory of gauge fields in four dimensions*. CBMS Regional Conference Series in Mathematics, American Mathematical Society, Providence, RI.

Lubke, M. (1982). Chernclassen von Hermite-Einstein-Vectorbundeln. *Matematische Annalen*, **260**, 133–41.

Lubke, M. (1983). Stability of Einstein-Hermitian vector bundles. *Manuscripta Mathematica*, **42**, 245–57.

Maciocia, A. (1990). Topics in gauge theory. D.Phil. Thesis, Oxford.

Mandelbaum, R. (1980). Four dimensional topology: an introduction. *Bulletin of the American Mathematical Society*, **2**, 1–159.

Margerin, C. (1987). Fibrés stable et metriques d'Hermite–Einstein. In: *Sem. Bourbaki* 683 (1986–87) Astérisque, Societé Math. France 152–53.

Markov, A.A. (1960). The problem of homeomorphy (in Russian). In: *Proceedings of the International Congress of Mathematics, Edinburgh, 1958*. Cambridge University Press.

Maruyama, M. (1977). Moduli of stable sheaves I, II. J. Math Kyoto University, **17**, 91–126 and 18 (1978) 557–614.

Matsushita, Y. (1988). Fields of 2-planes on compact, simply-connected smooth 4-manifolds. *Matematische Annalen*, **280**, 687–89.

Mehta, V.B. and Ramanathan, A. (1984). Restriction of stable sheaves and representations of the fundamental group. *Inventiones Mathematicae*, **77**, 163–72.

Milnor, J. (1958). On simply connected 4-manifolds. *Symposium Internacionale topología algebraica*, Mexico, 1958, 122–128.

Milnor, J. (1965). *Lectures on the h-cobordism theorem*. Princeton University Press.

Milnor, J. and Stasheff, J. (1974). *Characteristic classes*. Annals of Mathematics Studies, **76**. Princeton University Press.

Min-Oo, M. (1982). An L^2-isolation theorem for Yang–Mills fields. *Compositio Mathematica*, **47**, 153–63.

Mitter, P.K. and Viallet, C.M. (1981). On the bundle of connections and the gauge orbit manifold in Yang–Mills theory. *Communications in Mathematical Physics*, **79**, 457–72.

Moishezon, B. (1977). Complex surfaces and connected sums of complex projective planes, Lecture Notes in Mathematics, **603**. Springer.

Mong, K.C. (1988). Some differential invariants of 4-manifolds. D.Phil. Thesis, Oxford.

Mong, K.C. (1989). Polynomial invariants for 4-manifolds of type $(1, n)$ and a calculation for $S^2 \times S^2$. Preprint.

Mrowka, T. (1989). A local Mayer–Vietoris principle for Yang–Mills moduli spaces. Ph.D. thesis, Berkeley.

Mukai, S. (1981). Duality between $D(X)$ and $D(\hat{X})$, with application to Picard sheaves. *Nagoya Math. J.*, **81**, 153–75.

Mukai, S. (1984). On the symplectic structure of the moduli spaces of stable sheaves over abelian varieties and $K3$ surfaces. *Inventiones Mathematicae*, **7**, 101–16.

Mumford, D. and Fogarty, J. (1982). *Geometric invariant theory* (2nd. edn) Springer, Berlin.

Nahm, W. (1983). Self-dual monopoles and calorons. In: *Proc. XII Colloq. on gauge theoretic methods in Physics, Trieste*. Lecture notes in Physics **201** ed. Denado *et al.*, Springer-Verlag, Berlin, 1983.

Narasimhan, M.S. and Seshadri, C.S. (1965). Stable and unitary vector bundles on compact Riemann surfaces. *Annals of Mathematics*, **65**, 540–67.

Newlander, A. and Nirenberg, L. (1957). Complex analytic co-ordinates in almost complex manifolds. *Annals of Mathematics*, **65**, 391–404.

Newstead, P.E. (1972). Characteristic classes of stable bundles on an algebraic curve. *Transactions of the American Mathematical Society*, **69**, 337–45.

Newstead, P.E. (1978). *An introduction to moduli problems and orbit spaces.* Tata Institute for Fundamental Research, Bombay.

Okonek, C., Schneider, M. and Spindler, H. (1980). *Vector bundles on complex projective spaces.* Progress in Mathematics, **3**. Birkhauser, Boston.

Okonek, C. and Van de Ven, A. (1986). Stable vector bundles and differentiable structures on certain elliptic surfaces. *Inventiones Mathematicae*, **86**, 357–70.

Okonek, C. and Van de Ven, A. (1989). Γ-type-invariants associated to $PU(2)$ bundles and the differentiable structure of Barlow's surface. *Inventiones Mathematicae*, **95**, 601–14.

Palais, R. (1968). *Foundations of global non-linear analysis.* Benjamin, New York.

Palamodov, V. (1976). Deformations of complex spaces. *Russian Mathematics Surveys*, **31**, 129–97.

Parker, T. (1982). Gauge theories on four-dimensional Riemannian manifolds, *Communications in Mathematical Physics*, **85**, 563–602.

Quillen, D.G. (1985). Determinants of Cauchy–Riemann operators over a Riemann surface. *Funct. Anal. Appl.*, **14**, 31–4.

Quinn, F. (1986). Isotopy of 4-manifolds. *Journal of Differential Geometry*, **24**, 343–72.

Ramanathan, A. and Subramanian, S. (1988). Einstein Hermitian connections on principal bundles and stability. *Journal für die Reine und Angewendte Mathematik.* **390**, 21–31.

Ray, D.B. and Singer, I.M. (1973). Analytic torsion for complex manifolds. *Annals of Mathematics*, **98**, 154–77.

Rham, G. de (1984). *Differentiable manifolds.* Springer, Berlin.

Rohlin V.A. (1952). New results in the theory of four dimensional manifolds. *Dok. Akad. Nauk. USSR*, **84**, 221–24.

Rourke, C.P. and Sanderson, B.J. (1982). *Introduction to piecewise linear topology.* Springer, Berlin.

Sacks, J. and Uhlenbeck, K.K. (1981). The existence of minimal 2-spheres. *Annals of Mathematics*, **113**, 1–24.

Salamon, S.M. (1982). Topics in four dimensional Riemannian geometry. *Seminaire "Luigi Bianchi"*, Lecture notes in Math., **1022** (ed. Vesentini). Springer-Verlag, Berlin.

Salvetti, M. (1989). On the number of non-equivalent differentiable structures on 4-manifolds. *Manuscripta Mathematica*, **63**, 157–71.

Schenk, H. (1988). On a generalised Fourier transform for instantons over flat tori. *Communications in Mathematical Physics*, **116**, 177–83.

Schoen, R. (1984). Conformal deformation of a Riemannian metric to constant scalar curvature. *Journal of Differential Geometry*, **20**, 479–95.

Schwarzenberger, R.L.E. (1961). Vector bundles on algebraic surfaces. *Proceedings of the London Mathematical Society*, **11**, 601–22.

Sedlacek, S. (1982). A direct method for minimising the Yang–Mills functional. *Communications in Mathematical Physics*, **86**, 515–27.

Serre, J.-P. (1961). *Sur les modules projectifs.* Seminaire Dubreil-Pisot 1960/1961, Secr. Math. Paris exposé 2.

Serre, J.-P. (1973). *A course in arithmetic.* Springer, Berlin.

Sibner, L. and Sibner, R. (1988). Classification of singular Sobolev connections by their holonomy. *Journal of Differential Geometry.* In press.

Simpson, C.T. (1989). Constructing variations of Hodge structure using Yang–Mills theory, with applications to uniformisation. *Journal of the American Mathematical Society*, **1**, 867–918.

Smale, S. (1964). On the structure of manifolds. *American Journal of Mathematics*, **87**, 387–99.

Smale, S. (1965). An infinite dimensional version of Sard's theorem. *American Journal of Mathematics*, **87**, 861–66.

Soberon-Chavez, S. (1985). Rank 2 vector bundles over a complex quadric surface. *Quarterly Journal of Mathematics (Oxford)*, (2), **36**, 159–72.

Spanier, E. (1966). *Algebraic topology*. Springer, New York.

Spivak, M. (1979). *A comprehensive introduction to differential geometry*. Vols I–V (2nd edn). Publish or Perish.

Stein, E.M. (1970). *Singular integral operators and differentiability properties of functions*. Princeton University Press.

Sunderaraman, D. (1980). *Moduli, deformations and classification of compact complex manifolds*. Pitman, London.

Taubes, C.H. (1982). Self-dual Yang–Mills connections over non-self-dual 4-manifolds. *Journal of Differential Geometry*, **17**, 139–70.

Taubes, C.H. (1983). Stability in Yang–Mills theories. *Communications in Mathematical Physics*, **91**, 235–63.

Taubes, C.H. (1984). Self-dual connections on manifolds with indefinite intersection matrix. *Journal of Differential Geometry*, **19**, 517–60.

Taubes, C. H. (1986). Gauge theory on asymptotically periodic 4-manifolds. *Journal of Differential Geometry*, **25**, 363–430.

Taubes, C.H. (1990). Casson's invariant and gauge theory. *Journal of Differential Geometry*, **31**, 547–99.

Taubes, C.H. (1988). A framework for Morse theory for the Yang–Mills functional. *Inventiones Mathematicae*, **94**, 327–402.

Taubes, C.H. (1989). The stable topology of self-dual moduli spaces. *Journal of Differential Geometry*, **29**, 162–230.

Teleman, N. (1983). The index of signature operators on Lipschitz manifolds. *Inst. Hautes Etudes Sci. Publ. Math.* **58**, 39–79.

Trudinger, N.S. (1968). Remarks concerning the conformal deformation of Riemannian metrics on compact manifolds. *Ann. Scuola Normale Sup., Pisa*, (4) **3**, 265–74.

Uhlenbeck, K.K. (1982*a*). Removable singularities in Yang–Mills fields. *Communications in Mathematical Physics*, **83**, 11–29.

Uhlenbeck, K.K. (1982*b*). Connections with L^p bounds on curvature. *Communications in Mathematical Physics*, **83**, 31–42.

Uhlenbeck, K.K. and Yau, S-T. (1986). On the existence of hermitian Yang–Mills connections on stable bundles over compact Kahler manifolds. *Communications on Pure and Applied Mathematics*, **39**, 257–293; Correction: *Communications on Pure and Applied Mathematics*, **42**, 703–7.

Wall, C.T.C. (1962). On the classification of $(n-1)$ connected $2n$ manifolds. *Annals of Mathematics*, **75**, 163–89.

Wall, C.T.C. (1964*a*). Diffeomorphisms of 4-manifolds. *Journal of the London Mathematical Society*, **39**, 131–40.

Wall, C.T.C. (1964*b*). On simply connected 4-manifolds. *Journal of the London Mathematical Society*, **39**, 141–49.

Ward, R.S. (1977). *Physics Letters*, **61A**, 81–2.

Warner, F. (1983). *Foundations of differentiable manifolds and Lie groups*. Springer, New York.

Wells, R.O. Jr. (1980). *Differential analysis on complex manifolds*. Springer, New York.

Whitehead, J.H.C. (1949). On simply connected 4-dimensional polyhedra. *Commentarii Mathemataci Helvetici*, **22**, 48–92.

Whitney, H. (1944). The self-intersection of a smooth *n*-manifold in 2*n* space. *Annals of Mathematics*, (2) **45**, 220–246.

Witten, E. (1982). Supersymmetry and Morse theory. *Journal of Differential Geometry*, **16**, 661–92.

Witten, E. (1988). Topological quantum field theory. *Communications in Mathematical Physics*, **117**, 353–386.

Yang, C.N. (1977). *Physics Reviews Letters*, **38**, 1377–79.

INDEX